Upland Agriculture and the Environment

The NINEVEH Charitable Trust

ISBN 978-0-9570747-0-5

Printed by Badger Press Ltd
Longlands, Bowness-on-Windermere LA23 3AS

Upland Agriculture and the Environment

Lois Mansfield

Nineveh Trust

Badger Press

To farmers, sheep and family

Contents

List of figures

List of tables

List of boxes

List of acronyms

AAPS	Arable Area Payments Scheme
ABC	Abiotic-Biotic-Cultural
AD	*Anno Domini*
ADAS	Agricultural Development Advisory Service
AHDS	Agriculture & Horticulture Development Scheme
ALC	Agricultural Land Classification
AONB	Area of Outstanding Natural Beauty
ASSI	Areas of Special Scientific Interest (Northern Ireland)
BAP	Beef Annual Premium
BAP	Biodiversity Action Plan
BGMS	Broads Grazing Marsh Scheme
bp	before present
BSE	Bovine Spongiform Encephalitis
CAP	Common Agricultural Policy
CCW	Countryside Council for Wales
CEC	Cation Exchange Capacity
CEH	Centre for Ecology and Hydrology
CMS	Countryside Management Scheme (for Northern Ireland)
CoCo	Countryside Commission
Cr	'C' soil horizon on the regolith
CS2000	Countryside Survey 2000
CSS	Countryside Stewardship Scheme
CVM	Contingent Valuation Method
DA	Disadvantaged Area
DARD	Department of Agriculture and Rural Development
DEFRA	Department of Environment, Food and Rural Affairs
DM	Dry Matter by herbage
DMC	Dry Matter Content
DSH	Dwarf Shrub Heath
EAGF	European Agricultural Guarantee Fund
EAGGF	European Agricultural Guidance and Guarantee Fund
EAFRD	European Agricultural Fund for Rural Development
ECS	Energy Crop Scheme
EIM	Environmental Information Map
ELC	European Landscape Convention
ELS	Entry Level Scheme
ERDP	England Rural Development Program/ Plan (2000-2006)
ESA	Environmentally Sensitive Area
ESS	Environmental Stewardship Scheme
ESU	European Size Unit
EU	European Union
FAO	Food and Agriculture Organisation
FEOGA	*Fonds Europeen d'Orientation et de Garantie Agricole*
FEP	Farm Environment Plan

FER	Farm Environment Record
FFTS	Fell Farming Traineeship Scheme
FHDS	Farm & Horticulture Development Scheme
FMD	Foot and Mouth Disease
FPS	Farm Practices Survey
FRCA	Farming and Rural Conservation Agency
FWPS	Farm Woodland Premium Scheme
FYM	Farm Yard Manure
GAEC	Good Agricultural and Environmental Condition
GATT	General Agreement of Tariffs and Trade
GDP	Gross Domestic Product
HAP	Habitat Action Plan
HFA	Hill Farm Allowance
HGMM	Hill Grazing Management Model
HLCA	Hill Livestock Compensatory Allowance
HLMMS	Hill Lamb and Mutton Marketing Scheme
HLS	Higher Level Scheme
HNV	High Nature Value
HRA	Historic Reference Amount
IACS	Integrated Administration and Control System
IPA	Importance Performance Analysis
IPCC	Inter-Governmental Panel on Climate Change
ITE	Institute of Terrestrial Ecology (now CEH)
JCA	Joint Character Area (now NCA)
JNCC	Joint Nature Conservancy Council
LAG	Local Action Group
LBAP	Local Biodiversity Action Plan
LCA	Land Capability for Agriculture (Scotland)
LDNPA	Lake District National Park Authority
LEADER	*Liaisons Entre Actions de Developpement de l'Economie Rurale*
LFA	Less Favoured Area
LFASS	Less Favoured Areas Support Scheme
LKS	Local Knowledge System
LMCMS	Land Management Contract Menu Scheme
LRLIS	Livestock Rearing Land Improvement Scheme
LRR	Lagging Rural Region
LU	Livestock Unit
MAFF	Ministry of Agriculture, Fisheries and Food
MEA	Millenium Ecosystem Assessment
MLURI	Macaulay Land Use Research Institute
MTR	Mid Term Review
NAI	National Areas Initiative
NCA	National Character Area (was JCA)
NCC	Nature Conservancy Council (a pre cursor to Natural England)
NFFO	Non Fossil Fuels Obligation
NFI	Net Farm Income
NFU	National Farmers Union
NICMS	Northern Ireland Countryside Management Scheme
NNR	National Nature Reserve
NUMRP	Northern Uplands Moorland Regeneration Project
NYMPA	North Yorkshire Moors National Park Authority
NVC	National Vegetation Classification
NYM	North York Moors
OD	Ordnance Datum
OFS	Organic Farming Scheme
OGA	Other Gainful Activity
OLS	Organic Level Scheme
PDO	Protected Designation of Origin
PGI	Protected Geographical Indication
pH	Measure of acidity/ alkalinity
PMG	Processing and Marketing Grant
PSA	Point Score Analysis
PSA	Public Service Agreement
R	Regolith

RBST	Rare Breeds Survival Trust
RDPE	Rural Development Plan for England (2007 to 2013)
RDR	Rural Development Regulation
RES	Rural Enterprise Scheme
RICS	Royal Institute of Chartered Surveyors
RPA	Rural Payments Agency
SAC	Special Area of Conservation
SAP	Sheep Annual Premium
SAP	Species Action Plan
SCP	Suckler Cow Premium
SDA	Severely Disadvantaged Area
SES	Social Ecological System
SFSC	Short Food Supply Chain
SGM	Standard Gross Margin
SLF	Sustainable Livelihoods Framework
SLR	Standard Labour Requirement
SMD	Standard Man Day
SMRs	Statutory Management Requirements
SNH	Scottish Natural Heritage
SPA	Special Protection Area (designated under EU Birds Directive)
SPS	Single Payment Scheme (also known as Single Farm Payment)
SRC	Short Rotation Coppice
SSSI	Site of Special Scientific Interest
SWES	Sheep Wildlife Enhancement Scheme
SWOT	Strength Weaknesses Opportunities and Threats
TIFF	Total Income From Farming
TSG	Traditional Speciality Guaranteed
UELS	Upland Entry Level Scheme
UK	United Kingdom
UKROFS	United Kingdom Register of Organic Food Standards
UNEP	United Nations Environment Programme
UNESCO	United Nations Educational, Scientific and Cultural Organisation
UV	Ultra violet
VTS	Voluntary Training Scheme
WES	Wildlife Enhancement Scheme
WGS	Woodland Grant Scheme
WHO	World Health Organisation
WTA	Willingness to Accept
WTO	World Trade Organisation
WTP	Willingness-to-Pay

Chemical symbols

B	Boron
Ca	Calcium
$CaCO_3$	Calcium Carbonate
Co	Cobalt
Fe	Iron
H^+	Hydrogen
I	Iodine
K	Potassium
Mg	Magnesium
Mn	Manganese
N	Nitrogen
Na	Sodium
NH_{4^+}	Ammonium
OH	Hydroxide
P	Phosphorus
S	Sulphur
Se	Selenium
Ti	Titanium
Zn	Zinc

Acknowledgements

The author would like to acknowledge the following people for providing advice, guidance and information throughout the writing of this book: Bridget Andrews of the Museum of Rural Life; Adrian Banford of Cumbria Fells & Dales RDPE; Laura Banford; Dr Kevin Butt of the University of Central Lancashire; Mervyn Edwards MBE; the late Alan Fishwick; Robin Karfoot of DEFRA; Jean Johnston of Natural England; Nadeem Raja of DEFRA; Rosina Mansfield; Terry McCormick of ACT; Rachel Pickering of the Yorkshire Moors National Park; Hilary Wilson of High Carlingill; the Library staff at Newton Rigg and the University of Cumbria.

Particular thanks is extended to Sheila Ripper for drawing the many diagrams in this book, to Gerard Hill for his long suffering editing and the folks at Badger Press, especially Steve.

The author would also like to express her deepest gratitude to sponsorship of the Nineveh Trust without whom this book would not have seen the light of day.

Finally I'd like to thank Brian Ilbery and Julius Mage who first introduced me to agricultural geography.

Lois Mansfield
November 2011

1

The upland farming system

What's this book about?

The unique environment of the British Isles, on the western seaboard of Europe, has determined the way agriculture has developed. In global classifications it is recognised as a pastoral economy, but closer inspection reveals that the eastern part of the archipelago is dominated by arable production and the west by livestock. This divide is underpinned by higher land dominating western Britain (Figure 1.1), where the upland landscape has evolved over thousands of years through a combination of agriculture, mining and woodland management. The result is a landscape of high nature value (HNV), cherished by the public for its environmental and recreational qualities acknowledged in poetry and prose. It is the agriculture of the uplands and its relationship to the environment that this book aims to investigate.

Farming on the margins

Since the mid-19th century, agricultural activity in the uplands, like that elsewhere, has been transformed to meet the changing demands of society. Being at the margins of cultivation, many upland farming businesses have struggled to remain financially viable, despite a range of government interventions.

As a result, for most of the 20th century, upland farming has been labelled marginal at best, a problem at worst. Compounding this economic challenge is the tenuous balance between over- and under-exploitation of the land, a balance created by the long symbiotic relationship between farming and its environment. Whilst attempts to support the socio-economic core of the system have been partly successful in maintaining farm incomes and slowing rural depopulation, some polices have had a detrimental effect on the HNV qualities of uplands where overgrazing has degraded biodiversity.

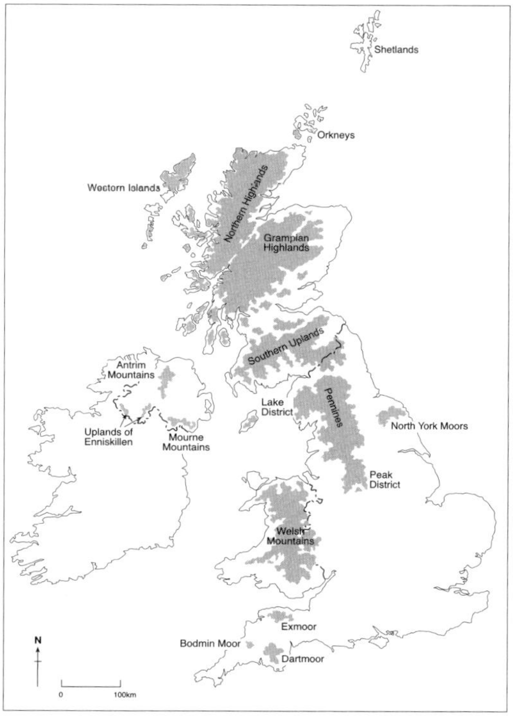

Figure 1.1 Uplands of the British Isles

The situation has recently become more complex as European Union (EU) agricultural support has shifted from food production to rural development and environmental management, devaluing the mainstay of the upland farming economy (meat and wool). On the one hand this switch is seen by farming communities as a betrayal of their role as food providers; on the other hand, it does finally imply political recognition of the role of extensive agriculture in maintaining HNV in upland environments. A fundamental thrust of these strategies is the reduction of stocking rates on uplands, to allow habitats to recover from inappropriate management in earlier periods. Unfortunately the pendulum can easily swing the other way, because destocking causes financial hardship for farming families and environmental undergrazing, which also reduces biological diversity.

In order to explore this balance between upland agriculture and its environment we need to answer several questions. First, what does an upland farm look like and how does it operate? We consider this later in this chapter. Second, we need to appreciate the extent of the resource base itself. Simply put, where are the uplands? Whilst it may seem rather an odd question to consider, it has far-reaching ramifications for people farming in upland-type environments; it can decide whether they are denied financial aid. We explore this issue in Chapter 2.

Factors in upland farming

We then consider the controls on upland farming patterns. This is much more complex to answer and, for agriculture in general, has generated protracted debates in academic circles (examples include Coppock 1965; Tarrant 1974, Ilbery 1985, Robinson 2004). The main factors are summarised in Figure 1.2, which shows that the fundamental control is the physical environment. Any type of farming that operates near to or on the physical margins of cultivation is severely constrained in this way. For upland agriculture it limits enterprise choice and demands sustainable management practices to avoid over-exploitation of the resource base. Chapter 3 explores the physical context and examines the various management challenges faced by contemporary upland farmers.

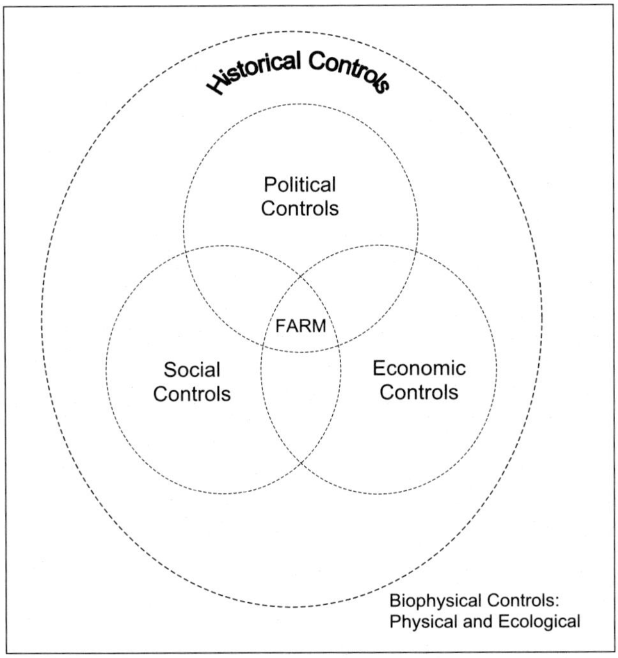

Figure 1.2 Controls on Upland Agricultural Land Use Patterns

Current upland agricultural land-use patterns are also a function of temporal landscape change. All landscapes are palimpsests,[1] but uplands more so because of their marginal character. Small changes in climate or population pressure over the centuries have seen an ebb and flow of

[1] A palimpsest is a vellum manuscript that has been scraped clean and re-used, but some of the original words still show through. So it is with landscapes where the land use of previous cultures shows through into the present – for example, medieval drystone walls or Bronze Age field lynchets (see Chapter 4) seen in the current landscape.

agricultural use up and downhill. Understanding the historical context of upland agriculture also helps to explain how the HNV interest evolved and why agricultural support then became essential for the sector. We explore this dimension in Chapter 4.

Underpinned by physical and historical processes, the current upland farming landscape is an inter-relationship between farming, ecological and social systems. Whilst Chapter 4 demonstrates how the HNVs developed, Chapter 5 explores the current status of these ecological systems and the effects upland farming has upon them. It becomes apparent that we need a delicate symbiosis between farming and ecological systems if both are to be sustained in uplands. For this to occur, upland farm businesses need to be economically viable without over-exploiting the environment, something which has proved difficult to achieve. The economics of upland farming are therefore dealt with in Chapter 6, where we also investigate opportunities to diversify the farm business.

The other important aspect of the upland farming landscape is the role of people. This book argues here that, for upland farming systems, people play a more central role than in other types of agriculture because of unique co-operative management processes that have developed over the centuries. Anything that destabilises the farming population will therefore destabilise farm business viability. This social dimension is contemplated in Chapter 7.

The final controlling factor for upland agricultural land-use patterns is politics. Whilst Chapter 3 shows what happened up until 1945, Chapter 8 unfolds the story of upland farm support in the UK from 1945 to 2000. The changing vagaries of political intervention demonstrate that British upland farmers need to re-invent themselves perpetually. The final chapter (Chapter 9) considers the role of this re-invention in turning upland agriculture and its environment towards a more multipurpose, multifunctional resource base for the 21st century.

The purpose of the rest of this chapter then, is to investigate the character of upland agriculture and operation of its farming system.

The structure of upland agriculture in Britain

The concept of the farm as a system has been conceptualized by a number of researchers (Figure 1.3). Systems are made up of inputs (things which go in), outputs (things that come out, both useful and not useful), throughputs (matter or energy flowing through) and storage (things that hold matter or energy for a while before passing it on).

On a farm, inputs may be variable, like seeds or forage, or fixed – those assets available continuously to the farmer, such as machines or land (see Figure 1.3a). The outputs are marketable products, such as cereal and meat. There are by-products as well, which can be desirable, like landscape (important for upland agriculture, so perhaps it should be an output rather than a by-product), or undesirable, like farm waste or excess chemicals. An example of a throughput could be nutrients derived from feed concentrates, which are eaten by livestock, pass through their gut, are deposited as cow faeces and then decompose into the soil. Nutrients may then be taken up by plants or be leached out of the soil into groundwater and out of the farm system. The storage element of the farming system can be seen as the livestock and crop enterprises developing on the farm, as well more recently valued CO_2 fixing.

The first real attempt to model the farm system was made by Morgan and Munton (1971), but this model was only concerned with internal movements within the farm itself (Figure 1.3a). Later work by Fowler (1978) expanded the farm system to include the importance of individual farmers, their knowledge and background, and the social organisation of their culture (Figure 1.3b). This is interesting because Fowler was writing for an archaeological audience and not a geographical one, but by doing so he took a more 'cultural' stance in his analysis of the farm system. It is important to do this in relation to upland agriculture because the model identifies traditional skills, such as hefted flocks, as being fundamental to the effective operation of upland farm systems (this is discussed again below and also in Chapter 7).

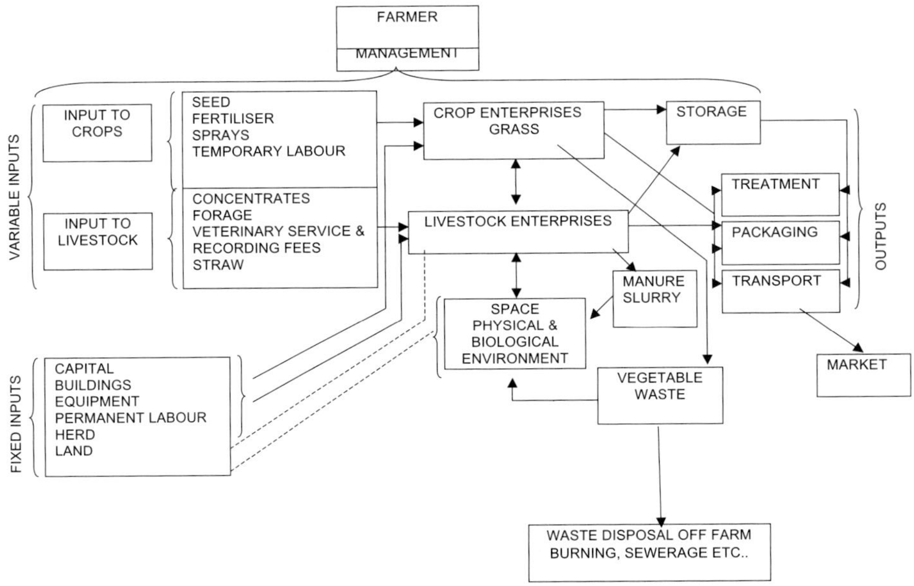

Figure 1.3a The Farm as a System according to Morgan & Munton (1971)
Reproduced from their book 'Agricultural Geography' Figure 1.2 on p22 with permission from Routledge.

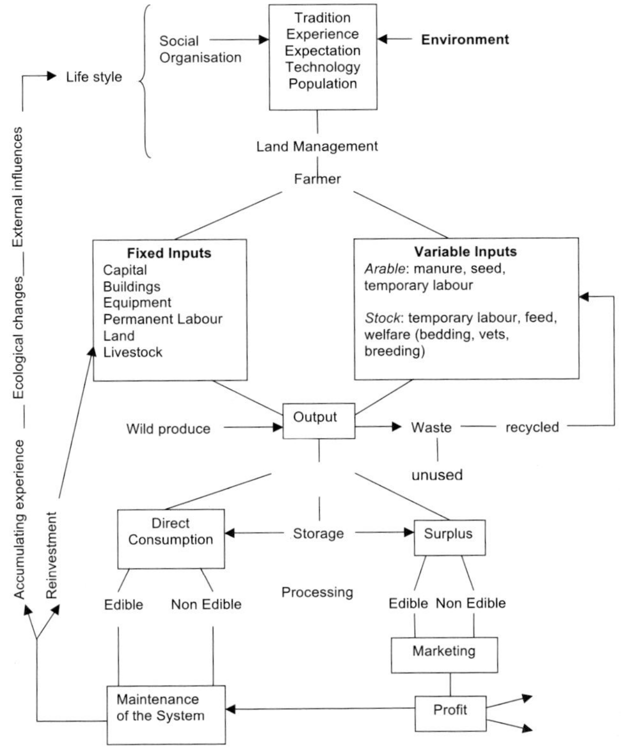

Figure 1.3b The Farm System according to Fowler (1983).
Reproduced with kind permission from '*The Farming of Prehistoric Britain*' Fowler PJ (1983) Cambridge University Press: Cambridge

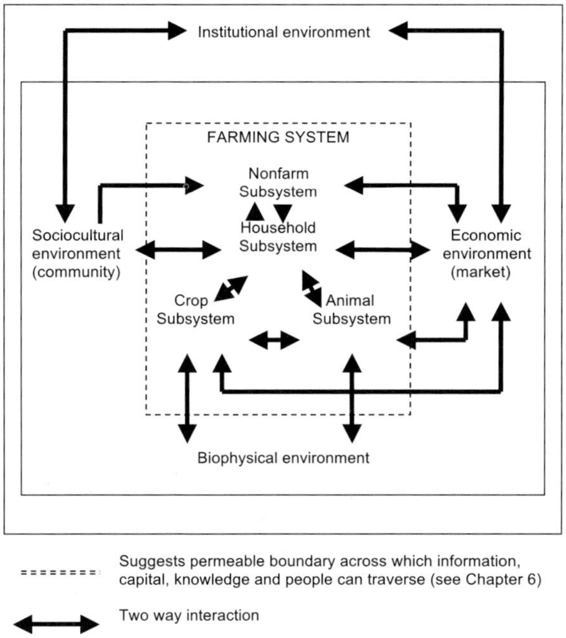

Figure 1.3c The Farm System according to P Caldwell.
This diagram was originally published in the *Encyclopedia of Agricultural Science* (1994:131) Copyright Elsevier (2011).

Caldwell (1994) developed the farm system concept still further by building in external influences to the farm system (Figure 1.3c). He added four subsystems: the biophysical environment, the marketplace where products are sold, rural communities and their effects, and institutional or political environments. Caldwell then identified several subsystems within the farm itself, each of which has been studied in isolation by different disciplines:

- a crop subsystem – typically studied by agriculturalists
- a livestock subsystem – studied by zoologists and veterinary surgeons
- a household subsystem – studied by sociologists
- a non-farm subsystem – studied by agricultural economists

In essence, Caldwell has extended the boundaries of the farm system to reflect the complex interactions of farming with other systems, thus making it a more realistic representation of reality (as becomes more apparent in chapters 6 to 8) beyond the bounds of the other two models. Agricultural geographers have also considered the internal and external relationships in farming systems (e.g. Evans 1990).

The various internal and external systems come together in different ways to produce farm systems that are unique to a particular location. One farm may be a dairy unit, another a grain farm and a third a hop farm. However, the complexity of variables interacting within systems can lead neighbouring farms to be unlike one another as well. Next door to the dairy farm may be another dairy farm, but the situation in the household subsystem may mean the farmer has an older child who can help with the milking thus providing an opportunity to develop other farm enterprises, unlike their neighbour. In terms of agricultural land use, it means that no two farms can be identical, which makes studying agricultural geography, as a topic, difficult.

Nevertheless, we can make some generalisations about farming systems, depending on the geographical scale of the investigation (e.g. Coppack 1971, Ilbery 1984). At a global scale our

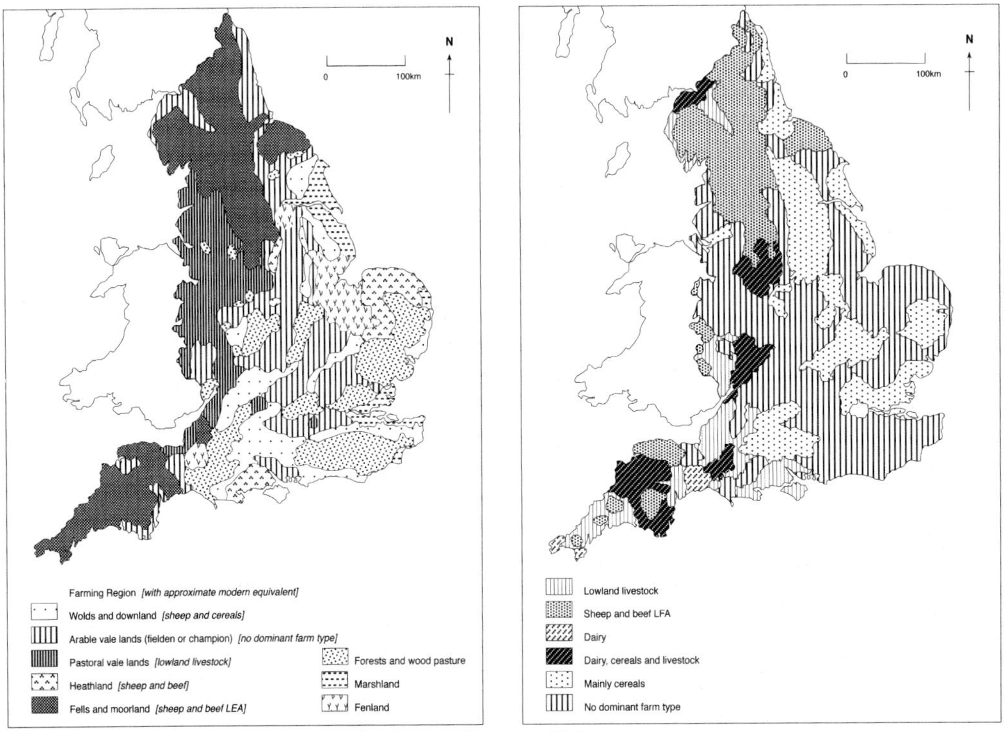

Figure 1.4 Agricultural Land Use in the United Kingdom 2006
(Adapted from DEFRA 2007, with permission from Thirsk J (1987) *England's Agricultural Regions and Agrarian History 1500 to 1750.* Macmillan: Basingstoke.

uplands are seen as being dominated by dairying, but they are in fact much more varied when we come down to the regional level (Figure 1.4). Agricultural activity has also varied over time (see Chapter 3), mutating from one dominated by mixed farming to the 21st century's more specialised pastoral system, where farming systems have developed comparative advantage.

Within the United Kingdom, the current structure of agricultural land use is skewed towards this pastoral economy (62% of the land area: Figure 1.5a). Roughly half of this 62% is grassland

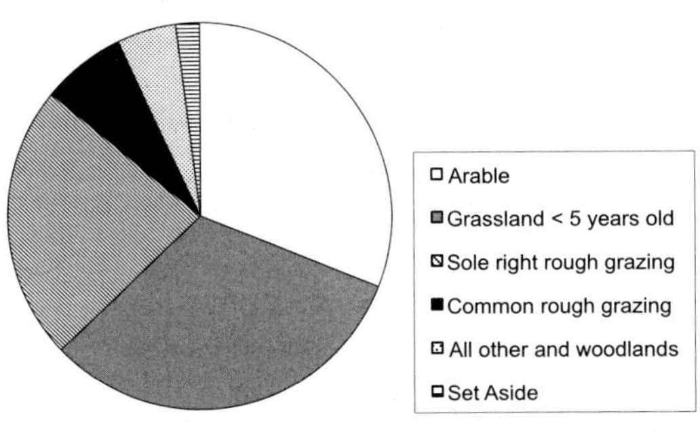

Figure 1.5a
(Adapted from DEFRA 2007)

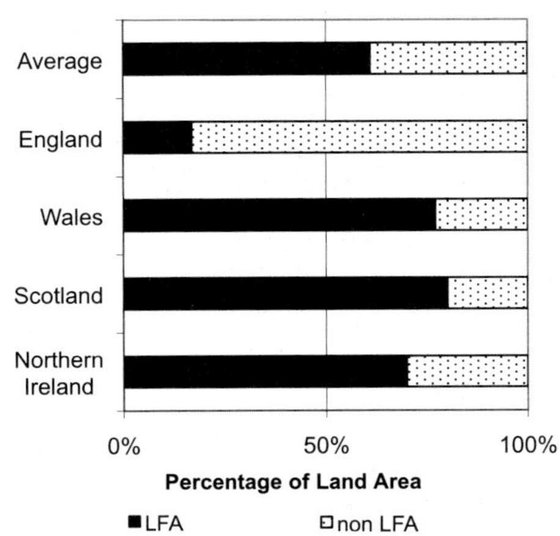

Figure 1.5b Less Favoured Areas in the United Kingdom 2007
(Adapted from DEFRA 2007)

under five years old, indicating that it is ploughed periodically to improve the grazing potential. The rest is rough grazing, of which three quarters is either sole-right grazing (one farmer's stock grazes the land) or common-right grazing, and the other quarter is open fell. The majority of grazing is in the uplands, where combinations of sheep, beef and dairy enterprise vary from farm to farm.

The upland area itself is roughly coterminous with the EU Less Favoured Area (LFA) designation. This accounts for 47% of the land area of the United Kingdom. England has only 16.3% classified as LFA in contrast to the other three parts of the United Kingdom, where it is between 70 and 79% (Figure 1.5b). We will see in Chapter 2 that this is somewhat misleading because the definition of LFA is not confined to uplands in the truest sense. The landscape of Northern Ireland is not 70% upland!

The upland farm system

Although the uplands of Britain vary substantially in character, as we shall see, the basic structure of an upland farm is universal. It is made up of the farmstead, inbye, open fell and intake (Figure 1.6). Farmers operate these as a management system to give them the flexibility to overcome the poor physical conditions of the upland environment.

Farmstead

First there is the central farmstead. This usually comprises the farmhouse, farmyard, sheep dipping area and several sheds, barns or byres. Many barns are used for storing equipment and machinery. Along with any sheds or byres, barns may also house livestock at certain times of year. The farmstead is sometimes referred to as the hereditament[2].

Inbye

Adjacent to the farmstead is the inbye land, made up of grass meadows and some occasional arable fields for the production of forage crops. This land is the best quality on the farm, but it is also improved by drainage and the addition of fertilisers and other products, and is therefore the most productive in terms of grass. These fields are either grazed by livestock or cut for hay, silage or haylage for winter feed.

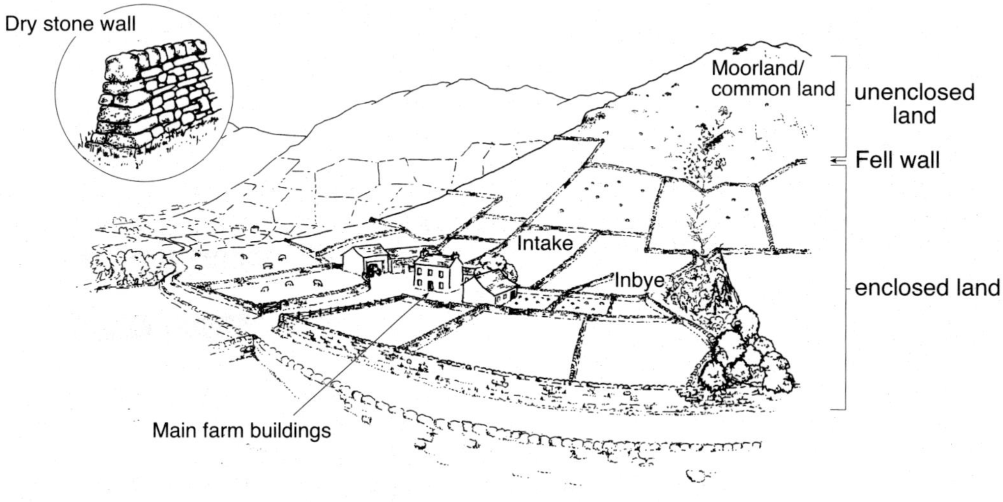

Figure 1.6 A Generalised view of an Upland Farm

[2] Hereditament - a legal term related to land as property rather than goods and chatels.

Historically the inbye fields produced hay, leading to floral-rich hay meadows. However, changes in farming practice since the 1960s have seen hay replaced with silage, which has little biodiversity value (see Chapter 5). Fodder crops such as turnips or beets may also be grown as supplementary feed for livestock in the winter months, although these are of fairly limited extent. Occasionally a farmer may grow a field of barley, maize or roots, again for fodder and not as a cereal product.

Open fell

The second type of land on an upland farm is unenclosed open fell. This land lies above what is known as the fell wall, which is the last drystone wall before the land opens out into an unwalled expanse (see Box I). The land here can be common land, owned by a single landlord and/or shared through common rights by the farms which graze livestock upon it (see below).

The fell is a mosaic of poorer-quality, semi-natural habitats, usually rough grassland, heather moorland and bog. It is this zone where biodiversity has suffered most from increased grazing (see Chapter 5) because those managers with grazing rights can graze as many livestock as their common rights allow, which may be beyond the ecologically sustainable carrying capacity of the land (see Chapter 3 and Boxes III and XXVI). The stock introduced onto common grazing in a certain area is often known as a heft or heaf (but see section 1.5 below).

Fells can sometimes be divided by a system of physical boundaries, and then typically each block of land on the fell has single ownership and is not common (sometime called sole grazing rights or sole occupancy). These fells are easier to manage because there is only one occupier who controls the grazing regime. The pattern of sole occupancy and common land varies across the English uplands (Figure 1.7). The Lake District is dominated by common land (65%) whereas the Peak District has none. Other uplands, like the North Pennines and the Borders, have varied amounts of the two. This makes national upland land management policies hard to apply because one could be dealing with one landlord owning thousands of hectares or many commoners, all of whom have to agree to any global change in the land management.

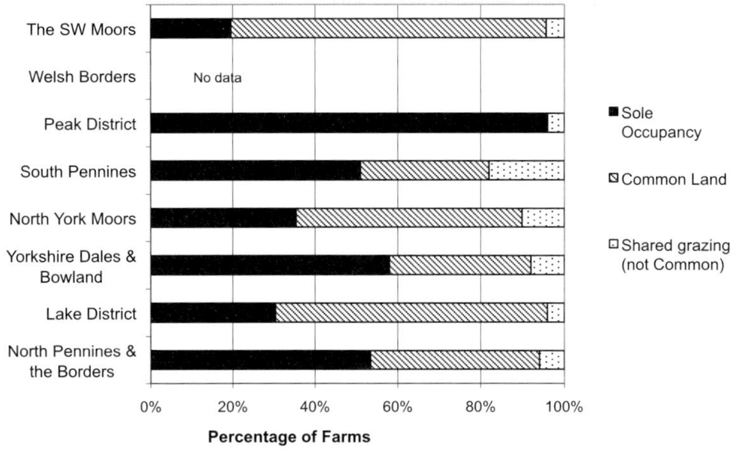

Figure 1.7 Sole Occupancy and Common Land in England and Wales 2009
(Adapted from DEFRA 2009a)

Intake

The third type of land found on upland farms is intake, or intack, which lies between the inbye and open fell. Intake is made up of pieces of common or other land which has been enclosed from the open fell, literally taken in. The quality of this land lies somewhere between inbye and open fell. Often farmers install (tile or plastic) drains to produce a semi-improved pasture of rush beds and some nutritious grasses.

Box I: The drystone wall

The drystone wall is an important feature of the British landscape. Farmers use these walls where local geology offers stone that can self-support without the use of cement. Drystone walls are not limited to uplands above 200m; lower ranges of hills, such as the Cotwolds and Lincolnshire Wolds, use this method of field division as well (Woodcock 1996).

A drystone wall is in fact two walls running near-parallel to each other but sloping gradually inwards from bottom to top. The two are held together with large stones known as throughs that stretch across the entire width. Throughs that protrude can be used to make steps up and over the wall, acting as a stone stile. As the two outer walls are built upwards, the gap between them is infilled with smaller stones, called heartings. On top of the two walls, capstones are set along the length, each leaning on its neighbour, their weight helping to bond the walls together.

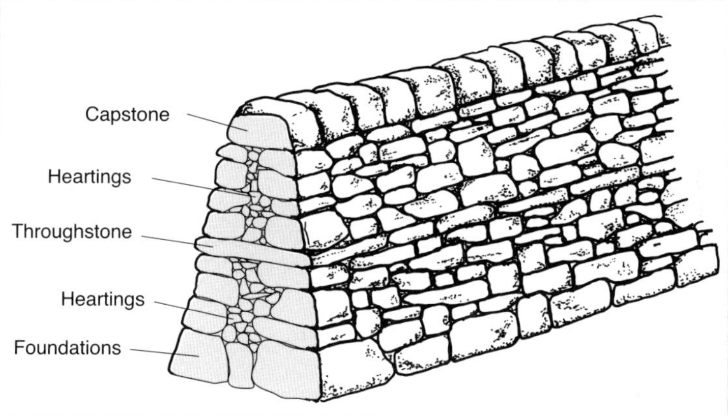

Anyone visiting for the first time an upland, where drystone walls are used, will be amazed by the thousands of miles of such walls that exist, often traversing difficult terrain and steep slopes. These walls date back many hundreds of years and some simpler versions date back even further, to the Bronze Age when the first animal enclosures were constructed (read more in Chapter 3). However, few of those walls are still in use, because after about 150 years the heartings inside the wall rot due to weathering and the two skins collapse inwards, giving the wall a distinctive snake-like appearance. The wall still has the same line, and most of its stone is recycled when it is repaired, but it is not the same wall.

Style of construction is a guide to the age of a wall. Before the age of parliamentary enclosure, walls were constructed in different ways in different areas, but many enclosure acts in the 17th and 18th centuries stipulated the construction and specification of walls; hence most walls from this time appear standardised.

Dating walls provides a glimpse into the evolution of a landscape. Hardcastle and Nisbet (2008) investigated wall evolution on a Mendip estate, that of Charterhouse near Cheddar. They based their analysis on the supposition that, where two walls meet as a T, the stem of the T is likely to be younger than the crossbar, as a result of the construction technique. Using this basis for field division and supported by historical documents and charters, they found that 36% of walls were early medieval (AD410 to 1066), 25% were late medieval (1066 to 1550), 28% post-medieval (1550 to 1750), 8% from parliamentary enclosure (1750 to 1950) and 3% post-1950.

A similar type of land is called 'allotments' in Cumbria and *fridd* (pronounced 'frith', with hard th as in 'that') in Wales. If it is specifically common land that is enclosed, it is sometimes referred to as stinted land. Stints are typical of Cumbria, which has a large area of common land.

These land types can be combined to form three distinctive farm systems in upland agriculture:[3]

1. Dairy farms – mainly inbye and intake, these farms are confined to the upland margins, typically at 200–300m altitude, where precipitation is enough to encourage high grass yields, but the climate is mild enough to allow dairy cows to flourish.
2. Upland farms – typically like that in Figure 1.6, with a mix of all three types of land, running sheep and cattle. Most farms in the uplands fall into this class.
3. Hill farms – mainly open fell, with one field of inbye or none at all, and some intake improved as far as possible, thus constraining operations to sheep only. These farms are confined to the higher upland areas, such as the heads of valleys or far up on valley sides just below the open fell. Farms such as these can be seen in the upper reaches of the Cheviots and the Yorkshire Dales.

Because of the variation in land types that exist across the three farm classes, farm businesses have developed different enterprise combinations (outputs) to make optimal use of the conditions.

Outputs on upland farms

Central to the operation of any farm is its enterprises, which usually consist of crops, grass (sometimes classed as a crop or fodder) and livestock products. In uplands, farmers have always tended towards livestock production in order to suit the restrictions of the physical environment (see Chapter 4).

Historically there were crops to feed the family and crops to provide fodder for the livestock (dominated by cattle and to a lesser extent sheep, goats or horses). However, as time has gone on, cattle have been seen less and less frequently alongside sheep on upland farms (the reasons for this are discussed in Chapter 8), though hill farms are naturally less likely to run cattle due to limited grazing and water supply.

Patterns of farming vary from upland to upland. Figure 1.8 shows that beef-only farms are more important in the Peak District than in other English uplands, and sheep-only enterprises are concentrated in the Lake District and Yorkshire Dales. Mixed sheep and cattle farms do however dominate (an average of 44% of farms) in the English uplands. This pattern is not replicated in Scotland, Wales or Northern Ireland. For example, in Scotland sheep-only farms are dominant (44%), and mixed-stock farms account for only 17.2% of upland LFA farms (Scottish Parliament 2010, accessed 5 Nov. 2010).

Livestock

Sheep and beef enterprises are managed by planning the movement of stock from one type of land to the next, fitting the needs of cattle and sheep around each other depending on time of year. A proviso is that, if upland farms do not have enough inbye land or sheds/barns, the size of the cattle herd will be substantially reduced.

Sheep

The sheep enterprise is based on a flock containing a range of ewes of various ages, which act as the breeding stock. Most farms also have one or two rams, usually from different flocks to

[3] The terms 'upland agriculture' and 'upland farming' are used from here onwards to represent both upland and hill farms together, unless the point under discussion requires them to be distinguished.

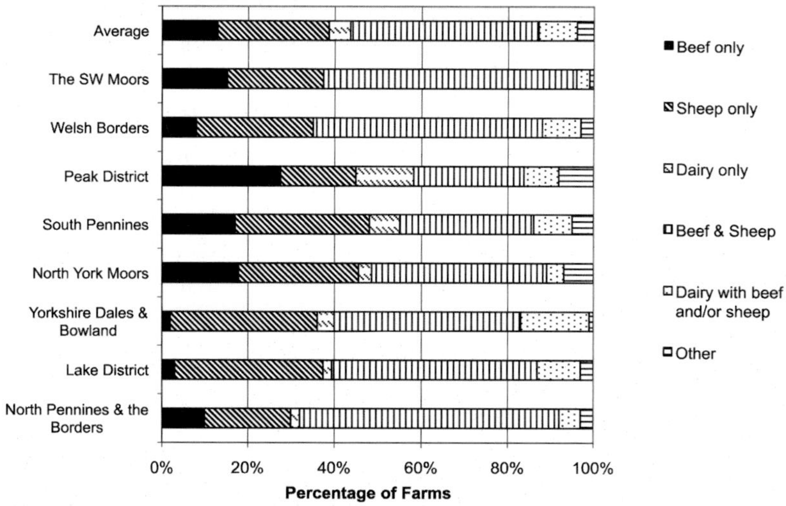

Figure 1.8 Enterprise Mix on English Upland Farms 2009
(Adapted from DEFRA 2009a)

avoid too much inbreeding. Lambs can be brought on to replace ewes that get too old to breed or can be sold on for fattening up in lowland Britain.

Breeds of sheep vary from upland to upland. Figure 1.9 shows some of the main breeds of England and Wales; most of them have been bred to deal with the conditions of their particular environment. Some upland flocks are made up of mules (cross-breeds) in an attempt to provide hardy stock with both good-quality fleeces and heavy carcasses for meat. Saul *et al.* (1992) investigated crossing hardy hill breeds with Australian Merino sheep, trying to raise both sale price and wool value, but this achieved only 6% to 9% increase in gross margins.

Cattle

Where cattle are kept, upland farms run livestock for beef, using suckler cows. The calves are reared by their mothers until they are moved off the farm for fattening in the lowlands. The suckler cows too are eventually slaughtered and enter the beef food chain. Herds are made up of one breed which is sired by certain breeds of bull.

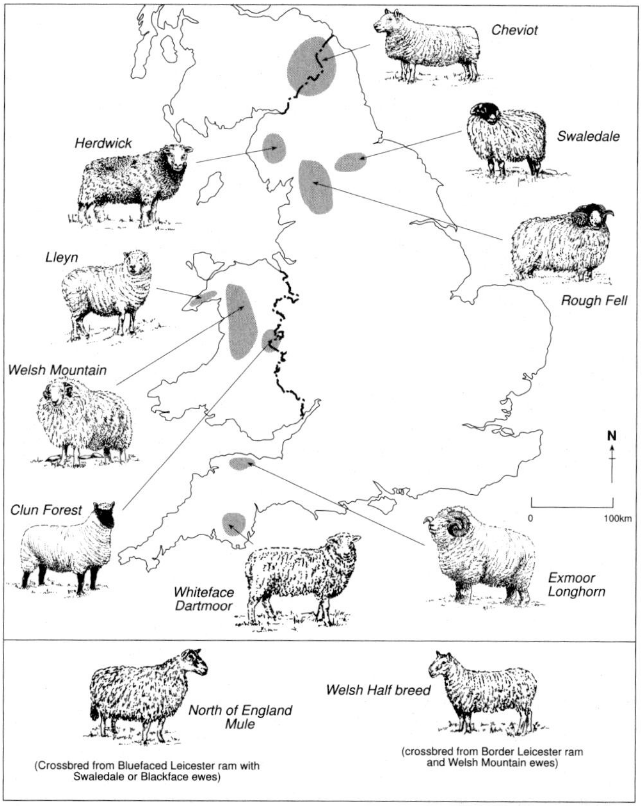

Figure 1.9 Upland Sheep Breeds in the United Kingdom

Currently, Limousin bulls are particularly popular as sires. Insemination may be either by natural means or through artificial processes. Rare breeds for specific purposes, such as Dexters or Belted Galloways, have become popular over recent years as farmers seek ways to add a premium to their meat to increase farm incomes (read more in Chapter 6).

Dairy herds on the upland margins are typically Holsteins, which have replaced Friesians because the Holsteins increase milk yields.

Ruminant anatomy

As the most important element of the upland farm system is livestock, it is crucial to understand a little about how sheep and cattle work. Both species are known as ruminants because they have a special chamber in the stomach known as a rumen, which has the ability to digest cellulose using symbiotic micro-flora (Figure 1.10). This means that ruminants are more efficient than, say, horses or ponies (which are monogastic – single-stomached) in digesting plant material, particularly mature vegetation of poor nutritional value (Bazeley and Hayton 2007).

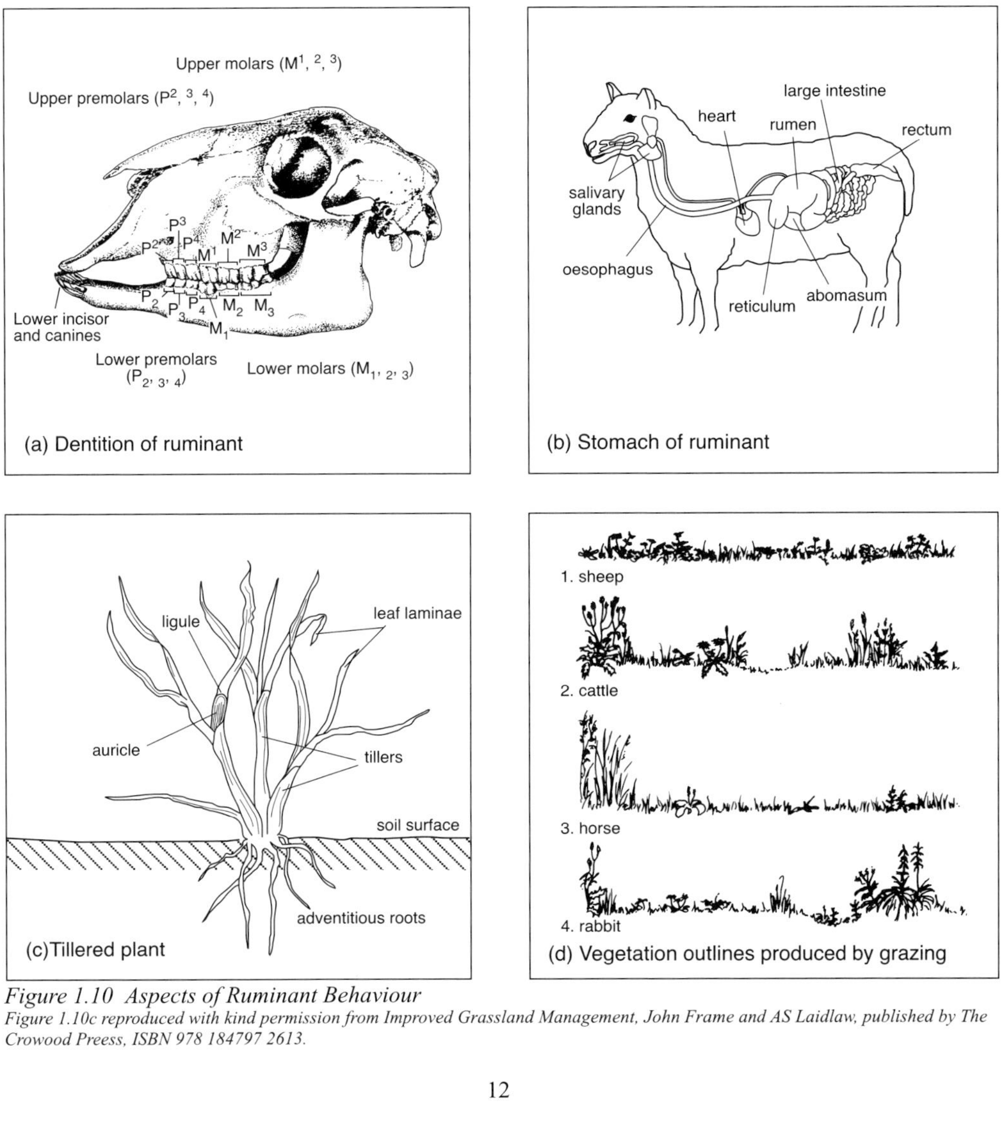

(a) Dentition of ruminant

(b) Stomach of ruminant

(c)Tillered plant

(d) Vegetation outlines produced by grazing

Figure 1.10 Aspects of Ruminant Behaviour

Figure 1.10c reproduced with kind permission from Improved Grassland Management, John Frame and AS Laidlaw, published by The Crowood Preess, ISBN 978 184797 2613.

However, not all ruminants are equally efficient at breaking down poorer swards. Those ruminants with a large rumen capacity are better; they can be spotted by their deep and wide body shape. This type of stock has gone out of fashion in many agricultural systems because the large offal quantity is rejected for human consumption at the end of the processing chain (Grayson 1997). Instead, current stock is bred to have smaller rumens, making them less efficient digesters of poor-quality swards. As a result, farmers have had to raise the nutritional value of grass (hence the explosion of silage and homogeneous swards) and rely more on supplementary feeds. Modern breeds of sheep and cattle therefore find it difficult to flourish on the poor-quality rough grazing of upland farms.

Livestock breeds

These fundamental facts of rumen physiology are important because they help to explain why upland farming systems rely on certain breeds of stock to graze the land.

Historically sheep and cattle were bred to fit in with their environment, so the stock on marginal land were adapted to survive cold harsh environments, to graze poor-quality swards and utilise the lower nutrition more efficiently. For example, Soay sheep will browse on many species and White Park cattle will specifically seek out coarser, older grasses. These 'native breeds' are often associated with particular environments where they can obtain enough nutrition to thrive. There are over fifty breeds of sheep and twenty of cattle that have evolved in this way in Britain (Evans and Yarwood 2000). The uplands are particularly well endowed with native breeds (Figure 1.9).

Notable breeds of sheep include the white-faced Cheviot and Welsh Mountain, who prefer to graze on white moors of mat grass and purple moor grass, and the black-faced Swaledale and Scottish Blackface, who prefer black heather moors (Alderson and Small 1997). Belted Galloways, White Park and Shetland are examples of cattle breeds typically found in upland areas.

Utilisation of native breeds is a double-edged sword. They put on weight more slowly, which delays sale; there is the aforementioned offal issue; some native sheep breeds do not respond well to shepherding with dogs; many have a low meat-to-carcase ratio, making them expensive to produce; and finally, decreased popularity has brought many breeds to the point of extinction. These are classified as rare by the Rare Breeds Survival Trust (see Box II).

Box II: Rare breeds

The range of livestock we now have has evolved over the millennia to fit in with its local environment. Figure 1.9 shows some of the different sheep breeds in England and Wales. However, in the 20th century many native breeds became increasingly rare or even extinct when they were abandoned in favour of breeds providing more lamb, beef, milk or wool.

The Rare Breeds Survival Trust (RBST) was set up in 1973 to protect British breeds of farm livestock and prevent any more from becoming extinct. Evans and Yarwood (2000) note that rare breed conservation can also:

- underpin our national heritage
- provide attractions at farm parks
- provide conservation management tools e.g. Hebridean sheep eat purple moor grass
- add a premium to sales as high-quality food
- act as symbols of success e.g. Dexter cattle are no longer an endangered breed

The RBST specifically defines 'rare' by three criteria (Alderson and Small 1997):

1. Genetics – requiring positive answers to three of these questions, including question a:
    a. Has the breed existed continuously for at least 75 years?
    b. Has there been an accepted stud/herd/flock for at least six generations?

c. Have other breeds contributed less than 20% of the genetic make-up of the breed in the last six generations?
d. Are the parent breeds used in the formation of this breed no longer available?

2. Numbers – a breed qualifies as rare if there are fewer breeding females than these:

Cattle	750
Equines	1000
Sheep	1500
Goats	500
Pigs	500

The effective population size can also be used as a numerical basis for rarity.

3. Current trends – is the breed's population declining significantly or is it found in fewer than four significant units greater than 50 miles apart?

Once rare status has been confirmed, the breed is placed in one of four categories (see Table 1), sometimes depending on its current trend, but usually based on the number of breeding females, as shown in Table 2.

Table 1: Rare breeds

	sheep	*cattle*
critical	Boreray	Aberdeen Angus (o.p.) Chillingham Wild Cattle Northern Dairy Shorthorn Vaynol Whitebred Shorthorn
endangered	Leicester Longwool North Ronaldsay	
vulnerable	Castlemilk Moorit Devon and Cornwall Longwool Hill Radnor Teeswater Whitefaced Woodland	Lincoln Red (o.p.)
at risk	Balwen Cotswold Lincoln Longwool Manx Loaghtan Norfolk Horn Oxford Down Portland Soay Wensleydale Whiteface Dartmoor	Gloucester Irish Moiled Shetland

Note: o.p. = original population.
Source: www.rbst.org.uk/watch-list/sheep and www.rbst.org.uk/watch-list/cattle, accessed 1 March 2011.

Table 2: Definitions of rare breed status

	critical	*endangered*	*vulnerable*	*at risk*
Cattle	150	250	450	750
Sheep	300	500	900	1500
Pigs	100	167	300	500
Goats	100	167	300	500
Equines	200	333	600	1000

Source: www.rbst.org.uk/watch-list/main, accessed 1 March 2011.

Grigg (1989) collated data for the structure of the national sheep flock and cattle herd. He demonstrated how Shorthorn cattle (dating from the 1780s) represented over 60% of the national herd before 1939, but by 1985 only 0.2%. The national sheep flock has always been more varied, with three dominant upland breeds, the Blackface, Cheviot and Welsh Mountain. During the 20th century, new breeds such as crosses, Clun, Kerry Hill and Masham have increased in importance.

There is also a European Union list of breeds at risk, which was created after the Foot and Mouth (FMD) outbreak of 2001 (EU Directive 2003/85/EC). The directive requires member states to create a register of holdings where stock is kept for the purposes of animal conservation and which are indispensable to the survival of the breed, if FMD were to threaten them. The idea is that special measures can be applied to these breeds at risk in such an event. At the time of writing, the RBST and relevant UK ministries and departments are drawing up a list of these breeds. The value of these indigenous breeds is also recognised in an EU proposal for a protocol on mountain farming, which suggests:

> livestock farming, including traditional domestic animal husbandry, must be maintained, with its characteristic variety of species and typical produce; such husbandry must be suited to local conditions and to the available land and compatible with the environment. …
>
> To this end, the necessary farming, pasture and forestry facilities shall be maintained, keeping a balance between pasture and livestock which suits local conditions, in the context of suitable, extensive grassland farming.
>
> Measures shall also be adopted to maintain the genetic diversity of livestock and crops, particularly in the field of agricultural research and advisory services. (EU Com 2006/059(CNS), Article 10)

The importance of rare breeds for conservation grazing has finally been officially recognised in the UK's agri-environmental strategies. For example, as part of the Higher Level Stewardship scheme in England, farmers can apply for a £70 per hectare supplementary payment under HR2 Native Breeds at Risk: Grazing Management (Natural England, 2008c).

In favour of native breeds is their ability to graze on poor swards, making them ideal for grazing semi-natural habitats managed for HNV conservation. Such grasslands can be organic in their management and this allows a premium to be added on the sale of organic meat or milk. However, some native breeds are more useful for this than others, depending on their environmental tolerances, grazing preferences and rumen efficiency. For example, a five-year trial by the Game Conservancy Trust in the North Pennines found that Hebridean native sheep were better conservation grazers than Swaledale sheep. The Hebridean sheep controlled purple moor grass and encouraged heather, whereas Swaledales did the reverse (Braithwaite *et al.* 1997). Native breeds also tend to be more resistant than other breeds to certain diseases prevalent in modern stock.

Grazing methods

Another important physiological difference between domesticated herbivores is that of grazing method. The morphology of the teeth and jaws is different in sheep and cattle (Figure 1.10).

A sheep's jaw allows it to bite off the vegetation. This produces an even sward, which can be grazed to as little as 3cm (1¼ inch) high. They also have nimble lips, which allow them to graze out selectively the most palatable species – a problem on open fells with a range of habitats on them (see Chapter 5). Complicating the grazing is the fact that young and old sheep do not have a full complement of teeth (which middle-aged sheep do), meaning that they do not graze very efficiently. The overall effect is the production of a uniform grass sward which over time can loose its plant diversity.

In contrast, cows bite, tear and pull at the vegetation by wrapping their prehensile tongues around a plant, leaving the sward with a more tussocky appearance. The design of their jaw means they can only graze to 6cm (2½ inch) above ground level and their unmanipulative lips mean that selective grazing is severely curtailed. The sward is therefore more structurally diverse – that is, more species are found and they grow to different heights. Cows are also more effective in controlling unwanted vegetation either by their ability to graze out dead plant material and litter or through the trampling effect of their hooves. Previously this has been useful for bracken control on upland slopes.

Fodder

The second most important enterprise on an upland farm is fodder. Fodder can be, in order of importance: grass, hay, legumes, root crops or cereal for feed.

Grasses

Grass is the dominant fodder because it is the most readily accessible over the farm system and least expensive to produce. It is estimated that there are over 10,000 species of grass worldwide. Of these about 160 are indigenous to the British Isles and adapted to the varying conditions of our natural and semi-natural habitats (Hubbard 1984). Although many of these are eaten by sheep and cattle, only very few have enough nutritional value to be the sole diet for modern livestock to survive on.

Thus grasslands used for stock grazing are often drained, ploughed and re-seeded with grass species of higher nutritional content. Lime and farmyard manure can be added to replace soil nutrients depleted or lost when a grass crop is harvested from the field. These processes are typical of inbye and to some extent intake land. Open fells are not treated in similar fashion due to the prohibitive cost of soil improvement; instead the stock feed off what they can forage, hence the use of native breeds adapted to these environments. Some attempts (generally unsuccessful) have been made to improve grazing quality through a process known as gripping (see Chapter 5). Attempts to improve grazing potential explain why many semi-natural grasslands have been lost in the lower areas of uplands rather than the open fells.

A further complication to the nutritional value of fodder is the palatability of a grass species. Palatability is a function of the taste, smell, digestibility, texture and leafiness of a plant. According to Crofts and Jefferson (1999), these features are influenced by:

- Chemical composition – alkaloids,[4] resins[5] and terpenes[6] make plants bitter
- Defence mechanisms – hairs, prickles and thorns
- Leaf: shoot ratio – leafy plants are more palatable than 'stemmy' ones

Favourable combinations of these characteristics are more prolific in grasses and legumes than other forms of plant, hence their use in pastoral farming systems. Many other plants, such as herbs or wild flowers, are less desirable in a sward and thus have been eliminated from the grassland composition in favour of more grasses.

Whatever the productivity and palatability of grass species, all of them behave in a similar way when grazed. The grass plant is defoliated above ground, either completely or partly, in a session of grazing. This activity stimulates grass plants to produce new buds and shoots from the base. This new shoot process is known as tillering; if carefully managed, it can actually improve the agricultural condition of a grass sward by thickening it up (see Figure 1.10c). At some stage in the growing season, the shoots switch from producing leaves to producing flowers. Grazing also increases soil nutrient status (by dunging and urine) and herbivore saliva has been found to stimulate plant growth (Miles 1987).

Though extremely good for agricultural purposes, excessive tillering can have a detrimental effect for nature conservation because a grass plant that is continually grazed never gets to

---

[4] Alkaloids are nitrogen-based compounds that are poisons or drugs.
[5] Resin is a semi-solid substance secreted from the sap of plants.
[6] Terpene is an aromatic hydrocarbon obtained from plant oils.

produce flower-bearing tillers. In extreme cases, excessive grazing pressure can lead to the tiller buds being eaten away completely, meaning the plant cannot re-grow. The permanent loss of sward reduces grazing capacity and eventually causes soil erosion. On top of this, different grasses have different tillering capacities. For instance, fescues (*Festuca* spp.) and bents (*Agrostis* spp.) are very good at tillering, whereas oatgrasses (*Trisetum* spp.) and sweet vernal grass (*Anthoraxatum odoratum*) are poor. Grasses that tiller poorly, and are therefore grazing-intolerant, can be eaten out or suppressed in a sward fairly easily – another way in which the conservation value of grassland can be removed (Crofts and Jefferson 1999).

Tillering is one mechanism by which a plant can resist grazing pressure. Plants can also tolerate grazing if they produce copious seeds or if their physiology compensates for the loss of production (e.g. by uptake of nutrients via roots). Plants also develop avoidance mechanisms (mechanical defence, chemical defence, escape mechanisms) to resist the effects of grazing. Thus, heavily grazed grasslands will eventually support only species that are grazing-tolerant, such as perennial ryegrass (*Lolium perenne*), eliminating diversity. Selecting only a few grazing-tolerant grass species results in the inbye land supporting the best-quality grazing, but having low species diversity from a biodiversity point of view.

On the intake and open fell, the converse is true. Here a wide range of grasses and other plants will flourish, providing high biodiversity but low grazing potential. A useful indicator of grazing potential is the dry matter content (DMC) of a grass sward. Frame (1992) describes how the potential maximum production of a grass sward has been calculated at 27–30 tonnes per hectare (t/ha) of herbage dry matter (DM). However, this was achieved on experimental plots with no limits to nutrient or water and the most nutritious grasses planted. In general, the maximum production for British farming practice has been calculated at around 10–13 t/ha DM.

Unfortunately, because of impoverished and unpalatable flora in upland grasslands beyond the fell wall, the maximum potential production averaged over all upland is around 3 t/ha, but lower for open fell because the grazing there contains more than just grasses (see Chapter 5). It is important to note that even this production potential is probably unachievable in many instances, since it does not take into account the other management challenges on upland and hill farmland, nor variations in weather. This explains why hefts are so extensive for relatively few sheep, in contrast to lowland situations. In practice, at least three times as much upland is needed compared with the same number of livestock on lowland.

Hay and silage

The grass produced on the inbye will either be directly grazed or cut at different times to produce hay or silage. Direct grazing on a pasture occurs at various times of year, usually in winter (see Figure 1.11a). The grass crop recovers for a first cut for silage, is then lightly grazed and later is given a second cut for more silage.

For a hay meadow, historically the field was briefly grazed (Figure 1.11b) and the stock taken off by midsummer to allow the plants to put on maximum growth ready for the hay cut in late August. Hay is defined by its moisture content being reduced from 80 to 20% of its weight (Frame 1992). The evaporation is mainly a field-based process, taking place over two or three days to minimise the loss of DMC and nutrients. The drying process slows as time goes by, as water is extracted first from the herbage wilting and then from stem material (a function of cell structure in various plant components). Thus, dry sunny weather with low humidity and a breeze gives ideal drying conditions. Any relative humidity over 65% will slow the process down considerably – a level that is often exceeded in the British uplands – so several days before cutting need to be as dry as possible too.

To enhance the drying process the hay crop was crushed or bruised and then tossed around (tedded). Once dried, the crop would be gathered into small heaps for final storage in barns or

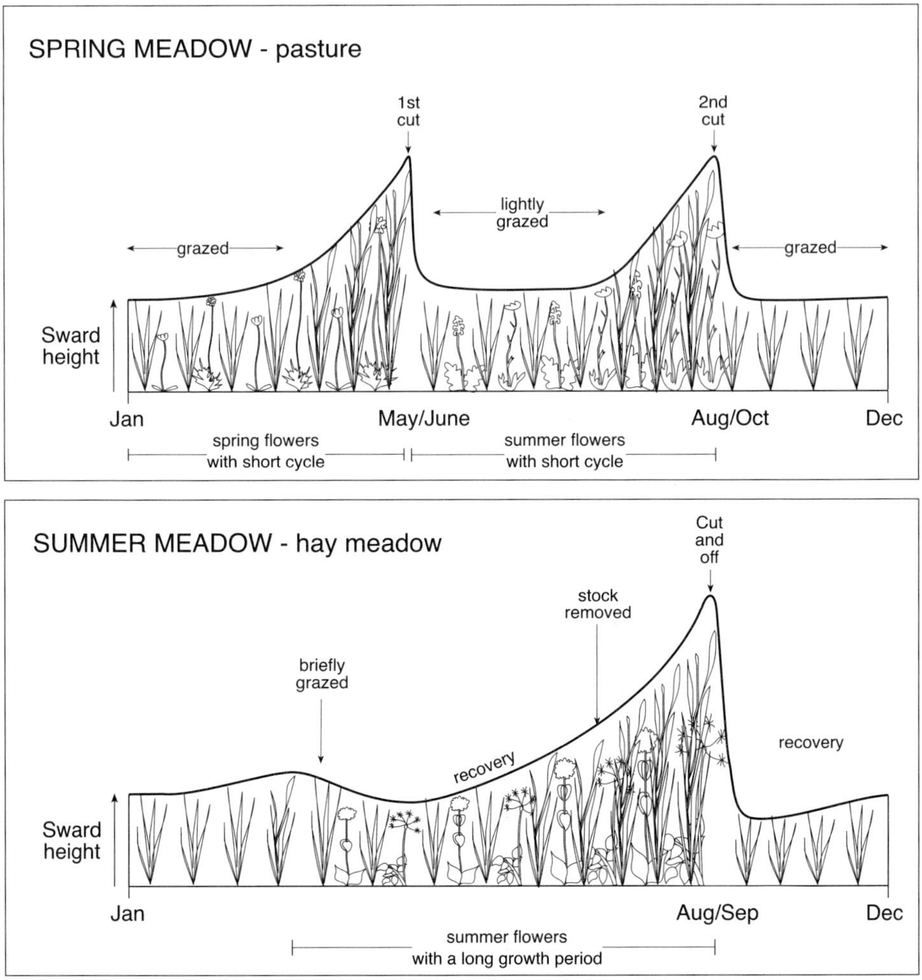

Figure 1.11 Cutting and Grazing regimes on grasslands

larger haystacks outside. Originally, this hay would be gathered and placed in barns out on the hillside, where several cattle would overwinter, so a farmer would have a number of field barns on their land (Figure 1.12). Before the advent of machines, haymaking was a labour-intensive task for the entire extended family. With the arrival of machinery, the hay was typically baled and then stored inside or out.

Most hay production – once typical of upland farms in Britain – has been replaced with silage (Brassley 1996), which is particularly important for the production of milk by dairy cows. Silage is grass that is harvested when its moisture content is at its peak; thus it is green and has about 35% dry mass by volume. Once the grass is cut, it is allowed to wilt in the sun, then baled and wrapped, usually in plastic. The bales are then stored either in the field in a stack or piled into barns back at the main farmyard. Another method of storage is to pile the loose silage in a three-sided bay known as a clamp. The clamp surface is covered with black plastic and weighted down, often with old tyres. With the air excluded (by either method) and an activator added, the grass ferments, releasing its sugars.

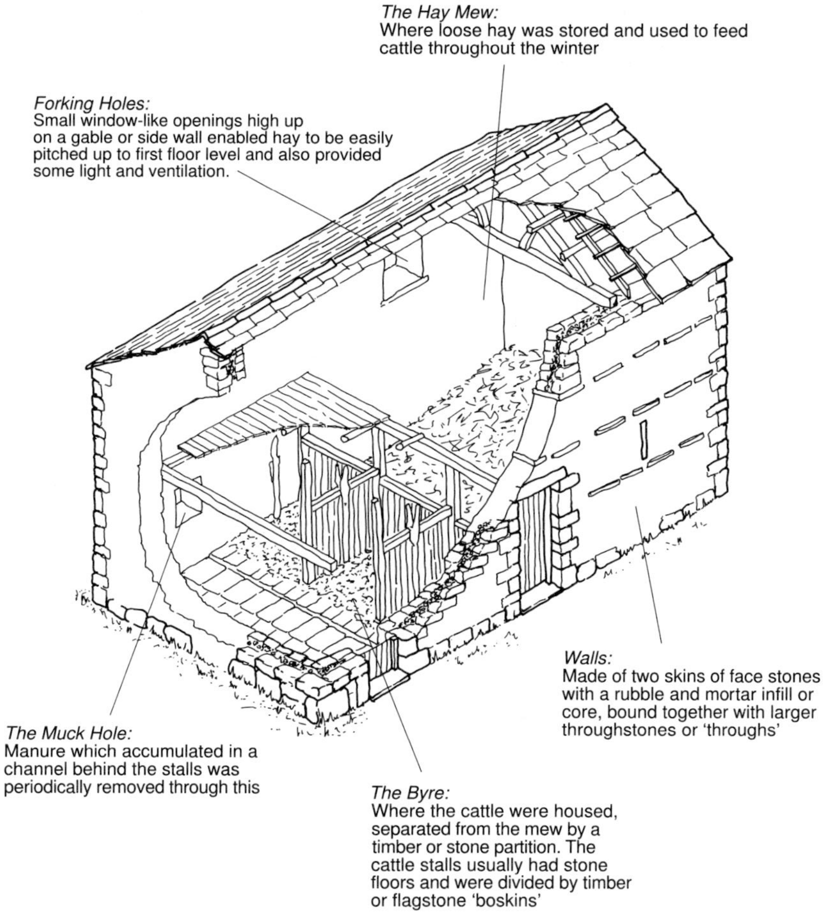

Figure 1.12 A Typical Yorkshire Dales Field Barn.
Reproduced with kind permission from Robert White, Yorkshire Dales National Park Authority.

The resulting silage is then used gradually over the winter months as a feed supplement for sheep out on fells or cattle inside in barns. Depending on location and weather conditions, several cuts a year can be taken from a single field. In the uplands this is usually two or just one. Frequent cutting is beneficial to cattle because grass that is younger when ensiled makes more digestible silage.

A particular concern for either hay or silage production is whether enough has been produced to feed livestock during the lean months. The length of supplementary feeding can increase considerably – from about 100 days to 130 days – if low temperatures, wet weather and/or poaching (see Chapter 4) delay either the turning out of stock or the beginning of grass re-growth. Thus two cuts of silage in the previous year help immensely, whereas a single hay crop would not. Lack of supplementary feed from hay or silage can force the farmer to look elsewhere for forage or even sell the stock in the worst case.

Other fodder

Roots like turnips or swedes tend to occupy an area for the whole growing season and thus can be expensive if a high yield cannot be achieved (Croston and Pollott 1994). They are fed to stock in late autumn and winter either by setting up a strip grazing system or lifting the roots and feeding them to livestock indoors. Forage can also be extracted from the green leaves of catch crops grown on land after a cereal harvest (June to August), before ploughing of an old pasture or as part of a land reclamation project. Brassicas are commonly grown for this purpose.

Maize silage is produced increasingly in some upland areas from lower lying fields. This is preferred by cattle farmers because it produces a highly digestible product after only one cut and the field can take high levels of slurry before planting (Phillips 2001). Leguminous (nitrogen-fixing) species are particularly beneficial for sheep where high protein levels are required. Croston and Pollott (1994) consider the inclusion of clover in upland grass swards as one way of allowing production systems that would not normally be possible in upland and hill areas. However, they do point out that a notable problem is retaining the clover proportion because competition from grass species, created by shading or high phosphorus content, can be difficult.

Concentrates are used when livestock quality needs to be high or there is insufficient grazing or hay/silage available on the farm. These feedstuffs are based on cereals or other high-energy protein plants, such as soybeans and oilseed rape. They are often fed as pellets or cakes bound together by molasses but they have to be purchased and brought to the farm at extra cost. Mineral and vitamin supplements can be added to the diet to correct any deficiencies.

The choice of livestock and the method of forage production are therefore important to ensure that economic success is underpinned by a sustainable management system. To do this, the upland farmer aims to operate a farm system which maximises the farm's potential while avoiding deterioration of the resources available. Most farmers reach this point through practical trial and error, experience and some training. We can now look at how this system actually operates.

Operating an upland farm system

At this point it is important to separate hill and upland farms from dairy units. In a dairy system the farmer is solely concerned with moving stock around between different improved grasslands at lower altitudes. The farmer ensures the grass remains as nutritious as possible by ploughing and re-seeding fields every few years. These temporary pastures are known as leys. Other inputs, apart from seeds, may include fertilisers, tile drains and fresh livestock.

Many dairy farms in upland areas overwinter their stock in barns because of the inability of modern breeds to tolerate harsh weather; this can lead to high levels of slurry production. It is not untypical for a dairy farmer to spread the slurry on vacant fields to recycle nutrients back into grass production, thus eliminating the need to buy artificial fertilisers. As the dairy cows languish in the barn over winter, it frees up the inbyes and leys for other stock. In this way, dairy farmers can graze sheep taken off the hills from upland and hill farms for the winter (known as off-wintering).

Hefting

Central to hill and upland farm systems is the hefting or heafing of sheep onto unenclosed land. Hefting ensures that sheep stay on a certain piece of land, either owned or managed through common rights bestowed on the farm unit (Box III). Initially, a shepherd and dogs show sheep the invisible boundary by herding them to this area or heft. Several hefts are known as a hirsle, but this term is now largely used only in Scotland. As time goes by, the sheep develop an instinct to remain within their virtual geographical boundary and do not wander through contiguous hefts (Figure 1.13).

Once the lambs are born, their mothers show them the heft and thus the knowledge of their grazing land is passed on from one generation of stock to the next. It is therefore important that the farmer maintains enough sheep in the flock to show the new generation the heft boundaries. Because there are no stock-proof borders between them, stock can wander into neighbouring hefts at will. They are contained on their correct heft either by active shepherding or by sheer

Box III: Common land

Common land dates back at least to medieval times. Land might be owned by one person, but access to and use of the land was bestowed on others. The right of common, as it is known, can include grazing rights (typical of upland areas), dead wood collection from the ground (termed estover), peat extraction (turbary), fishing (piscary) or pig grazing (pannage), as well as other rights such as bracken or gorse gathering, or mineral exploitation (Aitchson *et al.* 2000). Legally, a common is recognised as land that has rights attached to it, whereby

> A person may take some part of the produce of, or property in, the soil owned by another. (Aitchison and Gadsden 1992: 168)

Historically, turbary was particularly important in uplands because it provided fuel for domestic cooking and heating, and could also be used to produce charcoal for smelting lead and iron (Rotherham *et al.* 2004). A particular type of common grazing right is the stint, typical of Cumbria and the Yorkshire Dales (Winchester 2000; Raistrick 1984). This right is attached to only part of a certain piece of land, giving the right to graze a certain number of livestock (perhaps five sheep or one cow) within a known area (usually a particular hillside or watershed). Stinting has helped to underpin hefting and intercommoning.

From late medieval times, common land was enclosed by successive landowners and common rights gradually became eroded. In 1235 the Statute of Merton re-asserted the rights of commoners in an attempt to slow this erosion by stealth. In spite of the Statute, some lower-lying common land was taken into private ownership in late medieval and Tudor times. This was achieved through piecemeal enclosure or enclosure by agreement, either to improve the land for agricultural production or to stake claims for later, less scrupulous mineral extraction.

Land enclosed for farming in this period lay along the southern and eastern edges of upland areas where iron-pan podsol soils (see Chapter 4) only needed deep ploughing to improve their drainage (Williamson 2002). Agreed enclosure of stints and intakes for this purpose was more tolerated than the insidious piecemeal enclosure, because it helped to avoid direct confrontation between landlords and their tenants (Whyte 2006). On the high moors some piecemeal enclosure did occur before 1760, usually on forest land or where shieling grounds were converted to permanent hill farms (see Chapter 3).

Many commons were enclosed under acts of Parliament between 1720 and 1880. By then over 2.3 million acres had been enclosed (Overton 1996). The pattern was not uniform, the focus being on uplands, simply because they had not fallen to the earlier piecemeal enclosures. A series of commons acts in the late 19th century moved the emphasis away from their use to support people's livelihoods and instead towards recreation and access. This halted enclosure, but it created general misunderstanding about the legal status of commons (Staughton 2004).

Common management was eventually addressed in 1955 by the Royal Commission on Common Land, which pushed for a management system for commons, but the Commons Registration Bill (1965) only set up a register. This led to formal recognition of all commons in an inventory detailing their extent, ownership and rights. The register covers 554,409 hectares (1.37m acres), about 4% of the land area of England and Wales (Aitchison 1990). Unfortunately, there is evidence that during the registration process in the 1960s many rights were exaggerated. A typical example of over-registration was the grazing right of 10 ewes/ha recorded for the Long Mynd in Shropshire (Felton and Marsden 1990). Since registration, a number of commoners' associations have tabled and successfully enacted commons council bills (Hoskins and Stamp 1964).

Analysis by Aitchison *et al.* (2000) shows that there are large concentrations of commons in upland areas, particularly in Cumbria, North Yorkshire, Devon and Cornwall. The former two counties account for over half of the surviving commons in England (367,372 ha). The survey also revealed that 88% of commons are found in the lowlands (below 250m asl). However, they are generally smaller, accounting for only 27% of all common land.

The character of common rights varies from county to county. For example, the table over shows that cattle grazing rights are the most important in Cornwall (30% of commons), following the national trend, whereas in Cumbria sheep rights are most frequent.

Common land in England 1992 (Adapted from Aitchison and Gadsden, 1992)

A comparison of rights of common in 2000

	percentage of commons with such rights		
rights of common	*Cornwall*	*Cumbria*	*England*
Sheep	17.2	30.2	16.2
Cattle	30.0	29.8	19.9
Horses/ponies	19.3	20.8	12.9
Pasture (unspecified)	0.3	1.0	1.6
Other Livestock	6.6	8.9	7.3
Estover	8.3	13.2	10.3
Turbary	4.8	18.4	8.1
Common in the soil	3.8	14.9	4.7
Piscary	3.4	3.7	2.7
Pannage	3.1	1.0	2.0
Other rights	13.4	0	12.2

Note: Percentages do not add up to 100 because a common may have more than one right on it.
Source: Aitchison et al. 2000

The main aim of Aitchison *et al.'s* work was to record the character of the biological habitat on each common in preparation for a new Commons Act in 2006. National data show that the most typical upland habitats are acid grassland (27.2%), dwarf shrub heath (21.5%) and bog (18.4%), though their geographical distribution varies: acid grassland and bogs are more widespread in Cumbria than Cornwall (48.9% and 15.6% in contrast to 35.5% and 0% respectively). Dwarf shrub heath is greater in Cornwall (29.6%) than Cumbria (8.2%) but the ecological character is substantially different (see Chapter 5). Commons then are important resources for habitats identified under Biodiversity Action Plans and Habitat Action Plans (see Box XXI).

The recent Commons Act 2006 empowers commoners and landowners to manage commons more sustainably by regulating grazing and other agricultural activity. The act also strengthens protection against abuse, encroachment and unauthorised development. Commons registration authorities are required to bring the registers up to date and there are also provisions to prohibit severance of rights, that is, selling, leasing or letting common land separately from the property to which it is attached. Part 2 of the act provides the power for new commons councils to be formed. Three areas are currently piloting such projects in Gloucestershire, Bodmin Moor and Cumbria.

The right of common in the UK is not comparable with Garrett Hardin's 1968 concept of the 'Tragedy of the commons' (in *Science*, 162: 1243–8). He considered the impact of one person introducing an extra cow to a piece of land used jointly by a number of people, thus degrading everyone's grazing while the perpetrator gained from running another head of stock. Hardin considered whose responsibility it was to police such activities, when the common in question (which could include oceans, air and water) was not owned by anyone with ultimate authority for its management. In the British uplands, this does not happen as every piece of land now belongs to someone (see Box XXIX for further discussion of this concept). Commons are also not the same as waste, which is an historical term:

> although all commons were in law waste of some particular manor or group of manors, not all waste was necessarily 'common'. (Everitt 2000, quoted in Silvester 2004: 53)

In other words, waste lay beyond managed land; it was unoccupied, unclaimed and so accessible to all. After the Conquest in 1066 a lot of waste, such as central Dartmoor, was enforested[1] to secure legal ownership. In contrast, commons remained part of the village infield–outfield farming system (see Chapter 3).

[1] Enforestation brought land under the control of forest law (see Chapter 3).

grazing pressure interheft. A fell with common rights may therefore be made up of a number of hefts, which are shared between the farms surrounding the fell base. This is known as intercommoning.

Operating a farm management system in the uplands is more complicated than lowland pastoralism, for five main reasons:

1. Options are limited by how much of each type of land the farmer has access to, because inbye, intake and open fell are used for different activities in the farming year as an integrated system.
2. The poor nutritional value of the land limits grazing potential.
3. The proportion of enterprise mix between sheep and suckler cows can affect land-utilisation patterns and variable inputs.

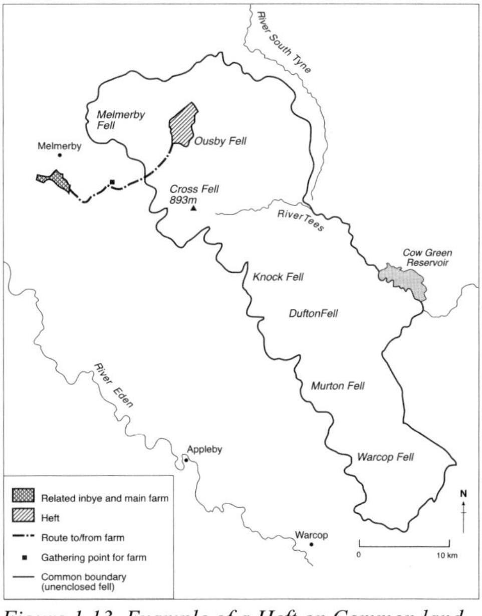

Figure 1.13 Example of a Heft on Common land on the North Pennines 2005

4. The erratic upland climate and weather patterns through any one year, and year on year, mean stocking rates may have to vary, leading to excess stock or underutilised land.
5. The impact of economic and political factors outside the farm can create imbalance between stock and land resource.

These farm-management issues and farmers' response to them combine to produce whatever agricultural land-use pattern develops in an upland area.

The first and second issues are inter-related. Hill farms have little or no inbye; this means they tend only to have sheep that can withstand the harsher environment and depleted nutritional base without excessive supplementary feeding (which costs money). The farmer is, therefore, totally reliant on the open fell and any semi-improved intake they might have. The problem is that in winter the only food available on the open fell tends to be heather (*Calluna vulgaris*), and the ewes particularly rely on this during their pregnancies and the early days of a lamb's growth.

The other challenge is that 75% of any grass growth occurs in just three months, fading away in later summer and autumn (Grigg 1989). Thus a farmer needs to avoid stocking densities that exceed the carrying capacity of their fell, adjusting stock levels to ensure that in the course of a year there are never enough sheep to permanently damage vegetation. The lowest density of sheep out on the fell is in winter, but this means there are too few stock there to take advantage of the summer flush earlier in the year. To solve this, historically farmers developed a strategy of maintaining only a breeding flock (or herd), selling on the new lambs (or calves) to keep winter stocking densities down.

Stratification

The sale of lambs to lowland farms for either fattening or overwintering has led to a process known as stratification, which is unique to the livestock industry of the British Isles.

Figure 1.14 shows how sheep are moved from hill to upland to lowland farms at different times of year and for different purposes. The system is also reversible, as when hoggs (weaned lambs) or first shearlings are returned back up the chain to the hill farm to be fattened further.

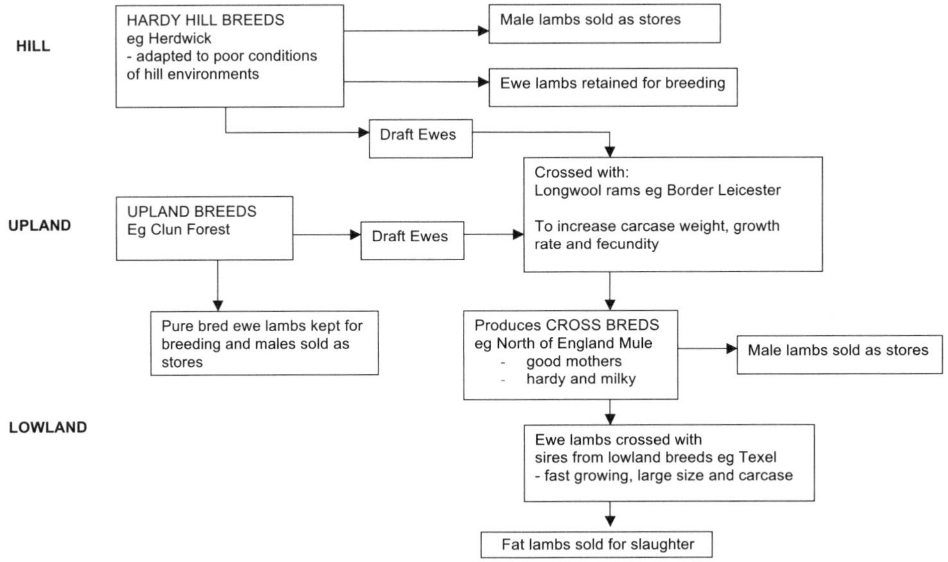

Figure 1.14 Stratification in the UK Livestock Industry

Suckler cows' movement follows a similar but simpler stratification system. The system is then modified in a variety of ways, depending on the local environment (hence the geographical variations of the upland farming system in the British Isles).

On Exmoor for example, upland farms tend to be only about 21% rough grazing and are smaller than in other English national parks. They also have a greater mix of stock types, with Exmoor ponies, cattle and sheep. Cattle are summered on the moor from May to June and the inbye is rested as ewes and lambs go onto the hill land in summer (June to October) (Miller *et al.* 1984). In Ireland, in a complete reversal of the system elsewhere in the British Isles, the wintering of stock on the limestone uplands of the Burren has been typical up until recently (Dunford and Feehan 2006).

The general farm management system is also altered in response to economic factors such as high production costs, low sale prices or lack of labour. The first option is to overwinter stock on a lowland farm, thus avoiding the loss of fell fodder. In this way, the sheep can come back to the farm, be fattened in a second year and then command a higher price when sold at market in the second autumn. The other option is to find ways to increase the nutrition available to the stock while overwintering on their own land. This can be achieved by improving the grass on intake or inbye, by planting fodder crops to supplement the animals' diet in winter or by buying in feed. Farmers will only choose the last-mentioned if wool or meat prices are very favourable; otherwise they will be out of pocket. However, increasing nutrition can cause physiological problems in the stock if they are native breeds (see above).

What the farmer opts to do is strongly limited by farm type and balance of land available. Hill farms, with no inbye and little intake, can rarely keep stock for a second year because the cost of overwintering or supplementary feeding outweighs the increase in sale price of lambs at market (unless prices are exceptional). For upland farmers there is greater choice because they can adapt management of their inbye and intake to support lambs on the land through to a second year (again depending on prices, though to a lesser extent).

Mixed livestock

The above suggests greater flexibility on upland farms in retaining sheep over winter, but this is countered by any suckler cows the farmer may have. At this point suckler herds and sheep flocks may be competing for the same resources, either those bought in or the use of inbye and intake land.

To understand the potential impact of this competition for resources, we need to appreciate the nature of the farming year for sheep and suckler cow production. Managing each enterprise is fairly straightforward if the farm contains either sheep or cows. However, if a farm contains both enterprises there are certain pinch points in the year when the stress on resources becomes difficult as sheep and cows are competing for the same land (see Table 1.1 for May/June and late September into October).

This is further complicated for a farmer who operates a hill farm, as they are unlikely to possess enough inbye to provide forage for the entire herd and/or flock when required between September and January. Thus stock are often sold off in November and gimmers[7] moved to the lowland to overwinter. To a certain extent this explains why hill farms tend not to have suckler cows, though many had a mixed economy in the past. Other reasons for the lack of mixed enterprises can be an overly harsh environment or specific financial incentives (discussed in chapters 4 and 8). For the upland farmer the stress on resources can be countered by moving stock (either sheep or cows) between the inbye and intake or by periodically selling animals. This also allows the farmer to produce hay or silage on some inbye fields for winter feed when stock have been moved elsewhere. Once again, specialisation in sheep or cows can help eliminate these problems of stock organisation at busy times of year on the inbye.

[7] Upland farmers use a range of traditional words to describe their stock, particularly sheep. Gimmers are female lambs, wethers are male lambs, draft ewes are older fell ewes; whereas first and second shearlings refer to sheep that have been sheared only once or twice, so are one or two years old. A hogg is either male or female and yet to be shorn at all.

Table 1.1: The Upland Farming Year

month	*sheep enterprise*			*suckler cow enterprise*			
	inbye	*intake*	*open fell*	*inside*	*inbye*	*intake*	*open fell*
November	Tupping of draft and hill ewes				Location dependent on weather		
December				Location dependent on weather			
January			Ewes with single pregnancies				
February		Ewes with twin pregnancies					
March					Calving, cows and calves		
April	Draft and hill ewes with twin lambs		Hoggs			Cows and calves	
May		Ewes with Twin lambs	Ewes with single lambs				
June							
July	Clipping and dipping				Bulls with cows		
August			Hill ewes		Cows and calves		
September	Surplus lambs sold or retained	Ewes for breeding or sold					
October					Calves sold		

(light shading) Pinch points	(dark shading) Single Occupancy of land

Weather

Unfortunately, there is one complication that neither hill nor upland farmers can avoid: variability of the weather through the year, and from year to year. For farmers, inclement weather can disrupt the entire system. If it is too snowy, wet or dry at certain times, the movement of stock can be delayed from one part of the farm to another, creating a domino effect elsewhere. Pastures that should have been vacated weeks ago can develop problems such as poaching or overgrazing as stock remain where they are; this happened during the Foot and Mouth crisis of 2001 (Box IV).

In extreme bad weather entire flocks can be wiped out, as in the winter of 1946 when three million sheep were lost in the uplands (Attwood and Evans 1961). Poor weather for a number of years can also exacerbate management problems on the farm and in some instances lead to financial ruin; this has been suggested as one cause of the Agricultural Depression of the 1880s and 1890s (Perry 1974; see Chapter 3). The full effects of global climate change will also have an impact (see Chapter 9).

Political and economic factors

The final factors manipulating the upland farming system are the impact of economic and political controls from outside the farm. The fluctuating price of meat or fleece can lead to huge swings in farm profits. Fifty years ago the June clip from sheep flocks would pay the land rent and wage bills for the year, now farmers are lucky to get 50p per fleece.

Governments have responded to this situation in the past by introducing mechanisms to guarantee farm incomes. However, if farmers are guaranteed financial support, it encourages them to increase the stocking of animals on their farms. This was particularly the case in upland Britain until 2000, when previous government policy encouraged stocking of sheep and

Box IV: Foot and Mouth Disease

Livestock can suffer from a wide range of diseases, which may kill them or substantially reduce their profitability. While stock cannot actually die from Foot and Mouth Disease (FMD), EU legislation requires the immediate extermination of stock infected.

FMD was first confirmed at an abattoir in Essex on 21 February 2001. The source was traced to a farm along Hadrian's Wall in Northumberland, but by then the disease had spread, finally infecting 44 counties by the end of the outbreak in January 2002. Over 6.5m animals were slaughtered, either from direct infection or as a result of the policy of culling livestock on over 7,000 contiguous farms. The news media showed graphic images of dead animals being burnt or buried and of bereft farm families, causing a public outcry about food safety, rural–urban equity and government ineptitude in dealing with the crisis. Initially, the government treated the outbreak as purely an agricultural problem, but as the situation deepened the wider rural economy was dragged into a social and economic crisis. Gradually the countryside shut down as local authorities closed footpath networks as a precaution, land-based charities shut their properties and the national parks asked people to stay away (Bennett *et al.* 2001).

The total economic cost of the outbreak has been calculated at £3 billion for the public sector, of which £1.1bn was compensation to farmers, and over £5bn for the private sector. Particularly affected in the wider rural economy were tourism – up to £3.2bn – and its allied industries, which lost £2.2bn (Franks *et al.* 2003). The transport sector also suffered as many haulage companies lost trade and laid drivers off; eventually some of the smaller ones went out of business. So, though the effect on farming was mainly linked to pastoral enterprises (notably upland ones), the impact on the rural economy was far more pervasive.

Many commentators could not understand why FMD spread so rapidly in 2001 and why the economic effects were so wide. The 1967 outbreak had been contained to 16 counties and only 442,000 stock were destroyed. The difference was simply that the livestock industry had restructured itself in many ways (Franks *et al.* 2003):

- There were more sheep and cattle.
- More stock moved around the countryside as a result of changing agricultural support.
- Stock movements covered longer distances as the number of livestock markets had declined.
- Production cycles had become shorter.
- Grass production had intensified.
- The farm and rural economies were more intertwined as a result of increased farm diversification.

Later inquiries recognised that the slow response of the Ministry of Agriculture, Fisheries and Food (MAFF), the ministry then responsible for agriculture, in not isolating the outbreak and not placing an immediate ban on the transport of livestock, had added to the speed and magnitude of the impact (Lessons to be Learnt Inquiry 2002).

For upland farms the most obvious effect was loss of income. Franks *et al.* (2003) estimated from a Cumbrian sample that an average farm lost about £41,000 net revenue. Farms directly affected or within the 3km contiguous cull zone received compensation; but the farm businesses beyond this suffered most financial hardship. They couldn't sell their stock, they had to continue feeding them and they received no compensation. As the crisis continued, the August lamb export market did not occur, leaving the remaining stock to be absorbed by the UK market (Barclay 2001).

Particularly acute were the social problems that emerged as the outbreak continued. Initially paid staff were laid off but were unable to get on and off farms, a lost variable cost that would help with cash flow. The personal trauma of culls is well documented; as movements and income became restricted, marital and domestic relationships suffered (Bennett *et al.* 2002). Self-policed farming and rural communities shut down, day-to-day movements were limited, farm children stopped going to school and some even boarded away for extended periods. A particular fear for upland farmers was the loss of hefted flocks. With no animals left to pass on the virtual boundary knowledge, the heft would either have to be abandoned or restocked using active shepherding or fencing. Many farmers had no experience of re-hefting because flocks often stay put when a farm changes hands. For others, the loss of cattle herds with long pedigrees was equally difficult. Even now, years later, many farmers still struggle to talk about what happened.

At the time opinions varied as to what would happen once the crisis passed. It was thought that some farmers would abandon land, some might extensify and others might restock to similar levels (Barclay 2001). Research in Wales suggested that many farmers who had previously contemplated intensifying livestock production would no longer do so, but only a small minority were considering giving up (Scott *et al.* 2004). This pattern was replicated in Cumbria, but a dichotomy emerged in increased diversification and pluriactivity by culled-out farms in contrast to farms not culled out (Franks *et al.* 2003).

The government, under pressure from all sides, set up the Task Force for the Hills under Lord Haskins to investigate how government could help English upland farmers develop more sustainable businesses that would also contribute more to the rural economy and environment (MAFF 2001). The ministry re-named itself.

discouraged cattle through a system of livestock subsidies (read more in Chapter 8). This also had indirect environmental consequences as rising sheep densities on open fells led to overgrazing, soil erosion and reduced biodiversity (see Chapter 5).

A further complication for farmers was that government then responded to these problems by introducing grant schemes, quotas and changes to the subsidy regime, which distort the economic market. With a farming system geared up to high stocking densities, the swift reduction of stock numbers has led to undesirable repercussions (Burton *et al.* 2005). Many farmers in the Lake District of Cumbria have reported problems of selective grazing on open fells as hefts have thinned out or disappeared completely. Other farmers have seen rapid encroachment of bracken (*Pteridium aquilinum*) across fells previously devoid of it. Furthermore, in the past, the drift of sheep off the heft would have been curtailed by a shepherd or by high stocking densities where competition for grazing kept sheep on-heft. The reduced densities mean sheep are free to wander off heft onto other hefts. Upland farmers therefore have a difficult balancing act to ensure their management system operates effectively enough to produce a profit at the end of the year. The proportions of land type and selection of stock are clearly important practical considerations for business success.

Chapter summary

Hill and upland farms are distinctive systems that exploit variations in topography, allowing stock to be moved around the land to make best use of resources available. Within this system, unique practices such as heafing and intercommoning exist. Hill farms focus on the production of sheep, which will either be overwintered at lower altitude to return in spring or be sold on to lowland farms. Upland farms have more flexibility in enterprise choice, often being able to combine sheep with suckler and/or dairy cows. Traditionally stock were bred to fit local conditions and a variety of breeds developed, many of which are now rare or less common.

The other important component of the hill and upland farming systems is production of fodder. This is derived either from grazing of unimproved fell land or from hay, silage and other fodder crops grown on valley floors. Productivity and palatability are essential to success, and thus farmers will employ any method to improve these two factors. The balance between stock numbers and food source can become critical and any imbalance can impair stock welfare or grazing potential. Farmers' ability to identify, manage and maintain this balance has developed through an accumulation of knowledge and experience over generations of farmers working with, and adapting to, the harsh upland environment.

Upland and hill farms rely fundamentally on sustainable exploitation of the upland physical and ecological environment, but other factors can control how farms operate, such as economics and politics. These two subjects are closely inter-related, as we shall see. But first we need to investigate more fully the definition of uplands and their physical and ecological character if we are to appreciate the physical context of upland agriculture in the British Isles today.

2

Defining the extent of uplands

To maintain any resource effectively, its extent and character need to be well understood. Mitchell (1989) sees resource inventories as critical to effective land use because they identify areas requiring special protection or controls, recognise dominant land uses and management potentials, identify immediate management needs and recognise data gaps and needs for further research.

Simple early land inventories, such as that developed in England by Dudley Stamp in the 1930s (Ilbery 1984), allowed the evolution of an agricultural land-quality system and proved useful planning tools as an initial phase of assessment. However, land inventories tend to be subjective (based on the knowledge and needs of the developer); even when replaced by landscape-evaluation techniques, subjectivity can still exist. For example, in England in the early 1950s the boundary of the Lake District National Park was decided by a group of 'specialists' looking at Ordnance Survey maps and conducting a variety of field visits (A. Fishwick, pers. comm. 2005). Instead, the mechanism of analysis needs to provide objective identification. In this way, there can be no cry of bias towards one user, suggestions of partisan management objectives or hidden agendas.

Where are the uplands?

The issue of subjectivity has long plagued inventory and appraisal of the upland resource. The precise delineation of uplands has been overlooked by many authors, keen to move onto the more meaty issues of describing their character and the challenges they face (for example Pearsall 1955; MacEwan and MacEwan 1982; Lowe *et al.* 1986).

There has been no collective agreement among them but, worse still, many have jumped onto a single erroneous bandwagon. The report *A Better Future for the Uplands* gave this broad definition of them:

> land predominantly above 240 metres (800 feet). … The 240 metre contour corresponded closely to the boundaries of the less favoured areas [LFA] of England and Wales as defined and approved under EC Directive 75/268, prior to the recent application to include marginal land. Consequently we have regarded the uplands as being synonymous with the LFA throughout this report. (Countryside Commission 1984: 9)

This definition directs our attention to two key features: first, the importance of altitude and second, the role of legislation. Much time has been spent in academic and professional literature defining mountains (e.g. Price and Butt 2000; Messerli and Ives 1997) and consequently uplands have been ignored somewhat. Smithson *et al* (2002: 501) in their physical geography text define mountains topographically as

> Land systems exceeding 600m in altitude with typically steep and often rocky slopes. The term is also applied to other high-altitude regions above 2km with more subdued relief which share other montane characteristics such as extreme climatic or geomorphic systems.

whereas they refer to uplands as

> less distinct elevated landscape with less extreme environments.

This vague definition is simply not adequate for policy makers dealing with land management in areas where uplands are commoner than mountains.

Legislation has done little to help. Although the Less Favoured Area (LFA) boundaries have their uses for upland farmers focusing on subsidies, grant schemes, other financial support or public perception, LFAs can label rural areas as 'marginal', 'lagging' or 'semi-natural', in contrast to 'prosperous', 'intensive' or 'planned'. The very phrase 'less favoured' instils little confidence that upland areas might ever prosper. However, the UK Government evidently finds the LFA meets its needs:

> Lord Henley asked Her Majesty's Government:
>
> Whether the Department for Environment, Food and Rural Affairs has a definition of upland or lowland Britain.
>
> Lord Whitty:
>
> There is no statutory definition of upland or lowland Britain. However, there are statutory maps of less favoured areas for the United Kingdom prepared under Council Directive 75/268/eec. These maps, *which are predominantly upland*, also include other farming areas which suffer disadvantage and are used for the purpose of targeting certain CAP support payments. (Hansard, 27 Jan 2003, Column WA137; author's emphasis)

Adopting the LFA definition raises several issues for the UK because:

- there is no formal boundary to uplands,
- the LFA, accepted as a substitute, contains areas that are not uplands,
- the definition is driven by legislative agricultural policy only.

Any of these situations would be unacceptable because, like all land-based resources, uplands are multifunctional; that is to say, they are used for a range of purposes, with a plethora of management objectives. Furthermore, as will we see, UK uplands are not synonymous with LFAs; if anything, the LFAs are a subset of the uplands, not the other way round. If a boundary has not been identified objectively, then the place where specific policy or management begins and ends may be inappropriate. Reliance on the LFA boundary also suggests a lack of dynamic, which does not reflect the reality of changing land management, whether intentional or not. The lack of a statutory definition has forced UK organisations to create their own, to suit their goals and objectives. Scottish Natural Heritage (2002a: 7) identifies uplands ('hills and moors') as:

> All land which lies above or beyond enclosed farmland or croftland, and which is covered by vegetation which is at least semi-natural in character.

Though still very vague, this does provide two more clues about the character of uplands: first, they have something to do with human activity and second, the type of vegetation plays a role in their definition.

As individuals we all have a notion in our mind of where uplands are and what they look like. Part of our 'world image', these ideas are shaped by our cultural background, education, race, religion, age, social class and so on (Walmsley and Lewis 1984). Generally the uplands in Britain are seen as part-tamed wild areas, free of the trappings of modern-day society, where people can experience freedom, peace and quiet, and a 'sense of place' (SNH 2002c). This is a dynamic concept, for even a hundred years ago uplands were seen as places of terror. This is how Dartmoor was described by Conan Doyle:

> Rank reeds and lush, slimy water plants sent an odour of decay and a heavy miasmatic vapour onto our faces, while a false step plunged us more than once thigh-deep into the dark quivering mire, which shook for yards in soft undulations around our feet. Its tenacious grip plucked at our heels as we walked, and when we sank into it it was as if some malignant hand was tugging us down into those obscene depths. (Doyle 1902/1986: 135)

A century earlier, in 1800, the uplands were romanticised by Wordsworth:

> There is an Eminence,- of these our hills
> The last that parleys with the setting sun;
> We can behold it from our orchard-seat;
> And, when at evening we pursue our walk
> Along the public way, this Peak, so high
> Above us, and so distant in its height,
> Is visible; and often seems to send
> Its own deep quiet to restore our hearts. (Wordsworth 1800/1888)

Unfortunately, such subjectivity is no help in resource management, but it does allow us to consider the sense of place – the tangible geographical idea of place along with the intangible concept of lived experience (Graham *et al.* 2009), which can effectively be used as a rural regeneration tool (see chapters 8 and 9). Nevertheless we have already identified four key parameters which can be used to define uplands more objectively:

- physical – topography, altitude and climate
- ecological – flora and habitats
- legislative – statutes and regulations
- cultural – human perspectives and activities

Each of these parameters is now considered in turn, before a short discussion of the results and what they mean for upland agriculture and the environment.

Physical parameters

In the British Isles we cannot afford to overlook our uplands and leave them undefined, because we do not have truly impressive peaks, such as those of the Alps, Pyrenees or Sierra Nevada.

Taking Smithson *et al's* 1994 definition, only about fifty peaks rise more than 600m above sea level (asl) in the British Isles. Instead, we have low-level mountains ranging from about 300 to 1344m asl, covering half of our land area. They have become important to us as our wilderness areas and our iconographic landscapes (Atherden 1992).

Not every state is lax over the physical definition of its upland areas. For example, France has specific laws on the *Creation of Massif Commissariats 1973* and the *Mountain Law with delimitation of massifs 1985* (Office of the Official Publications of the European Communities 1990). These two laws identify mountain (massif) areas as those generally over 700m, with slopes over 20% in over 80% of the area with two specific cases for the Vosges of 600m and Mediterranean area of 800m.

In Britain, one of the simplest physical definitions of uplands uses the Exe–Tees line (Figure 2.1), first proposed by Sir Harold MacKinder. It runs from the mouth of the River Exe in the south-west to the estuary of the Tees in the north-east. It is generally accepted – and British schoolchildren are taught – that the area north of this line is dominated by uplands; south of the line is mostly lowland (Millward and Robinson 1980). Mackinder's demarcation was based solely on his perception of the topography, which reflects the geology of these islands (see Chapter 4). However,

> Mackinder never thought of his line as a sharp 'frontier' but rather as a zone of transition where the lowland basins and lines of gentle hills of the south-east gradually gave way to the harsher contours of mountains and high plateaux of the north and west. (Ibid: 15)

This differentiation is geomorphologically correct and realistic, but it does not help policy makers and resource managers to identify uplands exactly for administrative and financial functions. One way to define uplands is by a contour height above sea level. Stapledon was the first to do this:

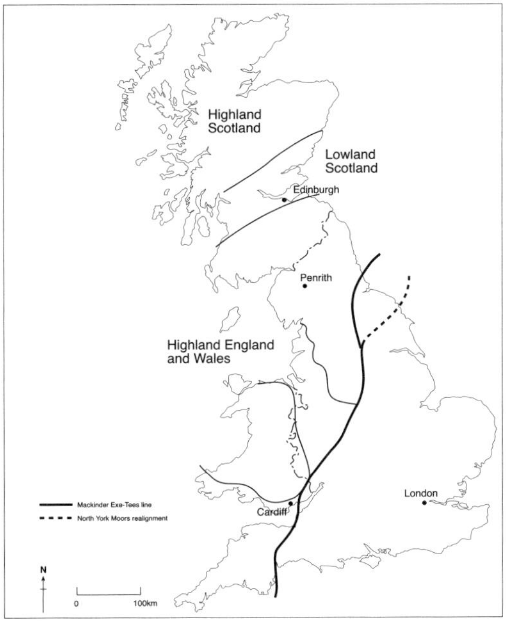

Figure 2.1 The Exe-Tees Line

> I shall define hill land as that part of the country which stands above the 700ft contour. ... in this country, however, farming practices, soil conditions, vegetation and the habitations of man all begin to show differentiating characteristics. (Stapledon 1937: 18)

The 700ft contour (215.4m asl) took in, at the time, most rough grazing land, hills and intakes. For the United Kingdom this represented 18m acres or one third of the country (Stapledon 1937). At another time he considered the hill areas as those above 1500ft (461.8m), about 3.5M acres (6.3% of Scotland, England and Wales), but viewed this land as 'utterly useless' in an economic sense. Since then, various contours have been stated in different text books, and even Stapledon himself chose different altitudes (800ft in Stapledon 1946; 200m by Hughes and Hunter 1988; 240m by Fielding and Howarth 1998; 250m by Backshall *et al* 2001; 300m by Atherden 1992).

These variations in altitude reflect the geographical scale of investigation and the environmental criteria used. The United Nations Environment Programme's World Conservation Monitoring Centre defines uplands as land above 300m asl (www.unep-wcmc.org/habitats/mountains/background.htm). Kapos *et al.* (2000) developed this typology by adding altitude and slope using GIS (Geographic Information System) data generated by the US Geological Survey in 1996. The results led to the division of mountains into six classes (Table 2.1), and most of Europe (99%+) lies under 3500m, with 41% of land being above 300m. For the UK various figures between 15% and 33% have been suggested above 300m (e.g. Ball *et al.* 1983; Ratcliffe and Thompson 1988). All land in the British Isles lies below 1500m, with the highest peak, Ben Nevis, reaching 1344m (Atherden 1992).

In a global context then, Britain and Ireland are not mountainous and nor are they perceived as such. Nevertheless, their 'low mountain' environments face similar agricultural issues to many higher parts of Europe. The reason for this is the British Isles' climate, which is influenced by their location on the western seaboard of Europe (see below). Climate as a criterion in defining uplands was explored by a European Union project (NORDREGIO 2004).[1] Initially the brief

[1] The NORDREGIO project defined Europe as including: Austria, Belgium, Bulgaria, Cyprus, Czech Republic, Denmark, Estonia, Finland, France, Germany, Greece, Hungary, Ireland, Italy, Latvia, Lithuania, Luxembourg, Malta, The Netherlands, Norway, Poland, Portugal, Romania, Slovakia, Slovenia, Spain, Sweden, Switzerland, UK.

Table 2.1: Classification of Mountains adopted by UNEP-WCMC

height asl	*global land area (10^6 km^2)*	*% global total*	*global mountain area ($10^6 km^2$)*	*% of global land area*	*European mountain areas (10^6 km^2)*	*% European land area*	*UK area (km^2)*	*% UK land area*
>4500	1.8	1.0	1.8	1.2	0.0	0.0	0	0.0
3500–4499	2.7	2.3	2.7	1.8	0.0002	0.004	0	0.0
2500–3499	6.9	7.8	6.9	4.7	0.50	9.2	0	0.0
1500–2499	11.9	9.5	5.3	3.6	0.16	2.9	0	0.0
1000–1499	15.1	9.9	6.2	4.2	0.36	6.6	402	0.2
300–999	53.5	33.6	13.0	8.8	1.22	22.5	34674	15.0
0–299	55.7	35.9	0.0	0.0	3.19	58.7	195453	84.8
Total	***147.6***	***100.0***	***35.9***	***24.3***	***5.43***	***100.0***	***231433***	***100.0***

(Sources: Kapos *et al.* 2000, p. 9; www.unep-wcmc.org 30 July 2004; Ball *et al.* 1983.)

followed UNEP-WCMC guidelines, but the report extended the mountain area definition, based on the premiss that the impact of altitude on temperature is relative (Box V). This is because temperature falls not only with high altitudes but also with higher latitudes; thus climatic conditions near the coast in higher latitudes are similar to those at higher altitude further south. A simple example of this is the Outer Hebrides, whose environment is similar to that of mainland highland Scotland. Some of the project's results are shown in Table 2.2.

Table 2.2: Example Areas pf Mountains classified by the NORDREGIO project

country	*country area (103 km^2)*	*mountain area (103 km^2)*	*mountain area % of country area*
Belgium	30.62	1.29	4.2
Germany	356.77	52.59	14.7
Estonia	45.23	0.00	0.0
Norway	323.90	295.86	91.3
Switzerland	41.30	37.46	90.7
UK*	245.49	65.01	26.5
Project total	**4760.68**	**1934.65**	**40.6**

*Note: The original report mistakenly placed Cumbria in Ireland; this has been recalculated here.

(Source: NORDREGIO 2004).

Part of the aim behind the NORDREGIO project was to overcome national variations in definition, a matter which plagues the EU (European Union) at many levels. For example, the French Government uses altitude in its legally accepted definition of mountains, as do other states, but defining altitudes can be as great as 1000m asl (Switzerland) or as little as 200m asl (Ireland). Table 2.3 shows that governments also use slope angles in defining mountains, and this is particularly relevant for farming. Steep slopes limit the use of machines, such as tractors (to 19° maximum), and accelerate soil erosion. Both problems limit the choice of enterprises for farmers, thus putting them at a disadvantage to their lowland neighbours.

Box V: Topographical criteria for NORDREGIO mountain areas

This classifies high land and steep slopes as mountainous, on the principle that the threshold for rough topography decreases as altitude increases; thus it includes steep-sided fjords and plateaux at high altitude. There are five classes, categorised by height asl (an example follows each):

- *Below 300m* – local variations in topography. Heights near the location are recorded at the eight points of the compass; if, when these are averaged, the standard deviation is 50m or more, the area is considered mountainous (e.g. Scottish sea lochs).
- *Between 300 and 1000m* – either class 1 criteria or have land between 300m and 1000m asl within a radius of 7km (e.g. most British Isles uplands).
- *Between 1000 and 1500m* – class 1 or 2 criteria plus slope angles of at least 5° (e.g. the uppermost peaks of the Grampian mountains and parts of Snowdonia).
- *Between 1500 and 2500m* – class 1, 2 or 3 criteria plus slope angles of 2° or more (e.g. Massif Central in France; none in the British Isles).
- *Above 2500m* – all land in this class (e.g. The Alps; none in the British Isles).

Each 1km square was allotted to one of these five classes, which slightly overlap.

Temperature variation was included by adding a contrast index, using two temperature variables: the average monthly maximum temperatures above 0°C and the average monthly minimum temperatures below 0°C. This index produces values between –1 (where average minimum temperatures never climb above freezing) and +1 (where the average minimum temperature never drops below freezing). For Europe, the range is between –0.36 and +1.0. To define mountains, the project chose +0.25 as the temperature contrast index threshold, and included areas with values worse than this.

Finally the project team examined the areas enclosed by municipal boundaries (in the UK census enumeration wards were used). A municipal area was classed as mountainous if 50% of its land fell within the topographical and climatic criteria.

Table 2.3: Definition of Mountain Areas in Europe

state	*min. height asl (m)*	*other criteria*
Austria	700	also above 500m if slope >20%
Belgium	300	–
Bulgaria	600	also >200m altitudinal difference/km2 or slope >12°
Cyprus	800	also above 500m if average slope >15%
Czech Rep.	700	–
France	700 (generally) 600 (Vosges) 800 (Mediterranean)	slope >20% in >80% of area
Germany	700	climatic difficulties
Greece	800	also 600m if slope >16% or below 600m if slope >20%
Hungary	600	also above 400m if average slope >10%; or average slope >20%
Ireland	200	–
Italy	600	altitudinal difference of >600m
Poland	350	or >12° slope for >50% of agricultural land in municipality
Portugal	700 (north of R. Tejo) 800 (south of R. Tejo)	slope > 25%
Romania	600	also on slopes of >20°
Slovakia	600	also above 500m on slopes >7o; or if average slope >12°
Slovenia	700	also above 500m if more than half of farmland is on slopes of >15%, or average slope >20%
Spain	1000	slope > 20%, elevation gain 400m
UK	240	–

Note: The other eleven states in the study had no formal definition. (Source: NORDREGIO 2004, p. 150).

The EU is now considering combining topography and climatic harshness to define mountains. Article 18 of Council Regulation (EC) no. 1257/1999 suggests that harsh climatic conditions could justify defining all lands north of 62°N latitude as 'mountain areas'. This is a function of climatic regimes in Europe. Moving from west to east across the continent, the climate becomes drier and less harsh, moving from oceanic to more continental (Barry *et al.* 2009). This is compounded by the ability of mountains to make their own climates to a certain extent, through orographic uplift (see Chapter 4). Also, the British Isles' location on the edge of the North Atlantic means their lower uplands are, in climate, like higher-altitude areas in continental Europe. This has influenced the character of vegetation and consequent economic activity unfortunately, Article 18's definition does not include any land in the UK, which reaches only about 58°N.

Using physical parameters is further complicated because we have two environmental gradients crossing the British Isles (west to east, getting drier; north to south, becoming warmer). The Countryside Commission considered this in their definition of uplands by combining 51 attributes of physiography, climate and soils to create eight classes (CoCo 1978). Classes 1 to 4 were described as marginal uplands – where altitude, temperature, slopes, rainfall and soils created milder conditions for agriculture – and classes 5 to 8 were the upland core where the environment was more severe. In this way the Countryside Commission could classify the upland national parks of England and Wales into two groups: Northumberland, the North York Moors, Exmoor and Dartmoor were marginal uplands, all others the upland core (Table 2.4).

Table 2.4: Countryside Commission's Definition of Uplands 1978

class	altitude (% land over 240m asl)	slope (modal* class in degrees)	rainfall (modal class in mm/yr)	temperature (modal class in annual sum of days over 5.6°C)	dominant soil types
1	28	12–22	760–1014	1375–1649	brown earth
2	16	0–11	1015–1524	1650–2000	brown earth
3	9	0–11	760–1014	1650–2000	brown earth and calcareous soils.
4	20	0–11	760–1014	1375–1649	gley, brown earth and stagnogley
5	76	12–22	1525–2284	1100–1374	podzol, brown earth and stagnogley
6	73	12–22	1525–2284	1100–1374	brown earth, podzol, stagnogley and peat
7	71	12–22	1015–1524	1100–1374	brown earth, gley and stagnogley
8	62	0–11	1015–1524	1100–1374	stagnogley, gley, brown earth and podzol

* Note: the most commonly recorded class in the entire sample. (Source: adapted from Countryside Commission 1978.)

Ecological parameters

A second way we can define uplands is by ecology. The occurrence of specific ecological zones at certain altitudes has long been recognised (Humboldt 1805, cited in Messerli and Messerli 2007). As we go higher, climates become harsher, more precipitation falls, temperatures are lower throughout the year, and wind chill and exposure increase. Growing seasons get shorter and aspect affects the amount of sunshine a slope receives.

These variables combine to influence soil development and plant communities (Chapters 4 and 5). Globally, the resulting natural variations are typically classified into four main ecological zones: nival, alpine, sub-alpine and colline (Figure 2.2). Because of the earth's angle of tilt and local insolation levels, these zones occur at higher altitude as we get closer to the equator. Figure 2.3 shows the zonation and slightly different terminology used in North America and Eurasia.

Each zone has different ecological attributes. In Europe they are as follows:

- Nival – the highest zone, where snowfall exceeds snow melt all the year, is dominated by glaciers and snow. This zone has no biodiversity save for transient birds and mammals. It does not exist in the British Isles.

- Alpine – a zone where snow melt can exceed snowfall, so ground can be intermittently exposed to form simple soils supporting fragile plant and animal communities. It is divided into three sub-zones:

  – Upper Alpine: low temperature and thin soils allow discontinuous open vegetation to develop. Not found in the British Isles.

  – Middle Alpine: low temperatures encourage snow to lie long into summer, and harsh scouring winds limit plants to prostrate forms. These environments are found at 900m+ in the British Isles and include grasslands, moss beds and lichen heaths (see Chapter 5). It covers about 113,000 ha of Britain (Thompson *et al.* 1987).

  – Lower Alpine: with snow melts for large parts of the year, plant and animal communities evolve in low-temperature, high-precipitation environments. This zone, found above 600m in the British Isles, consists of grassland plus bilberry (*Vaccinium myrtillus*) and ling (*Calluna vulgaris*) heaths, representing about 554,000 ha of Britain (Thompson *et al.* 1987; see also Chapter 5).

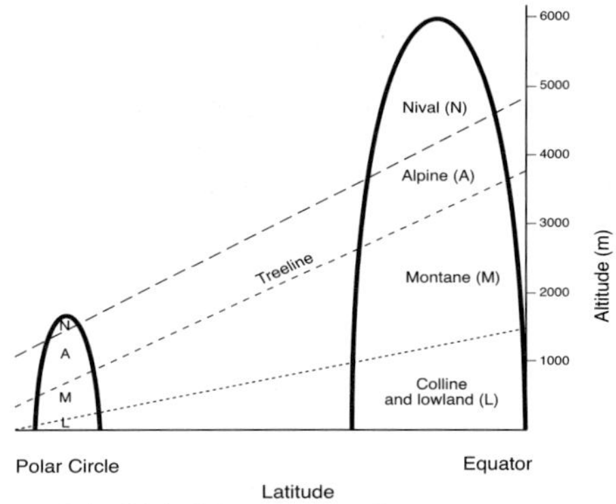

Figure 2.2 Global Bioclimatic Zones.
Based on materials first published Scottish Natural Heritage in Thompson DBA, Price MF & Galbraith CA (2005:45) *Mountains of Northern Europe*

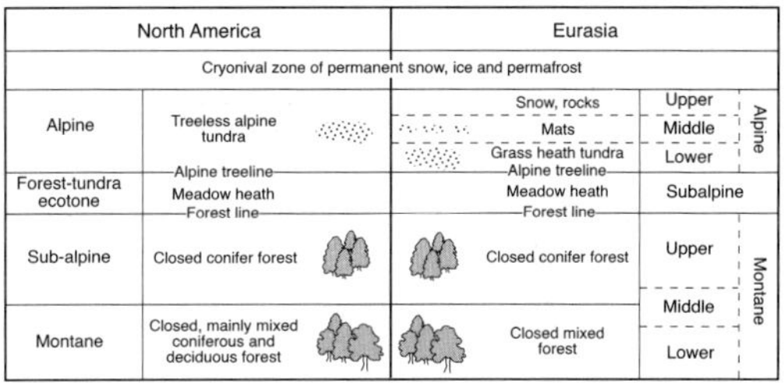

Figure 2.3 Changing Vegetation with Altitude.
Reproduced from Smithson P *et al.* (2002) 'Fundamentals of the Physical Environment' 3rd edition Figure 25.12 with permission from Routledge.

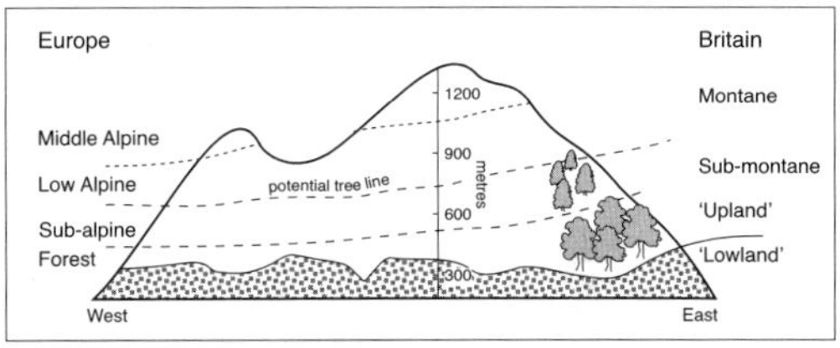

Figure 2.4 Ecological Zones in Northern Hemisphere Mountains.
Reproduced from Fielding AH & Howarth PF (1999) '*Upland Habitats*', Figure 2.2 on page 9 with permission from Routledge

The middle and lower alpine vegetation is sometimes known collectively as montane heath. In Britain, 'montane' has been traditionally used to describe the uppermost zone of upland (Poore and McVean 1957), but it is probably better to use the continental term 'alpine' as Horsfield and Thompson (1996) suggest, for straightforward comparison with mainland European research (Figure 2.4).

The zonation continues downwards:

- Tree line – the ecological division between alpine and sub-alpine zones (Box VI). In the British Isles much of our natural tree line has been lost to clearance. Only in a few places can it still be seen, in inaccessible gullies or on upland areas where human interference has been minimal, such as north-west Scotland. Referred to as montane scrub in Britain, it has been mainly replaced by various types of grassland and shrubby heathland through management (see Chapter 5).

- Sub-Alpine Zone – runs from the tree line down to the level where field boundaries begin to appear. Depending on the level of human activity, this zone is made up of various more natural habitats like deciduous woodland or pine forest. In the British Isles, most of these have been cleared and replaced by semi-natural (*plagioclimax*) vegetation to form grassland or dwarf shrub heath (heather moorland). Among the exceptions where ancient woodland remains are Wistmans Wood on Dartmoor and Young Wood in the Lake District. These have survived because they were inaccessible and did not provide useful timber.
- Lowlands – the zone down to the tidal high-water mark, where human activity has had most influence on biodiversity through various forms of management. Zone sometimes called 'colline' (Nagy *et al.* 2003).

The terms 'sub-alpine' and 'colline' are typically used in continental Europe. In the British Isles 'sub-montane' has traditionally been the equivalent of 'sub-alpine', but once again it is probably more appropriate to follow Horsfield and Thompson (1996) and use the continental term for ease of comparison.

Box VI: The tree line

The tree line is the contour above which no more trees can grow. Its altitude depends ultimately on a combination of latitude and continentality. Higher latitudes and more oceanic conditions lead to lower tree lines because environmental lapse rates are steeper in these conditions, being particularly affected by higher wind speeds and more cloud (see Chapter 4). Thus temperatures are higher on the general tree line in the UK than at comparable altitudes in continental Europe.

There is no single treeline because every tree has different ecological requirements. Thus, beech trees are limited to lowlands, whereas sessile oaks and aspen can grow at much higher altitudes. Generally, coniferous trees can grow at higher altitudes than deciduous ones. Tree growth on mountains is primarily limited by the tree's ability to deal with winter desiccation (gradual drying out of foliage). Growth can also be limited by air temperatures at night that are too cold for canopy foliage to survive at a certain height above the ground. Lack of nitrogen-fixing nodules or *mycorrhizae* (symbiotic fungi) in certain species can also inhibit growth.

Trees at the tree line tend to be stunted, their limbs are twisted and annual growth rings very narrow. Typically they grow to 1m to 2m high at most. Being so close to their physiological limits, trees growing here can in some years fail to set seed, waiting for warmer years to reproduce (Mackenzie 2000). There are two main types of tree line:

- *Abrupt* – where the occurrence of trees stops abruptly as you move uphill because trees no longer have the appropriate conditions to flourish, or sometimes because of high grazing pressure (Hester and Brooker 2007).
- *Gradual* – a function either of the surrounding topography or lower grazing pressure. A concave slope encourages trees to grow further up slope in what would normally be the alpine zone; gullies typically allow this. A convex slope encourages cold air to descend lower down slope and thus the tree line descends into what would normally be the sub-alpine zone.

Tree-line communities are examples of marginal habitats, where the upper altitudinal limits are controlled by climate and the lower ones by competitive displacement by other species (Hester and Brooker 2007). As a consequence these habitats can be limited in extent, as is the case with the montane scrub found in upland Britain, which can easily be reduced by climate change or land-use pressure (see chapters 4 and 5). The situation is complicated because tree and shrub species react more slowly than flowers and grasses to changing conditions, an effect exacerbated by the phenological strategies of individual species adjusting to periodic events.

Another term of importance is the timberline. This is always lower down slope than the tree line. It is important for farmers and foresters because it is the altitude above which trees are no longer able to provide marketable timber. Confusingly, the term 'timberline' is sometimes used to define the natural tree line, particularly in North America (Horsfield and Thompson 1996).

The spatial inter-relationship between ecological variations with altitude and the changes in abiotic (physical) factors is known as biogeographical zonation. This can be used on any spatial scale – the British Isles as a whole are classed as 'Atlantic' under the EU 1992 Habitats Directive (see Box VII) – and in the UK such zonation has been used as the basis for inventories of land resources since the 1980s. At this scale, three biogeographical systems impinge on our discussion of uplands definition:

- The ITE Land Classification of Great Britain
- The Countryside Survey of Great Britain
- The Natural Areas Initiative in England

The ITE land classification of Great Britain

The Institute of Terrestrial Ecology (now Centre for Ecology and Hydrology) developed the ITE land-classification system in the 1980s. It identified 13 types of upland out of a total of 32 land classes (Table 2.5; Bunce and Barr 1988; Bardgett *et al.* 1995).

Table 2.5: ITE Upland Land Classes

land class		*description*
17	MARGINAL UPLAND	Rounded intermediate slopes, mainly improvable permanent pasture
18		Rounded hills, some steep slopes, varied moorlands
19		Smooth hills, mainly heather moors, often afforested
20		Mid-valley slopes, wide range of vegetation types
28		Varied lowland margins with heterogeneous land use
31		Cold, exposed coasts with variable land use and crofting
21	UPLAND	Upper valley slopes, mainly covered with bogs
22		Margins of high mountains, moorland, often afforested
23		High mountain summits, with well-drained moorlands
24		Upper, steep mountain slopes, usually bog-covered
29		Sheltered coasts with varied land use, often crofting
30		Open coasts with variable land use and crofting
32		Bleak undulating surfaces mainly covered with bogs

(Source: Bardgett *et al.* 1995).

As with all biogeographical classification systems, the ITE inventory was based on physical features rather than land use (which is a function of human management or impact). The survey allowed for long-term trend analysis; it was completed in 1978 and first revised in 1984. The ITE classification divided uplands into two main groupings (landscape types), which Bunce *et al* (1996) refer to as marginal upland and upland (see Figure 2.5). The upland group (six classes) was identified mainly in Scotland and northern England and covered 30% of Great Britain. The uplands were typified by high mean altitude, very low winter temperatures, low sunshine hours and most snowfall (Table 2.6); the geology was predominantly igneous and metamorphic. The landscape consisted of hill farming with semi-natural vegetation cover (bog, moorland and grassland), woodland and open countryside with little infrastructure.

Box VII: The Atlantic Biogeographical Zone

The character of the Atlantic zone is influenced by climate, topography and animal and plant migrations (Hopkins and Buck 1995). Climate is controlled mainly by air masses developing over the Atlantic, where the circulation of the Gulf Stream brings warm water to Europe's western seaboard. The effect is to produce warm winters and cool summers with high precipitation, which does not have any marked seasonal variation. Local climatic conditions provide many ecological niches for habitats and species that have uniquely adapted.

The landscape of the Atlantic Region can be divided into two distinct topographic units:

- *The Northern European Plain* – an area of relatively flat land dominated by sedimentary rocks cut through by rivers forming a series of escarpments. Intensive land use had a profound impact on its biodiversity in the second half of the 20th century.
- *The Celtic Fringe* – consists of the western seaboard of mid-latitude Europe dominated by higher ground. The geology covers all three main rock classes. Some uplands contain alpine vegetation, and other sub-alpine habitats are extensive.

Plant and animal movements give the Atlantic zone an ecological coherence. Contemporary seasonal migration routes are typical, particularly for birds. Since the retreat of the ice (10,000BP) sedentary organisms have gradually migrated northwards. Many species which were once adapted to cooler conditions now reside in small refugia, which are rare though widespread.

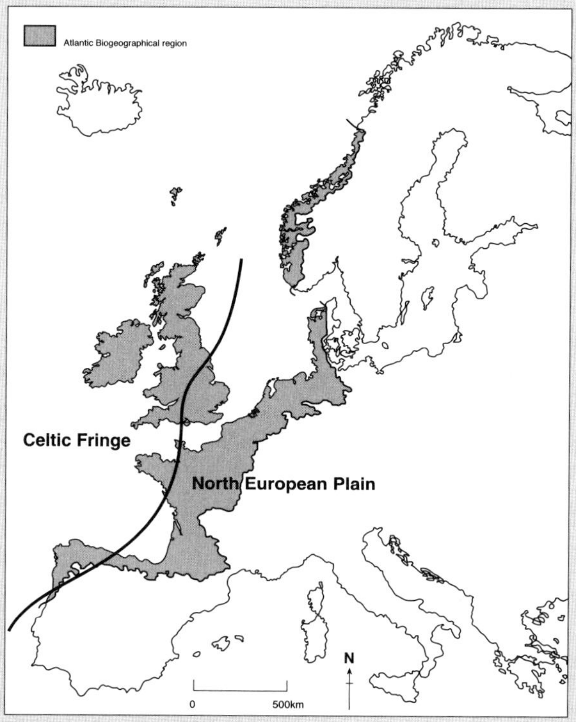

The Atlantic Biogeographical zone
Adapted from Hopkins and Buck, 1995

The inter-related factors of climate, topography and movement have produced distinctive flora in the Atlantic Region. Roisin (1969; cited in Hopkins and Buck 1995) divided the flora into three broad groupings:

1. *Eu-Atlantic* species, which are confined to the Atlantic Region;
2. *Sub-Atlantic* species, which are at a maximum within the region but spread beyond into other regions, diminishing with distance;
3. *Eury-Atlantic* species, which are confined to areas with Atlantic climate in both Europe and North America.

Roisin estimated a rough split of 49/45/6% for these three groupings. But, as Hopkins and Buck (1995) point out, his list of species was incomplete, missing out UK endemics and bryophytes for example. There are 12 bird species that breed almost totally within the Atlantic Zone (e.g. Scottish Crossbill *(Loxia scotica)* and Red Grouse *(Lapogus scoticus)*) and various species that overwinter within it (e.g. Barnacle Geese). Other birds use it as a stop-off point on their north–south migration routes.

Table 2.6: Comparison of the Four Main Groupings under the ITE Land Classification

	upland	marginal upland	arable	pastoral
Mean altitude asl (m)	313	244	76	87
Maximum altitude asl (m)	1225	985	280	340
Sun hours (per day in July)	4.4	4.9	5.8	5.7
Mean minimum January temperature (°C)	0.3	0.9	0.7	1.4
Average days of snowfall each year	48.0	36.9	26.0	22.1

(Source: Air Ministry Data 1941 to 1970, cited in Bunce *et al.* 1996.)

In contrast, marginal uplands covered 16% of the land area and occurred on the fringes of the upland grouping in western and northern Britain. These areas had a more moderate environment than the upland group, with medium altitude, low winter temperatures, and near-average amounts of sunshine and snowfall (Table 2.6). The geology was dominated by softer metamorphic rock. Land cover was more diverse with low-intensity agriculture, forestry and semi-natural vegetation (Bunce *et al.* 1996). This classification for the first time gave some generic guidance on the typical physical features of uplands and their related human activity.

The Countryside Survey of Great Britain

In 1990, the ITE classification was replaced by the Countryside Survey of Great Britain, whose purpose is long-term environmental surveillance of biodiversity in relation to negative activities and amelioration measures (Smart *et al.* 2003). Thus it is more than an inventory; it is an analytical tool for management decisions.

Based very much on the ITE land classification groups, the Countryside Survey has been refined to use remotely sensed imagery to measure a range of other variables, such as soil, breeding birds, and hydromorphology (Haines-Young *et al.* 2000). It was completed in 1990, first revised in 1998 (published as CS2000) and again in 2008. As Firbank *et al.* (2003) note, the challenge for the Countryside Survey was to ensure continuity of data from 1978 to allow for comparative monitoring. With new parameters measured and administrative devolution of parts of the UK, the land classes were re-grouped into six environmental zones (Figure 2.5), replacing the landscape types (Haines-Young *et al.* 2000). For the uplands, three environmental zones were identified:

- Zone 3 – uplands of England and Wales
- Zone 5 – intermediate uplands and islands of Scotland
- Zone 6 – highlands of Scotland

All three zones are typified by extensive areas of semi-natural habitats and low levels of intensively managed enclosed landscapes.

The Natural Areas Initiative in England

Designed by English Nature in the 1990s, the Natural Areas Initiative (NAI) was an attempt to develop joined-up thinking about local nature conservation sites at regional landscape level. English Nature hoped that other organisations would use it as a framework for management and resource allocation, thus developing a more integrated partnership approach to nature conservation and land management. The idea was also that the system would translate national and international nature conservation targets into a workable framework.

ITE Land Classification Marginal Upland and Upland areas	Upland Natural Areas in England (English Nature)	Countryside Survey 2000 Environmental Zones

Figure 2.5 Mapping the Uplands: ITE, Natural England and CS2000.
Figure (a) reprinted from *Journal of Environmental Management, Volume 4,* Bunce RGH, Barr CJ, Clarke RT, Howard DC & Lane AMJ 'Land classification for strategic ecological survey' p 54 (1996), with permission from Elsevier. Figure (b) taken from Backshall J *et al*, 2001, 'The Upland Management Handbook' p1:6 © Natural England 2001 material is reproduced with the permission of Natural England, http://www.naturalengland.org.uk/copyright. Figure (c) reprinted from *Journal of Environmental Management* Volume 67, Haines-Young R, Barr CJ, Firbank LG, Furse M, Howard DC, McGowan G, Petit S, Smart SM, Watkins JW (2003:269) 'Changing landscapes, habitats and vegetation diversity across Great Britain.*', with permission from Elsevier*

Each natural area (NA) is unique because its distinctive character is a function of the inter-relationship between physical and ecological parameters. English Nature delineated 18 upland NAs (Figure 2.5) covering a total of 23,485 km^2. Typical vegetation includes semi-natural grassland, blanket bog, dwarf shrub heath, exposed rock, and montane heath and scrub. Detailed profiles exist for each NA, providing inventory-type information, discussion of management issues affecting biodiversity and management objectives by habitat (see www.naturalengland.gov.uk, accessed 21 November 2008). There has been criticism that some NAs are too large to reflect distinctive local character (Porter 2004); this is very much the case in the Cumbria High Fells NA, where every valley is unique.

These three biogeographical classifications have been constructed using similar physical and ecological data. As a result they recognise the same semi-natural habitats as defining an upland environment. However, they do not classify the same geographical areas as uplands. For instance, the Natural Areas Initiative classifies the Quantocks as upland, but the ITE system does not. Under the NAI, the whole of Bodmin Moor is recognised as upland, but under the ITE and CS2000 it has only a tiny marginal upland core (Class 17) and the rest is not considered upland at all (Class 6).

We may also note that the ITE is a resource inventory, the CS2000 a resource appraisal tool and the NAI a resource management tool. Ideally, inventories inform appraisals, and appraisals inform management. If these systems fail to recognise the same areas as upland, not only will work be replicated between systems, but also – depending on which classification is adopted as the baseline inventory – upland policy makers may be overlooking part of their remit.

Of course it could be argued that the three systems' aims differ and thus all are equally valid and fit for purpose. The NAI is focused on the development of specific nature-conservation objectives for each area. The CS2000 (and its predecessor, the ITE LC) is more of a watchdog, monitoring and recording change.

Legislative boundaries

A third way to identify the extent of upland areas is by using administrative boundaries developed through various forms of legislation. In continental Europe, as has been noted above, some countries have specific legislation to identify 'mountain' areas; in the United Kingdom, no such legislation exists. Nevertheless, our uplands have been identified indirectly through two levels of legislation:

- European Union legislation
- national administrative definitions

European Union legislation

In the British Isles, financial support for hill farming has a long history, dating back to the early 1940s (see Chapter 8). If subsidy was to be paid, there had to be some way of deciding whether a farm was a 'hill farm'. Thus, agricultural subsidy is where the definition of uplands really became centred in the UK. It evolved in 1940 for England and Wales as a Hill Sheep Subsidy, and in 1941 for Scotland as a Hill Cow Subsidy. Two commissioned reports underpinning the Hill Farming Act 1946 specifically identified hill sheep areas:

> At present day there are some 10,700,000 acres of 'mountain and heath land used for grazing'. This area, along with a comparatively small acreage of land classified in the agricultural statistics as permanent grass and arable land, is the basis of the hill sheep industry. (Balfour 1942, para. 42, p. 10)

The De La Warr Report for England and Wales narrowed down the idea of uplands:

> Most farm crops can be grown with reasonable prospects of success up to an altitude of 600/800ft [185/246m]. Between this altitude and 1400/1500ft [431/462m], where arable cropping and intensive livestock management become uneconomic, are transitional zones known as 'uplands'. (Cmnd 6498, para. 10, p. 5)

From this point on, agricultural support for the hill farming sector allocated funds under the 1946 Hill Farming Act and then the Hill Livestock Rearing Act 1951, whereby hill land eligible for grant aid was defined as:

> Land situated in an area consisting predominantly of mountains, hill or heath ... suitable for the use of breeding, rearing and maintenance of sheep or cattle but not for carrying ... of dairy farming, the production of ... fat sheep or fat cattle or the production of crops in quantity naturally greater than necessary to feed the number of sheep and cattle capable of being maintained by the land. (Bowler 1976: 274)

Whether farms were eligible was decided by government officials making farm-by-farm assessments, but to all intents and purposes the boundary contour came out at about 244m asl for hill sheep and 183m asl for hill cattle farm units (Bowler 1976).

The influence of these early pieces of legislation was far reaching and was used in the designation of less favoured areas in the UK, following accession to the European Union in 1972. The EU Directive 'on mountain and hill farming in less favoured areas' (75/268), known by its short title as the Less Favoured Areas Directive, had many purposes (see Chapter 8), but pivotal to this debate on the definition of uplands in the UK are two aims of the legislation:

- To counteract large-scale depopulation caused by declining farm incomes and poor working conditions,
- To ensure the conservation of the countryside in mountainous and other less favoured areas. (LFA Directive 75/268: 3)

The directive allowed EU member states to identify regions that needed extra financial support to overcome natural handicaps to economic development. Either aim was sufficient, and thus less favoured areas were not necessarily mountainous. The United Kingdom Government drew up LFA boundaries under the criterion of 'other areas' (also called 'intermediate') and not 'mountain areas' (MacEwan and Sinclair 1983), using economic criteria rather than altitude (Table 2.7). Nonetheless, from this point on 'less favoured areas' gradually became (mistakenly) synonymous with the uplands of the UK.

Further expansion of LFA boundaries in 1986 compounded the problem by dividing LFAs into a core of 'severely disadvantaged areas' and a periphery of 'disadvantaged areas'. For example, in north-west England (Figure 2.5) the Disadvantaged LFA stretches into the lowlands, well below the 240m contour. With its 1975 boundaries the LFA could more accurately be described as mountainous (from 1986 an SDA), but the extension undermined any validity that had.

Other EU states have used the directive to much better effect to identify their uplands or mountainous areas, though not all states. As Table 2.7 shows, member states define LFAs using a range of criteria with different thresholds, reflecting the variations in their physical and ecological environments. This is not to say that other member states use LFA designation only for mountainous areas, because the directive also allowed for areas of low economic activity or poor infrastructure – sometimes called specific handicaps (see Figures 2.6 and 2.7, and Tables 2.7 and 2.8).

In the 1990s the delineation of LFAs came under further scrutiny:

> The original justification for the LFA scheme – to offset natural handicaps – has become ambiguous, and the legitimacy of the whole approach today has been largely eroded. On the one hand, the farm income situation has changed considerably since the mid-seventies (when LFA support was first introduced). With the notable exception of Ireland, the 'simple' LFAs of the northern and central member states are in a better position than the EU average, while southern member states, regardless of the type of area, are in a less-favoured situation as concerns farm incomes. (DG Agriculture, July 1997).

This echoes the Hill Livestock Compensatory Allowance evaluation of 1997 in England, which identified a range of farms that no longer seemed to fulfil the LFA criteria of disadvantage (Drew Associates 1997).

Table 2.7: Criteria Employed to Define Less Favoured Areas in the EU 1975

member state	*defining altitude*	*slope angle*	*land use*	*livestock density*	*agricultural population (%)*	population density (sq/km)	other factors
UK	–	–	rough grazing >49%	<0.78LU/ha	19 high (national average 3.1)	36 low (national average 228)	farmer's rent <48%/ annual income <75% of national average
France	600 –700m	>20%	–	–	–	–	none
Italy	700 –800m	15–20%	rough grazing >50%	<0.65LU/ha	15>	75 low (national average 181)	wheat yields <16.5q/ha
Eire (Ireland)	-	-	rough grazing >38%	<0.54LU/ha	50 high (national average 26)	24 low (national average 46)	rateable value <65%/ salary <68% of national average

(Source: europa.eu.int/smartapi/cgi/sga_doc, 09/01/04)

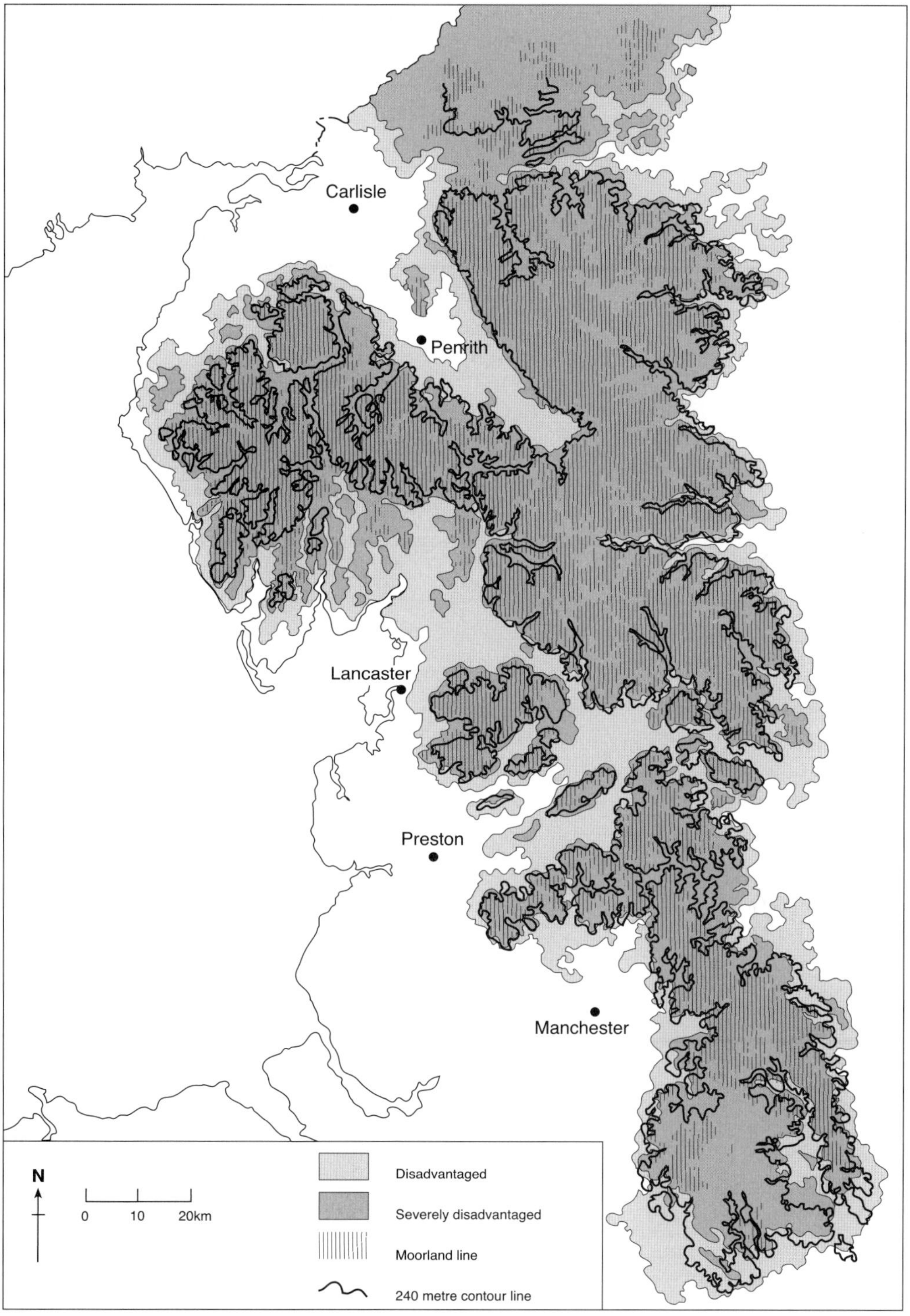

Figure 2.6 Less Favoured Areas in North West England
Adapted from DEFRA (1999) ERDP 2000 to 2006 NW volume p16 reproduced by permission of Ordnance Survey on behalf of HMSO.

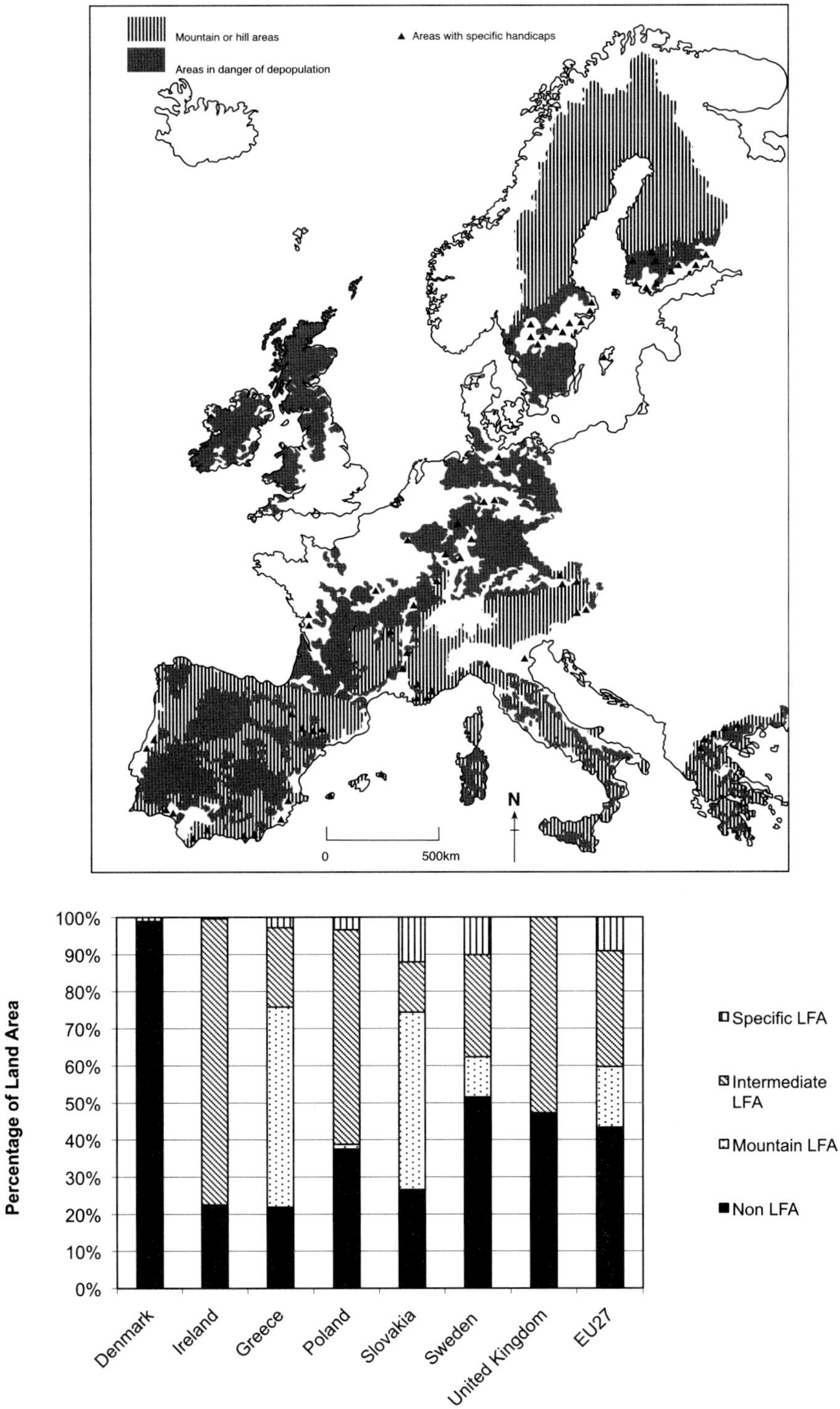

Figure 2.7 Less Favoured Areas in the European Union
(Adapted from: NORDREGIO, 2005:11, CEC 2009b:4)

Table 2.8: Percentage of UAA classified as LFA in Various EU-27 States in 2005

member state	*non-LFA*	*mountain LFA*	*intermediate LFA*	*specific LFA*
Denmark	98.9	0	0	1.1
Ireland	22.5	0	77.1	0.4
Greece	21.9	53.9	21.4	2.8
Poland	37.5	1.3	57.9	3.4
Slovakia	8.7	69.5	19.8	17.5
Sweden	51.6	10.8	27.6	10.1
United Kingdom	47.2	0	52.8	0
EU27	**43.3**	**16.4**	**31.2**	**9.1**

(Source: europa.eu.int/smartapi/cgi/sga_doc, 09/01/04)

The issue arose again in 2005, when the LFA scheme was revised. The socio-economic objectives were removed and more emphasis was placed on land management (CEC 2009b). Council Regulation 1698/2005 redefined LFAs as:

> areas affected by significant natural handicaps, notably a low soil productivity or poor climate conditions, and where maintaining extensive farming activity is important for the management of the land. (Council Regulation 1698/2005. OJ L277, p. 1)

In preparation for this event, DEFRA contracted the University of Reading to develop a new classification of upland (DEFRA 2003). This project used a range of physical parameters to define the uplands (Table 2.9), but the authors did not foresee any problem in adding to the mix social variables such as distance from markets or distance to nearest A-road. However, the European Council did not achieve agreement on a community-wide classification and set in motion a new review (CEC 2009a; see Chapter 9).

Table 2.9: University of Reading Definition of LFAs in the UK

general characteristics →	*classification of disadvantage* →	*weighting of variables*
Temperature	Accumulated temperature	Each variable is weighted 1 to 7 to represent increasing disadvantage to farming. Scores are combined to give an overall score.
Precipitation	Annual rainfall	
Geology	Rock type Elevation Slope Aspect Landform	
Soils	Soil characteristics	
Remoteness	Distance to market	
Population Density	Population by radii	
Socio-economic status (per capita GDP)		

(Source: adapted from DEFRA 2003)

National administrative definitions

LFA designation is not the only way of identifying uplands for the purposes of hill farming financial support in the United Kingdom. In 1992 the Ministry of Agriculture, Fisheries and Food (MAFF) set ADAS (Agricultural Development and Advisory Service) the task of drawing up the moorland line.

This boundary was developed for the Moorland (Livestock Extensification) Regulations 1995, which articulated EEC Council Regulation 2078/92 'on agricultural production methods compatible with the requirements of protection of the environment and maintenance of the countryside'. The national grant aid scheme that developed from this was called the Moorland Grant Scheme. It was designed by MAFF to encourage those farmers who farmed moorland outside any other environmental initiative area to reduce stocking densities to regenerate heather moorland (Soffe and Hetherington 1996/7). The moorland line was used to identify whether farmers had land eligible for the scheme.

Later the moorland line was adopted by the England Rural Development Programme 2000–2006 (ERDP 2000: 138), where it marks out

> Land within England that has been defined as predominantly semi-natural vegetation or rocky outcrops and semi-natural vegetation, used primarily for rough grazing.

The moorland line was used to separate rates of hectarage payment under the Hill Farm Allowance scheme – a subsidy for agriculture in less favoured areas – and, from 2005, the Single Farm Payments (more on this in Chapter 8). As the example of Bodmin Moor in south-west England (Figure 2.8) shows, the moorland line lies very much within the boundary of the Less Favoured Areas SDA. It tends to follow the fell wall, beyond which lies un-enclosed land but not much else. The line bounds about 33% (798,896 ha) of the total LFA in England and 46% of the total SDA (ERDP 2000, main document; www.magic.gov.uk accessed 16 August 2004).

Though the moorland line has been legitimately designated by trained ground surveyors and the use of aerial photography (ERDP 2000), it is hard to ignore the fact that it has led to reduced payments for many farmers whose land falls inside the line (see Chapter 8). A cynic may suggest this has been an attempt by the UK Government to reduce farm payment costs. It might also be criticised for not following any topographical feature, being based on where rough grazing and semi-natural vegetation starts to occur as one goes upslope. Thus the moorland line encloses only a very small part of Bodmin Moor (Figure 2.7). As with any such line, there is also the problem of periodic redefinition as land use changes, agri-environment schemes take effect and nature conservation targets are met under the Natural Areas Initiative and the Public Service Agreement with Natural England (more on this in Chapter 5).

Another way that uplands have become delineated is the designation of national parks in England and Wales (MacEwan and MacEwan 1982; 1987). A cursory inspection of Figure 2.9a shows that most are found in 'upland' areas. This was not accidental, as the preparatory reports by Dower for the 1949 Act (Cmnd 6378, National Parks in England and Wales 1945) and Hobhouse (Cmnd 6628, Report of the National Park Committee (England and Wales) 1947) demonstrate. Dower specifically defined national parks as:

> An extensive area of beautiful and relatively wild country (Dower 1945: 6)

By default these areas should be our uplands because these are our 'wild' areas, farmed at low intensity, where farming adapts to the environment and not the other way round (see Chapter 4).

The 1949 National Parks Act defined this 'open country' as including mountain, moor, heath, down, cliff or foreshore, and was extended in 1968 to encompass woodlands, rivers and canals (MacEwan and MacEwan 1982). The boundaries were decided subjectively by Hobhouse and a committee of specialists, who simply drew a line on maps, following recognisable features on

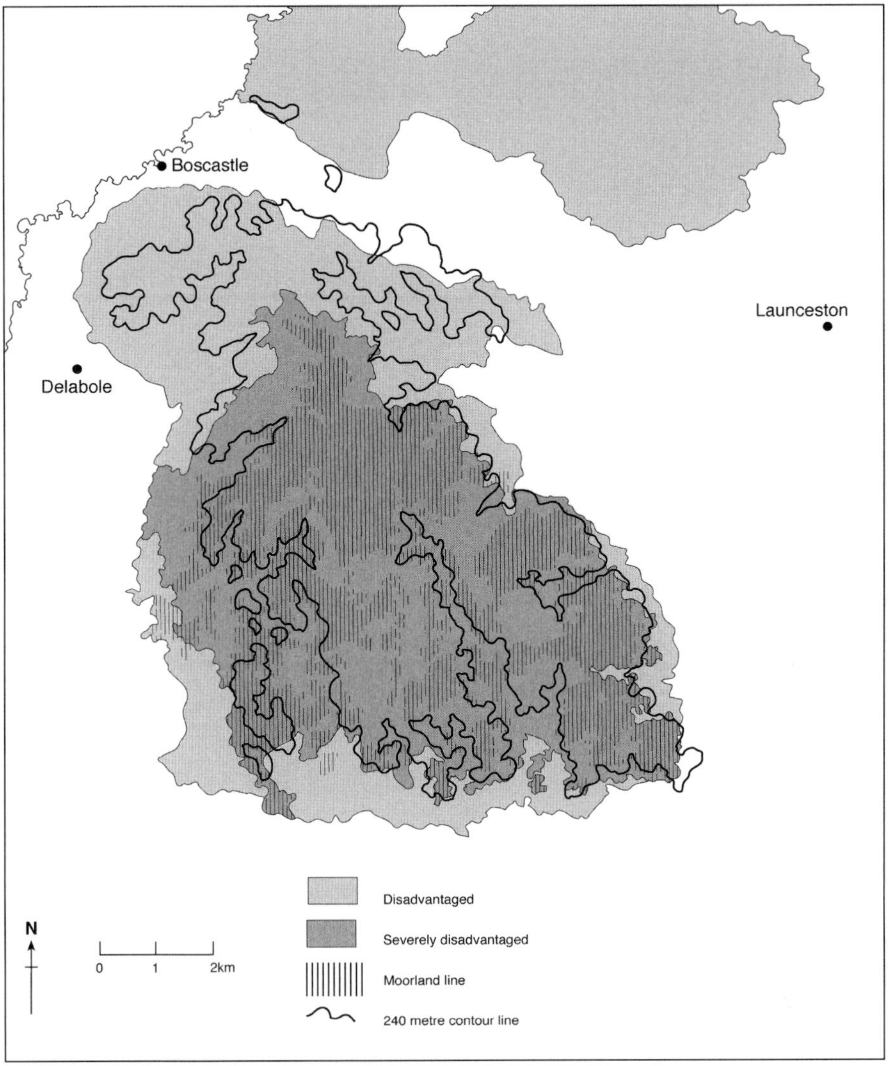

Figure 2.8 Bodmin Moor LFA and the Moorland Line
Adapted from DEFRA, Accessed: 24/11/05 www.defrgis.defra.gov.uk/moorland/src reproduced by permission of Ordnance Survey on behalf of HMSO.

the ground judged on the basis of landscape quality and recreational value (pers. comm., A Fishwick). After the parks were designated between 1951 and 1957, it was apparent to a casual observer that many of the boundaries roughly followed the 240m contour (Figure 2.9b). That the national parks were largely defined by uplands was confirmed by the Countryside Commission in 1984:

> One approach has been to use the 240m contour line to separate 'lowlands' and 'uplands' …. These areas correspond closely to those defined under EEC Directives as Less Favoured Areas. They also include the major part of the land designated as national parks. (Countryside Commission 1984: 3)

In summary, it would seem that the legislative development of hill farm support has driven the agenda for upland designation, albeit not deliberately, but over time it has become the accepted norm (see Lord Whitty's response, quoted earlier p.30). Organisations like the

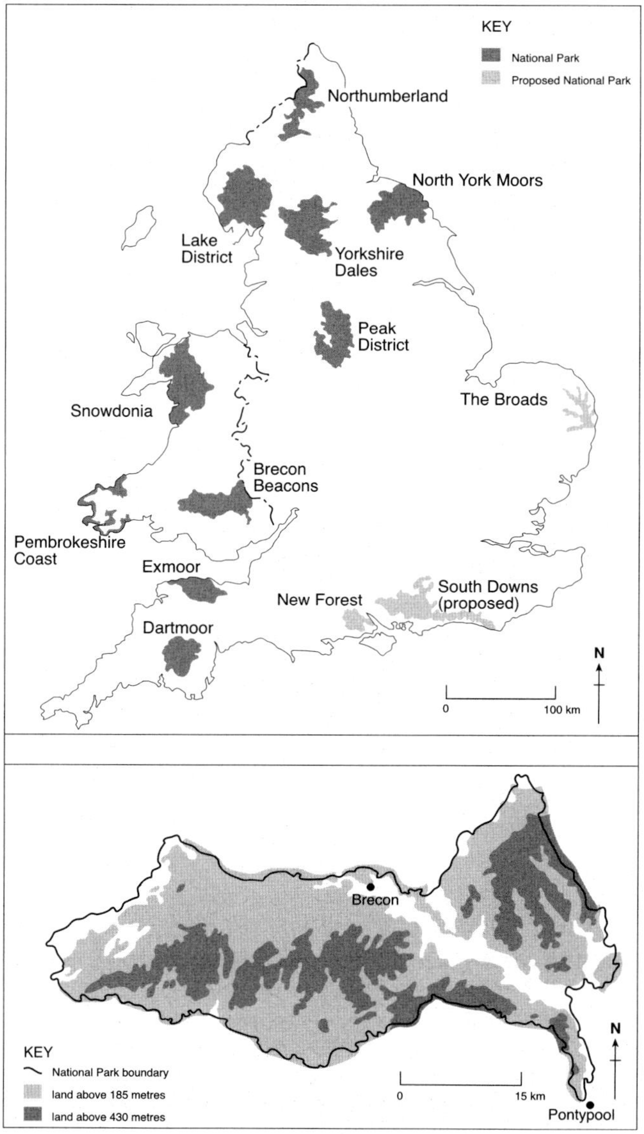

Figure 2.9a National Parks in England and Wales
Figure 2.9b Brecon Beacons National Park Boundary with respect to 185m contour

Countryside Commission have solidified this by suggesting that the LFA boundary and 240m contour are to all intents and purposes identical. These legislative approaches to definition have moved the terms of the debate from basing upland identification on natural phenomena to identifying how people utilise the land and the impact this has on the actual character of a landscape. Integrating this cultural dimension into the discussion is where we turn next.

Cultural boundaries

We can consider a cultural definition in two ways: first by 'feel' and second by 'cultural imprint', which some refer to as sense of place (Graham *et al.* 2009).

> The upland zone feels quite different from lowland Britain, both in the physical make-up of the land and in the cultural imprint which people have made upon it. (Atherden 1992: 1)

If we consider 'feel' first, it seems essentially subjective, reflecting the way each individual perceives their environment. Our world images are dissimilar, and different cultures – for instance, Dutch, Swiss, Tibetan – identify mountains and uplands in different ways. Visitors to the Yorkshire Dales will see mountains if they are from Holland, but foothills if they are from Tibet. It is all a matter of perception (see Box VIII). Likewise, each user group can have its own perception of what an upland might be. An ecologist will see an upland landscape differently from a farmer or a recreationalist (Meinig 1979, cited in Walmsley and Lewis 1994).

Even so, Walmsley and Lewis (1994) suggest that most people share some inherent meaning of a concept, much of which is derived from their cultural construction of landscapes. This is particularly relevant for agriculture in Britain. Burton (2004) considered this by investigating what makes farmers in Scotland respect their neighbours and develop a bond of identity with one other, expressed as their sense of place (discussed further in Chapter 8).

Individuals' perception of an area can play a pivotal role in the way its resources are utilised. Many uplands are typically seen as economically marginal, with poor infrastructure and non-existent services, because they are no longer centres of commerce (see Chapter 3). If these views persist, they can undermine upland development by two opposite views: either preserving the upland landscape in aspic, strangling development, or marginalising the role of agriculture and other long-established economic activities in shaping the upland environment. Winchester articulates this admirably when he says, referring to J. Thirsk's *Agrarian History of England and Wales, Vol. IV: 1500–1640* (Cambridge University Press: Cambridge, p. 184):

> It is striking that the maps of early modern farming regions in the *Agrarian History of England and Wales* classify the northern English uplands first in terms of cultivation: they were characterised by 'subsistence corn' with cattle, sheep and rural industry. The result is an historical paradigm which tends to see upland areas as deficient in terms of lowland norms, concentrating on what they lack, rather than on their central activity, the breeding and rearing of domesticated animals. What is more, this view has tended to re-enforce this assumption that grain-growing was the main determinant of wealth and that such 'marginal' areas were necessarily poor. (Winchester 2000: 3)

Box VIII: Landscape perception studies

People have long searched for a technique to define landscapes of high amenity or heritage value. Mitchell (1989) identifies three main approaches: consensus, description and preference.

Landscape consensus is where a group of experts delineate an area, basing their decisions on field reconnaissance and aerial photography. This approach was used to identify the national parks of England and Wales, and latterly the Cairngorms National Park in Scotland. Although this process seems accepted as reliable, it lacks the capability of replication and has no systematic evidence to justify decisions.

Landscape description is a cluster of techniques that aim to either identify and measure specific attributes of a landscape or else delineate areas by assessing landscape quality against some standards or criteria. They are, in effect, a resource inventory. One of the best-known descriptive studies in the UK was that of Scotland by Linton (1968). To identify areas he used landform and land use on a grid, developed a scoring system and then allocated areas to the matrix developed. Critiques of this technique questioned the scoring system and the way he dis-aggregated landscapes to score them. However, on a positive note, Linton was one of the first to draw land use into the debate on landscape evaluation.

Landscape preference aims to get visitors and users to identify the landscapes they prefer, by indirectly eliciting their preferences in landscape qualities, typically by showing respondents a range of photographs of landscapes and asking them to rank them by quality. However, even where experts and non-experts choose the same views, it is hard to judge what it is they like about the landscape (Walmsley and Lewis 1994).

We could suggest that such perceptions remain, when we consider what we mean by marginal or almost second-class. This is further demonstrated by Dodgshon's (1988) view of Highland peasant farming between 1500 and 1800: he recognised that their strategy was to sustain their limited arable resource by exploiting labour and land only suited to grazing – they were making the best of what they had. In contrast the 19th-century 'improving' Scottish landlords perceived the Highlands as impoverished and uneconomic. Even earlier, from the Bronze Age to the 16th century, Davies (2007) considers, through her analysis of the pollen record in north-west Scotland, that many upland farmers exploited land on a micro basis, cultivating the pockets of higher-quality land for arable production and using the rest for grazing. Today's sheep stratification system, whereby sheep are moved from upland to lowland at different times, is a manifestation of this adaptation to the resource base. The cultural dimension is therefore fundamental to identifying uplands.

Subjectively, areas such as the Lake District or Dartmoor could be included here. The NORDREGIO report refers to these as 'massifs':

> These areas are designated by name, and are generally perceived as culturally and economically specific. As such they are one of the cornerstones of people's perception of how their territory is structured, together with major rivers, islands etc. (NORDREGIO 2004: 33)

However, Table 2.10 shows an objective method, developed by the NORDREGIO team, to delineate massifs, so there seems to be a double classification system operating here - massif by cultural and then use of physical parameters and enumeration words to tighten it up (Box V).

Table 2.10: Massifs identified by NORDREGIO in the UK

NORDREGIO name	*national name*	*area (km²)*	*% of total area*	*total population*	*% of total population*	*population density (per km²)*
Cumbria	Lake District	696.3	1.08	6356	0.25	9.1
Dartmoor and Exmoor		953.8	1.48	83,132	3.26	87.2
Highlands and Islands	Cairngorms Grampians	37539.9	58.35	414,408	16.21	11.0
Mourne Mountains		236.2	0.37	20,547	0.80	87.0
Northern English Mountains	Pennines Yorkshire Dales North York Moors Peak District Cheviots	9164.9	14.25	968,286	37.93	105.7
Northern Ireland Mountains		1032.2	1.60	89,109	3.49	86.3
Southern Uplands		4317.9	6.71	68,286	2.67	15.8
Welsh Mountains	Snowdonia Brecon Beacons Cambrian	10393.5	16.16	902,852	35.36	86.9
Totals		**64334.7**	**100.00**	**2,553,976**	**100.00**	**mean: 61.1**

* Note: the original NORDREGIO report placed the Cumbria massif in the Republic of Ireland; that has been corrected here.

(Source: NORDREGIO 2004.)

Despite this the NORDREGIO names are somewhat misleading as they are not commonly used terms; for example 'Northern English Uplands' which is really five distinctive upland massifs, each of which should be managed very differently.

Countryside Character Areas Initiative (CCAI)

The CCAI blends abiotic, biotic and cultural information – to classify areas (the ABC approach to resource inventory, Mitchell 1989). It was developed by the Countryside Commission of England in the mid- to late 1990s. The Countryside Character Areas project emerged when professionals recognised a need to move away from characterising special landscapes, such as national parks, to look at the wider countryside (Swanwick 2002a). The project's underlying principle was the concept of landscape, defined as:

> The relationship between people and place. It provides the setting for our day-to-day lives … It results from the way that different components of our environment – both natural (the influences of geology, soils, climate, flora and fauna) and cultural (the historical and current impact of land use, settlement, enclosure and other human interventions) – interact together and are perceived by us. (Swanwick 2002b: 2)

Swanwick (2002) summarises this interaction in a diagram (Figure 2.10), which brings together many points discussed in this chapter, demonstrating that many phenomena create landscapes, including those of the uplands.

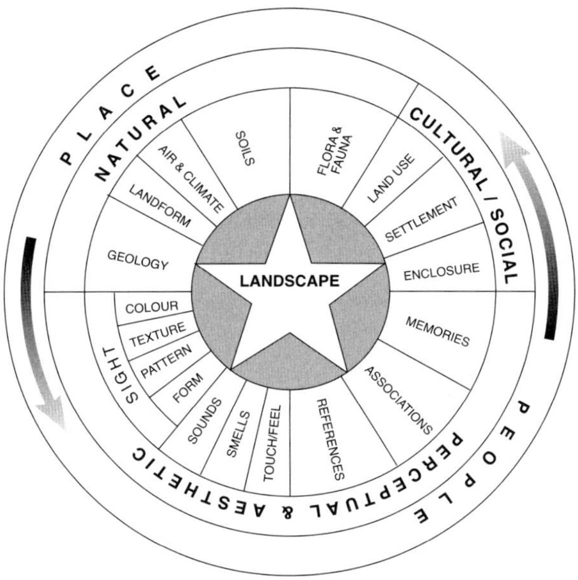

Figure 2.10 Swanwick's Interpretation of Landscape
Based on materials previously published in Swanwick C (2002:2) *'Making sense of place - landscape character assessment - summary guidance for England and Scotland'* :Scottish Natural Heritage

The initial work for the Countryside Commission was carried out in the south- west peninsula in 1993-4 in collaboration with English Heritage (now developing their own Historical Characterisation Map; see Porter 2004). Later the CCAI project was expanded to cover the whole of England, with support for landscape conservation and enhancement among its aims. The programme ran alongside the English Nature NAI project, and in 1996 the two schemes were merged to produce the Character of England Map, from which emerged Joint Character Areas, now central to agri-environment grants (see Chapter 8).

The map, with its 159 character areas, was supported by literature on their physical character, historical and cultural influences, buildings, settlement and land cover (Figure 2.11). The Countryside Commission scheme also allowed for a hierarchical classification of landscape (see Box IX). There are 23 upland character areas and they account for 18,184 ha of England, but they do not match the 'natural areas' of English Nature's NAI. For instance, the Cumbria Fells and Dales Natural Area covers the equivalent of five character areas (Figure 2.12). Given that the five character areas are very distinct from each other geologically and (by inference) ecologically, it seems odd that the NAI does not officially recognise this (Countryside Commission 1998). The Cumbria Fells and Dales Natural Area profile report identifies these variations at sub-NA level, but only at the point of practical management. In other instances, the two schemes use the same boundaries, as with Bodmin Moor.

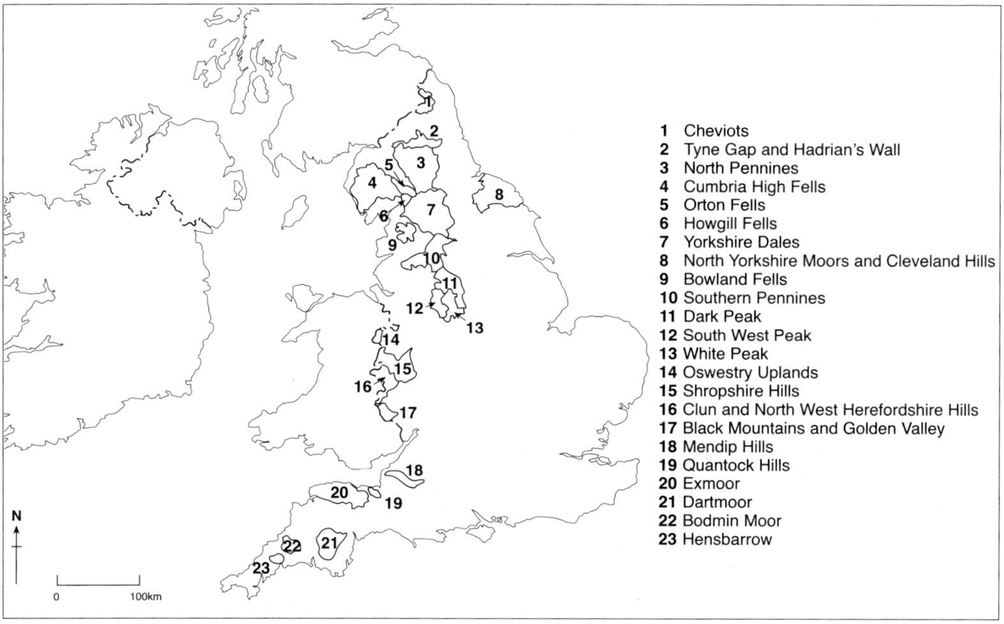

Figure 2.11 Upland Countryside Character Areas of England
from p7 of: *'The Character of England's natural and manmade landscape'* (CGP535) © Natural England 1999 material is reproduced with the permission of Natural England, http://www.naturalengland.org.uk/copyright.

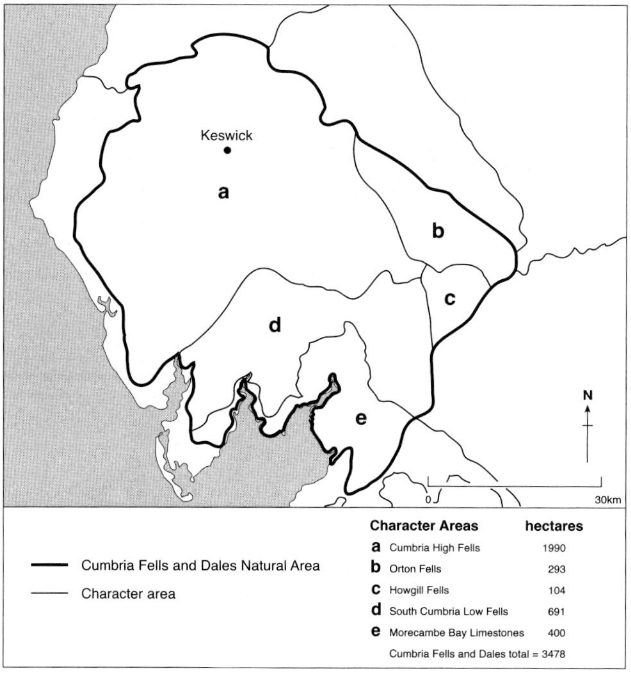

Figure 2.12 A Comparison of Countryside Character Areas and Natural Areas Initiative in Cumbria
Reproduced with permission from © Natural England 1999, http://www.naturalengland.org.uk/copyright.

Natural Heritage Zones

These form another ABC resource inventory approach. The Scottish equivalent of England's Countryside Character Areas Initiative, the scheme aims to

> Enable discussion and agree action to address the needs and opportunities facing the natural heritage in different parts of Scotland. (SNH 2002: 13)

To do this the 'parts' are identified using three main criteria:

- attributes that form the basis of the natural heritage – species, habitats, landform, topography, land cover, settlement patterns and so on
- additional descriptors of the natural heritage itself – particular species or habitats, peoples' experience of the landscape and a 'sense of place'
- issues that can influence future options for the natural heritage – how human activity can shape the future direction of an area

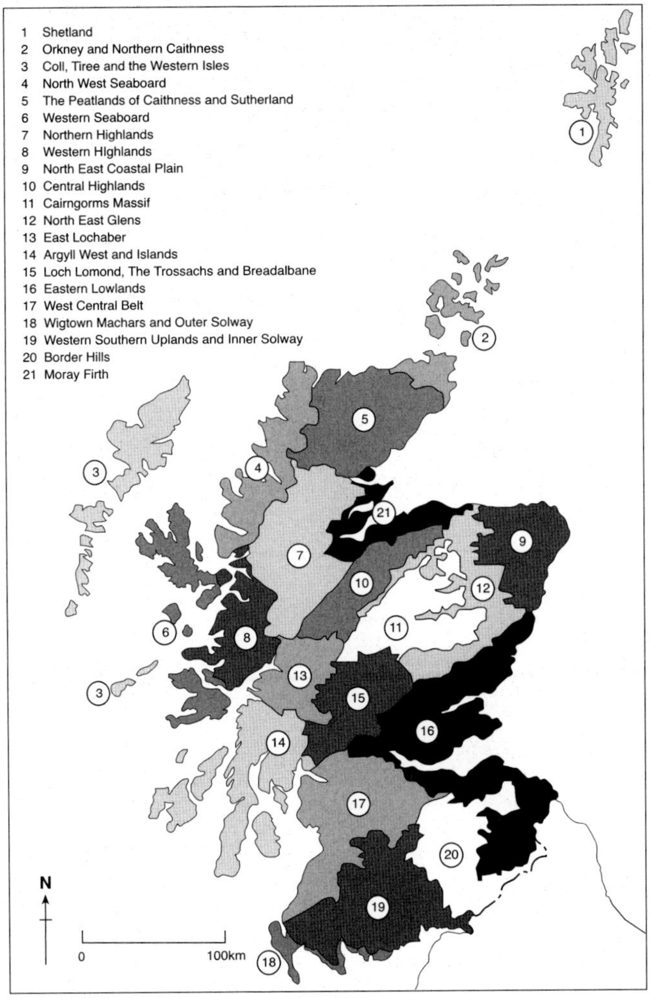

Figure 2.13 Natural Heritage Zones in Scotland
Reproduced with kind permission of Scottish Natural Heritage 2011

Box IX: The Countryside Commission's landscape character assessment

The landscape character assessment was developed using a number of spatial datasets:

- natural features – geology, landform, rivers and drainage systems, soil types, vegetation cover (phase 1 habitat survey, Land Cover map of Scotland, Natural Area profiles, land cover map 2000, Natural Heritage futures) and tree cover;
- cultural features – land use, settlement patterns and patterns of field enclosure;
- historical features – English Heritage's historic landscape characterisation GIS package and the historical land use assessment developed by Historic Scotland and RCAHMS.

Desk work was underpinned by field survey to ascertain how people see the landscape and to record perceptual and aesthetic elements. A field sheet with key words to ring helped surveyors express their views. The completed form was supported by annotated sketches and photographs. These were then put onto maps as a series of overlays to produce coherent character areas.[1]

The system can be applied at national, regional and local level. At each level, there can be landscape character types and areas. The figure below shows a typical hierarchical analysis. Landscape character types are distinct kinds of landscape with a generic character, occurring in different parts of the country as a result of similar combinations of variables, for example chalk river valleys or upland moorlands. Landscape character areas are unique and thus take the name of a specific place. They may be similar in landscape character type, but each has its own distinctive 'look' derived from different circumstances. Consider Exmoor and Dartmoor: both have upland moorlands, but each is unique, Exmoor with its dry summer-haze moorland and Dartmoor with a wild landscape of wet heath bogs and tors.

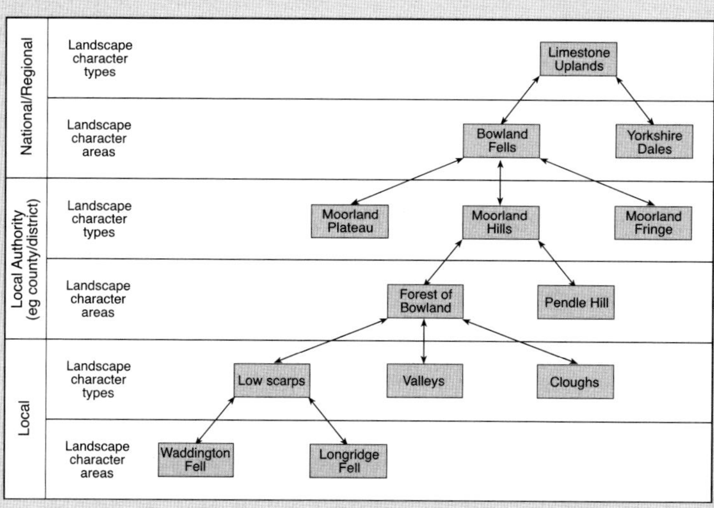

Landscape character assessment hierarchy; CoCo
Based on materials previously published in Swanwick (2002-11) 'Making Sense of Place - landscape character assessment - summary guidance for England and Scotland': Scottish Natural Heritage.

This method of assessment is a brave attempt to standardise how people perceive cultural landscape. It has become a benchmark for many types of planning activity. An example of this is the Cumbria Landscape Strategy by Cumbria County Council (1995; 2001), who have used the classification to realise the strategic economic ambitions of the local population while sustaining the high-value landscapes for which the county is known. The classification (it does not cover the Lake District National Park, which is a separate planning authority) identifies 13 landscape types, of which five are 'upland'. These are divided into sub-types, each of which has landscape guidance for dealing with changes in land management (farming, forestry) or construction (housing, commerce, services, communications, infrastructure, manufacturing). In the example shown in the table below, it is evident that agriculture plays a defining role in the cultural landscape.

[1] This technique is generally now computerised and referred to as GIS (Geographic Information System).

A landscape type as defined by Cumbria County Council

Landscape Type 12	*Higher Limestone* – an upland landscape where limestone has a strong influence on patterns of drystone walls, buildings and villages. Remnants of heath and limestone pavement on moors and commons
Sub Type 12A	*Limestone Farmland* – rolling pastoral landscape evolved from a long history of habitation with a mix of enclosed and unenclosed grazed land
Vision	Conservation and enhancement of the landscape
Guidelines specifically for:	*Field boundaries* – discourage field enlargement and maintain boundaries using traditional methods *Unimproved grasslands* – restrict agricultural improvements and encourage species-rich grassland/meadow *Woodland/trees* – encourage afforestation of small blocks not in conflict with other guidance *Cultural features* – conserve the prehistoric and historic heritage *Farm Buildings* – conserve traditional farm buildings, and new builds to be in sympathy with surroundings *Villages* – developments to respect village integrity and character avoiding settlement fringes

Source: Cumbria County Council 2001.

Using these criteria, SNH (2002) identified 21 areas, of which 14 can be described as having hill or upland character (Figure 2.13). Each area has its own set objectives, with specific actions over the short (0 to 5 years), medium (5 to 15 years) and long term (15 to 25 years). The actions are based on a synthesis of research and practitioner knowledge that can aid the maintenance and recovery of high plateau habitats (see Chapter 5). Table 2.11 shows objectives and actions for the Cairngorms Massif.

Table 2.11: Objective 1 of the Cairngorms Massif Natural Heritage Zone

Objective 1: to enhance existing upland habitats, including the diverse inter-grading plant communities of the high plateaux, and expand key habitats such as alpine tall herbs	
Summary of Actions:	
Modify deer management by developing deer management plans that identify damage and population levels that will achieve habitat restoration	Short term
Ensure agri-environment initiatives include enhancements for the habitats of this zone	Short term
Promote the Muirburn Code	Short term
Develop and promote upland management training programmes	Short term
Include policies for conservation and upland habitat management into local authority plans	Medium term
Reduce greenhouse emissions	Medium term
Reduce nitrogen deposition	Medium term
Continue to monitor montane species with respect to climate change	Ongoing
Promote positive management agreements on SSSI and SACs	Ongoing

(Adapted from: SNH 2002b: 2)

Cultural definitions matter because humans play a role in shaping landscapes. They are important more in countries where little true wilderness remains, such as the UK (see Chapter 5). They are perhaps even more critical for upland areas because they direct our attention to the fact that land management needs to continue if we wish to maintain the landscapes we treasure.

Summary

We can see that uplands can be and have been defined in many ways, ranging from straightforward altitude to more complex administrative and cultural classifications. It is evident from Table 2.12 that these various systems lead to different amounts of land being classed as upland. The restrictive moorland line encloses 7,898 km^2 in contrast to the extensive CS2000 Environmental Zone 3 of 25,900 km^2 for England, a difference of some 17,911 km^2.

Table 2.12: Comparing English definitions of Uplands

Parameter	*variables*	*area (km2)*	*originator*
Physical	Exe–Tees line	19,560	estimate based on Mackinder
	300m asl and local elevation range of 300m	n/a	UNEP-WCMC
	300m asl, temperature index, topography, 50% of ward	10,815	NORDREGIO 2001
Ecological	*Land Cover:* Topography, climate, geology, habitats, land cover	unknown	ITE
	Environmental Zone 3 – a development of the ITE system	25,900	Countryside Survey 2000
	Natural Areas Initiative: Geological maps, landscape accounts, agricultural treatises, Habitat and species distributions. County floras and the Land Utilisation Survey	23,485	English Nature
Legislative	*Subjective* – roughly 240m contour for 6 'upland' parks	9,502	National Parks Act 1949
	Less Favoured Areas by parish: Rough grazing <49%, <0.78LU/ha, % agricultural population and population density < National Average. Farm rent <48% of National Average Annual income <75% of National Average	23,800 (total) 17,464 (SDA) 6,337 (DA)	LFA Directive 72/268
	The Moorland Line: Rough grazing and semi-natural vegetation	7,989	Agricultural Advisory Service for England
Cultural	*Subjective*	n/a	individual perception
	'Massif' identification	11,088	NORDREGIO 2001
	Countryside Character Areas: The physical character, historical and cultural influences, buildings and settlement and land cover	18,184	Countryside Commission

These discrepancies, which we could dismiss as a function of the varying aims and objectives of the classification systems, hinder clarity in policy and management operations. What is worse, when one of these classifications is used out of context from its original objectives and becomes the inventory for another resource appraisal or resource-management exercise for uplands, if there is no real consensus among actors and agencies of where the uplands are, inappropriate surrogate definitions creep in with unforeseen results.

We have seen that policy literature increasingly uses LFAs as roughly equivalent to UK uplands. This has been further entrenched by the statistical data on upland farming available for analysis, largely limited to LFAs and national parks, as the data sources used in this book demonstrate. It is clear that definition by LFA is not really appropriate, unless only the SDA is used. Recent moves to redefine LFAs at European level by 2013 suggest that the UK Government may be adopting the SDA officially. Though a cynic might view the reduced area as simply a way to restrict rural and agricultural support payments, it may better delineate the uplands in line with other classifications used by various land-management organisations in the UK.

Whilst LFA SDA has some merit in these ways, it may not be the most appropriate in identifying and managing coherent environmental landscape units. Upland massifs may become dissected by land not in the LFA, making policy development and scheme operation almost impracticable. Furthermore, by using the new LFA boundaries, UK uplands will still be defined through agricultural policy, even though the LFA directive has been subsumed into the general rural development aspects of the Common Agricultural Policy (see Chapter 9).

For resource management, it may be more effective for the UK Government to look elsewhere for its definition of uplands. This change in direction is being driven by the growing official recognition that upland agricultural landscapes are not monofunctional but multifunctional, and farming can produce a range of public goods, including ecosystem services – for example, landscape, timber production and nature conservation (see Chapter 9 and Box XXXII) – which have been among the targets of agri-environment initiatives since the mid-1980s (more in Chapter 8).

Among those public goods is the contribution that upland farm landscapes can make to integrated river catchment management. A catchment is the natural drainage area of a single river system, within which a range of land uses will operate (Figure 2.14). Because changes in the management of one land use will impact on another, it is more effective to link their management to avoid negative impacts. This land-management approach has been much easier by the EU Water Framework Directive of 2001, leading to a series of nested integrated management plans shared between stakeholders (Figure 2.15). It is argued here that a catchment approach makes demarcation of upland from lowland redundant, because a change in upland management affects lowland areas too. A simple example is felling trees in an upland area, which increases the water run-off from hillsides (Nisbet *et al.* 2004). In a storm, water runs directly into the river system, causing problems downstream in the lowlands, such as flash flooding or soil erosion, creating sediment build-up in any reservoirs.

Integration of upland and lowland management would also help to dispel the ingrained view that uplands are economically marginal, as well as placing farming centre stage. In one fell swoop, management of the uplands would change from an administratively demarcated system led by a single land use to one that is integrated through natural processes adapted by human activity. These ideas are already employed in some regional UK schemes, such as the Stillwaters Partnership[2] which manages the Bassenthwaite Lake catchment (Lake District Stillwaters Partnership 2010).

---

[2] Stillwaters Partnership is a consortium of organisations and individuals concerned with Bassenthwaite Catchment, in North West England including Environment Agency, National Trust, Natural England, Lakes District National Park, Forestry Commission, parish councils, farmers, private landowners and local communities.

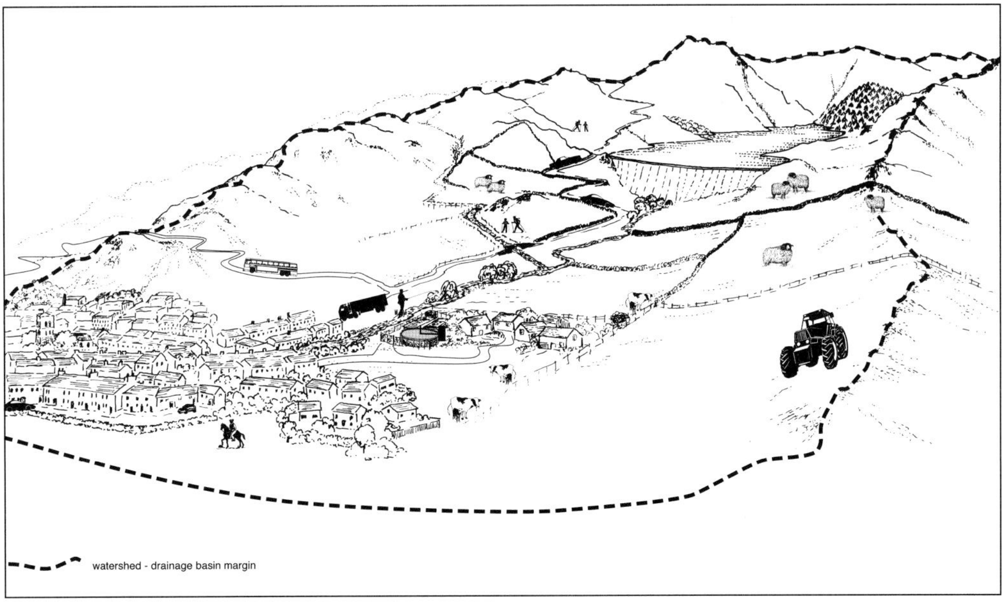

Figure 2.14 The River Catchment

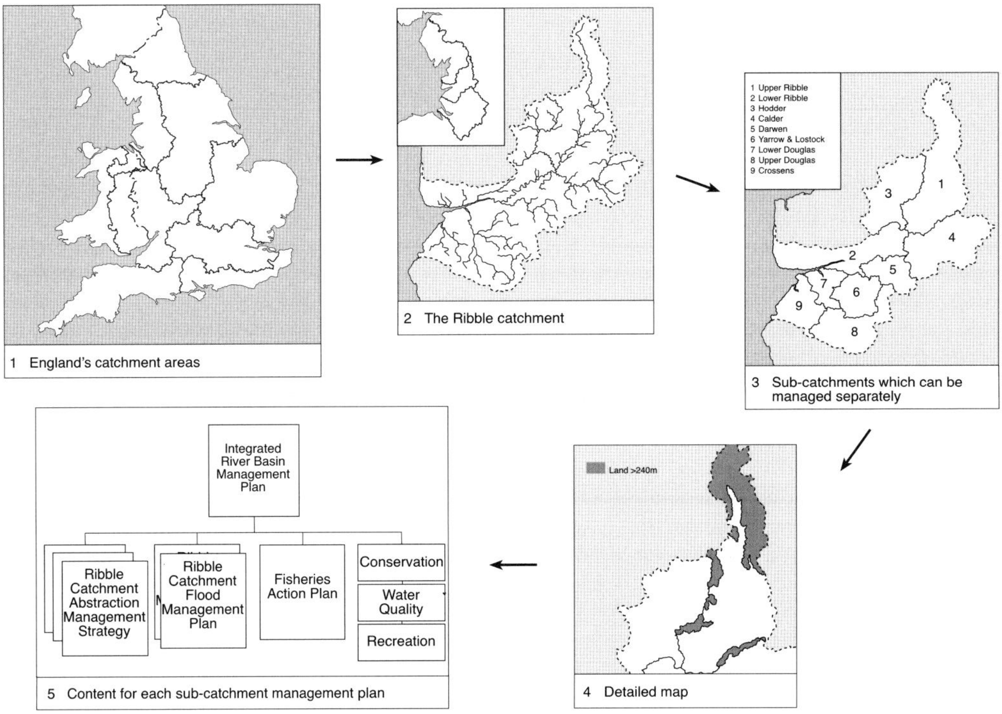

Figure 2.15 Integrated Catchment Management

The issue of delimiting the uplands may seem a purely academic debate, but it can be vital for the farming community that live and work in these areas. Since the 1940s much of the support for agriculture and rural development has been allocated to the uplands and increasingly, since the 1970s, based on LFAs. These support mechanisms, the official response to the growing marginality of upland agriculture, have intensified that marginality by only supporting monofunctionalism – namely, livestock production. It is true that protected areas, such as national parks, have allowed farmers to exploit tourism, and agri-environment schemes like Environmentally Sensitive Areas have tried to redress economic marginality, but these have only tinkered at the edge, not really addressing the core issues (see chapters 6 and 8). By being addressed in isolation from other upland activities, the decline of upland agriculture has become a self-fulfilling prophecy – we created it ourselves.

All is not lost. The emerging agendas of ecosystem services and integrated catchment management can help restore the balance. By abandoning LFA definition and embracing multifunctionality, upland agriculture can shrug off its marginal, anachronistic image. This chapter has demonstrated the flaws in using the LFA and other approaches to defining the uplands, but we need to explore further why upland farming has become monofunctional. Historically, upland farming systems were far more multifunctional than they are now. So what happened to force them into this monofunctional, marginal image? It is this we investigate in the next chapter by unfolding the historical development of upland agriculture.

3

An evolving upland farming landscape

Agricultural land-use patterns are not fixed and unchanging. Landscapes are a product of the way successive (or simply different) cultures have used the land in the British Isles over many thousands of years. Uplands are perhaps more of a palimpsest than any other landscapes, because so often they have been on the physical and economic margins of human activity. A small change in circumstances might lead to land being abandoned, to be only lightly re-used by later generations or perhaps never used again. Good examples include the Bronze Age sheep folds at Ewe Close in eastern Cumbria, medieval field systems above Malham Cove in the Yorkshire Dales and the Dartmoor reaves (Darvill 1986; Raistrick 1986; Muir 1987).

However, we must remember that uplands are not solely a product of agriculture. They bear the marks of resource exploitation – timber, minerals and water, as well as food – making them multifunctional landscapes supporting farming communities, in contrast with the monofunctionalism of 20th-century agriculture.

The move from multifunctionality to monofunctionalism is the theme of this chapter. We investigate how this shift has come about, by tracing the evolution of upland farming from the first hunter-gatherers to its pre-eminence in the late medieval period, followed by its slow decline to a marginal economic activity by the beginning of the 20th century.

The first uplanders: 14,000 to 6500BP

The British Isles have undergone a series of climatic changes over the last two million years, alternating between cold glacial conditions and warmer interglacial periods. During interglacials, vegetation flourished and supported diverse fauna. Evidence of small Palaeolithic populations has been found near Clacton-on-Sea, Essex, from the Hoxnian interglacial (250,000 to 200,000BP), but these early peoples had no impact on upland landscapes because repeated glaciations forced vegetation to re-establish itself each time (Ingrouille 1995). It is only about 14,000BP that the story of our upland landscapes truly begins.

With the general retreat of ice sheets in north-west Europe, about 14,000BP, the climate improved and the land was colonised by flora and fauna through processes of ecological succession (Box X). The conventional view is that climatic climax here was reached around 8000BP when woodland replaced open country in most areas. What is now the British Isles was still connected to mainland Europe because sea level was lower, and woodland spread rapidly northwards to develop two main ecological communities. Scotland was dominated by boreal

Box X: Ecological succession

Communities can be characterised as a group of interacting populations (plants and/or animals) living together in a defined area (Morin 1999). We recognise these communities in two ways: either from the environment or habitat in which they occur, for example a sand dune or montane heath, or through recognition of the dominant species such as an upland ashwood.

Because a community is a living system it will change over time in cycling nutrients, soil, food webs and chains, and the character of the interacting populations. These changes coalesce to form a process known as ecological succession, which charts alterations in structure and species composition of a community (Chapman and Reiss 1998).

At its most extreme, ecological succession occurs when new bare rock is gradually colonised by successive communities of plants replacing each other. Usually we consider sand dunes, volcanic rocks, landslips and glacial debris as the initial land surfaces for this process of primary succession, which began in northern Britain around 14000BP as the ice retreated.

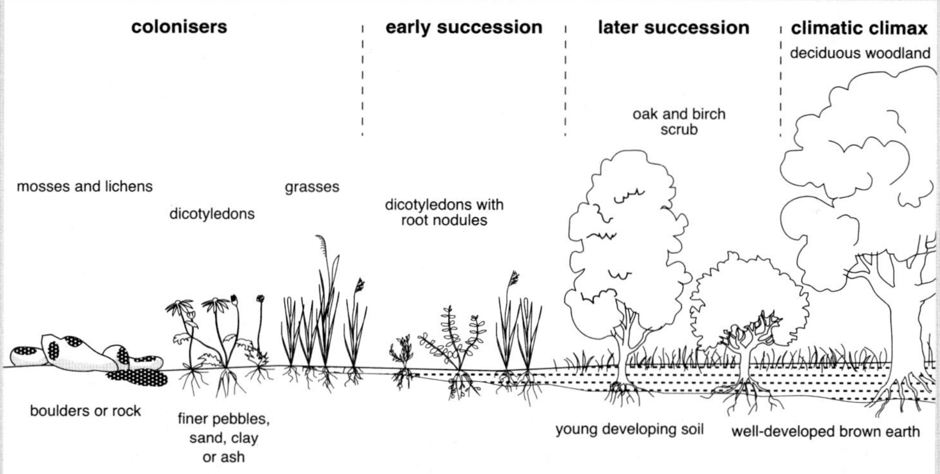

Initial colonisers on bare ground will include mosses and lichens, which do not need soil to grow. As these die they donate organic matter, which coalesces with weathered mineral matter to form a primitive soil that can support more sophisticated vascular[1] plants. Typically, leguminous plants begin to dominate, allowing nutrient cycling to initiate. As living plant matter (biomass) increases, so soils become deeper and more sophisticated. As succession continues, larger shrubs replace grasslands. Eventually trees grow and a woodland emerges, which classic succession theory states is at equilibrium with its environment. It is generally accepted that this final habitat, the climatic climax community, was reached in the British Isles about 8000BP.

Other forms of ecological succession occur. Typical on abandoned farmland is secondary succession, where not all life has been eliminated by a disturbance. Agricultural weeds spread, eventually followed by shrubs and trees. In other instances autogenic succession can occur, whereby changes in the soil or plants themselves activate succession, for example the invasion of birch (Betula spp.) onto a lowland heath. Allogenic succession, in contrast, is driven by processes external to the habitat such as climate change or the addition of domesticated grazing animals. Another type of succession, cyclic, develops where two species are inter-related and one ebbs as the other flows. Watt (1955) showed this happens when ling *(Calluna vulgaris)* dies off at around 25 years and bracken invades, with the former fighting back later and dominating the bracken.

[1] Vascular plants contain lignified conducting tissue (xylem for water transport), whereas non-vascular plants cannot extract food and water from the surrounding soil. Instead, like *Sphagnum* they gather nutrition from the atmosphere.

A particular issue is whether climatic climax actually exists: that is, do habitats find a stable end point at equilibrium with their environment? Climax was a concept promoted by Clements (1916) whereby different climax communities were the product of different environments. This idea is known as monoclimax theory because climate is the over-riding factor controlling the end point. Clements' ideas have been challenged by the concept of polyclimax, which suggests that many factors prevent climatic climax from being reached, for example fire or grazing. As a result some areas never reach a climatic climax but are a mixture of different polyclimax communities, sometimes referred to as subclimax. On top of this we all know that climate is dynamic itself and thus ecological populations will be continually adjusting to new conditions, so climax may well be unachievable.

One final aspect worthy of consideration is the plagioclimax community, an important concept for conservation management in many parts of the populated world, particularly upland Britain. Most of our landscape is cultural – a product of people's relationship with their environment – as I show in this chapter and in chapters 5 and 7. Areas that have undergone human intervention, halting or re-directing the natural succession of communities, are known as plagioclimax. Upland examples of plagioclimax include upland moorland, limestone grassland and coppice woodland. It is these that our national nature conservation aims and objectives are currently directed towards (see Kirby 2005).

Thus we are not promoting natural habitats, but semi-natural ones manipulated by people's activities. This is really why agriculture and nature conservation are such uneasy bedfellows, because the latter is often a product of the former but at an extensive level. Thus any increase in intensity of agricultural activity negatively affects the nature conservation value of the habitat, yet nature conservation needs agriculture to perpetuate many desirable semi-natural habitats.

forest of scots pine and birch above an understorey of bilberry and ling (Gimingham 2002). This is also referred to as the Caledonian Pine Forest. Southern Britain was dominated by deciduous woodland forming a series of subclimax communities, with upland woods (Figure 3.1) dominated by oak *(Quercus spp.)*, elm *(Ulmus spp.)* and hazel *(Corylus spp.)* in northern and western England (Ingrouille 1995).

The view that the climatic optimum was reached around 8000 has been challenged by Birks (1988), who suggested that it came earlier for the uplands, around 9000 to 8000BP, and therefore by 7000BP large areas of upland in the British Isles were already above the tree line (see Box VII), which varied depending on location. To substantiate her hypothesis Birks developed a multi-proxy technique using pollen analysis (Box XI) and tree-stump data to estimate crude tree lines on specific upland massifs (Table 3.1). The data show that the tree lines descend going northwards and westwards. Therefore large areas of the Lake District, upland Scotland and

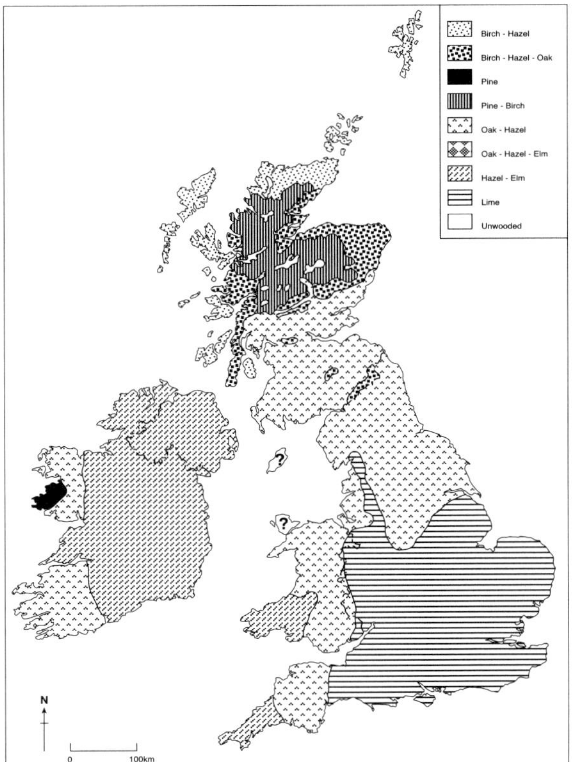

Figure 3.1 Climatic Climax in the British Isles 8500bp
Adapted from Rackham O (1986) *'The History of the British Countryside'* JM Dent & Sons Ltd: London p70 and Holmes G & Crofts R (2000) *'Scotland's Environment: the future'* Tuckwell Press: East Linton p6.

Table 3.1: Maximum Treelines on Various Upland Massifs during the Holocene

upland massif	*tree line estimate (m)*	*highest point (m)*	*tree line as % of highest point*
Bodmin Moor	419	419	100
Dartmoor	551	621	89
South Wales	>715	886	>81
North Wales	>635	1085	>59
North York Moors	454	454	100
Southern Pennines	595	634	94
Northern Pennines	893	893	100
Lake District	760	978	78
Galloway	>457	843	>54
Western Grampians	716	1214	59
Eastern Grampians	700	1068	65
Cairngorms	793	1311	60
N.W. Highlands	520	1081	48
Isle of Skye	457	1008	45
Average	**618.9**	**892.5**	**73.7**

(Source: adapted from Birks 1988: 42.)

north Wales were montane treeless environments well before the influx of larger populations around 5000BP. It is an important shift in interpretation that should influence our understanding of what is natural and semi-natural habitat in uplands (Box X). MacDonald (2004) calculated that about 29% of Scotland could have been already montane before people made any direct impact, rather than the 4% previously assumed (Figure 3.2). Thus woodland was not completely dominant in Scotland and montane habitats (treeless landscapes) played a greater role. It seems that campaigns to afforest the Scottish uplands to return them to their natural state may be simply misleading.

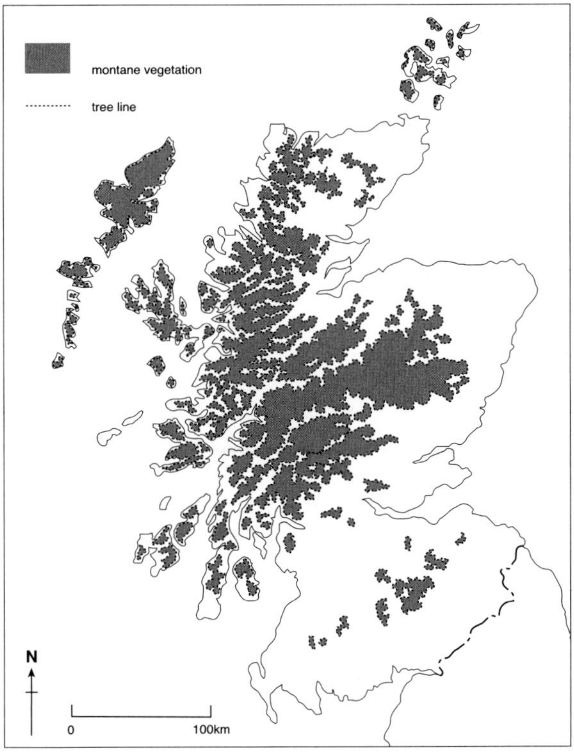

Figure 3.2 Possible Tree line in Scotland 5000bp.
Based on materials previously published in MacDonald A (2004) 'New Developments at the tree line' Look to the Hills Vol 12

Both Palaeolithic and the later Mesolithic people (down to 6500BP) were hunter-gatherers, relying on what animals they could hunt and the plant or vegetable matter they could gather. Flint implements and the bones of their prey are all that remains of their activity. The impact on vegetation of Mesolithic people in upland areas was

Box XI: Reconstructing past environments

Learning about past environments relies on our ability to seek out materials that were present then and date them. In this way plant and animal communities can be reconstructed and related to periods in the past. There are various techniques for doing this, which can be used separately or (increasingly) together. When used together, they are known as multi-proxy:

- morphological – landforms, sea-level changes
- lithological – sediment variations, oxygen isotopes
- biological – pollen, macrofossils (wood and bone), beetles, snails
- historical – diaries, paintings, tax returns, charters

Multiple techniques have the added advantage of cross-verification.

A tried and accepted technique uses pollen and spores expelled from plants as part of their reproductive process. Made from a protein known as sporopollenin, these microscopic grains can survive in anaerobic (oxygen-free) conditions for thousands of years with little or no deterioration. Peat is a particularly effective environment for preserving these grains.

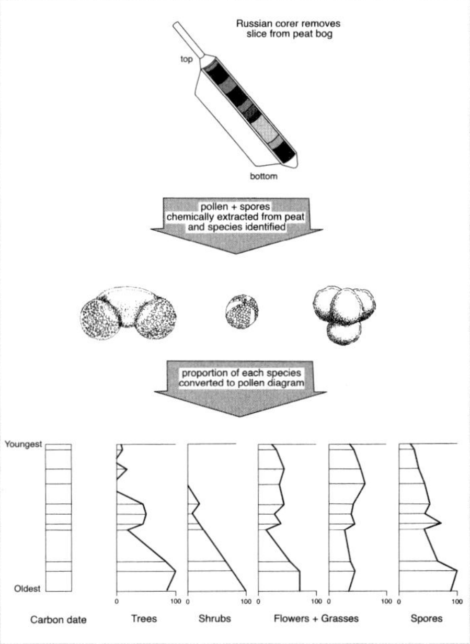

Samples of sediment are extracted from a core, with the uppermost layers being the youngest and the lower ones consecutively older. Taken at set distances down the core, pollen and spores are extracted, quite simply, from the sediments they fell into at dispersal and identified under a microscope. Identification of species or genus can be made because the grains come in different three-dimensional shapes with their surfaces covered in combinations of textures, protuberances and holes (pores). For a set number of grains, the proportion of each species is counted and from this a graph known as a pollen diagram is constructed.

The left-hand side of the diagram represents depth from the surface to further and further down into the sediment deposit. The horizontal axis lists the species or genera found, ordered thus: trees, shrubs, herbs (flowers), grasses and then spore-bearing plants. A range of techniques can be used to illustrate the pollen and spores on a diagram (Moore and Webb 1983), but the general principle is that the thicker the bar for each species, the greater the proportion of that plant in the vegetation's composition at a set time. At the right-hand end lies a series of pollen assemblage zones, which mark notable changes in the proportions of grains as we go up the column; for example, a decline in total trees or a sudden rise in aquatic plants, which could mark important events in the landscape's history.

A great deal of palaeoenvironmental research in uplands has been undertaken using this technique because peat is widespread there, but there are problems with interpreting the pollen counts. For example, some species such as hazel produce far more grains than others, making it appear disproportionately numerous; other plants produce very little, such as holly. Some grains decompose more readily than others and some species cannot be distinguished from other plants of the same genus (Lowe and Walker 1984). A particular criticism of pollen analysis is the fact that grains are light and can blow for many tens of miles, so they may not be representative of local vegetation. We should, in fact, see pollen diagrams as showing regional vegetation trends; this is why multi-proxy developments are important, because they confirm local interpretation. Despite these issues, much of our understanding of what the British countryside looked like in the past has been based on this technique.

limited to woodland clearance and summer-season deer hunting (Ingrouille 1995). These types of activity have been found in the pollen records in the Pinswell area of Dartmoor (Caseldine and Hatton 1993) and in changes from trees to grasses and herbs on Bodmin Moor (Gearey *et al.* 2000).

Given Mesolithic levels of technology, their clearances were limited to areas where trees were naturally thinner (uplands and equivalent areas, such as chalk hills and coasts) and fire could destroy all in its path. However, the purpose of this burning was not to grow crops, but to open up vistas to flush wild animals into clearings to allow them to be killed easily (Simmons 2005). A recent contrasting theory is that these clearings may not have been created by people. Instead, Vera (2000) suggests that large herbivores could have been the driving force behind the creation of this open-canopy landscape (Box XII). The ensuing discussion of his ideas has again raised important questions about what exactly we are trying to conserve in the uplands, as well as how (Hodder *et al.* 2005; Mitchell 2004; Kirby 2005). The reality may be a combination of both human and large herbivore activity, for sheep and then cattle were domesticated between 10,000 and 6000BP (Evans and Yarwood 1998).

By the end of the Mesolithic era, the uplands were wooded up to about 600m asl, though in places thinned out either by large herbivores or by people (depending on your point of view). Above this zone of clearings and thinning woodland lay the open montane environment of natural grasslands, moors and exposed rock, which could have covered as much as 25% of uplands in some areas.

The arrival of cultivation (6500 to 3900BP)

The next major impact on uplands was the arrival of cultivation. Spreading out from the Middle East, the concept and practice spread across Europe between 8000BP and 6000BP, when it reached the British Isles (Bell and Walker 1992; Simmons 2005).

The ensuing transformation of the uplands was quite dramatic, demonstrated by changes in the pollen record. Rapid drops in tree pollen mirrored an increase in grasses and other herbaceous species, notably plantains, and the arrival of cereal pollen is linked to an increase in charcoal fragments (Ingrouille 1995). Certain species, particularly elm (*Ulmus*), became rarer and tree stumps now beneath the peat record demonstrate the move from woodland to moorland in various uplands (central Wales, Moore 1972; North York Moors, Dimbleby 1961; south Pennines,[1] Tallis 1964). These shifts were previously believed to be the result of climatic deterioration (Pearsall 1950; Conway 1954), but there are instances where these changes do not occur simultaneously in one area (Edwards 1988), which implies that people may have been responsible for some of these environmental changes.

As the population increased from 6500BP, so landscape change accelerated in both upland and lowland Britain. Neolithic cultures lasted from around 6500BP to 4500BP (Adkins and Adkins 1982), though there is some argument over the actual timing of transition in different parts of the country (for south-west England, see Gearey *et al.* 2000; Edwards 1988). Neolithic cultures were typified by an agricultural economy and development of the axe. Stone-axe factories existed in many northern uplands, notably in the Lake District, where the Langdale Pikes volcanic tuffs were made into axe heads, which were traded throughout the British Isles and beyond (Millward and Robinson 1980).

Fowler suggests that this was a time of high exploitation of the natural vegetation, soil structure and chemistry, and it resulted in an upland landscape that limited any later agricultural activity. He comments:

> In that thousand years or so (5500 to 4500bp) the first critical steps were taken towards what in today's terms is considered scenic amenity, and wrongly, a natural phenomenon of unspoilt beauty; but in pedological and vegetational terms much of northern and western Britain is a landscape of dereliction. (Fowler 1981: 247)

---

[1] By the southern Pennines I mean the hills of Derbyshire and Staffordshire; the central Pennines are the section in Yorkshire and east Lancashire; and the northern Pennines are in east Cumbria and west County Durham.

Box XII: Vera's hypothesis and the uplands

Frans Vera works for the National Forest Service of the Netherlands. His theory seeks to explain the role of large herbivores in the dynamics of vegetation in the Atlantic period[1] (Vera 2000a). He starts from the fact that light-loving tree species, such as oak (*Quercus* spp.) and hazel (*Corylus avenula*), persist even in a closed-canopy forest of pre-Neolithic times, as shown by palynological records, which is an ecological contradiction. Furthermore, these tree species struggle to regenerate in forest reserves in western Europe today, even when any grazing livestock have been removed. So how could this be?

To answer this dilemma, Vera theorised that the trees were there as a result of a four-stage vegetation cycle, with large herbivores acted as controlling agent. He demonstrated that in wood pastures oak and hazel seem to regenerate quite happily when spiny scrub acts as a nursery crop to prevent the young trees from being browsed out. As the new trees grow, they shade out the underlying spiny scrub, which then dies, creating a grassland grove with trees growing randomly. The scrub survives by migrating onto new grassland where there is no shade, and its protection role begins again. On the original plot the tree canopy develops, so young trees cannot grow there either because they are shade-intolerant. Over time the canopy trees die and are not replaced since any seedlings that do manage to germinate are grazed out by herbivores. These changes combined with grazing pressure create grassland, which is eventually reinvaded by spiny scrub, ensuring the entire cycle begins again.

Vera (2000b) called this process 'the cyclical turnover of vegetation' with grazing livestock as the driving factor. The overall effect is to produce a mosaic landscape of open grassland, scrub, solitary trees and groves, which he believes closely resembles the landscape of pre-Neolithic Europe in the Atlantic period.

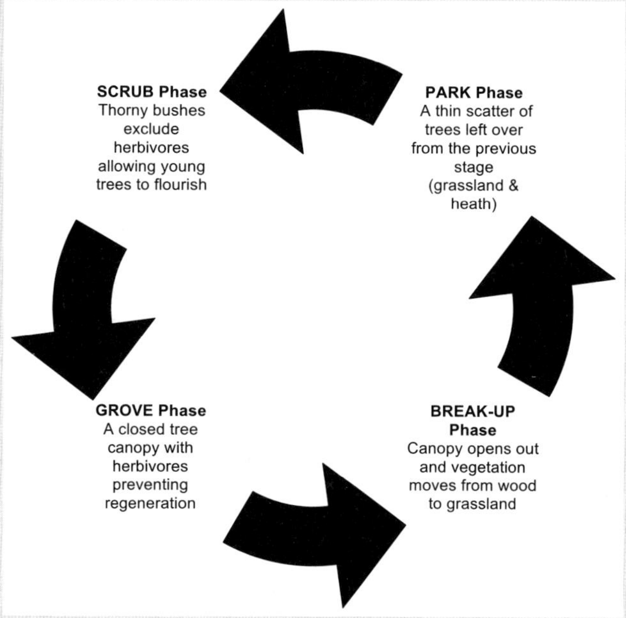

Kirkby's interpretation of Vera's model of the cyclical turnover of vegetation
(Adapted from: Vera 2000a, b; Kirkby 2005)

[1] The Atlantic period is 9000 to 5000BP, before people influenced the vegetation of western Europe. It is often used to demarcate the era of natural landscape from that of cultural landscapes. It is these which have been maintained by nature-conservation techniques in the 20th century.

Although Vera's model specifically addresses the evolution of lowland landscapes, it still has useful lessons for upland environments. First, herbivores do not self-regulate themselves to stay in the uplands or lowlands; they move across the upland–lowland divide when they migrate (Bullock 2005). Vera's model can, therefore, be applied to upland landscapes, where indigenous red deer *(cervus elaphus)* and the now-extinct migrating aurochs were present. The question is whether or not these two species were able to drive Vera's cycle of vegetation change in upland environments. Bullock (2005) feels that red deer did so directly through their browsing behaviour and indirectly through their predation by wolves, which led to deer control allowing a more closed canopy. The aurochs used uplands more as summer grazing, thus influencing vegetation seasonally. Bullock considers that fire played an important role, both directly as a management tool to flush out livestock and also to encourage gaps in woodland, something which Vera overlooked.

Bullock's partial rebuttal of Vera's ideas is not the only one. English Nature in 2000 instigated a review of the model to understand the role of Vera's theory and also the possible function of large herbivores today for nature conservation (Hodder *et al.* 2005). It concluded that the role of large herbivores in vegetation patterns has been underestimated, but their position as the dominant driver is weak.

Furthermore, nature conservation in Britain is focused, through the UK Biodiversity Action Plan, on maintaining cultural landscapes of the last three thousand years and not those of the earlier Atlantic period (Kirby 2005). If we were to adopt the Atlantic landscape as our natural vegetation, we would need to identify areas that could be re-wilded in line with Vera's ideas (see Chapter 9 for re-wilding). However, this would create many practical challenges to current land use needs and ownership, buffering from economic land use, public perception and loss of species already at equilibrium with the extant environment (Hodder *et al.* 2005).

In other words, he is suggesting that, first, Neolithic activity in the uplands led to a landscape from which all other activity developed, and second, these early peoples over-exploited the environment, leading to peat-ridden moors and damp, rough grassland. We in turn, have misinterpreted this in the past as the original natural landscape and made it central to our upland image (more evidence that a cultural rather than natural landscape is pivotal to nature conservation).

Pioneering palynological work by Iversen (1941, cited in Edwards 1993) and Troels-Smith (1954, cited in Edwards 1993) led to a proposal that, to exploit upland woodland for arable production, Neolithic peoples used two clearance processes: *landnam* (Old Norse for 'land take') meant woodland was cleared using fire to open land up for browsing by ruminants and, less so, for cereal cultivation; *swidden* was fire-cleared woodland that was cultivated and then abandoned as soil fertility fell (Adkins and Adkins 1982). Vegetation burning was crucial as it produced potash, helping to fertilise the soil. A typical example of this cycle was recorded by Simmons and Ines (1996) at North Gill (354m asl) on the North York Moors, where wheat was cultivated on a site that was temporarily cleared of woodland. The land recovered its forest canopy after abandonment.

Another possibility is that abandoned land evolved into moorland. Moore (1988) has outlined the possible pattern of events (Figure 3.3). His theory focuses on the role of waterlogging brought on by either *landnam* or *swidden*. Experiments carried out in the US and UK have shown that, as trees are felled, water run-off increases, simply because the trees are no longer there to intercept water and store it. Moore describes how Mallik *et al.* (1984) demonstrated that fine charcoal particles from fires block soil pores, thus preventing the drainage of water out of the soil. They found the soil permeability dropped by a massive 74%, so water remained in the soil, causing waterlogging and allowing moorland to form. The whole process would be exacerbated by livestock grazing, helping to suppress any regeneration of trees. Moore (1988) postulates that this waterlogging triggered the conditions for peat development in the uplands during Neolithic times.

In other parts of the uplands, clearance for agricultural activity is shown in the pollen record by the appearance of ribwort plantain (*Plantago lanceolata*) and other agricultural weeds (Smith in Simmons and Tooley 1981), indicating the opening up of grassland and the development of arable cultivation. Typical cereals grown at this time included einkorn and emmer wheat, both of which grow in cool temperatures in exposed areas with short growing seasons.

All this clearance for agriculture led in many places to temporary open areas, later becoming re-afforested or turning into permanent moorland mire. Other temporary clearances became more permanent, as in parts of the Lake District (Pennington 1970), central Wales (Moore 1973), Bodmin Moor, North York Moors and Pennines (Dark 2004). Above the tree line were montane grasslands and heaths, and below a belt of woodland. Valley bottoms and lowland areas were a patchwork of cleared and uncleared vegetation, depending on local conditions (Simmons 2005). The Neolithic peoples must have used the open upper areas for continued wild-game hunting as well as arable activity (Figure 3.4). They herded livestock in a semi-domesticated sense, using natural migratory instincts from lowlands in the winter to uplands in summer to develop early transhumance of wild aurochs (*Bos primigenius*) and deer (Fowler 1981; White 2002).

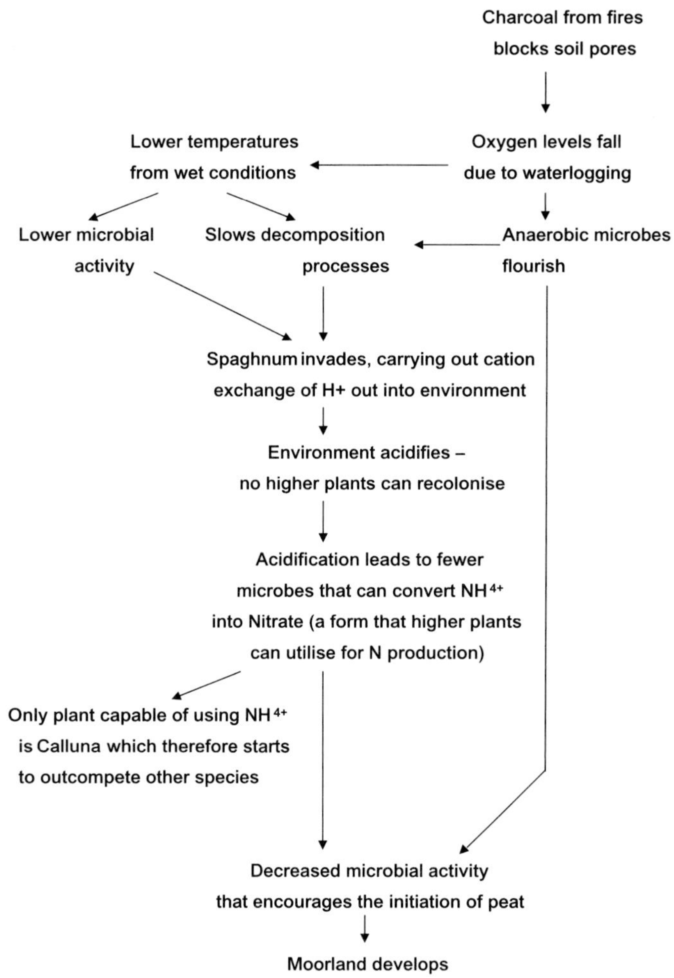

Figure 3.3 Waterlogging and Moorland Evolution

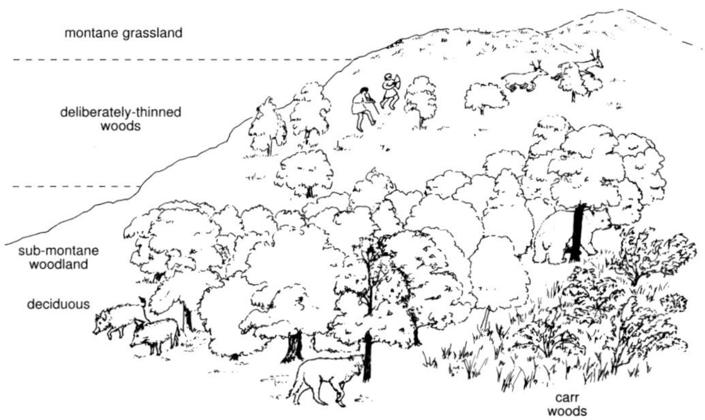

Figure 3.4 The Neolithic Upland Landscape

The field arrives (3900 to 2040BP)

With the Bronze Age culture, and then the Iron Age, the British uplands underwent another substantial shift. Previously, humans had fitted in with the environment, making only temporary inroads to change it (a form of environmental determinism). From the Bronze Age onwards, fields were created, a system of sedentary agriculture developed and landscape changes became more permanent; people began to adapt the environment to suit their needs. Nature was now less the determining factor in shaping uplands than it had been.

Another trend was a gradual deterioration in climate (Pennington 1969; Godwin 1985). Increased precipitation and strong winds heralded a more oceanic climate as Atlantic storms swung closer to the British Isles (Birks 1988); seasons became wetter and colder. It is thought the deterioration began in the extreme west and north-west (Tinsley 1981; Dark 2004). This hypothesis is supported by palaeoenvironmental reconstruction for the southern Pennines, where high-altitude forest was replaced rapidly by the growth of peat (Conway 1954), a trend replicated in the North York Moors (Dark 2004). Higham (1986) records how increased precipitation in areas like western Cumbria led to high levels of soil wash, making many agricultural areas unviable. Until the 1990s the accepted view amongst archaeologists and landscape historians, supported by palynological interpretation, was that such situations led to many upland areas being abandoned wholesale as the population retreated downhill (Parry 1981; Burgess 1986).

The thesis of climatic deterioration as a determining factor in vegetation change and depopulation has latterly been contested by various researchers (e.g. Jobey 1985; Young and Simmonds 1995; Tipping 1997, 1999). Using chronological settlement and re-interpretation of palaeoenvironmental data, they have instead proposed that uplands were continuously settled right through the Bronze and Iron Age. The most persuasive pieces of evidence relate to varied time frames of woodland clearance in British uplands. On an archipelago scale, Birks (1988) collated a wide range of pollen studies and found that deforestation occurred in four distinct phases in different parts of the British Isles over a long time scale, starting in the Bronze Age and continuing through to the present day:

3900 to 3700BP – confined to north-west Highlands and eastern Skye

2600 to 2100BP – all uplands in Wales and England (except the Lake District) plus Sutherland and northern Skye

1700 to 1400BP – the Lake District, Galloway, the Ardnamurchan peninsula and southern Skye

400 to 300BP – the Grampians and Cairngorms

On a regional scale this pattern of variable clearance, indicative of human activity, is evident. Dumayne-Peaty (1999) summarised the work of a range of researchers in the English and Scottish borders (Figure 3.5). Of the 19 sites she correlated, eight record gradual clearance of woodland into open landscapes and a further eight show woodland regenerating only to be cleared again (Table 3.2). Earlier palynological investigations elsewhere in the British Isles back up this view, particularly in the Howgills (Cundill 1976), Pennines (Tinsley 1976) and North York Moors (Atherden 1976). Atherden's work also revealed a number of iron forges, which would have used massive amounts of charcoal, thus encouraging wholesale woodland clearance, though forges could also be fuelled by peat charcoal (Ardon 2004). Furthermore, palynological work by Davies (2008) in a single upland valley in Scotland revealed that, instead of wholesale retreat downhill, Bronze Age and then Iron Age people were being selective over the use of land. Patches of very fertile soil were used for cereal cultivation, whereas anything of poorer quality was used for grazing.

While the climate deteriorated and then later partly recovered, the first functional, enclosed fields appeared. Most archaeological evidence suggests these fields were used for arable cultivation. Land beyond the enclosed field was used for stock grazing, managed by shepherds

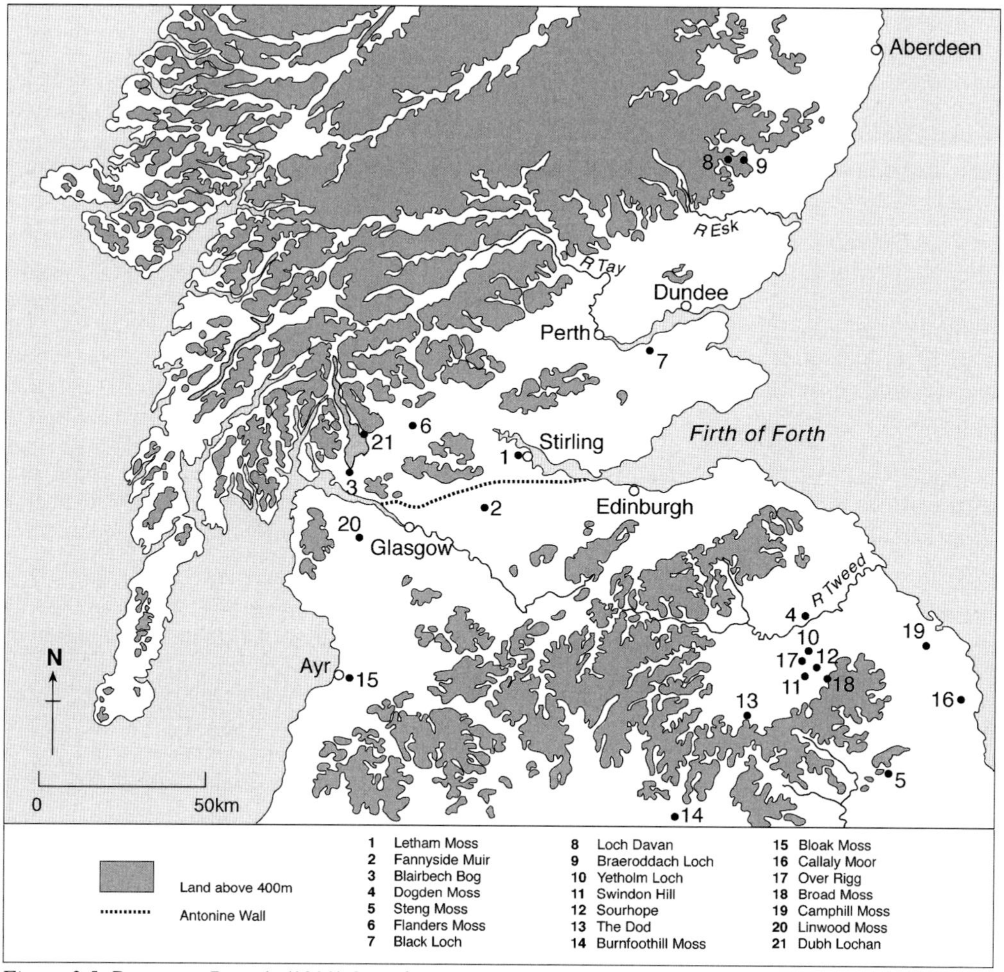

Figure 3.5 Dumayne-Peaty's (1999) Sampling sites for Northern England and Southern Scotland.
Reprinted from *Review of Paleaobotany & Palynology* Vol 105, Duymayne-Peaty L 'Late Holocene human impact on the vegetation of south eastern Scotland: a pollen diagram from Dodgen Moss, Berwickshire' p130, 1999, with permission from Elsevier.

protecting the livestock from wild animals. Any cultivation was based on the crook-ard, a forked implement which was dragged behind a person walking forwards. This created a groove in the ground into which cereal grains could be planted.

These first fields have been recorded all over the uplands, from Dartmoor to Northumberland (Taylor 1987). They have survived to the present day because they lay (and lie) in land considered marginal and, in later periods, not worth cultivating. These areas are littered with cairnfields (Figure 3.6) where stones were cleared from an area of land and piled up to form a cairn, leaving open land of a higher productive value (Higham 1986). Higham suggested that the archaeological evidence indicates that these areas were rapidly exhausted of their fertility and thus people moved on, clearing the next segment of forest; this ties in with Fowler's observations of the earlier Neolithic. Higham believed the technique was not really appropriate for uplands, but the vastness of the upland forest made its productivity when cleared unimportant. It may also be that lowland people were moving gradually into the uplands as the climate ameliorated. These activities led to some of the first permanent clearances of forested uplands, notably in north-eastern and western Britain. Dark (2004) has catalogued a range of palaeo-ecological data to support this for uplands in Wales, Exmoor, Dartmoor and the Trough of Bowland (Lancashire).

Table 3.2: Vegetation Changes in North Eastern England & the Scottish Borders

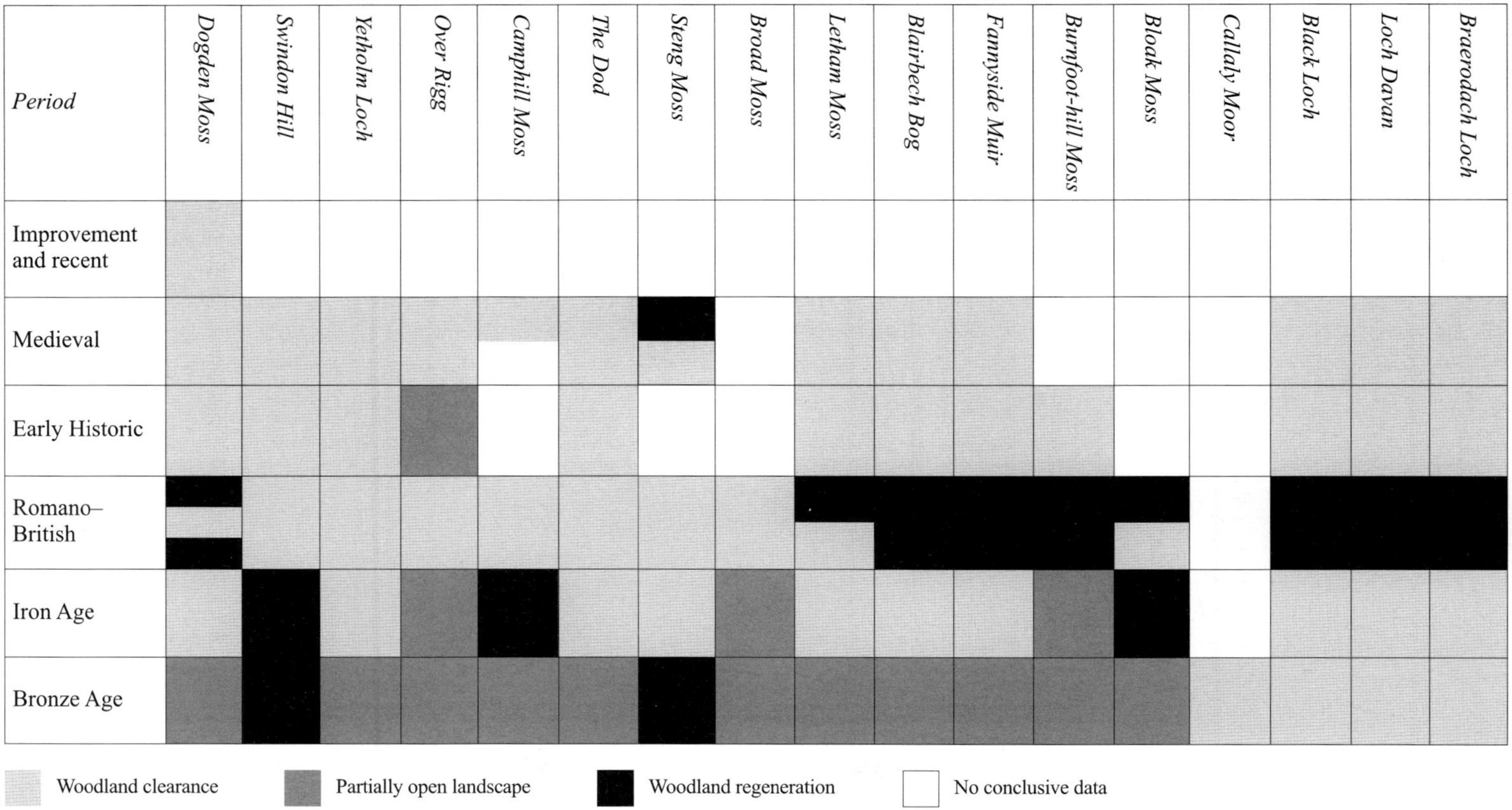

N.B: The vertical axis is not an absolute chronological scale since radiocarbon dates are imprecise – the table is designed to give a general impression of vegetation change. Location map shown in Figure 3.5.

Source: Adapted from Dumayne-Peaty (1999: 134)

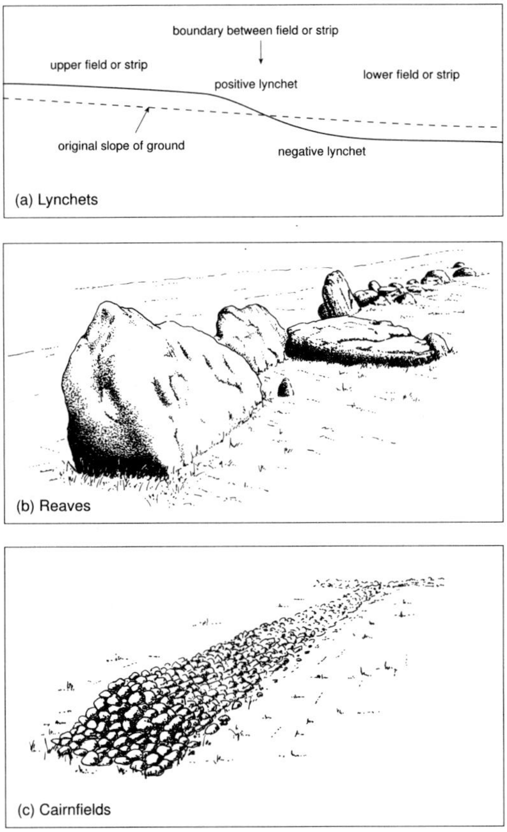

Figure 3.6 Bronze Age Field Structures

The characteristic slopes of uplands forced Bronze Age people to adapt their agricultural practices by forming scarps or lynchets, created by soil moving downhill as a consequence of ploughing. Following the contour, natural field boundaries were created, good examples of which can still be seen today around Malham Cove in the Yorkshire Dales (Figure 3.6). In other parts of the country, for example on Dartmoor, field boundaries were marked by stone constructions known as reaves (Muir 1987). The reaves of Dartmoor form part of enclosed fields of at least 200 hectares in size. Fleming (1983) describes two types: parallel reaves and radial single reaves (Figure 3.6). The parallel reaves occur in groups to form sets of enclosed fields running at right angles to the contour. The radial reaves occur singly and run away from the parallel reave blocks, up onto the higher ground in a radial pattern. Some run along the watersheds. Palynological evidence suggests that these field systems supported arable crops and high-quality pasture, which was abandoned around 3000BP as the climate deteriorated and blanket bog encroached.

Collectively then, cairnfields, lynchets, terraces, reaves, rectangular fields and fossil pollen provide evidence that Bronze Age people were conducting arable agriculture as high as 400m asl (Burgess 1985). The climate in the British Isles at the beginning of the Bronze Age was also warmer and wetter than now, with summer temperatures 2–3°C higher (Tinsley in Simmons and Tooley 1981). The ability to farm at higher altitudes was aided by the adoption of more climatically tolerant cereals; emmer wheat (*Triticum dicoccium*) was replaced by spelt (*Triticum spelta*) (Dark 2004). In modern trials, spelt has been shown to be productive even in heavy, damp soils prone to frosts (van der Veen and Palmer 1997).

Livestock management in the Bronze Age focused on semi-domesticated cattle; sheep were kept in smaller numbers, producing wool for cloth. Early cattle were generally smaller than modern animals, typically the size of Dexter cattle today (up to 1.5m tall), and are referred to as Celtic shorthorns (*Bos longifrons*). Sheep were already fully domesticated, but their wool seasonally moulted; thus no shearing was required. Instead, the wool was plucked in pieces from the fleece (Fowler 1981). No individual breeds existed, but it is believed most were reminiscent of the Soay sheep now found feral on St Kilda, which have coloured (pigmented) fleeces.

The era of transition from bronze to iron was also a time of climatic deterioration, which some consider to be the reason behind the movement of people into hill forts as resources became scarcer and were fought over (Burgess 1986). However, Davies (2008) seems to show that the uplands were abandoned a piece at a time and thus depopulated gradually, not in 'Burgess' rush'. This work also suggests that resource utilisation in uplands then was far more sophisticated than previously thought.

During the Iron Age, farmers deliberately developed two significant characteristics in sheep. First, selective breeding led to loss of pigmentation, giving whiter fleeces. Second, fleeces now grew continuously, requiring the introduction of shearing (Fowler 1981). The first domesticated sheep moulted at certain times of year. Their fleece was made up of two types of hair: kemp, an outer coat of hairy fibres, and the true wool, forming the undercoat. In the Iron Age a third type of wool evolved, the heterotype, which is long and grows continuously. It lies somewhere between kemp and true wool, being coarse and hairy in summer but thin and woolly in winter. Shearing meant stock management had to be more formalised, with an identified clipping period in early summer, structuring the farming year on the uplands.

From 300BC the climate ameliorated and another major episode of woodland clearance began (Higham 1987). Upland margins were increasingly exploited as the population expanded, particularly after Rome annexed Britannia in AD43 in search of mineral wealth. For example, the Clywdian range in north-east Wales was exploited for lead, which opened up the land for agriculture (Brown 2004). In northern England, the Romans cleared swathes of woodland beside the line of Hadrian's Wall, where military activity was at its height in the first and second centuries AD (Dumayne and Barber 1994). Much of the fell sandstone of Northumberland, the Cheviots above 300m, the Cumbrian mountains and central Pennines was switched to a mix of cereal crops and grazing (Tipping 1987). However, west of a line from the Quantocks to Exeter the villa economy was possibly absent as the Romans failed to take this land from the Cornish tribes (Fyfe *et al.* 2004).

Cattle were still the mainstay of society, evidenced by scant archaeological remains and stone stock corralls that litter upland landscapes from this time. Sheep were less utilised as they were more susceptible to attack from wolves, a fact which kept the Lancashire uplands devoid of sheep well into the 13th century AD (Higham 1986). The original semi-wild cattle were now domesticated, leading to a more formalised transhumance, with high-level summer pastures and winter grazing on low ground suitable for taking off a hay crop. The Romans introduced new, larger cattle, not unlike the surviving semi-domesticated Chillingham cattle, which may date from this time (Hall 1979; Evans and Yarwood 1998).

As demand for fodder increased, so new practices appeared, notably hay meadows on inbye land (Dark 2004). The grasses making up hay meadows were, at this time, those natural to the locality. For instance, limestone uplands produced grasslands with lime-loving species such as quaking grass (*Briza media*) and blue moor grass (*Selesia caerulia*). Dry acid grasslands had other dominant species like sheep's fescue (*Festuca ovina)* and various bents (*Agrostis* spp.), whereas wet acid grasslands were dominated by mat grass (*Nardus stricta*) (Duffey *et al.* 1974).

In the third and fourth centuries AD, plough technology began to improve. The Romans also took more interest in sheep for meat and milk, as well as their saleable by-product, wool. Northern Wales remained wooded, or benefited from woodland regeneration, while Southern Wales developed a more intensive arable economy (Dark 2004). In other upland areas, typically the light-soiled limestones (Yorkshire Dales and Lincolnshire Wolds), the local Romano-British started to develop long fields (200m by 20–40m). Gradually land switched to arable production as the population rose and settlements expanded uphill (Brown 2004).

This mixed economy did not last, because by AD300 overgrazing of the upland core had led to another cycle of resource deterioration, soil impoverishment and moorland expansion (Higham 1986; Ingrouille 1995). The Romans simply exploited these new environments, taking heather for browsing, thatch and bedding, along with peat and gorse for fuel (in northern England) and bracken for bedding (Ardon 2004). By the end of the Roman period, British society was in closer contact with more advanced ones in continental Europe, and this helped emphasise the divide between the lowlands, where the denser population and an influx of merchants and traders encouraged the evolution of a tamed patchwork of fields and woods, and the uplands, which remained patchily exploited, with few resources that were valued.

From the Bronze Age through to the end of Roman rule, the uplands underwent change. Fields were introduced as a technique to manage arable cultivation more effectively. Population pressure led to clearance of more upland woodland, which either regenerated or initiated the expansion of blanket mire, exacerbated by overgrazing and periodic shifts in climate. Later upland exploitation for mineral wealth led to a more multifunctional landscape, but at the same time these areas became marginalised from the main mercantile activities of the lowlands.

The medieval period (AD410 to c.1550)

Years of unrest followed the collapse of Roman rule in Britain in AD410. Waves of invasion by Germanic tribes, such as the Saxons and Jutes, along with bouts of plague and a deteriorating climate, led to extensive abandonment of agricultural land (Ingrouille 1995). Plague entered Britain about AD550 and had a profound impact on the population's age structure. The next generation, unable to replace these losses, had to abandon less favoured agricultural land (upland and wetland areas) and switched to less intensive pastoralism (Higham 1986). As a result the population nosedived from six million to two, leaving upland communities isolated.

In this devastated landscape the Anglo-Saxon kingdoms emerged. The political geography of England and southern Scotland began to develop in the 6th century; over time chiefdoms coalesced into kingdoms like Bernicia and Deira, which merged to form the 'super kingdom' of Northumbria in AD603. In the 9th century, Viking invasions and power struggles between super-kingdoms led to a loose peace, with the north and east controlled by Vikings (the area known as the Danelaw), the south and west controlled by the English under Alfred, and Scotland largely separate under the clan system with a king as overlord. England was eventually united under Athelstan (Alfred's grandson) in AD927 and, with a few hiccups, remained under English rule until the arrival of the Normans in 1066.

In this political landscape, upland areas were still controlled by the Celts alongside the later Danes and Norsemen, whereas Angles, Jutes and Saxons gravitated towards the lowlands of the south and east. Lowland English society was controlled through a strict hierarchy, both secular and non-secular (Roberts and Roberts 1980). The thegn – the system's lynchpin – could accrue land through gift from the king, for services rendered, and maintained law on his manor. Farmers were mainly tenants of the thegn, providing various services for him while working the land. In the Danelaw, the structure of society was different. Here farmers enjoyed freedom of tenure of their land, though still subject to an overlord. Settlements were sparser, forming a pattern of isolated homesteads and hamlets, in contrast to Saxon Britain where settlement was more nucleated. The Vikings introduced new breeds, including polled British white cattle and sheep such as Hebridean and Manx Loghtan (Evans and Yarwood 1998). Whichever side of the line a farmer lived, Danelaw or Saxon, the manor system became central to the economic life of the countryside.

The impact on upland landscapes of this dual Anglo-Saxon society and those which followed (Norman and Plantagenet) was complex. Unlike Iron Age people and the Romans, who focused on livestock farming, Anglo-Saxon agricultural activity was mainly arable, so they exploited lowlands rather than uplands. The detail of their agriculture is poorly understood, because there is a lack of evidence from the pollen record or from artefacts and structures (Dark 2004). The gradual decline in agricultural activity in the uplands was exacerbated by political unrest, poor climate and plagues. In some places secondary woodland regeneration occurred, particularly in the central Pennines (White 2002), northern England and southern Scotland (Dark 2004). Similar abandonment has been documented on Exmoor, with land reverting to Ling moorland or woodland (Fyfe *et al.* 2004; Dark 2004). Fyfe *et al.* (2004) also show how a more intensive agricultural production system was developed on lowland in the south-west: known as convertible husbandry, this is a sort of crop rotation of grain and grass crops interspersed with long grass leys, with rough grazing beyond in the outfield (see below).

In later Anglo-Saxon economies, a subtle shift in agricultural activity occurred. Depleted populations, war-torn and plague-ridden, switched back to less time-consuming enterprises. Livestock production became central, favouring the uplands, and sheep became more important, along with goats; breeds such as Orkney, Shetland and Norfolk Horn appeared (Evans and Yarwood 1998). Cattle were used more as beasts of burden to draw ploughs and consequently highly valued on an individual basis, but it was sheep that were central to later Anglo-Saxon farming. Saxon documents record them everywhere in the landscape (Rackham 1985), as do village names in *sceap*, corrupted later to forms like Shap, Skip, Ship or Shep.

As well as political change, several other factors had a substantial effect on medieval landscapes (see Table 3.3):

- the royal forests – eventually converted to common land
- the infield–outfield system – expanding common rights with transhumance
- monasticism – developing sheep farms for wool
- the Black Death – emphasising the marginality of upland farming

We now look in turn at each of these interdependent factors, which helped shape the three-tier landscape of inbye-intake-open fell that now typifies upland agriculture.

Table 3.3: Historical Landscape Changes

		extensive agricultural activity	*infield–outfield*	*transhumance*	*monasticism*	*Black Death*	*royal forests*
Time line	5th C 6th C 7th C 8th C 9th C 10th C 11th C 12th C 13th C 14th C 15th C 16th C 17th C 18th C 19th C						
Main characteristics		fall in population abandoned land extensification	infield–outfield waste	sheiling grounds hafod–hendre	desmesne farming grange farming	fall of desmesnes wool from 1450 more peasant farming	forest law
Main effect on upland landscape		increase in sheep and cattle	development of common land marginal piecemeal enclosure	summer pasturage hefting rudimentary stratification	cattle until 1450, then switch to sheep	extensification of land use abandonment in some places	agistment rudimentary stratification

Royal forests

An important feature of secular estates in upland areas from the 11th to the 17th century was forest. The concept of a forest then was different from ours: they were areas of mixed land use, marked out only by being under forest law with the right to keep deer. Generally they were the prerogative of the monarch, who might bestow forest rights on a nobleman.

Over time, this right was relaxed and private land was afforested as well. In upland areas most forests were moorland, run by large baronial estates and technically known as chases. Here the local lord had the freedom to hunt, but could not exercise forest law; offenders were punished under common law (Winchester 2004). Actual forest law gave the king powers to fine people for various misdemeanours, such as taking venison, abuse of grazing or illegal felling of timber and to restrict how land was farmed, what dogs could be kept and what fences could be constructed, so as to keep an open landscape suitable for hunting (Winchester 2000). The forest was run by several tiers of officials (Neville-Howins 1976). The king's representative was known as the Master Forester, Warden or Constable. Beneath him were verderers, who enforced forest law and held the forest court, and subforesters whose duty it was to implement the orders of the Master Forester, for example by felling timber for certain purposes. The subforesters were supported by woodwards or haywards, who were essentially gamekeepers. The verderers were supported by agisters, who collected rents for authorised grazing in the forest.

In upland areas, forest land was a combination of open moorland for grazing and valley bottom enclosed for stock farming. Good examples include Macclesfield Forest in east Cheshire, the Peak District, the Forest of Bowland, Inglewood and Skiddaw in Cumbria, Weardale in Durham, and Dartmoor. Throughout the medieval period much of the North York Moors came under the Forest of Pickering (Hey 2002), which were remnants of the land laid waste by William I's harrying of the north in the 1080s. Large areas of upland in Shropshire were also forest or chase in the 13th century: the Wrekin, Clun, Clee, Long Mynd and Stiperstones. In the south-west both Dartmoor and Exmoor are believed to have been forests well before the Normans arrived (Winchester 2004).

Lesser nobles might lease open fell to monastic orders, who developed granges, or peasants might pay them rent *per capita* for agistment, by which livestock that did not need milking – horses, gelded cattle, wether and/or hogg flocks –were driven up onto open fells and left to their own devices for the summer. Agistment therefore differed from shieling grounds, which included a temporary settlement for those managing the stock (Winchester 2000). Agistment rental raised substantial income for the owner, and agreements developed between lowland farmers and hill farmers without the need of shieling settlements. Reverse agistment also developed in the monastic era, whereby hill flocks could be overwintered off the fells. This pattern of 'off and on' the fell is probably the forerunner of the sheep stratification system that now operates in the UK (see Chapter 1).

The infield–outfield system

It was also in the Anglo-Saxon period that the infield–outfield system, commonly found in uplands, emerged (Astill 1992). The infield comprised a number of small enclosed fields, intensively farmed for arable crops. Beyond them lay the outfield, larger fields used for more extensive farming. Beyond the outfield might lie forest, controlled by agistment, or waste or common land, these two exploited through managed transhumance[2]. Contrary to popular belief, the infield–outfield system was not brought to Britain by invaders from continental Europe, but developed here over time (Taylor 1987).

It is at this point that archaeologists and landscape historians begin to disagree. A field system sometimes known as the 'Anglo-Saxon shuffle' developed, whereby each villager cultivated a strip in each of two or three large open fields; this minimised ownership conflict and allowed everyone a fair share of the resource base (Astill 1992). Taylor (1987) suggests that population pressure

---

[2] Transhumance - Movement of stock and people from lowland to upland and back to exploit summer grazing at higher altitude.

forced the infield to expand, engulfing the outfield. The resulting pressure on grazing land could be resolved only by communal agreement, leading to the development of common grazing rights.

The counter-argument is that common land more probably developed as a result of the economic marginality of uplands, linked to the loss of low-level communal grazing on better soils because of population expansion and enclosure. The strip element of the open-field system may then have evolved on the remaining outfield as population pressure increased or inheritance laws developed. This Midland open-field system covered much of central England (Astill 1992), though not everywhere else. As Fyfe *et al.* (2004) have shown, the convertible-husbandry system had spread in south-west England and managed transhumance was emerging in upland areas.

A third option accounting for the open-field sysem is the new plough technology. Re-harnessing a team of eight oxen after turning at the end of a small enclosed Iron Age field wasted too much time, so larger open fields developed, up to 22 yards (20m) long (Roberts and Roberts 1980). As no peasant owned eight oxen, ploughing became a co-operative effort, each peasant farmer being awarded a strip of land in the open field.

At the same time, the evolution of common land and agistment allowed for transhumance, whereby people and their herds and/or flocks migrated from a winter settlement to summer pasture. Before the Early Medieval period, the seasonal behaviour of animals was harnessed in natural transhumance, with people following behind (Ryder 1981). Not only did this ensure the availability of fodder, but it also reduced the risk of disease wiping out the stock as they roamed freely from summer grounds to winter and vice versa. The first documented evidence of artificial transhumance is that of Gildas the monk (6th century AD), who noted uplands in Wales suited to summer pasturage (Fowler 1981).

A dual system emerged in Wales, known as *hafod–hendre*. The lowland *hendre* was linked to an upland *hafod* or *lluest,* whither stock were traditionally moved on May Day (*Dydd Calan Mai*) and brought back down in November (Figure 3.7; Thomas 1992). Water meadows (*gwairglodd/dol*) were cut for hay (*gwair*), which was fed as fodder to cattle. A little further upslope, the drier pastures provided winter feed along with some arable land (*cyfer, llain, talar, dryll* or *erw*). Beyond lay the *ffridd* – land enclosed from the open fell – which was used as pasture or, in some cases (e.g. in parts of the Berwyn Mountains) as arable strip fields around 400m asl (Silvester 2000).

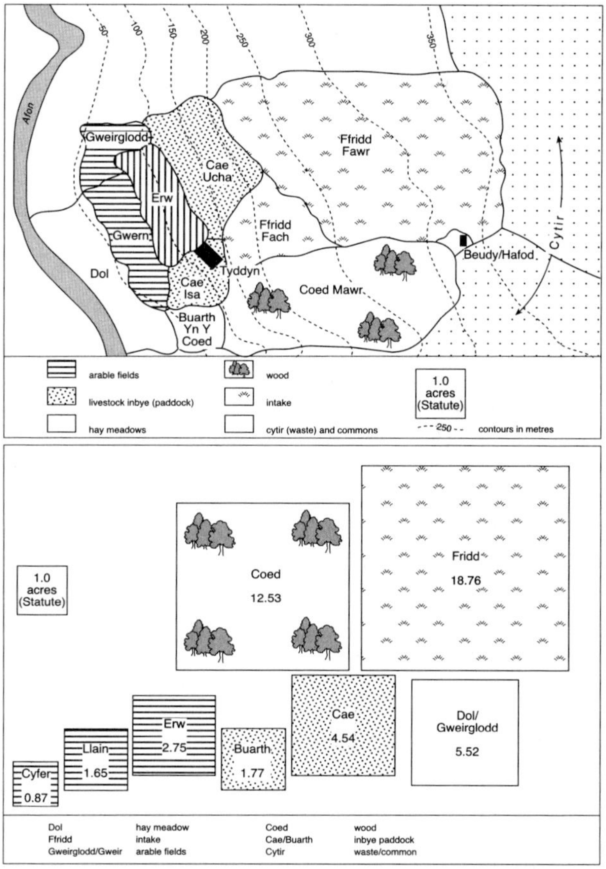

Figure 3.7 Agricultural field patterns on a Medieval Welsh upland farm

Reproduced by permission of the author and Society for Landscape Studies from *Landscape History*, Vol 14, Figures 2 and 3 pp.44 and 46

Finally, above all the other land mentioned, lay open moorland, used in summer for transhumance and often shared with neighbouring farmsteads, using hefting to separate sheep pastures. The *hendre* farm also contained intensive grassland, or inbye (*cae/buarth*), and typically areas of woodland (*coed*). Brown (1992) describes how Clywdian upland farming families often earned cash by grazing other people's stock. Unfortunately, many pastures became overgrazed and soil erosion ensued.

In Scotland and northern England there was a similar system known as shieling, shielding or summering. The shielings were summer dwellings lived in by herdsmen and their families, and the land was known as the shieling ground (Woodside and Crow 1999). Shieling allowed for stock that required milking (Winchester 2000). Of such importance were shieling grounds that they were written into charters of land ownership, family or clan. Woodside and Crow (1999) show how a number of civil parishes beside Hadrian's Wall still reflect this twofold land management in the way they stretch from the Tyne valley up to the waste of the Great Whin Sill (Figure 3.8).

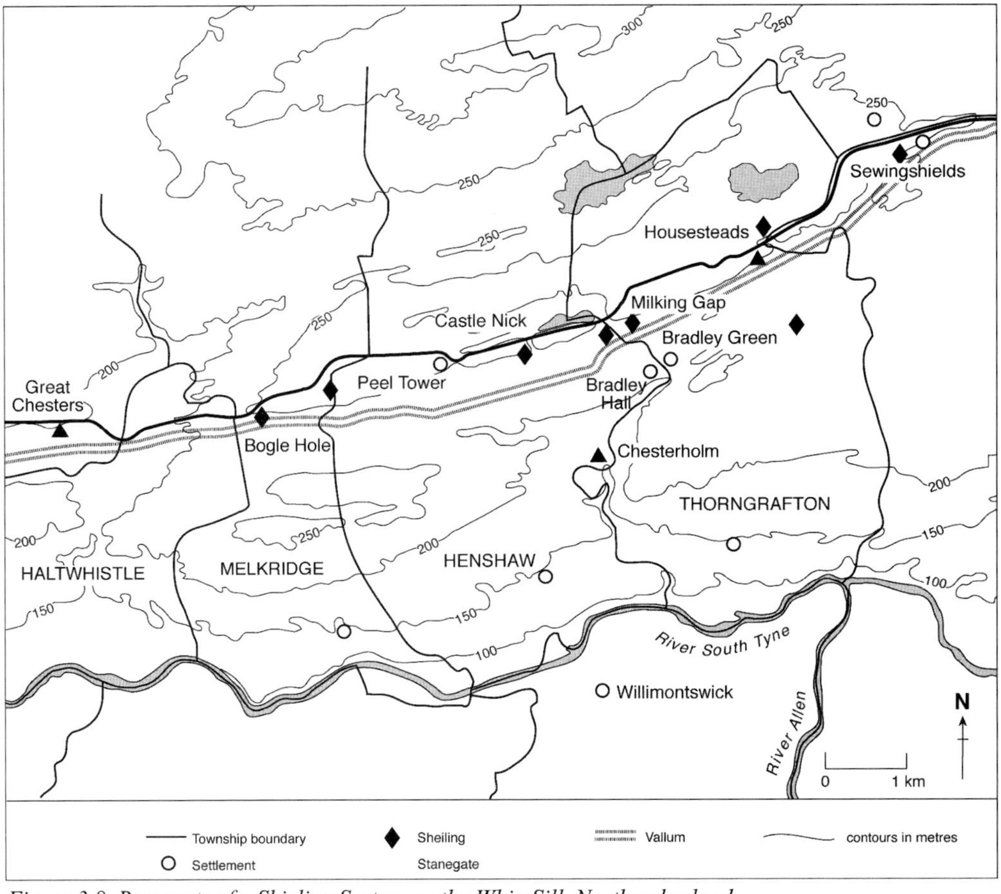

Figure 3.8 Remnants of a Shieling System on the Whin Sill, Northumberland
Adapted from Woodside and Crow (1999:65) *Hadrian's Wall: an Historic Landscape* © National Trust.

Atkin (1994) used Ightenhill Estate documents to interpret the system of secular upland farming in England at this time. The farm enterprises comprised oats, barley and cattle, the latter being herded on vaccaries (cattle farms) to produce ploughing oxen. These vaccaries ran from the valley bottoms up onto the fellsides. Stock was moved off the inbye onto upland summer pastures after May to allow for hay production. In late September the stock returned to the valley, to be divided into:

- stock to be sold on, either as oxen for work or because they had murrain[2]
- stock too old or barren (the geldherd), put out to pasture in winter or summer (agistment)
- the rest of the herd, which the estate steward reallocated back to the vaccaries

In this way herd numbers were reduced to match the 'hungry gap', the time when stored hay ran out and spring grass had yet to grow enough to support the herd.

The shieling grounds and *hafod–hendre* system gradually disappeared in the 16th and 17th centuries with the expansion of permanent settlement into the uplands, the last occurrence being on Snowdon in 1862 (Roberts 1949). With it went the custom of transhumance:

> Its long-term survival since the early Middle Ages reflects not only its economic viability, but also with what difficulty change came to this isolated area [Hadrian's Wall], so long devastated by wars and raiding. The final act that may have brought about the demise of shielding may have been the result of Crown Estate policy, since the Crown had long regarded transhumance as unsatisfactory in terms of revenue and defence against raids. (Woodside and Crow 1999: 66)

Some shielings evolved into permanent farms, such as Steel Rigg on Hadrian's Wall. This is symptomatic of an increase in population, where more marginal land was pressed into use and summer pasturage became permanent pasture (Williamson 2002). This also happened in Wales when the *hafod–hendre* system collapsed, with many *hafodau* becoming separate stock farms or simply falling into disrepair (Brown 2004). This was exacerbated by Welsh law, which awarded 4 acres of land to every 14-year-old boy, creating a dispersed population (Roberts 1949). Once dominated by cattle, which were driven to markets in London, the Welsh uplands gradually became populated by sheep, which became more profitable as wool demand increased. By the end of the 18th century the upland *hafod* became almost exclusively a sheep farm. Flocks developed their own *cynefin* whereby sheep passed on their heft knowledge to the next generation.

Monasticism

Christianity reached the British Isles in Roman times, but Anglo-Saxon England was converted by Irish monks (AD450) in the north and by missionaries from Rome (AD597) in the south. Monks were granted land and their communities became self-sustaining. Not much is known about the impact of these early communities on the landscape (Aston 2000), but it has been suggested that Christianity reduced the level of slavery (Roberts and Roberts 1980), and this allowed the gebur or tenant-farmer workforce to expand. In AD787 Viking raids began and by AD870 monastic life had all but ceased north of the Humber.

There was a resurgence of monasticism in western Europe in the 10th century in Burgundy, Flanders and Lorraine, and in Britain in AD940 probably at Glastonbury, where the Cluniac Order founded a monastery. The main impetus for landscape change in uplands came after the arrival of the Cistercian Order from Normandy in 1105. This order began as a response to the laxity of Benedictines and Cluniacs, who were now seen as decadent! The first Cistercian foundation was at Furness (now in southern Cumbria) in 1113 and from this one monastery many of the most influential foundations developed, including Fountains, Tintern and Rievaulx (Figure 3.9).

Since their rule did not allow the monks to leave the monastery precinct, the Cistercians used the manual labour of an army of lay brothers to run their outlying estates or desmesnes (pronounced 'domains'). According to the monastic rules of charity, obedience and poverty, individual wealth was not allowed. However, as Burton (1994: 233) points out, 'poverty was

[2] The exact nature of murrain is unknown. Atkin (1994) suggests it may have been a collective term for a range of cattle diseases including foot and mouth, rinderpest and pneumonia. On the Ightenhill Estate he estimated that 5% of one year's herd fell foul to murrain.

individual and not corporate' and so monastic houses were able to accrue wealth in land ownership and management. As perpetual corporations they could take a long-term view, developing sheep pasture and assarting[4] new land (Greene 1992). Until 1279, the land was farmed by monastic staff. From 1279 onwards, land began to be leased out to secular tenants.

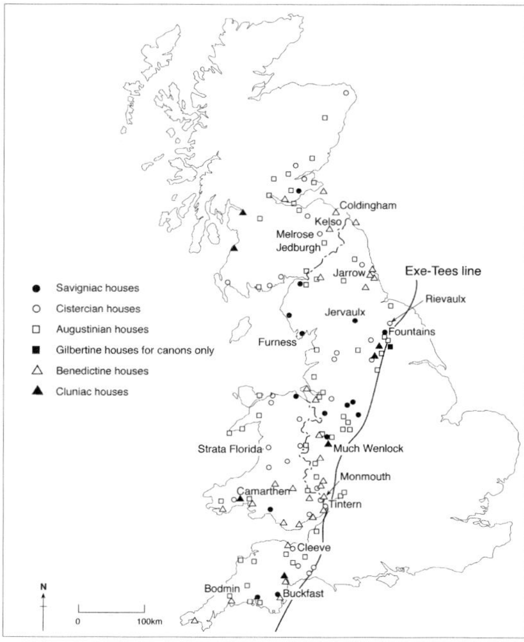

Figure 3.9 Medieval Monastic Foundations in the British Uplands
(Adapted from: Greene 1992)

The Cistercians selected isolated locations for their foundations, and about 40% of their land was waste[5] in the 12th century (Burton 1994). They rejected the open-field system that had developed in the Midlands and parts of Wales and instead re-used enclosed Iron Age or Roman field systems. Aston (2000) notes how many historical records from this time show the reclamation of upland areas by monks, particularly large tracts in the Lake District, Yorkshire Dales and Cheviots. To be successful in a predominantly agrarian economy, the monasteries had to exploit the land efficiently. The Cistercians in particular became expert in watercourse engineering and invented the concept of outlying farm units known as granges. They were also more innovative than other orders in their livestock management, which aimed to improve breeding, nutrition and disease control (Bond 2004). Finally, Cistercians graded their wool, unlike the Benedictines, who sold mixed wool sacks, and thus their high-quality fleeces commanded the best prices.

From the 10th century to the 15th, the enterprise mix of monastic farms went through several changes (Bond 2004). Up to the 12th century most of them focused on crop production for subsistence with cattle herds for milk production, particularly in mid-Wales. The cattle also extended the limits of occupation upslope as scrub was grazed off and trampled. Gradually this system was replaced by desmesnes based on sheep. By the 14th century most upland foundations were sheep enterprises; those in the lowlands concentrated on cereal production. At this point sheep farming became more specialised, and geographically limited to the regions with the best pasture – coastal marshes, chalk downs, the Cotswolds, Mendips, the Pennines and other northern moorland areas. Flock composition became more sophisticated, with some ewes for milk and breeding, and over half being wethers kept for wool (Bond 2004). In the 13th and 14th centuries, when the transhumance system developed further, the wethers were often left on the upland pastures in winter and the ewes and hoggs moved down to the valley bottom to overwinter. Transhumance could cover exceptional distances: Fountains Abbey pastured its cattle 60 miles away in Cumberland (Bond 2004). The flocks themselves probably varied considerably in size year on year, owing to diseases like sheep rot and murrain. As stock mobility was so extensive, there was no real breed development other than coarse-haired fleece in the north, south-east, Dorset and Wiltshire, and softer fleeces in the Welsh marches, Cotswolds and Lincolnshire Wolds.

[4]An assart was an area of trees and undergrowth cleared for agriculture and enclosed by hedges and ditches. Some hedges were a relic of the woods; hence ancient woodland species (e.g. dog's mercury) can be found in a hedgeline.

[5] Waste was undeveloped or underdeveloped land, often used as common land. Much had been previously abandoned through William I's vicious suppression of his revolting barons (the harrying of the North).

Putting into perspective the inroads of monastic land management in the uplands, Bond says:

> the role of the Cistercians as pioneer settlers should not be exaggerated. In reality, England had little virgin wilderness left by the twelfth century, and much of the Cistercian effort was directed towards resuming cultivation on previously abandoned land. Some 40 per cent of twelfth-century granges occupied lands described as 'wasted' in 1086. (Bond 2004: 71)

More significant was the grange system used for upland sheep pasture. This land could be extensive: for example, in 1242 Furness Abbey had at least 5700 ha (14,000 acres) of upland fell managed by a web of granges (Aston 2000). A network of drove roads emerged, to allow sheep to be moved between granges. In the valley and dale heads of the north, vaccaries were established, using inbye land for hay meadows as in the secular system (Bond 2004).

About 1321, an Italian merchant, Pegolotti, produced a guide to the wool trade in Britain (Aston 2000). He listed the amount of wool available for purchase from the top seven foundations in Britain (Figure 3.10a). Of all the houses in the wool trade, 85% were Cistercian (Figure 3.10b), Fountains Abbey being the largest supplier with 76 sacks. From this the size of the flock can be calculated, because a sack held 200 fleeces, so at this time Fountains Abbey owned and ran 15,000 sheep! Unfortunately the export market did not last; by 1457, markets had all but collapsed, with Fountains reduced to supplying only four sacks and the entire Cistercian Order providing about 4% of the wool market in England (Bond 2004; Aston 2000). Rival orders and secular sheep farming had become dominant.

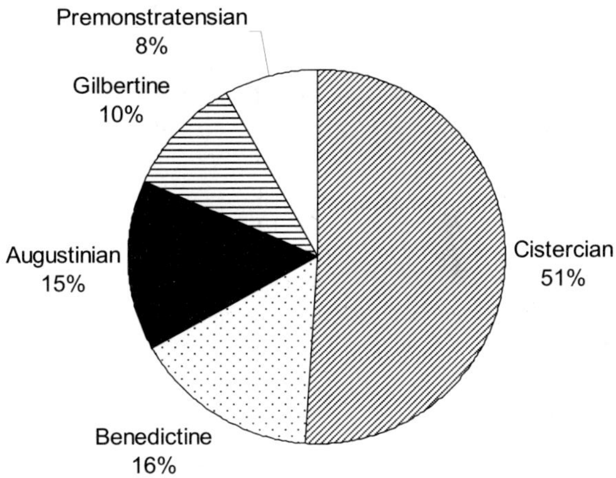

Figure 3.10a Monastic Wool Production by Foundation in Britain 1321
(Adapted from: Bond 2004)

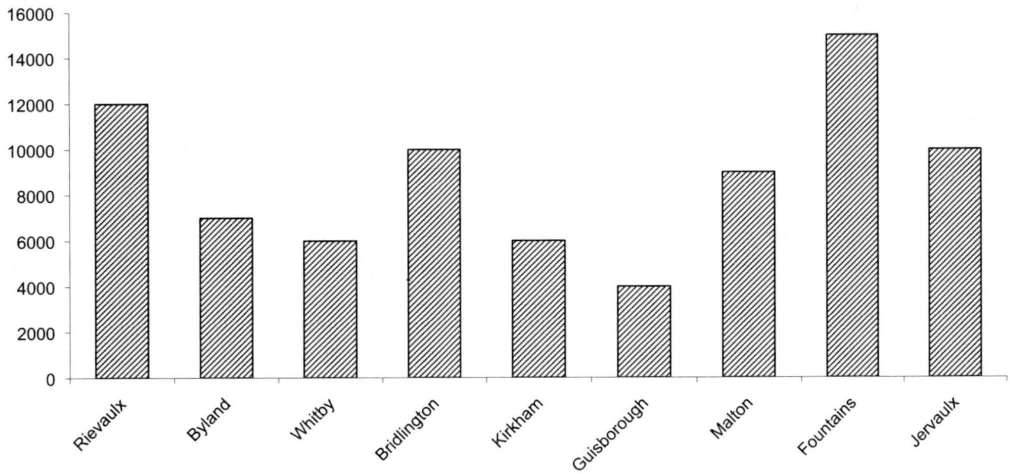

Figure 3.10b Quantity of Wool Production by Foundation in Britain 1321
(Adapted from: Bond 2004, Aston, 2000)

It would be easy to think that the monastic orders were pre-eminent in sheep farming and the wool industry in upland Britain. However, historical records, such as Pegolotti's, show that their economic role must not be overestimated. Much (later, most) British sheep pasture and wool production was managed by secular estates. In 1328 free-trade laws were introduced in England, which led to a rapid increase in wool exports and secular producers (Glassock 1973). What is important, however, is the overall impact of monastic activity on large swathes of upland landscape in terms of maintaining semi-natural grassland. With the dissolution in 1537, ownership passed to a small number of secular overlords, leading to the control of upland land management by a few, which is still the case today.

Impact of the Black Death

The Black Death reached the south coast of England in 1348. This date is often used to divide Early and Late Medieval periods because in one generation the population fell by a quarter or even a third, to around 3m by 1377. As a result, most marginal uplands were abandoned or extensified. A severe drop in land values was paralleled by a labour shortage and wage rises. Prices for products fell, but by 1450 had begun to rise again. This situation had three consequences (Baker, cited in Darby 1973):

- Decline in demesne farming – as the lay workforce disappeared, tracts of land were leased to tenants who became yeoman farmers. The least profitable land returned to waste, as landlords sought to cut their production costs.
- Growth of peasant farming - in parallel with the rise of tenancy on desmesne lands was the opportunity for peasants to buy their own land from secular landlords. The land market became fluid and land began to change hands more frequently. Parcels were subdivided and many small farms emerged increasing the area of private land holdings.
- Changing land use - with the loss of labour, production extensified, switching from arable to pasture. From 1450 grass became one of the most desirable crops because it allowed farmers to move into sheep production for wool. To increase efficiency in wool production, much of the 3 fold infield-outfield was fenced to control straying stock. This was particularly true in limestone upland, where resistance to enclosure was weak owing to a sparse population.

Early Modern uplands

During the medieval period these four new factors – forest law, infield–outfield, monasticism and the Black Death – helped to create the upland farming system we see today. Common land, transhumance, overwintering, stratification and a sheep economy all have their roots in this period, when people were not just adapting to and working with the uplands, but also exploiting the advantages of uplands.

This led to the distinctive three-fold upland landscape of inbye, intake and fell. Inbye was developed for subsistence crops – usually oats, rye or bigg (a type of four-row barley) – or for hay as fodder, but it still contained less nutritious grasses such as Yorkshire fog *(Holcus lanatus)* and herbs like yellow rattle *(Rhinaanthus minor)*. Intakes evolved where land could be partly improved, but was still of poor quality dominated by bent. The wetter areas on slopes were infested by rushes and sedges. Fells were used for agistment as forest law gradually collapsed or for shieling grounds where common land existed. Fells were dominated by grasses, particularly purple moor grass, used for hay for cattle, and cotton grass, indicative of wet peat.

By being used in new ways and exploiting their strengths, the uplands came to specialise in livestock production. Emery (in Darby 1973) suggests three distinct upland regions emerged in England and Wales by the end of the Middle Ages:

a) Northern England – a landscape of summer transhumance to higher land. In dales, enclosed land was developed for arable crops and hay meadows.

b) Wales and the Welsh Marches – predominantly a landscape of cattle, dairy and pig enterprises, supported by a mix of enclosed and unenclosed land.

c) South-West – the upland areas developed as summer rough grazing, using transhumance underpinned by some shifting cultivation.

In highland Scotland a different system emerged – that of the clan or extended family. Small townships grew up, like Polmaddy in the Southern Uplands (Forestry Commission 2006). Settlements had several communally tilled arable fields divided from the open moor by a head dyke or main wall. Land management was tightly regulated by the local barony court, which issued dates for putting stock onto the shieling grounds, allocated timed windows for muirburn and designated areas of heather–grass mix, which could not be burnt at all (Dodgshon and Olsson 2006). They kept black cattle and sheep that were rather like North Ronaldsays. Stocking densities were extremely low (0.0625 to 0.125 LU/ha) but based on hundreds of years of practical knowledge recognising the limits of the uplands. Shieling grounds were at best used for only four months per year (early June to late September, with a staged return to the valley). As a result of these restrictions, pressure on winter fodder and wood was extreme, keeping society at a subsistence level. Doubtless those who organised the later Highland clearances perceived their activities as improving.

The Agricultural Revolution and high farming

By the mid-17th century the grange system had disappeared and most land was held by a landlord–tenant relationship, moving British agriculture from subsistence to a capitalist economy, characterised by four processes affecting the uplands:

- adoption of innovations
- rise of the entrepreneur
- parliamentary enclosure
- Highland clearances

Adoption of innovations

Before 1700 most farmers saw corn as the most important crop because it was a staple food for the population (Beckett 1990); even in upland areas, more tolerant rye, oats and barley were grown. Hardly anyone produced just livestock. The dominance of cereals began to weaken as new production methods evolved.

For livestock enterprises the big leap forward came in the late 17th century with improved fodder crops (roots, nutritious grasses and clovers), selective breeding and the import of cattle sires. Unfortunately for upland farmers, physical constraints of the terrain limited their ability to exploit the new fodder crops unless they employed drainage or liming.

Poorly drained pasture was a major problem on their enclosed land. Not only did it lead to soil poaching and sedge and rush infestations, which lowered productivity, but it could cause a variety of foot-rot diseases amongst livestock. One solution was to put in drains beneath the sward. As early as 1686, under-drainage was developed in Staffordshire by digging a ditch, filling it with pebbles and brushwood faggots, and then covering it with soil. This practice spread slowly during the 18th century, helped later by the use of clay tile drains. There were attempts to drain high moors by digging open channels (grips or gripes). These had some effect in draining surface water, but the underlying peat still supported undesirable vegetation (see Chapter 4).

Liming was a more widespread improvement in uplands, often based on the lime-rich geology of many such areas. Liming was effective because it favoured the spread of white clover, which on alkaline soils out-competed the non-nutritious bent. Unfortunately, as land was improved and in some cases ploughed and re-seeded, grass species became more homogeneous (see chapters 4 and 5). Limestone had general economic value too, as a high-grade product commanding a higher premium than other such materials like marl or chalk. Its use was supported by the canal revolution of the 18th century, expanded production of coal (for the kilns) and finally by rail and road developments, which lowered transport costs. Lime could be extracted and burnt on site using a lime kiln. It was best used in a process known as 'pare and burn' (Williamson 2002): old grass swards were scarified off and burnt in heaps, ash was spread over the fields with lime and the mixture was ploughed in. Fields could be re-seeded with grass after a few years of arable cultivation. This technique eventually merged into convertible husbandry; swaling is a remnant of this (Russell and Harris 2008) (Box XXIII).

Farmers could also improve their livestock. One of the best-known innovations in this period is the selective breeding of the New Leicester sheep by Robert Bakewell. The resultant cross-breeding allowed many farmers to produce sheep that put meat onto the carcass rapidly; thus sheep were ready for slaughter in two years rather than four. Bakewell's work, starting in 1745, was not unique or original; he simply was affluent enough to give more attention to breeding and be more selective in the animals he chose to breed from (Beckett 1990). Improvers such as Bakewell worked in small groups and it was not until well into the 19th century that their influence spread widely. Today some upland areas benefit from crossbreeds such as Clun, Border Leicesters and Mashams (see Figure 1.5), whereas other areas retain breeds like Herdwicks, Cheviots and Welsh blackface that evolved to fit their environment, well before selective breeding began.

Rise of the entrepreneur

As farmers adopted new techniques, production increased and they produced a surplus for the first time. The agricultural economy shifted from subsistence to capitalism – what is termed 'high farming'. More efficient production also meant that more food was available to buy and fewer people were needed to work the land. This led to an increase in subsidiary occupations around the uplands such as quarrying, mining (Whyte 2004) and eventually mill work. It also led to mass emigration to the cities and into manufacturing and so began a gradual spiral of decline in the rural population, slowed only by the development of waterpower.

Supporting this capitalist agriculture was a change in the character of the agrarian class (Beckett 1990). In medieval times, land was owned by the lord of the manor and local people worked for him, receiving food in payment. Now farmers either owned or rented their land. Contrary to popular belief, the landlord-tenant relationship was not one of extortion. In fact, many landlords charged fair rents that allowed tenants to invest in their land. This was, of course, in the best interests of the landlord as it improved the quality of their land. The industrialisation of agriculture was also aided by the rise of the large estate, the decline of the small landowner/owner-occupier and an increase in farm size in certain regions (Grigg 1987), gradually eroding the shieling/*hafod–hendre* system as the combined units separated (Brown 2004).

Parliamentary enclosure

In England most open fields in valley bottoms disappeared before the 18th century, mainly enclosed piecemeal or by private agreement, particularly in the south-west (Emery 1981). In contrast, fell common land in (what are now) Northumberland, Cumbria, Clwyd and Powys was subject to parliamentary enclosure in the late 18th and early 19th centuries. Overton (1996) estimates that overall 24% of all enclosed land (4.5m acres of open arable field, 2.3m acres of waste) was enclosed by private agreement in this period, but there was no effective parliamentary enclosure in the uplands before about 1760 (Figure 3.11a).

Parliamentary enclosure began with a petition for an area of land (for instance, a parish) to be enclosed. Once an Act of Parliament awarded enclosure, surveyors identified ownership of every strip of land in the parish. Amounts were totalled and the land was then re-apportioned as amalgamated blocks. Pieces were allocated automatically to the lord of the manor and to the Church (replacing tithe land). Unfortunately farmers could end up with less land than they had before, though sometimes this loss was outweighed by the greater efficiency of having all their land in one block (Whyte 2004). Roads were added to ensure access and then the land was enclosed by hedgerows (Box XIII) or, in the uplands, drystone walls (Box I). The cost of enclosure was borne by the beneficiaries and this led to many farmers and families leaving the land. Large landowners often won from this exercise, but small farmers could lose out because their block might have none of the best land. These farmers were forced to improve what land they had. Conversely, Whyte (2004) notes that enclosures in Cumbria were often a relief to small farmers, because much of the common land had previously been overgrazed by drovers moving stock from Scotland.

Figure 3.11a Rate of Parliamentary Enclosure in Britain 1730 to 1860
(Adapted from: Overton 1996:151)

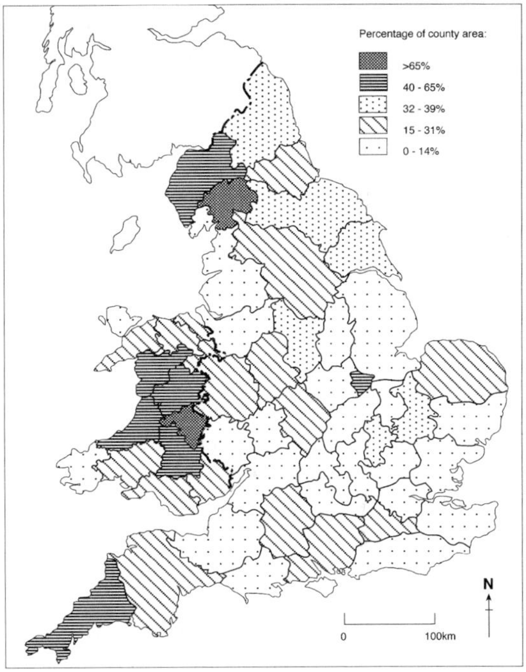

Figure 3.11b 'Waste' in England and Wales circa. 1800 (Board of Agriculture 1808)

Enclosure was a way of increasing efficiency and thus output, a particular concern during the wars with France (1793 to 1815), when the army needed more fodder and Britain could no longer rely on grain imports from the continent. In 1793 the Board of Agriculture was set up to investigate utilisation of upland waste (Overton 1996). In 1808 the board reported (Figure 3.11b) that some counties still had very high levels of waste, for instance 83% of Westmorland and 73% of Montgomeryshire (Darby 1981).

Cumbria had the most unenclosed waste in England before 1760, but by 1830 a quarter of this had been enclosed, most of it above 300m asl (Whyte 2000). North Westmorland, for example, had over 101,000 acres enclosed in three distinctive phases (Figure 3.12). In addition, Whyte (2000) calculated that 1,788 km of new boundaries were created within a 130km^2 area, together with 277 km of new public roads and 46 new farms. The magnitude of the task led to a standardised technique for walling, which traversed all terrain geometrically and employed

Box XIII: Hedges

Hedges are constructed from a range of woody species, most commonly hawthorn (*Crataegus monogyna)* and blackthorn (*Prunus spinosa*). Many other species can be used in a hedge as long as they are indigenous, do not shade out lower growth, are not poisonous to stock and can be managed by the process of laying (see below). Soil type also controls hedge species; for instance, alder (*Alnus glutinosa*) prefers damp soils.

The hedgerow is made up of a number of features: the hedge itself, intermittent fully grown standard trees that act as perches for owls and other raptors, the hedge bank at the hedge foot that may support a wide diversity of herbaceous species and acts as animal cover, and sometimes a ditch running the length of the hedge. A well-managed hedge will be thick enough to ensure stock cannot squeeze through any gaps while growing no more than 2m high to avoid negatively affecting any adjacent crop growth. In upland areas, height is not such an issue for the pastoral economy, but with height the hedge can become 'leggy', eventually allowing stock to escape.

If it is to control livestock, the hedgerow requires regular maintenance. There are two main techniques for this: trimming and laying. Trimming tidies the hedge faces and, if not overzealous, thickens up the hedge surface. Traditionally it was carried out twice a year but now, to minimise time and cost, most farmers trim once a year using a tractor-mounted flail. Overtrimming can eventually kill the hedge; poor trimming (typically trying to cut material too thick for the machine to cope with) can damage the stems and encourage disease. Flails are typically used for trimming because they are a more versatile tool for the farmer; they can also be used to cut grass verges, for example. More effective is a reciprocating cutter bar, but these are limited to only one or two jobs around the farm.

The second way to maintain a hedge is to lay it. Laying is a centuries-old technique that encourages the hedge plants to re-sprout, thus re-thickening the hedge (Agate 2002). It can only be done manually and its cost in labour and time discourages many farmers from doing it. Each woody stem (or 'pleacher') must be cut almost through, near ground level. The pleacher is then bent over onto its pre-layed neighbour. Where the pleacher is still attached to the rooted stem, new shoots appear over time to replace the layed one. The pleachers form an impenetrable dying barrier until the new shoots are thick enough to dissuade stock from escaping. The whole layed hedge is then staked every few metres to 'hold' it together, until the new growth establishes itself. Some layed hedges also have a top binding of woven wooden stems as additional support, but this depends on hedge style. To be effective as a stockproof barrier, most hedges need to be layed regularly.

Newly laid hedges are susceptible to animal grazing, so usually they are temporarily fenced until the hedge has become established. Those with a ditch and bank may not need fencing. The style of laying varies with the region. The defining feature is the angle between the pleacher and the ground. A common style is the Midland hedge, also found in the Cheviots, Yorkshire Dales and Peak District, in which the pleachers sit at about 30°. In contrast, the Cumberland style is for the pleachers to be laid almost along the ground; the whole structure is then pegged down using small stakes and baler twine. In Wales the style of hedging varies slightly from county to county.

Most farmers today know how to maintain their hedges. It is a traditional skill that has been either passed down through the family or learnt at the local agricultural college. Intensification of agriculture, lack of labour and the extra cost, along with financial support to remove hedges in order to make farming more efficient, meant that from the 1950s to the early 1980s hedgelaying become much less common. In upland areas this trend was most noticeable on dairy farms in the upland margins, where a switch to silage production led to larger fields to allow machinery to be used more effectively (Shoard 1980). Between 1946 and 1974 about 120,000 miles (75,000 km) of hedgerows were lost (Westmacott and Worthington 1974).

Since the late 1980s this trend has been reversed by financial support for re-establishing hedges (see Chapter 8). This may seem both ridiculous and beneficial, since in the long run the stock of hedges will return to its original level, but that statement hides the conservation reality. Quite simply, the older the hedge, the more species it supports. Thus newly planted hedges have less wildlife value than older ones. The older grubbed-out hedge may also have held historical value, for instance, as a parish boundary dating back to Anglo-Saxon times (Maclean 1992).

After some time the hedge species will become too woody to lay; at this point the plants need to be grubbed out and replaced. The restoration of a hedge can also be necessary as a result of dereliction or redundancy, whereby the hedge has lost its functionality. Restoration combines several techniques: replacing old pleachers with new plants in the gaps; coppicing larger woody stems; planting new standards; and some laying. The hedgeline will need protecting until it has become established, so temporary fencing is required along the side(s) that will be grazed.

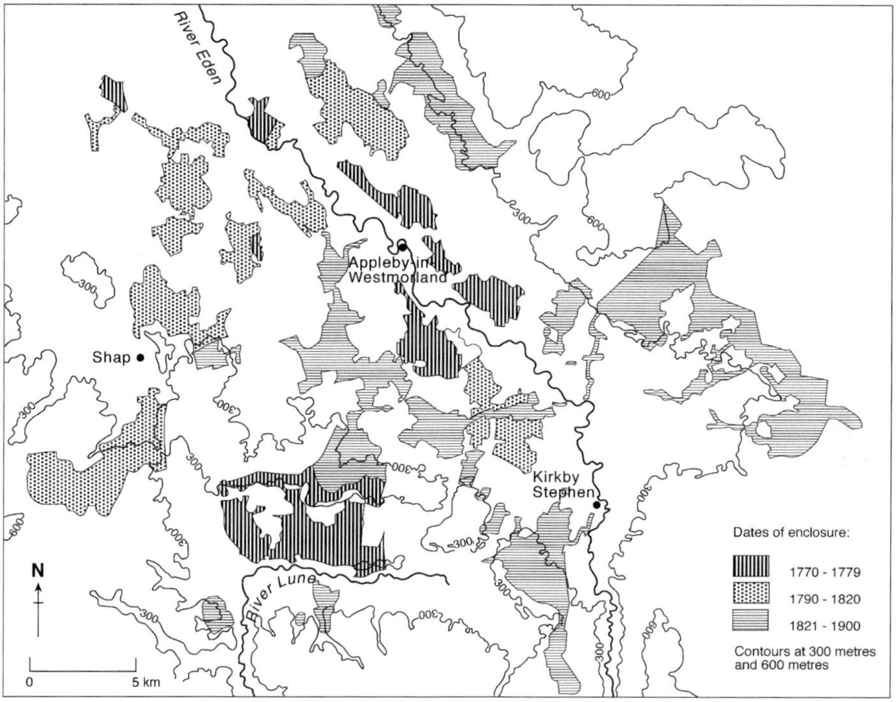

Figure 3.12 Enclosure in Northern Westmorland 1770 to 1890
Reproduced with kind permission from Whyte I (2003:23) 'Wild, Barren and Frightful' – Parliamentary Enclosure in an Upland Country: Westmorland 1767-1890. *Rural History* Vol. 1 Cambridge University Press

double-wall construction. In some cases, as Raistrick (1983) notes, the wall specification was even laid down in the enclosure award documents.

Lower fell fringes were also enclosed and reclaimed (pare and burn was typical, along with liming, re-seeding and in some cases drainage) allowing replacement of heather and rough grazing land with improved grass pasture.

Attempts were made to reclaim the higher, wetter moors around the turn of the 19th century, but these were often short-lived, designed only for owners to make a quick profit from the inflated prices caused by the Napoleonic Wars. Enclosure in these instances often had a hidden agenda because it was a way for landowners to tighten their legal claims to mineral rights on the land. It also allowed them to extend control over vast areas of moorland to exploit the new fashion for shooting, a pastime emerging among the *nouveau riche* of the Industrial Revolution. It did lead to more systematic burning of heather, which in time allowed the maintenance of dwarf shrub heath (see Chapter 5). A smaller agricultural labour force was also needed and this led to a drift from the land. However, many farmers could not afford to improve the enclosures, which simply perpetuated marginal farming.

Despite these drawbacks, the uplands were needed for the production of meat and milk for an expanding urban population. In 1805 the sheep population had reached 2.6 million.

Unimproved hill pasture carried about one sheep for every 3 acres (0.8/ha), whereas in lowlands the density could be as high as 5 per acre (12.3/ha) (Beckett 1990). For sheep, mutton production was dominant, with wool a secondary consideration. The best-quality land was allocated to cattle, typically valley bottoms where dairy farmers continued to use cow vaccaries dating back to monastic times. Gradually, sheep became a more valued commodity, replacing cattle, and as a result transhumance and shieling/*hafod–hendre* systems were abandoned. Farms began to be subdivided as the upland population increased (Thirsk 1987).

Highland clearances

The Highland clearances warrant attention as an example of the emerging capitalist economy. Thousands of people were forcibly ejected from their traditional lands by new landowners. Many emigrated to the New World, never to return. Their carefully developed land management system was seen by improvers as under-use of a resource (Dodgshon and Olsson 2006).

In the Highlands, sheep replaced black cattle as the mainstay of the farming system, reinforced by the clearances (Dodgshon 2004), breakdown of the clan system, industrialisation of the central lowlands and outbreak of the Napoleonic Wars (Collins 1978). The township system was abandoned as shared hill grazings were divided up into enclosed blocks attached to named farms. Sheep were sent to the hill much earlier, in April, and stocking densities of sheep quadrupled in some areas, leading to a switch from heather to grassland (Stevenson and Thompson 1992). Though this was seen as an improvement by commercial farmers, it led to less well managed muirburn (see Box XXIII). In other areas pasturage deteriorated through overgrazing and bracken expansion. Both environments suffered further as profits from mutton, wool and lamb spiralled upwards, encouraging conversion of valley bottom arable land to pasture. This increased the quantity of winter feed available, expanding off-wintering and summer capacity on hills. Finally, the absence of cattle and selective grazing by sheep left unpalatable mat grass and purple moor grass of low ecological value. What had been perceived as raising the economic viability of vast tracts of upland Scotland should, in retrospect, be seen as an overexploitation of resources.

High farming

From 1700 to 1850 was the high farming era in Britain, a time of improvement, increased production and better stock nutrition. Farmers became wealthy, benefiting from the misery of the Napoleonic Wars as the market for home-grown produce increased. The Corn Laws aided this. They were introduced between 1805 and 1815 to protect British wheat growing from foreign imports after the end of the wars. This increased bread prices rapidly which the public hated leading to riots in many towns and cities.

During this period sheep numbers peaked at nearly 27 million, a figure not reached again until 1985 (Figure 3.13). Simultaneously, farmers began to develop enterprise specialisation at the expense of corn,

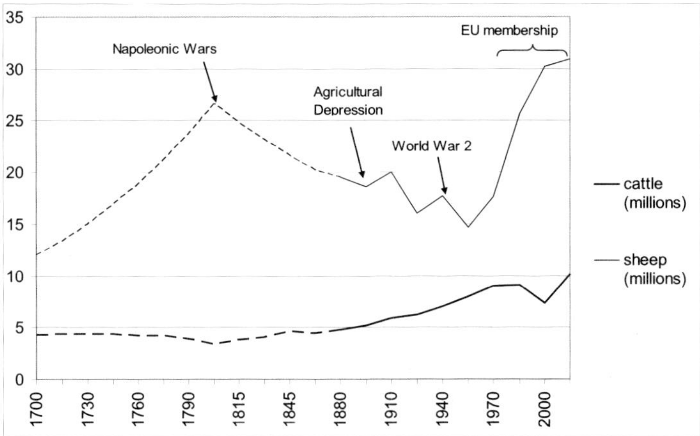

Figure 3.13 Changing Structure of British Agriculture 1700 to 2010
Adapted from: Grigg 1989; Marks and Britton 1989: 208-9; DEFRA 2000/2010

spurred on by technical innovations improving financial returns. Upland farming focused on livestock-only enterprises supported by a consolidating stratification system (see Chapter 1). With rising living standards and an expanding rail network, urban demand for meat and dairy products escalated, supporting upland livestock production.

The Agricultural Depression (1875 to 1945)

The utopian period for farming was unfortunately not to last in the uplands. In 1846 the Corn Laws were repealed, allowing free trade. The negative impacts of repeal were not felt until much later, in the 1870s; by then a series of factors depressed home product prices, most notably developments in oceanic and rail transport, the invention of ship refrigeration and the opening up of the grain lands in he USA and the colonies of Australia, New Zealand and Canada (Grigg 1989). Merino wool became sought after in the cloth market, old animal fat was no longer wanted for cooking (being replaced by imported vegetable oils) and the public's taste for younger lamb and smaller joints had an effect too (Balfour 1944; De Ware 1944). British agriculture had entered a phase of depression, and it moved from economic dominance to minor significance in terms of GNP (1850s 20%, 1890s 6%: Perry 1974) and labour (1851 20%, 1901 9%).

Between 1870 and 1901 land rents fell by 29%, wool prices fell by 50%, cattle by 33% and sheep by 25%. Interpreting the effects of these price movements on farming has led some researchers to suggest that lowlands suffered the most because of their reliance on cereal crops, suggesting that hill farms were more depression-proof because livestock farming was more resilient, with prices of winter feed dropping and lower labour costs (Collins 1978). Conversely, Perry (1974) argues that upland areas ended up suffering more, the depression exacerbated by their harsh physical environment, poor communications, distance from market and deep-rooted cultural traditions. For instance, between 1871 and 1942 the number of sheep fell by 25% in the Scottish uplands, but increased by 27% in the lowlands (Balfour 1944). When the depression started, hill farms were livestock businesses selling reared store cattle and sheep to the lowlands, for fattening. Most farmers grew a few crops to use as winter feed. Thus hill farmers were very much dependent on what went on in the lowlands, which controlled prices.

However we interpret the situation, upland farmers had to overcome the loss in income as the depression tightened. The responses available to them, roughly in order of preference (Perry 1974; Balfour 1944), were:

- increase the number of stock they reared
- reduce their standard of living
- reduce hired labour to a bare minimum and use more (free) family labour
- de-intensify use of marginal land – for example, switch from permanent pasture to rough grazing, leading to management neglect typified by bracken encroachment (Moore-Colyer 1998)
- change of use to forestry – this happened in Scotland, where the Forestry Commission acquired 0.6m acres between 1919 and 1944
- replace sheep farming with sporting estates – these increased by 55% to 3.5m acres in Scotland between 1883 and 1912
- abandon the land and migrate elsewhere

There is also evidence that flock structure changed: older wethers (3 or 4 years old) were replaced by breeding ewes; younger wether lambs were quickly sold on for fattening to lowland farmers, who were adjusting to the economic crisis. Cattle numbers also declined. From the 1870s farmers tried silaging, but the process was still unreliable and was largely abandoned until the 1960s (Brassley 1996).

In the lowlands the response was entirely different: farmers shifted from arable to pastoral farming. Thus the effect of depression varied geographically: for a grain-producing county like Essex it was catastrophic; for a mixed-farming county, such as Lancashire, the financial impact was slight. Rents fell in the south and east, but not in the traditional livestock areas. The upland farmer became caught in a cost–price squeeze of someone else's making.

As the depression continued, the effect on upland agriculture increased. Perry shows this clearly in his analysis of records of farm bankruptcy. The first phase, from 1871 to 1883, mainly affected south-east England and served to

> Eliminate the weaker brethren, the idle and incompetent, the undercapitalised, the overcapitalised and the profligate spenders. (Perry 1974: 37).

Deepening depression resulted in a wave of bankruptcies sweeping north and west into the uplands as the cumulative effect of twenty years' financial crisis took hold. By now, wool prices had collapsed and upland store cattle were not needed, because lowland farmers began to produce their own to escape financial ruin. Aided by lower rents and able to produce their own fodder on superior arable land, they rapidly out-competed the upland farmers.

The crisis was not helped by some poor weather patterns. In the lowlands a number of wet summers led to disastrous cereal harvests; in the uplands, hay harvests failed, making silage temporarily more popular than hay (Brassley 1996). Wet summers also led to a sharp rise in livestock diseases like pneumonia, rinderpest and liverfluke (rot). Between 1879 and 1881, 10% of the sheep population succumbed to liverfluke. These colder years also hastened the retreat downhill as marginal land was abandoned. Farmers and officials alike often incorrectly blamed the depression on poor conditions and disease infestations, and thus wasted time by not treating the real causes – price collapse and external competition.

The outbreak of the Great War in 1914 did have some positive effects, increasing demand. At the end of the war many lowland farms needed restocking and this gave upland farmers a slight reprieve (De La Warr 1944). Between 1920 and 1930 meat prices continued to decline, but upland farmers were compensated by improved wool prices (Balfour 1944) and some turned to dairy production. The removal of farm subsidies in the form of deficiency payments in 1920 after only one year (known as the Great Betrayal) instilled little confidence in the government's support for farming.

Then in 1931 agricultural prices collapsed as the British market became a dumping ground for international products (27% of world wheat, 44% of meat, 59% of butter). Moore-Colyer (1998) reviewed the depression in this period in Wales, and showed that price collapse was one of a range of factors that coalesced to deepen the economic crisis. Paramount were the tenancy arrangements in Wales at this time. Not only were tenancies annually renewed, but many landlords, recognising the economic crisis, sold off their tenancies at inflated prices, leaving the new owners little incentive or money to improve the management systems and techniques used on farms. The Agricultural Wages (Regulation) Act 1924 led to minimum wages, which forced costs up by 30%. Rural Wales also suffered massive depopulation, which led to the collapse of co-operative working – a lynchpin in upland areas. By the early 1930s Wales contained 4.5m acres of grass, of which half was unimproved bent, mat grass or purple moor grass and the other half had drainage problems.

The government finally acknowledged that hill farming was suffering. The evidence was undeniable: reports citing poor sheep prices, assistance to buy bracken-cutting machines and increases in sheep disease research pointed to the need for support. Influential writers and researchers like Stapledon (1937), then director of the Cahn Hill Experimental Station in Wales, recognised that the uplands were under-utilised. Between 1931 and 1933 marketing boards were set up, deficiency payments introduced for wheat, barley and oats, and subsidies given for cattle rearing (Howkins 2003) (see Chapter 8). The plight of hill farming did not go unnoticed and measures were put in place to improve the resource base and gross margins, for example the 1937 Liming Subsidy (Attwood and Evans 1961). The only overwhelming desire expressed by farmers – financial support for sheep rearing – was ignored until the outbreak of war in 1939. Despite these support mechanisms, the spiral of decline brought on by agricultural depression did not really halt until after the Second World War.

During World War II, land was reclaimed to increase home production, much of which was abandoned in the 1880s (Grigg 1989). Wartime policy focused on supporting enterprises that gave a quick calorific return – milk, cereals and potatoes – which disadvantaged hill farming. Only potatoes had some value in marginal uplands, where hay meadows could be ploughed for their production (Rodwell *et al.* 1992). The Ploughing-Up Campaign led to a decline in lowland sheep, which sent a ripple back up the stratification system, affecting upland agriculture as the market contracted.

In 1940, therefore, the government introduced in England and Wales the Hill Sheep Subsidy Scheme, offering a deficiency payment of 2s 6d per head of stock. By 1942 this had risen to 8s 0d, a rise of over 200%. A similar system for cattle emerged in Scotland in 1941, to ensure hill herbage was maintained and to reduce the spread of bracken, and in 1943 the scheme was expanded to cover England and Wales.

The emergence of the hill farming problem

It was during the Second World War that the UK Government eventually grasped that hill farming in Britain had particular issues. They are typically referred to in the literature as 'the hill farming problem' – a concept that had been gathering ground through the 1930s (e.g. Stapledon 1937).

In the severe winter of 1940/41, many Scottish hill farmers lost 50% of their flock. Increasing rents and labour costs (up 25% in 1942) compounded the problems of the catastrophic numbers of lamb deaths (Balfour 1944). On top of this the cost of off-wintering sheep rose steeply as lowland grassland was lost to the Ploughing-Up Campaign. Consequently, true hill farmers suffered more than other upland farmers because they could not take advantage of ploughing-up or fertiliser subsidies, nor the fixed payments for cattle, milk and crossbred lambs. In response the UK Government set up two committees (De La Warr for England and Wales, Balfour for Scotland) to investigate the problem, how it had occurred and how it could be tackled. The issue was revisited later by the Zuckerman Committee, which investigated sensible ways to utilise marginal land (Zuckerman 1955).

The two hill farming committees reported back in 1944. Their findings and recommendations (Table 3.4), particularly those of the more comprehensive Scottish report, resonated through the rest of the 20th Century, influencing hill farming policy until the 1990s. The Scottish report identified two over-arching aims:

- restoring fertility on hills and inbye
- encouraging the return of skilled labour to hill lands

from which three objectives evolved. The England and Wales Report had multiple aims, but in the end the reports emphasised similar recommendations, bar the state control suggested by De La Warr. How the recommendations were articulated for the upland farming sector is discussed in more detail in Chapter 8.

With the benefit of hindsight, we can see that the problem was essentially that the financial returns of hill farming were lagging behind those of other sectors. Attwood and Evans (1961: 12) saw the problem as related to the poor combination of factors of production (inputs) rather than the level or type of output:

> The present combinations of land, labour and capital do not generally produce returns to these factors on a scale comparable with those which are earned in other sectors of British agriculture.

Table 3.4: The Recommendations of the Two Hill Farming Committees 1944

England & Wales		*Scotland*	
Main topic	*To include:*	*Main Topic*	*To include:*
State assistance for 10 years to put farms in good working order with good husbandry	Improve & extend intakes Reseed Drainage, liming Fencing, shelter belts Bracken & scrub control Heather burning Repair & adapt hereditaments Replan farm layouts Construct & repair roads	*Financial assistance for long-term improvement of conditions & equipment of farms and livestock*	As per England & Wales plus: Machinery Hire of implements Livestock Sera and vaccines Control of disease Pest destruction
Develop state-owned hill land as an overseeing authority	Improve intake quality More overwintering on intake Reseed inbye Drainage – grypping Liming & potash Summer grazing & agistment Bracken control Regular heather burning	*Long-term financial assistance from the state*	Research and development Develop vet services Improve marketing organisation Modify transport charges Improve social conditions in hill areas
Improve the health and hardiness of sheep	More disease investigation Good dipping facilities Develop hardiness Improve meat for consumer demand Committees to approve 'ram' quality	*Develop administrative and legislative reforms*	Sheep stock valuations Heather burning Deer Afforestation
Recognise the importance of cattle and act upon it	Increase cattle on upland farms More agistment More breeding herds Develop dairy herds and beef stores		
Other factors needing action	Commons legislation Landlord–tenant safeguards Manage water supplies by catchment Farm amalgamation Fencing of public roads Limit stocking to set maximum Proper breeding control Agricultural training and education Post-war housing needs Communal services & amenities Need for shepherds, wallers etc.		
Increase research	Set up research stations in hill areas		
Afforestation	Transfer land to forestry Set up state-run organisation to adjudicate		

(Source: Balfour 1944; De La Warr 1944.)

These economic concepts are considered in more depth in Chapter 6, but suffice it to say here that hill farming began to be seen as the poor relation of British agriculture. Wartime agricultural policies, magnified by events during the depression, made this even clearer.

A more recent review of why and how the hill farming problem emerged indicates a range of circumstances that coalesced to marginalise the industry. Grigg (1989) suggests that it was a result of loss of population, increasing costs of service provision and loss of other economic sectors (e.g. mining, fulling and cloth production) to other, more profitable parts of the country. Until this time many farming families had survived through various forms of pluriactivity, where different members of the family farmed, mined or wove cloth (Tranter 1978). Having to fall back on farming as the sole occupation, compounded by the agricultural depression, left many upland farming businesses economically marginalised. Once this happens, it is almost impossible to climb back out of the financial quagmire, because there is no cash to invest in innovations to improve the farm's productivity, as Moore-Colyer (1998) demonstrated in the Welsh uplands in the 1930's.

Summary

This chapter has sketched the historical evolution of upland agriculture. Initially, uplands in Britain were used as natural hunting grounds where wild animals were stalked and killed. As time went by Neolithic peoples began to settle these areas and introduce cultivation, while continuing to rear livestock. With the development of field systems from the Bronze Age onwards, substantial adaptive management of the landscape began, which gradually intensified in the medieval period as the uplands were exploited. Large-scale cattle and sheep production became dominant through the monastic and secular farming practices of the 11th to 15th centuries.

With the dissolution of the monasteries, land redistribution and disafforestation, upland landscapes developed into a system of landlord–tenant and small holdings. The adoption of innovations like liming, drainage and selective breeding enabled some upland farmers to become financially independent in the 17th or 18th century. Underpinned by large-scale land ownership reforms driven by parliamentary enclosure and abandonment of the sheiling system, a landscape of inbye, intake and open fell evolved in the uplands during the later 18th and early 19th centuries. From the 1870s to the 1940s there was an agricultural depression, whose aftermath left upland farms in the rearguard of British agriculture. Marginalised economically, politically and physically, upland farmers found themselves unable to climb out of the poverty trap in the early 20th century. This led to the stigma of being labelled the 'hill farming problem', a legacy from which farmers still struggle to free themselves today. To understand the hill farming problem more fully, we need to appreciate the physical constraints within which upland farmers operate. It is to this we turn next.

4

The physical environment

At the most fundamental level, any agricultural activity on this planet is controlled by the physical environment. This direct relationship is a form of environmental determinism in that the environment controls the scope of our actions. In other words,

> geographical conditions affect the details of agriculture as definitely as they do larger problems of crop distribution … relationships [exist] between slope, exposure, drainage, groundwater, weather, climate and plant and animal production. (Dodge 1911: 144)

Central to this relationship are the physiological needs of any organism that forms part of a farm enterprise. These needs broadly explain the distribution of agricultural land-use patterns, for example, arable cultivation in the warmer and drier south-east of England and pastoral activities in the cooler, wetter northern and western areas.

Limits and balances

In uplands, the key organisms are fodder crops and animals. Farmers need to appreciate the physiological needs of both to ensure a balance to maintain effective production. For a plant these factors include light, temperature, water (not too much nor too little) and nutrients. For an animal the controlling factor is food, with enough nutrition to support its physiological functions without producing stress in some form. The limiting point of any factor is not necessarily that the organism will die, but that its physiology works less efficiently, causing stress and reducing its growth and reproductive potential (Cox and Moore 1993). Each limiting factor operates along a continuum known as an environmental gradient (Figure 4.1). Every organism has adapted so that it functions most efficiently within a certain section of the environmental gradient ofany one factor, and for every species that section is different. Each organism will also have a unique combination of environmental gradients for the factors that matter, so a farmer needs to manage many conditions to produce sufficient food for subsistence or, preferably, a surplus.

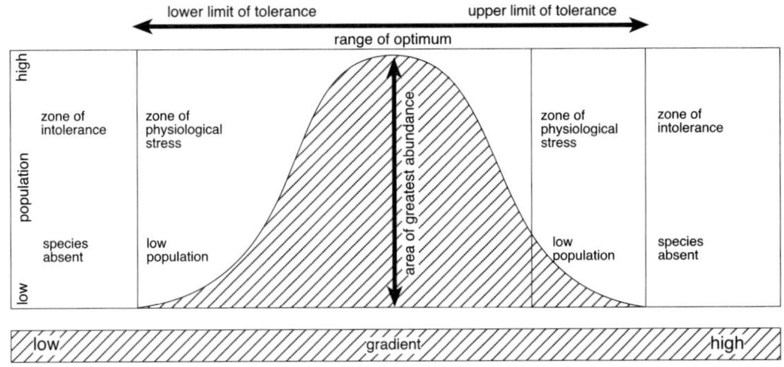

Figure 4.1 Environmental Gradients

We saw in Chapter 3 how clan-based farmers in the Highlands worked closely with nature to avoid overstocking, but managed this only at subsistence level. The clearance landlords tried to achieve a surplus, but their lack of understanding of the environment and its limits eventually led to overexploitation. These limits are known as the physical margins of cultivation, and they can be mapped – for example, to show the optimal range of fodder crops or the potential maximum extent where it can grow, albeit under duress. Grigg (1986) has modelled this concept as a simple diagram.

Figure 4.2 shows how a plant is limited by the two factors of temperature and water availability. As a plant moves out of its optimal zone, a farmer can compensate artificially for the loss of ideal growing conditions; for example, drier conditions can be overcome by irrigation. However, there will come a point when no amount of irrigation will compensate enough to allow the plant to grow satisfactorily (usually because of the combined effect with other variables). It is unusual for capitalist agriculture to operate on the extreme margins of cultivation because the levels of input required to compensate usually cost more than the profits gained from the enterprise; these are the economic margins of cultivation shown on Figure 4.2 (see Chapter 6). The other option is to switch to an alternative crop which can tolerate the new conditions, such as changing from wheat to barley cultivation.

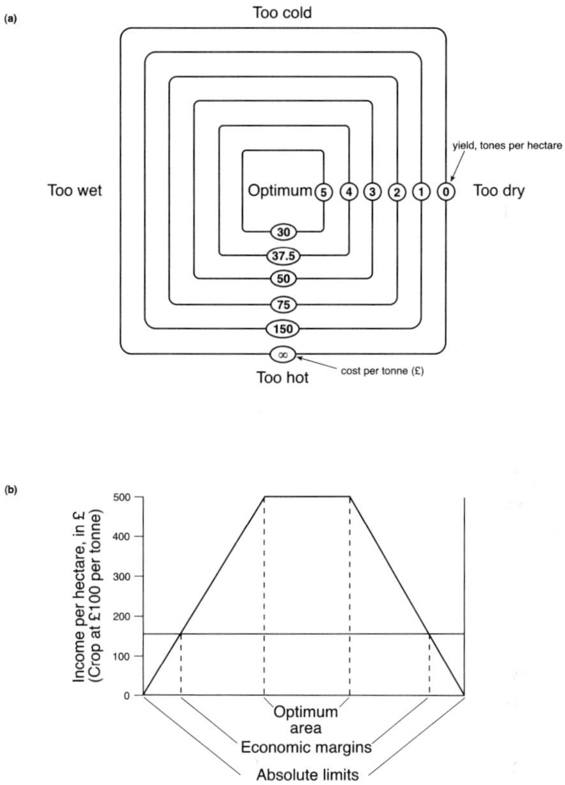

Figure 4.2 Margins of Cultivation.
Reproduced from Grigg D (1984) *'An Introduction to Agricultural Geography'* Hutchinson Figure 28, p115 with permission from Routledge.

The physical controls that matter most to agriculture are climate and soil, with geology exerting an indirect influence through topography and pedology. Climate controls the macro level location of agriculture in uplands, whereas soil, slope and aspect underpin local levels of cultivation (Parry 1975). For upland agriculture, these physical factors are more of a controlling influence than for lowlands. Higher altitudes combine with wet climates and thin or unproductive soils to limit enterprise choice at the physical margins of cultivation. This in turn pushes upland farmers closer to the economic margins. However, the climate in the British uplands was once warmer – allowing more enterprise variation in the past – and it may be so again in the future (see Chapter 9).

The role of climate

For agriculture the most important elements of climate are temperature and precipitation (rain, hail or snow). Our climate is particularly variable because of Britain's position at the edge of a continent with an ocean beyond. Our weather is dominated by maritime air masses (Figure 4.3), giving high levels of precipitation. Thus even at sea level there can be upland-like environments, with their own soil type, plant communities and land use, and this fact influences (as we saw in Chapter 2) some mountain classification systems.

The effects of topography

Mountains are known to create a climate of their own and modify weather (Barry *et al.* 2009) in different ways at various geographical scales. At a continental scale, major mountain ranges such as the Himalayas or the Western Cordillera of the Americas create their own climate, causing vegetation to change with altitude (Figure 2.3). Even regional massifs like those in the British Isles affect climate and weather in comparison to the surrounding lowland.

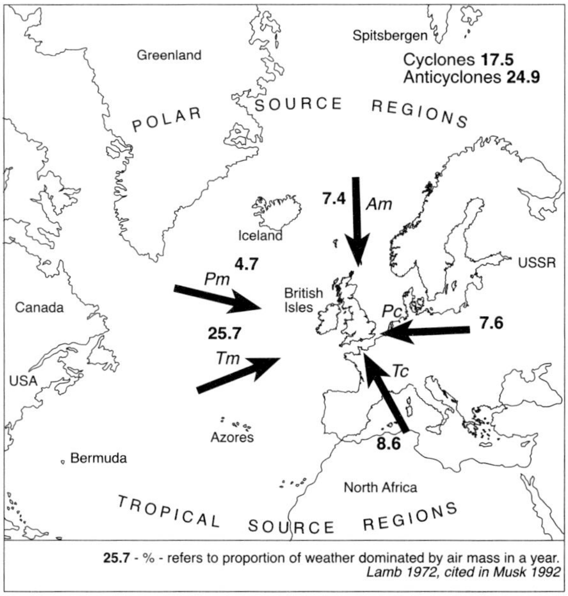

Figure 4.3 Air masses Affecting UK Climate.
Adapted from Musk 1992.

British uplands have slopes that encourage steep lapse rates, forcing temperature to decline faster with altitude than it would above a flat plain, because the air is forced to rise quickly (Chandler and Gregory 1976). As the air rises, moisture condenses from the atmosphere. Eventually it is released as precipitation or directly onto the vegetation as dew. This phenomenon has been used to fill dew ponds, which were built to water livestock on hills not served by springs or natural ponds.

The steep upland lapse rate[1] leads to higher humidity, more cloud and water surpluses. This is the orographic effect, which causes high levels of precipitation on the uplands themselves. Soil erosion and landslips are more frequent due to flash flooding in these areas. The uplands also cast a rain shadow over lowlands in their lee, protected from the prevailing wind. These lowlands will experience warming air, increased evaporation and periodic moisture deficiencies in summer (Figure 4.4). The Lake District, Cairngorms and Brecon Beacons regularly exhibit these characteristics.

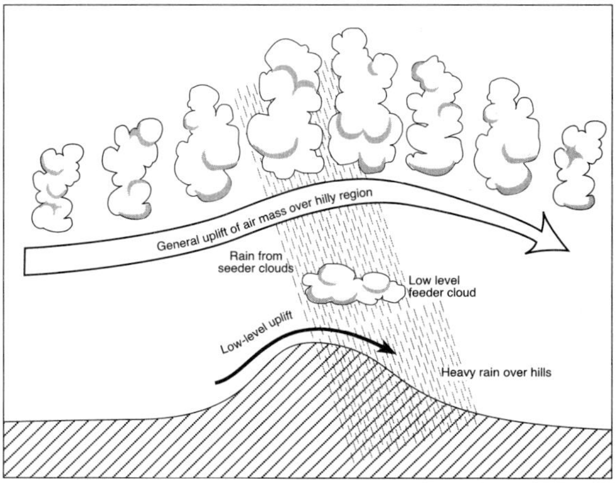

Figure 4.4 The Orographic Effect of Uplands

Part of the effect is orographic enhancement, created by a feeder–seeder mechanism. Precipitation falls from seeder clouds through the lower feeder clouds, which scavenges out water to enhance the droplet size. This can increase rainfall events from 2mm per hour to 10mm per hour (Browning and Hill 1981, cited in Grace and Unsworth 1988). The feeder–seeder

[1] lapse rate: decrease in temperature as altitude increases.

mechanism can also enhance the ability of lower clouds to adsorb pollutants from industrial areas as they cross to mountains, the corollary of which is higher levels of acid deposition in comparison to adjacent lowlands. Orographic effects can be enhanced further by the geography of upland massifs, particularly when:

- there is a rapid and large change in elevation
- such a change is at right angles to the predominant winds
- the upland blocks are compact and extensive at the greatest heights

However, not all British uplands lie on the western side of the archipelago. Uplands in eastern Britain, such as the North York Moors and Lammermuir Hills, have lower annual precipitation than their western counterparts, which in turn influences vegetation growth and farming patterns. In Scotland a west–east divide exists, with precipitation levels rising above 2000mm per annum in the west compared to eastern areas receiving about 800mm p.a. (Gregory and Essery 2006).

Local elevation on a massif can influence precipitation type and temperature regime. As altitude increases, the probability of snow, blizzards, prolonged snow lie and frost rises. Snow cover can encourage penetration of frost deep into the soil, leading to frost heave, disintegration of soil, rapid thaw and eventually slope soil failure. Gregory and Essery (2006) estimate that snow cover generally increases by 15 days for every 100m of ascent in Scotland, but the snow lie varies from one upland to another. The North York Moors average 40 days' snow lie per year (NYMNPA 2006), but higher plateaux, like those in the Cairngorms, which are open and exposed, have hardly any snow lie, though further downslope in the lee over 250 days' snow lie can be recorded in a year (Rodwell *et al.* 1992).

Temperature also decreases with altitude, typically by 0.6°C per 100m, but the humidity of rising air can result in temperatures up to 1°C higher or lower, depending on the massif. Solar radiation affects uplands too. It can be 30% of the norm in upland margins, only 25% in our central uplands and as little as 0.2% in parts of the Scottish Highlands. On the highest peaks above 1000m, cloud cover reduces and so radiation levels increase again, forcing plants to adapt to higher UV levels.

Plateaux also suffer from katabatic effects, caused by cold air sitting on a plateau that drains into a valley, creating either a cold wind or pockets of cold air, possibly frost, in the valley bottom. These conditions affect farming, for example in the Eden Valley at the edge of the Pennines and in the Ochil Hills north of Stirling (Harrison and Harrison 1988), and influence precipitation. On the North York Moors, the higher ground records 1016mm pa in contrast to the valley bottoms with 762mm pa (NYMNPA 2006). Snow lie also varies, from 40 days on high ground to 15 or 20 on low ground. Microscale temperatures also vary. Grace and Unsworth (1988) showed how temperatures varied on a small patch of heathery vegetation on a scree slope (Figure 4.5).

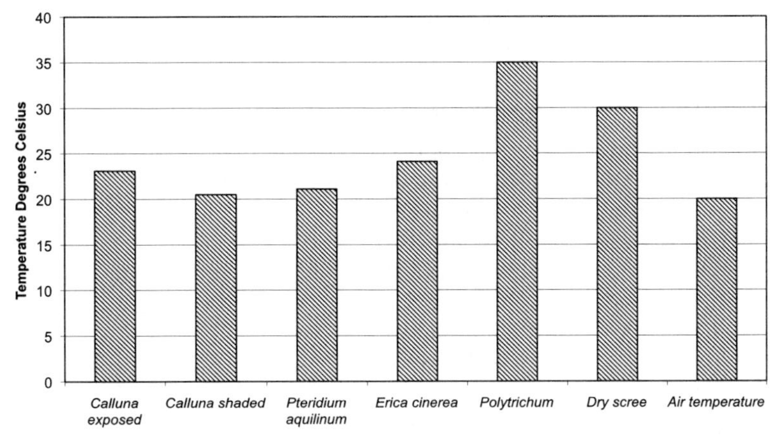

Figure 4.5 Microscale temperature variations on an upland slope.
Adapted from Grace & Unsworth (1988:143)

Growing and grazing seasons

Variations in precipitation and temperature combine to control plant growth, the length and intensity of the growing season, growth rates and related crop yields. A common measure of production potential is the number of days in a year that the temperature rises above the growth point, defined as the point when the top 30cm of the ground reaches 6°C (Harrison and Harrison 1988). Above this temperature, plants cease to be dormant and begin to grow. In contrast, the growing season is the time between the average date when temperature rises above 6°C (the start of the season) and the average date when it falls below 6°C (the end of the season). The length of the growing season varies with the soil temperature, and thus height above sea level, slope and aspect all play a role. The growing season varies most in the extreme south-west of England and Wales and least in northern Britain (Table 4.1a; Figure 4.6), but the growing season can vary over much shorter distances as well. In the Ochil Hills of southern Scotland, Harrison and Harrison (1988) found that at a height of 35m asl the growing season averaged 160 days, but at 352m asl it was 110 days (less than 16 weeks). In turn the growing season affects dates for stocking land and has been used to limit grazing seasons for agri-environment schemes (see Chapter 8).

Table 4.1a: Decrease in the Growing Season with Increasing Altitude

area	*days/100m*
North, Midlands and East Anglia	15 to 20
Wales	20
Southern England	20 to 25
Devon and Cornwall	30 to 35

Table 4.1b: Decrease in the Grazing Season with Increasing Altitude

area	*days/100m*
North-East England & Midlands	45 to 50
North-West England & Wales	33 to 40
East and South-East England	40 to 45
South-West England	65 to 75

(Source: Smith 1984:19.)

The grazing season starts later than the growing season, to allow the grass to put on sufficient growth to be of benefit to grazing animals. The delay increases with altitude and has been calculated variously between 3.6 and 5 days per 30m (Smith 1981; Harrison and Harrison 1988). The end of the grazing season is determined more by the weather and ground conditions than by the continuance of growth. The length of the grazing season depends on height, slope, aspect and rainfall, being longest in Britain in the south-west (Table 4.1b). Growing and grazing seasons vary from year to year.

Other relevant aspects of climate are transpiration rates, excess winter rain, drought factors, soil moisture deficits and irrigation needs (Smith 1984). The cumulative effect of these factors is that upland growing seasons are longest in western and northern Britain, which is why dairy farming is concentrated in these areas (Figure 1.4). However, upland cores have shorter grazing seasons because other physical factors, such as altitude, limit the length. This pattern may alter with future climate change (see Chapter 9).

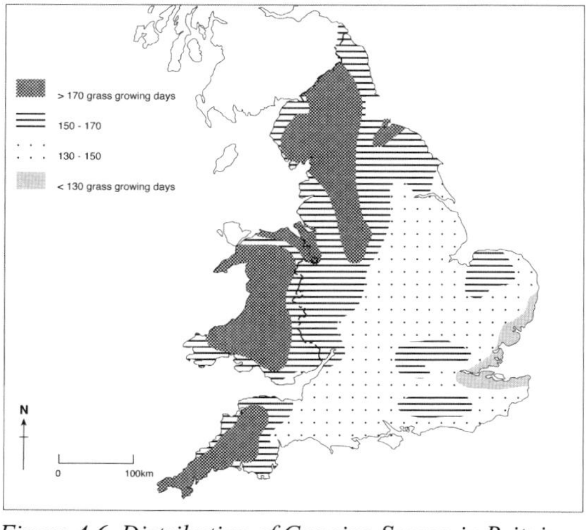

Figure 4.6 Distribution of Growing Season in Britain. Reproduced from Grigg D (1984) *'Introduction to Agricultural Geography.'* Hutchinson Figure 32, p121 with permission from Routledge.

Climate therefore fundamentally controls upland agricultural activity and it is enhanced by orographic effects. This has led to a distinct division in the British Isles between west and east, with uplands in the west dominated by high precipitation, cooler weather and low moisture deficits, whereas high ground in the east has lower precipitation, warmer weather and higher moisture deficits. Growing seasons are thus shorter in northern and western uplands.

The role of geology

Geology indirectly controls agricultural land-use patterns because it determines:

- the parent material of soils and thus their physical and chemical properties,
- topographical variations in altitude, aspect and angle of slope,
- precipitation, enhancing it through altitude and the orographic effect.

For a small island, our geology is extremely complex, mainly because our land mass has gradually drifted northwards during geological history, from 30°S to our present 52°N. In the process sedimentary rocks formed, such as limestone, which can support fertile soils (for example, in the Yorkshire Dales and Mendips) and vegetation that can be quite diverse and palatable to livestock (see Chapter 5). Unfortunately, many of our uplands are higher and more rugged, and are made up of hard, crystalline, igneous rocks (Figure 4.7).

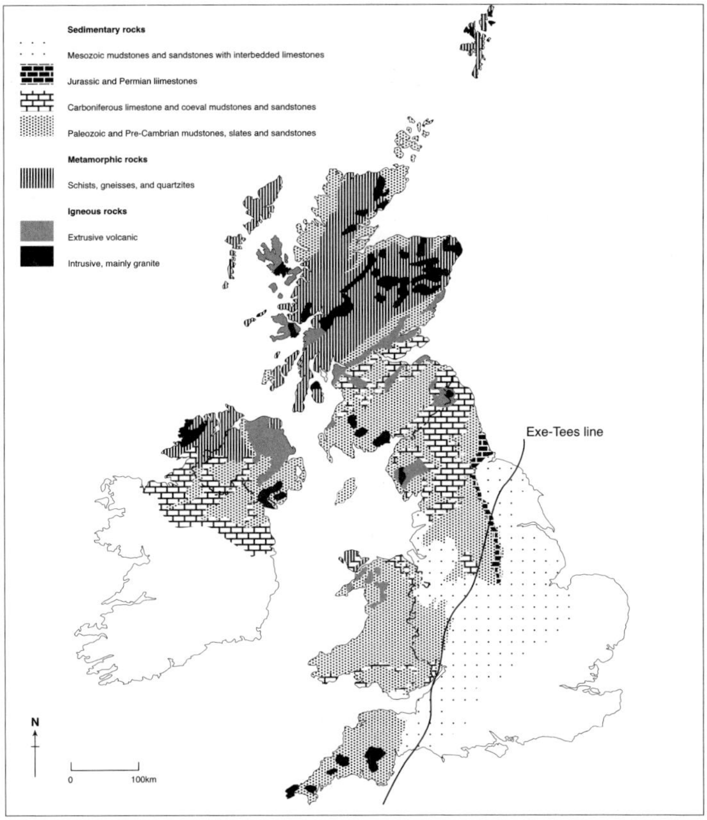

Figure 4.7 Generalised Solid Geology of the British Isles.
Adapted from Webster 1993: 125

Igneous rocks are typically the result of orogenic (mountain-building) processes caused by continental plates colliding on a geological timescale. One plate is subsumed (forced under) the other, creating friction, which leads to igneous activity. Huge plumes of hot magma either gush out of volcanoes (extrusive igneous rocks) or gradually cool before they reach the surface as a large plutonic mass (intrusive igneous rocks), only to be exposed by weathering later on. Large expanses of andesitic lava and acidic granites are characteristic of central Scotland, Dartmoor and the Lake District core. As these weather, the decomposing rock yields material for soil, but their acidic properties offer little nutrient, resulting in impoverished vegetation and poor grazing land (see Chapter 5).

Table 4.2: Geological Evolution of Upland Massifs in Britain

time line (mya)	*geological period*	*main activity*	*upland massif affected*
0.1 ↑ 1.6	Quaternary	repeated glaciations sculpted landscape	all areas except SW Peninsula, which was affected by extreme periglacial activity
1.6 ↑ 195	Tertiary/ Cretaceous	igneous intrusions Alpine orogeny changes the drainage pattern	Scottish Islands eg Rum North York Moors uplifted all tilted eastwards
195 ↑ 280	Jurassic/ Permo-triassic	warm seas creating marine sediments	North York Moors
280 ↑ 395	Carboniferous/ Devonian	sea level rise warm seas creating limestone igneous intrusions Variscan orogeny	Pennines, Yorkshire Dales, Peak District and Forest of Bowland SW Peninsula pluton – Dartmoor & Bodmin etc. Pennines & Peak District Great Whin Sill, Cheviots
395 ↑ 570	Silurian/ Ordovician/ Cambrian	supercontinent collision Caledonian orogeny	Central Highlands Southern Uplands Lake District Wales
570 ↑	Pre-Cambrian	formation of Iapetus Ocean Grampian, Athollian and Camodian orogenies	Welsh Borders Western Isles of Scotland

Igneous areas are typically flanked by metamorphic rocks created by:

- contact/thermal metamorphism, as hot magma slowly rises from the mantle to the surface, 'cooking' the adjacent rocks (much of Wales, northern Scotland);

- regional metamorphism, from the collision of two tectonic plates back in geological history (central Lake District, Shropshire Hills).

These geological processes have occurred at several periods (Box XIV), resulting in distinct upland massifs (Table 4.2). Throughout history the rocks have been exploited for minerals, by inhabitants and invading armies (see Chapter 3).

The last two million years gave our uplands their final shape. In each cold period, glaciers gouged out river valleys, leaving steep-sided, flat-bottomed (U-shaped) valleys, which have given rise to the classic upland farming system of valley floor inbye, side intake and open fell top.

Box XIV: The British Isles' geological history in brief

In geological history we know most about the last 600 million years because field evidence (the rocks) exists for us to examine and interpret.

Before this, the record is sketchy at best, because rocks have been re-worked by weathering and erosion and are no longer in their original form or location. Alternatively they have been subsumed back into the mantle as one plate descends beneath the other in the process of plate tectonics. This box summarises the main events in British upland geology. More detailed, easy-to-read accounts can be found in Webster (1993), McLeish (1992) and Toghill (2000).

Pre-Cambrian: over 590 million years ago
In this period what are now the British Isles existed in two parts on two continents divided by the Iapetus (pronounced 'Yapetus') Ocean at about 30°S. What is now Scotland and the north of Ireland was affected by the Grampian and Athollian orogenies on the northern continent, while Wales, England and the south of Ireland was uplifted by the Cadomian Orogeny on the southern continent. The only rocks still outcropping at the surface from this time are gneisses, sandstones and schists forming the core of the Scottish Highlands and Western Isles (Figure 4.7).

Cambrian/Ordovician/Silurian: 590 to 415 million years ago
The next three geological periods saw a range of other rocks form, which were then reshaped by the second main orogeny to affect the British Isles, the Caledonian. The sediments of the Iapetus Ocean (which was closing up) were metamorphosed into slates. These were interspersed by plutons, which had risen up through the crust. Outcrops of these can be seen as the uplands of the Central Highlands, Southern Uplands, Lake District and Wales (Figure 4.7). By the close of the Silurian period the British Isles had become one landmass for the first time.

Devonian/Carboniferous: 415 to 280 million years ago
Granitic pluton formation continued well into the Devonian period, especially in the Scottish Highlands, notably the Cairngorms, and the Southern Uplands. This Variscan Orogeny gained momentum in the Carboniferous period, when three distinctive massifs were formed:

- The Pennines – limestones and millstone grits underwent gentle uplift out of the Rheic Ocean that then covered the south.
- The South-West Peninsula – here a colossal pluton of granite was injected into the sea-floor sediments. Subsequent erosion has led to exposure of the pluton to form a string of granitic uplands: Dartmoor, Bodmin, Hensbarrow, Carnellis and the Lizard (Figure 4.7).
- The Whin Sill – an igneous injection of dolerite through Carboniferous rocks. It has been exposed through weathering and erosion to form steep, craggy uplands running through northern Cumbria and southern Northumberland and out to the Farne Islands in the North Sea (Figure 4.7).

Most of upland Britain had been formed by the end of the Palaeozoic era (630 to 280 million years ago). After this point, one last orogeny affected the British uplands.

Permo-Triassic: 280 to 195 million years ago
In the Permo-Triassic period the British Isles lay north of the equator, roughly where the Sahara is today. As a result, there were many desert-like landscapes, which eventually compacted and lithified to form the sandstones of marginal upland northern England. Further south, a salty sea (the Zechstein) evaporated to create extensive rock-salt deposits later in the Cheshire area.

Jurassic/Cretaceous: 195 to 65 million years ago
Much of what is now northern Britain was a land mass, but southern areas were submerged. Large deltas that snaked into the southern ocean eventually formed sandstones and shales in north-western Scotland and northern England. This era saw the beginnings of the Atlantic Ocean, spreading from the south northwards. Later, in the Cretaceous period, sea covered most of the British Isles depositing extensive beds of chalk, which survive only in south-east England.

Tertiary: 65 to 2.5 million years ago
The Atlantic Ocean continued to expand, the North York Moors were uplifted (Figure 4.7) and further igneous activity created many of the Inner Hebrides, notably Skye, Rum, Mull and Arran.

Finally, it was during the Alpine Orogeny that most of the present drainage system of the British uplands evolved and, as the land was tilted down towards the North Sea, most rivers diverted to flow from west to east in the direction of continental Europe.

Quaternary: 2.5 million years ago to the present
Two million years ago the geology of upland Britain had reached its current configuration. Until about 10,000 years ago, the world was plunged into a series of cooler glacial periods. These ice ages, by a series of glacial and periglacial processes, had a profound effect on the landscape of our uplands, in three main ways:

1. Formation of erosional features by glaciers, such as corries and over-deepened valleys typical of the Cumbrian Lake District.
2. Deposition of eroded material (known as drift deposits) in valley bottoms and marginal uplands. Large areas can be found in the Vale of Pickering, south of the North York Moors.
3. Periglacial activity in areas not covered in ice but exposed to severe freeze–thaw processes, such as the Cairngorms.

These processes are relevant to upland agriculture because they have left their mark on the relict landscapes: steep-sided valleys with high erosion potential, flattened summits with sudden breaks of slope, valley bottoms filled with impermeable drift deposits which impede field drainage. The combination of summit, slope and valley bottom are the basis of upland farming, because they provide contrasting situations which can be managed as inbye, intake and open fell.

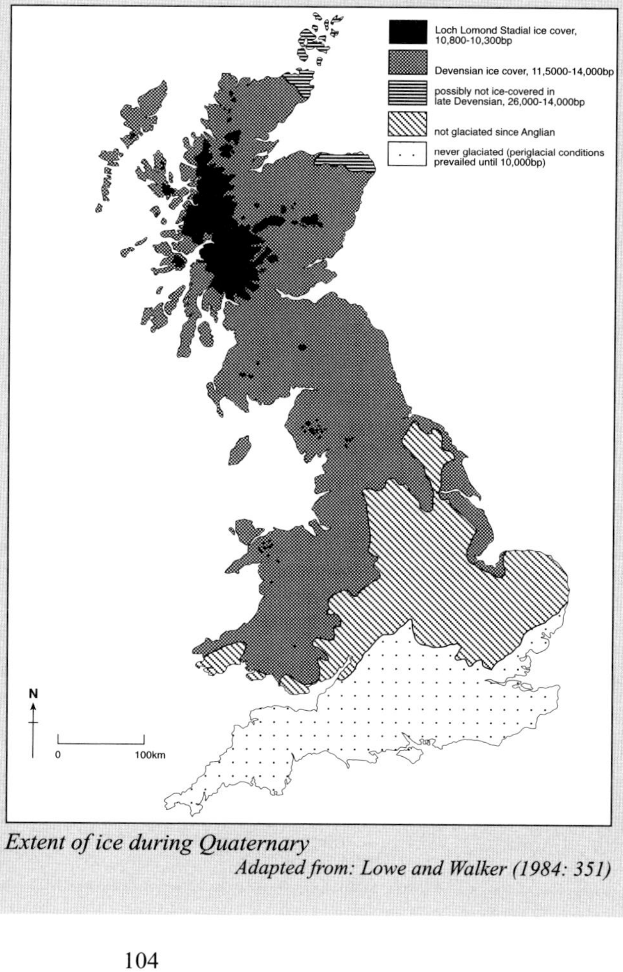

Extent of ice during Quaternary
Adapted from: Lowe and Walker (1984: 351)

Contemporary geomorphological processes

Current natural landform development is dominated by frost action. In the uppermost areas of Snowdonia, the Lake District, Pennines and Scottish Highlands we can see these periglacial processes at work (Ballantyne and Harris 1995). This is particularly the case in the Cairngorms, where the annual temperature range on the plateau around Cairngorm itself is only about 11°C, from −5.0 to +6.9°C (Gordon *et al.* 1998). These higher uplands are subject to freeze–thaw, frost heave and soil creep, which create distinctive periglacial landforms supporting thin soils and sparse vegetation.

Research suggests that these landscapes are particularly susceptible to human activity, and this is exacerbated by the extreme climate. Gordon *et al.* (2001) classify these geo-ecological landscapes as sensitive because of their susceptibility to trigger events that can lead to a chain of irreversible geomorphological change (Figure 4.8). Very limited overgrazing can be one such trigger as well as acid rain (see Chapter 5 under montane habitats).

Winter temperatures often oscillate around 0°C, so water can freeze, thaw and refreeze, especially in the higher reaches of the sub-montane (sub-alpine) zones. As water freezes in cracks in the rock, its volume expands by up to 4%, widening the crack and causing the rock to disintegrate after repeated cycles. In soil, this expansion and contraction weakens the bonds between particles, leaving them open to wind or water erosion. Blanket peat falls into this sensitive category (see below and Chapter 5). The high rainfall in sub-alpine zones accelerates erosion, a particular challenge for those farming in the uplands (see below).

Other upland environments are considered more robust, in that they will return to their original state after a trigger event. Gordon *et al.* (2001) argue that understanding whether a landscape is robust or sensitive will influence our nature-conservation strategies, but we should also use this concept in upland farm management by blending it with other types of resource management like integrated catchment or ecosystem management (see chapters 2 and 9).

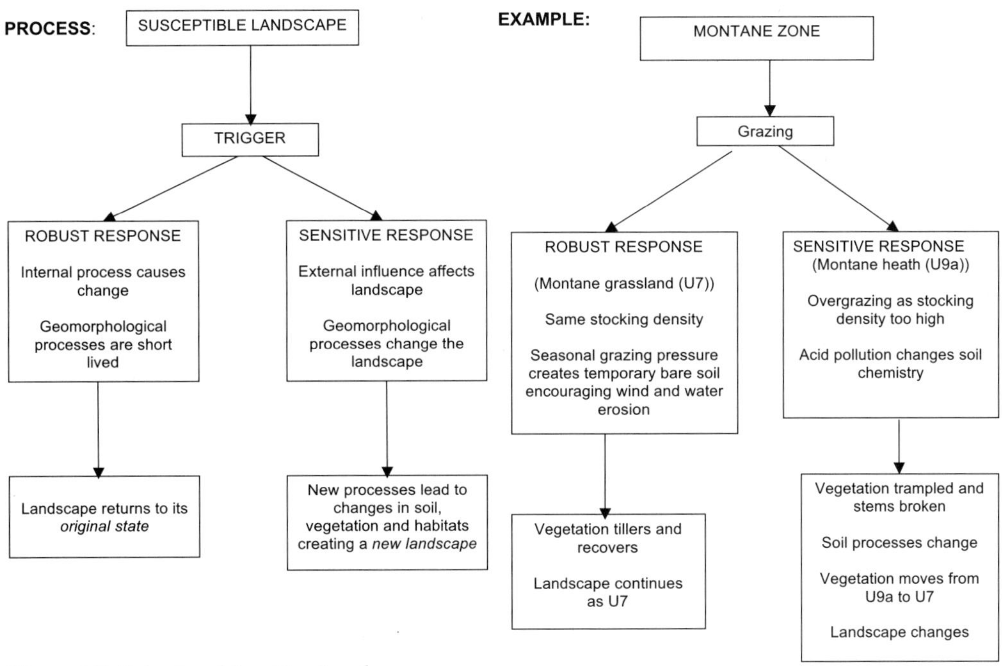

Figure 4.8 Robust and Sensitive Landscapes.
Adapted from Gordon *et al.*, 2001: 325

The combined effects of the British Isles' varied geology and subsequent geomorphological activity have formed distinct upland landscapes (Millward and Robinson 1980), a link which some notable 19th-century observers began to see all around them:

> Not only were characteristic landforms found to occur time and time again with a particular rock type, but there was also a distinctive human response reflected in the whole appearance of the landscape. (Millward and Robinson 1980: 18)

These inter-relationships are now formalised in classifications such as the National Character Areas Initiative and Scotland's Natural Heritage Futures (see Chapter 2), which are used as the basis for nature and landscape conservation management.

Aspect

Geology is also responsible for local features of aspect and angle of slope. Aspect refers to the direction of a slope with respect to the sun. In the northern hemisphere, south-facing slopes receive the most sunshine, warming the soil and encouraging plant growth and soil organisms.

North facing slopes tend to support cooler, damper soils which are in shade for large parts of the day. Their shady disposition also means the processes of freeze–thaw are more apparent, giving rise to steep-sided slopes. Though it might seem that south-facing slopes are more agriculturally viable, in Britain they suffer from the spread of bracken as it exploits the higher insolation levels (Box XV).

Box XIV: Bracken management

Originally a plant of woodland glades, bracken (*Pteridium aquilinium*) is a well-known fern that has become a management problem, especially in the uplands of England and Wales (Countryside Survey 2000). It escaped from the woods and has thrived on open south-facing hillsides, which fulfil its environmental needs.

Historically the fern was cut and harvested to be used in upland economies for bedding, thatch, compost, soap and potash (Rymer 1976). In the past, these activities controlled its spread, together with more cattle grazing (Ministry of Agriculture and Food 1934). Unfortunately, these management practices have almost ceased due to changes in the upland economy, the discovery of bracken's carcinogenic properties and the decline in cattle grazing over the last 70 years (see, for example, De La Warr 1944; Balfour 1944).

The ecological value of bracken is constantly being reconsidered and debate rages whether it has a high enough conservation value or not (e.g. Backshall 1999; Pakeman *et al.* 1996). Depending on the management objectives of a site, bracken is usually considered a weed requiring some form of control. The challenge with bracken results from its unique characteristics (Marrs *et al.* 2000):

- Ability to spread by rhizomes – though able to reproduce sexually, it prefers to spread vegetatively through rhizomes below ground from which new fronds unfurl above ground. A single plant can cover a whole hillside.
- High productivity – giving a dense frond canopy that eliminates most competition.
- Litter accumulation and lack of biodegradation – shading out other species.
- Toxicity – produces a range of toxins that prevent it being eaten or decomposing effectively.
- Tick environment – ticks hide in the vegetation ready to infect stock and humans.
- Height – so tall that sheep become lost and gathering is nigh on impossible.

In most cases, the control of bracken requires multiple techniques. The chemical Asulox™ is commonly used and is applied in a number of ways (Hawkings-Byass 1999; Robinson 1999):

- Hand-operated back packs – for steep slopes and small areas, but labour-intensive;
- Helicopter spraying – extensive areas, but very expensive;
- Weed wipers mounted on tractor/quad bike – where spray drift needs to be minimised.

Crushing, bruising and rolling are also possible if herbicides would result in water body pollution or destroy other fern species, but rollers work only on slopes of 19° or less; bracken usually occurs on steeper slopes than this. Outlying clumps are eliminated first, followed by the main infestation from one front. The whole process can take as long as ten years, alternating one technique after another year on year (Brown and Robinson, n.d.) until the infestation is eliminated:

year	*treatment*
1	Aerial application of Asulox™
2	Scarifying/ breaking and hand-pulling of fronds
3	Spot respray
4	Surface raking and application of heather brash
5	Temporary fencing
6	Spot treatment and general maintenance
7	Spot treatment and general maintenance
8	Spot treatment and general maintenance
9	Remove fence and monitor

There is concern that recent changes in EU policy away from production will unleash another wave of expanding bracken. Whilst past fears of overgrazing threatened to allow bracken to expand through the removal of competitors, the recent switch from headage to hectarage payments could also encourage re-expansion through undergrazing (Pakeman *et al.* 1996).

The role of soils

Geomorphological processes break down rocks to produce mineral matter, which typically makes up 45% of soil (apart from peat, which contains none). Soils are a product of the parent material, relief, climate, biodiversity and time. Because soil science is fundamental to agricultural systems (except for hydroponics) and food production, it is well understood and there are plenty of excellent texts on the subject (e.g. Gerrard 2000; Courtney and Trudgill 1984; Bridges 1984). Here we need to consider the nature and properties of soils and the challenges they present for upland farm management.

Technically, there is no such concept as a distinctly upland soil. Similar soils will form at any altitude as long as environmental conditions are identical. In the British Isles, montane environments can be seen at sea level, for example in the Western Isles off the Scottish mainland (see chapters 2 and 5). Allowing for this, five main soil types are typical of the British uplands (Table 4.3); unfortunately their varied nature and properties mean farmers may face a range of management challenges even within a single farm (Box XVI).

Table 4.3: Typical Upland Soils in Britain

soil type (major group)*	*main soil types*
Lithosols	non-calcareous calcareous
Lithomorphic soils	rankers rendzinas
Podzols	brown podzolic soils humic cryptopodzols podzols gley podzols stagnopodzols
Gley Soils	surface water gleys groundwater gleys
Peat Soils	raw oligo-fibrous peat raw oligo-amorphous peat

*Box XVI outlines the Soil Survey of England & Wales system

Lithomorphic soils

These soils are characterised by their shallow depth. Lithomorphic soils are typically 30cm deep and rest on bedrock or little altered regolith (Cr or R on soil-profile diagrams). They are limited to surfaces affected by erosion or deposition and other locations where soil development has been restricted by the nature of the bedrock. They account for 7% of the surface area of England and Wales and about 15% of Scotland (Mackney *et al.* 1983).

Lithomorphic soils can occur on bare rock, unvegetated screes and raw coastal sands, so they are not restricted to the uplands. Many lithomorphic soils are uncultivable because of:

- shallowness,
- stoniness/ rockiness,
- low water-storage ability or low water availability,
- liability to flood or erosion.

Because of these properties they are of particular interest to ecologists, because they demonstrate the earliest phase of ecological succession, support unusual biodiversity and can help us understand the rehabilitation of waste tips, quarries and industrial sites.

Within the Lithomorphic Major Group two main types of soil are found in upland areas: rankers and rendzinas. Rankers are non-calcareous lithomorphic soils usually having bedrock or rocky debris within 40cm of the surface (Figure 4.9). The term ‘ranker’ was coined by Kubiena (1953; cited in Duchafour 1982) from the German word *Rank* meaning a steep mountain slope. In the British Isles they are confined to mountainous areas and glacially eroded landscapes. Rankers are found in association with lithosols (see below) and occasionally other deeper soil types. They develop on acid geologies such as granite and are mostly uncultivable; as a result, they are dominated by semi-natural vegetation and their use is limited to extensive sheep grazing and/or recreation. Any attempt at cultivation often leads to accelerated erosion.

The other main lithomorphic soil represented in the British uplands is the rendzina, which forms over carbonate bedrock or extremely calcareous unconsolidated material lying 10–30cm beneath the surface (Figure 4.9). They occur over the Carboniferous limestones of northern England, Wales and Ireland, as well as on a few small areas of limestone and Dalradian Schist in Scotland. These soils support distinctive woodland communities, scrub and grassland, including various calcicolous (calcium-loving) species not found elsewhere, such as blue moor grass (*Seslaria albicans*) and thyme (*Thymus* spp.). The beneficial properties of rendzinas are that they have high levels of organic matter, pronounced faunal and microbial activity leading

Box XVI: The Soil Survey of England and Wales

In England and Wales soils are classified in a hierarchy, not unlike that used for biodiversity. The system was developed over a number of years by a team of soil scientists working under the direction of Brian Avery at the Soil Survey at Rothamsted, Hertfordshire (Avery 1980; 1990). There are five levels to the system: major groups, groups, sub-groups, series and phases.

Major groups
There are ten classes, coded from 1 to 10: terrestrial raw soils, raw gley soils, lithomorphic soils, pelosols (clay soils), brown soils (coded 5), podzolic soils, surface water gleys, groundwater gleys, manmade soils and peat. The divisions are based on the presence or absence of major diagnostic horizons that have agronomic, hydrological, ecological or engineering significance.

Groups
There are 43 of these, an example being an argillic brown earth (coded 5.7). Each major group will contain a range of groups which are subdivided from each other by general properties inherited from the soil's parent material and a number of key diagnostic horizons. In the case of argillic brown earths, these are loamy or clayey brown soils which have a clay-enriched horizon.

Sub-groups
There are 118 of these of which a gleyic argillic brown earth is an example (coded 5.73). It is distinguished from other sub-groups in the group by the existence of further diagnostic horizons and conditions. In the case of 5.73, this has evidence of gleying through mottling (orange and blue soil peds adjacent to one another) and permeable B horizons.

Series
Series, of which there are about 700, are named after the place where a particular soil profile was first described. For example in sub-group 5.73 is the Waterstock series. This will have a set of horizons and properties which differentiate it from all other gleyic argillic brown earth series.

Typically these horizond and properties include: parent material, texture(s) of the profile and presence or absence of materials with a distinctive mineralogy or colour (Clayden and Hollis 1984). Once a soil is found and named in one location, any identical profiles elsewhere are given the same name. Series are the most commonly used soil-type descriptors in land-management texts.

Phases
Soil types at phase level are differentiated from one another by identifying a particular feature in the profile that occurs over a significant part of the soil series, but not all of it – for example, a stony layer. Phases are rarely used in land-management information.

Every soil in England and Wales has been classified down to series level. Maps are available at varying scales. Particularly useful are the 1:63,360 (1 inch to a mile) and 1:10,000 series maps, which are accompanied by booklets ('bulletins') describing the local soils and their management issues. Bulletins are also published for smaller-scale maps and these provide detailed information on each soil series found in a particular region (e.g. Jarvis *et al.* 1984 for northern England). The bulletins contain data about geology and climate of an area, soil formation and classification, and land use management.

A useful function of these reports is to identify geographical combinations of series that typically occur together. These are known as associations. Other work by the Soil Survey focused on particular management issues, such as field drainage and soil–water regimes, which has led to the mapping of predicted underdrainage treatments for arable land use (Thomasson 1975) and land-capability maps.

As with any classification system, the Soil Survey of England and Wales has often been criticised (Ellis and Mellor 1995). A main issue is that the divisions between classes are purely arbitrary so that, by using different criteria, soils can end up in completely different divisions – as Webster and Oliver (1990) demonstrated for the soils of the Wyre Forest on the England–Wales border. Other industries have simply adapted the system to suit their own needs – as, for instance, the Forestry Commission in the United Kingdom (Kennedy 2002).

to rapid incorporation and decomposition of plant residues, stable soil aggregation and good aeration. However, they also have small available water capacities and are generally stony and hence unsuitable for deeper root systems (Avery 1990). The presence of $CaCO_3$ in the soil is a mixed blessing because, though it curbs weathering, with little iron being freed to move around and cause problems, the humification of organic matter is slowed down, which leads to the distinctive A/C profile of a rendzina soil. (A relates to the upper horizons of the soil, which are usually eluviated; B the lower illuviated horizons; and C the parent material.)

There are two main types of rendzina soil in uplands. Humic Rendzinas exhibit the classic A/C profile of rendzina in soil textbooks. In many upland environments, high precipitation (>1500mm pa) continually leaches any weathered carbonates from these soils (Box XVII), keeping them acidic and so promoting acid-loving plants. The overall effect can be to have calcicolous and calcifugous (calcium-hating) communities of plants growing next to each other on Carboniferous limestone! This is archetypal on and around limestone pavements. In contrast, brown rendzinas develop a simple weathered horizon beneath the A horizon, with a total soil depth of about 25cm.

Lithosols

Some even thinner soils occur in upland areas, and these closely resemble the parent material beneath them. These lithosols or 'raw skeletal soils' lie on bedrock only 10cm below the surface. They are found in the Scottish Highlands and islands and at the uppermost levels in England and Wales. Avery (1990) notes that lithosols can be non-calcareous, developing over acid geologies, for example those on the tops of the Cuillin on the island of Rum (Figure 4.9). Calcareous lithosols are typical of limestone pavements and pavement margins, especially in the grykes (Bullock 1971).

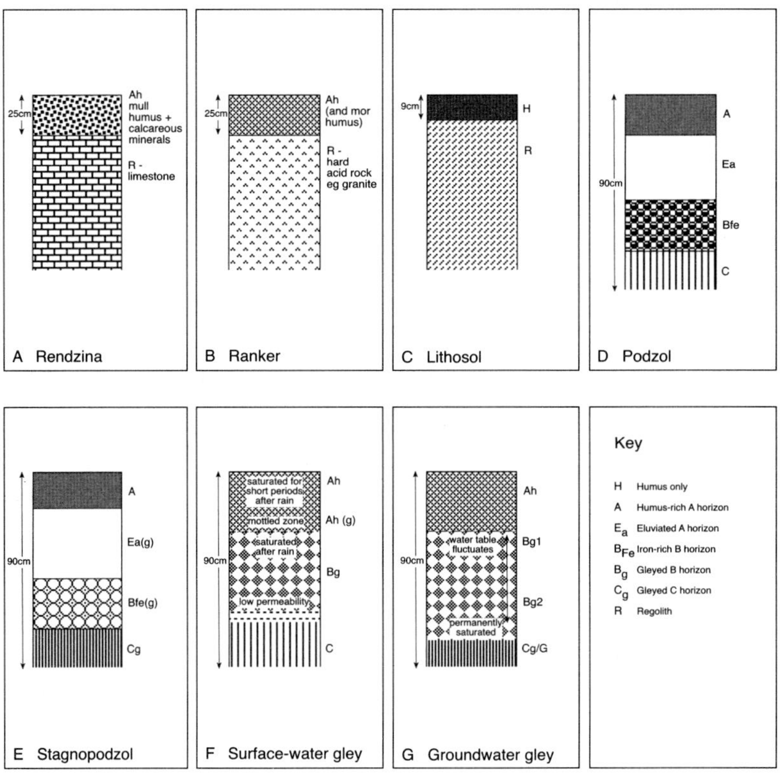

Figure 4.9 Examples of Upland Soil Profiles

Box XVII: Soil processes in temperate environments

A range of soil processes occur in temperate environments, where precipitation generally exceeds evaporation. Soil processes can be divided into transformations and translocations.

Transformations

Transformations are some form of chemical change which takes place *in situ*, for example:

- *Mineralisation* – the breakdown of complex organic structures into simpler, inorganic products. It happens slowly in uplands because the cooler, wetter soil results in lower levels of soil fauna.
- *Humification* – the decomposition of organic matter to form humus is also slower in uplands owing to the lack of faunal decomposers.
- *Hydrolysis* – the disassociation of water molecules into their constituent ions (hydrogen and oxygen) and their separate adsorption into the structure of chemicals; this makes the mineral fragments more susceptible to decomposition. Common in uplands.
- *Hydration* – the adsorption of the entire water molecule into the chemical structure of compounds, again encouraging breakdown. Common in uplands.
- *Carbonation* – the dissolution of calcium carbonate by rainwater leaching through the soil, owing to the existence of carbonic acid. Typical of upland areas dominated by limestone parent material.
- *Oxidation and reduction* – oxidation occurs when oxygen molecules are adsorbed into the chemical structure of Fe minerals in the soil. Reduction is the process in reverse, making iron more mobile. Typical of any soil that is waterlogged at some point, hence so common in uplands.
- *Solution* – the breakdown of ionic solids in water to produce a stable ionic solution. Very common in uplands.
- *Chelation* – metal ions, such as Fe, form covalent bonds with organic molecules to produce chelates. In this state the metal ions become soluble in soil water. The organic component is derived from ericaceous or coniferous leaf leachates in upland areas.

Translocations

Translocations occur where matter (certain chemicals especially) are moved around the soil profile. In many instances a specific transformation is needed before a particular translocation will occur – for instance, chelation for cheluviation. Common translocations include:

- *Leaching* - soluble material is washed down profile. Commonly bases or soluble salts such as Mg, Na or K move in this way.
- *Lessivage (mechanical eluviation)* - particles of clay are mechanically washed down a soil profile. Typically they are deposited as a clay-rich (argillic) layer in the B horizon, which can lead to an impermeable plough pan. This in turn can cause gleying if not managed.
- *Cheluviation* – the net downward movement of chelates through a soil profile. Parent material tends to be permeable to encourage good drainage.
- *Podsolisation* – a combination of leaching, lessivage and cheluviation which results in the removal of most cations and clay from the upper levels of the soil. It forms a distinctive bleached horizon just below the humus surface layers.
- *Gleying* – reduced conditions develop in a soil where water fails to drain away and waterlogging ensues. Iron will enter its reduced state (Fe^{2+}). If the soil dries out, the iron can oxidise back to its other state (Fe^{3+}).

In upland environments, what types of process dominate is controlled by climate in combination with parent material. High precipitation encourages processes that remove chemicals and other materials from the soil profile (e.g. leaching and cheluviation). Parent material affects the drainage capacity of the soil from good (podsolisation) to poor (gleying):

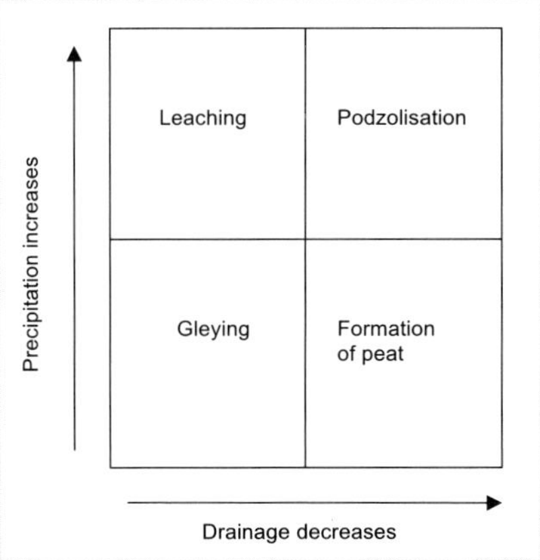

Surface vegetation plays a secondary role by encouraging certain processes such as chelation. Fuller accounts of soil processes can be found in Gerrard (2000).

Lithomorphic soils and the closely related lithosols are thin, fragile, sensitive systems that support little vegetation. Their immature development does not allow for complex multi-horizoned soil profiles. Many are consequently impoverished from the point of view of fertility. In totality, they are susceptible to wind (mainly lithosols) and water (rankers and rendzinas) erosion. Soil processes are limited to mineral weathering and the slow decomposition rates of any existing organic material. The exception is the rendzina soil, which demonstrates carbonation.

Podzols

Podzols are a direct contrast to lithomorphic soils. Their defining characteristic is an eluviated E-horizon[2] that has lost many of its nutrients, organic matter and clay (Figure 4.9). The upper E horizon takes on a typical bleached colouring after various soil transformations (e.g. solution and chelation) and translocations (leaching, lessivage and cheluviation) have left just silica grains behind. The lower B-horizons can have distinctive red staining indicative of re-deposited iron, speckles of brown illuviated organic matter and clay skins pasted onto soil peds (Figure 4.9).

Podzolic soils cover about 5% of England and Wales and about 25% of Scotland. Their distribution in the British Isles is a function of climate, parent material and biotic factors. They require: a parent material that encourages eluviation, usually on free-draining slopes or plateaux; a climate dominated by high levels of precipitation that exceed evapo-transpiration, encouraging the downward movement of excessive water; and specific types of vegetation cover, notably coniferous trees, lowland heath or ling moorland. This is because these plant types produce polyphenols, chemicals that trigger the process of cheluviation (the leaching downwards of soluble metals and organic acids).

Of particular note is the speed at which these soils can form. Birkland (1988) suggests tens of years rather than hundreds, as with other soil types, or thousands, as with histosols (peats). Inspection of any under-forest soil profile in the coniferous woodlands planted between 1920 and 1970 will show podzolisation.

[2] E-horizon - the layer just beneath the top soil from which chemicals are removed.

A wide range of podzol soils occurs in the British uplands. They can be roughly divided into non-hydromorphic and hydromorphic podzols, the latter tending to become waterlogged. The first group develop over free-draining parent material, usually under coniferous or ericaceous vegetation. Such soils are poor in available nutrients, with carbon and nitrogen restricted and locked up in the organic surface layers. The macro-nutrient phosphorus and the micronutrients manganese, cobalt and copper are immobilised in the B-horizons (Box XVIII). Sometimes liming and marling can unlock these nutrient reserves for agricultural purposes. Non-hydromorphic podzols can also be found under montane vegetation, typically lichen heath, where they are very sensitive to grazing and recreation pressure (see Chapter 5).

Hydromorphic podzols develop in valley bottoms where the water table fluctuates. They exhibit gleying (waterlogging, with reduced/oxidised iron mottling), supporting wet heath and purple moor grass. They have inherent chemical and physical limitations related to their properties. A common type of hydromorphic podzol is the stagnopodzol, a podzolised soil that shows evidence of waterlogging above or within the B-horizon, with iron pans and build-up of peat in the A-horizon (Figure 4.9). They are archetypal upland soils, but they occur at all altitudes and in all climatic zones (Birse 1971). Once formed, the peaty surface acts as a sponge and retains water, which encourages iron eluviation directly beneath, only to be illuviated as an iron pan in the B horizon. This iron pan can slow water percolation. From an economic point of view, these soils offer poor grazing potential; but on upland they have been drained and used as rough pasture or intake land as a compromise.

Generally, podzolised soils in the British Isles are not susceptible to erosion. Many of their problems arise from their impoverished fertility and the cost of improving them, which may not be economically viable. Whereas podzolisation does occur naturally, it is exacerbated by plantation forestry (Grubb 1969, cited in Northcliff 1984; Gerrard 2000).

Gleys

The main soil type of the British uplands is the gley, characterised by the soil being (permanently or temporarily) waterlogged. Field observation of gleying shows the soil contains blue/grey horizons which may be mottled with red areas. The waterlogging develops an anaerobic situation where iron is reduced from its oxidised state, hence the colour change from red to blue/grey. The mottling is caused by fluctuations in the soil water level, as parts of the soil dry out and the iron oxidises back to a red colour.

Gleys are estimated to cover about 40% of England and Wales, and about 30% of Scotland (Avery 1990). Their distribution is governed mainly by topography and how this affects either impeded drainage on plateaux or valley bottoms, or water inputs developing where precipitation is low to moderate (up to about 1500mm pa). Thus, in upland areas, gley soils often develop a peaty upper horizon, particularly under moorland as altitude and precipitation increase.

Ther are two main types of gley. Surface water gleys (stagnogleys) evolve when an impermeable barrier forms in the soil. The water can no longer drain naturally into the groundwater and begins to pond up above the impermeable layer. Typical impermeable barriers include layers of stones, iron pans and plough pans (Figure 4.9). The second type is the groundwater gley. Here the gleying results from an increase in the height of the soil's natural water table, usually a seasonal occurrence. This leads to mottling because of the distinct seasonal variation in precipitation patterns in Britain (Figure 4.9).

In contrast to podzols, research into gley soils in the British Isles is limited. This could be explained by the fact that many podzols can be improved more cheaply than gleys and then planted up with conifers, encouraging research on improvement of timber yields. Gleys, on the other hand, are typical of upland farming areas which command much less economic value, and it is more costly to improve the soil to give a similar financial return. Consequently they have not aroused so much research interest.

Box XVII: Macro- and micro-nutrients

Plants need twelve essential nutrients to grow, and six of these are macro-nutrients – ones required in large amounts (1000 parts per million or more). The macro-nutrients are: N, P, K, Mg, S and Ca. The micro-nutrients are required in smaller quantities (less than 1000 parts per million). They include: Mn, Cu, Fe, Zn, Mb, B and Co.

There are also other nutrients – selenium, sodium and iodine, for instance – that animals need but plants do not. For every nutrient there will be concentrations where it is insufficient (limiting growth), excessive (becoming toxic) or sufficient; the exact levels will depend on the recipient species' requirements. The main deficiency and excess issues for plants and animals are shown in the table.

nutrient	*function*	*deficiency issues*	*excess issues*
macro-nutrients			
Nitrogen (N)	converted to nitrates to produce amino acids used to improve photosynthesis	stunted grass growth roots less able to seek water and nutrients	leaves become sappy and blueish, which stock abhor crops will fail slower drying for hay/silage eutrophication of watercourses
Phosphorus (P)	energy component of ATP aids plant metabolism used for nucleic acids	seedling establishment poor, leaf growth limited	inhibits floral diversity for conservation purposes creates Fe and Zn deficiency
Potassium (K)	regulation of cell water content aids ion transport	restricts root development poor transpiration control wilting in droughts lack of winter hardiness	over-application can reduce Mg and Ca uptake from the soil, which is then passed on to livestock
Magnesium (Mg)	component of chlorophyll and growth enzymes	causes 'staggers' in livestock slows photosynthesis	excess creates Ca and K imbalance reduces growth
Sulphur (S)	part of amino acids	affects the efficiency of nitrogen use reduction in yields	if over-applied to Cu-deficient swards, affects livestock Cu metabolism causes 'staggers' in cattle causes premature leaf fall
Calcium (Ca)	required for cell division	stunted growth and leaf distortion in plants	halts availability of Mn, Zn, Fe or I in livestock Mg and K deficiency in plants
Micro-nutrients			
Manganese (Mn)	used in photosynthesis and enzyme systems	crop failure above pH yellowing of leaves	brown spots on leaves
Copper (Cu)	used in enzyme systems and photosynthesis increases yields aids haemoglobin production	diarrhoea and poor growth in livestock causes 'swayback' increases Mb, making grasses less palatable	can poison stock (toxic)
Iron (Fe)	manufacture of chlorophyll	causes anaemia	leaves 'bronze'
Zinc (Zn)	activates enzymes aids haemoglobin production	reduced leaf size leaves yellow causes anaemia	creates Fe deficiency in some plants
Molybdenum (Mb)	used for enzymes that control nitrogen metabolism	accumulation of nitrates in leaf tissues	halts availability of Cu in stock
Boron (B)	root and shoot growing points used for cell wall development	root and shoot growing points fail	leaf tips yellow leaf necrosis, followed by fall
Cobalt (Co)	aids N fixation in legumes aids vitamin B12 production in stock	anaemia may result poor appetite: livestock fail to thrive, become emaciated and can die	carcinogenic dermatitis in stock
Required by livestock only			
Selenium (Se)	cellular function production of enzymes	white muscle disease infertility in stock ill-thrift (similar to anaemia)	poor reproductive performance alkali poisoning toxic
Sodium (Na)	nerve impulse operation fluid control	masks other deficiencies reduces stock productivity	inhibits enzyme functions
Iodine (I)	effective thyroid function	abortion stillbirth birth of weak calves	goitre development in cattle toxic

Adapted from: Frame, 1992; Bazeley and Hayton, 2007

Peat soils

The fifth main class of soils found in the uplands is the organic soils, also called histosols, which derive from partly decomposed plant remains that accumulated under waterlogged conditions (Soil Survey Handbook 1980). To be recognised as a histosol by the Soil Survey of England and Wales, the soil needs to have:

- more than 40cm organic material in the first 80cm (excluding leaf litter or moss)

or

- more than 30cm organic material resting directly on R or Cr

and

no overlying mineral horizon of colour value 4+ and extends below 30cm depth.

Histosols are a difficult group of soils to investigate because different disciplines describe them in different ways. Ecologists often refer to them as mires (Table 4.4) and classify them into five broad types based on the source of their water and nutrients (Heathwaite and Gottlich 1993; Table 4.5).

Table 4.4: Ecological Definition of Peat soils

1) MIRE : bodies of organic material that have accumulated over the last8000 years (Godwin, 1975)	a) Ombrogenous Bogs - Raised bog (a dome) - Blanket bog (flat land or basins) - Intermediate bog (half way between two)
	b) Fen – form in a basin and are under the direct influence of ground water charged with mineral plant nutrients
2) PEATLANDS: peat areas no longer accumulating	

(Sources: Avery 1990; Bragg and Tallis 2001; Bragg 2002.)

Table 4.5: Mire Typology

Basin Mires
• Often found in glaciated cirques • Peat is deep, as much as 10m and some vegetation may be floating • Generally no outflow of water
Blanket Mires
• These carpet hill and mountain sides and tops to a depth of 5m • Typically acidic peat intermixed with other soil types • Angle of slope and precipitation are key
Floodplain Mires
• Develop on alluvium built up from rivers flooding • Typically alkaline peat
Raised Mires
• Mire develops by rising above water table • Typically acidic peat builds up owing to reliance on atmospheric water • Fringes can support alkaline peat
Soligenous Mires
• Found where water moves, flushes or springs • Alkaline peat builds up very slowly
Valley Mires
• Occur in the direction of water flow • Peat builds up at varying rates • Nutrient status very variable depending on surrounding geology

(Source: adapted from Heathwaite and Gottlich 1993.)

A common characteristic of mires is a high water table and thus they are typical of the uplands. Some mire types are acidic since they gain their water from the atmosphere only; for example, blanket mire is rain-fed. Other mires form in depressions and their acidity or alkalinity (with the related biodiversity) is determined by the properties of the water that drains into them (see Chapter 5).

Typically peat soils are identified by their lack of an earthy topsoil and visually identifiable remains of bog moss, cotton grass and ling, a pH (acidity level) less than 3.5 and their top 20–30cm being waterlogged (Mackney *et al.* 1984). Fibrous peat is found in northern England and Scotland; more amorphous peat, where plant remains are not visible, is typical of Wales and south-west England (Figure 4.10). These properties result in a number of economic problems with upland peat soils:

- low-quality grazing potential,
- wetness leading to poor load-bearing strength,
- extreme infertility with little improvement potential,
- need for intensive drainage (below 400m asl) with only afforestation viable.

Drainage and deep ploughing can overcome these challenges, as over time the peat will dry out, improving the aeration; but such activities remove much of the ecological interest (see Chapter 5) and can cause erosion.

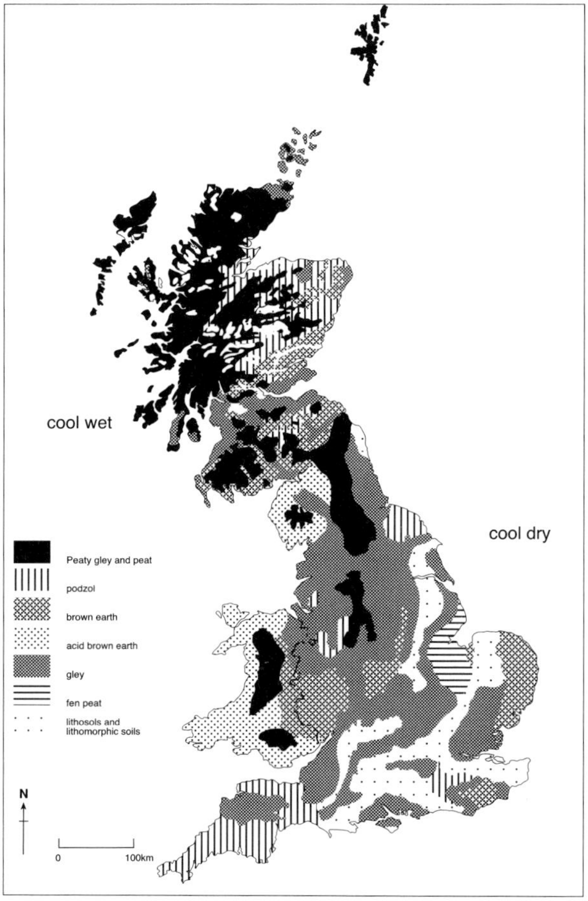

Figure 4.10 Generalised Soil map of Britain.
Adapted from Avery, 1990: 14

Agricultural potential of upland soils

Having characterised the main soil types in upland areas, we can now consider the agricultural potential of soils by reviewing their general properties and then relating these to the management challenges.

Soils of whatever type have certain physical and chemical properties that are important in terms of agricultural potential. The five most important physical properties of any soil are: structure, texture, porosity, mechanics and water availability/content. These are closely related to three chemical properties of soils: acidity, organic content and nutrient status (Figure 4.11). Some explanation is needed of these properties and how they develop in upland environments.

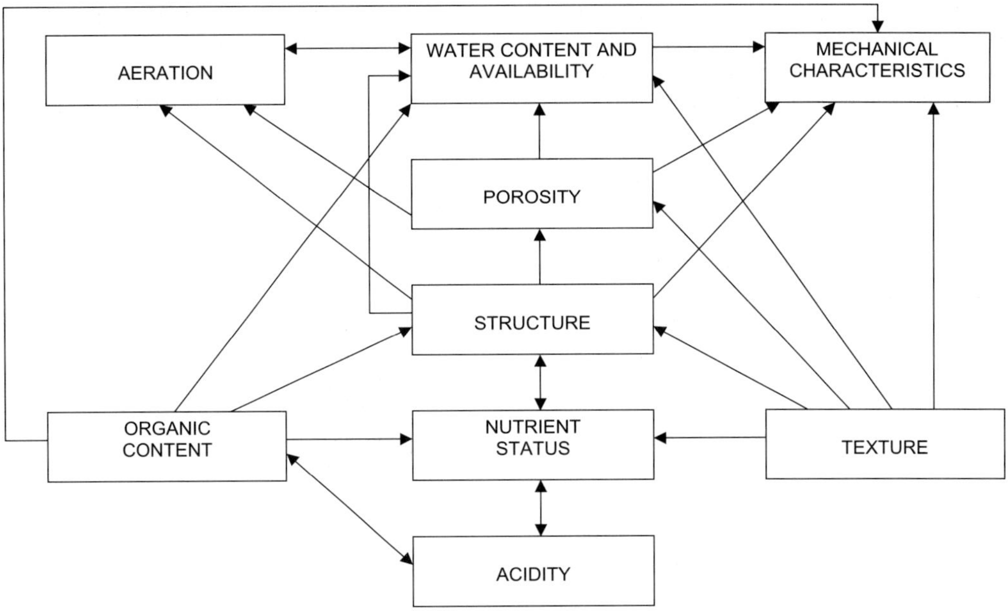

Figure 4.11 Physical and chemical properties of soils.
Reproduced with kind permission from Ellis S & Mellor A (1995) '*Soils and Environment*' Routledge: London Figure 2.2, p11.

Texture

Texture is a physical property defined by the proportions of mineral particles – of sand (granule size 0.2 to 0.02mm), silt (0.02 to 0.002mm) or clay (under 0.002mm/2 microns) – found in a sample of the soil. These proportions are used to determine an overall soil texture using a standard triangular plot (Figure 4.12). So fundamental is this soil property that, under the GAEC (Good Agricultural and Environmental Conditions) rules (RPA 2005), DEFRA requires a farmer's Soil Conservation Management Plan to identify soil 'type' as clayey, sandy or silty. As Figure 4.11 shows, texture is so closely related to other soil properties that it acts as a cheap, straightforward way for farmers to assess the soil-management problems on their land and then rapidly develop sustainable resource strategies.

There are, in essence, two extremes of soil texture: sandy and clayey. Sandy soils such as podzols are problematic because they cannot hold moisture and nutrients. Furthermore, sand does not form complex chemical structures with other components (such as humus or clay), which makes it more susceptible to erosion. Soils dominated by clay, such as gleys, also create management conundrums. Clays, by their chemical nature, have the ability to hold moisture and nutrients, but they may hold too much water and encourage waterlogging, slaking, poaching or pan formation. For agricultural potential it is best to have a roughly equal amount of clay and sand, thus balancing their worst and best properties, allowing the retention of a certain amount of water and nutrients. Such soils, known as loams, are unfortunately rare in the uplands. Loams are usually found in valley floors where inbye fields have been intensively managed over years deliberately to develop this type of texture.

For peat soils (and peat horizons within mineral soil profiles) there is no soil texture, because by definition peat does not contain mineral matter. This lack means that soil structure and nutrient retention are weakened because there are few clay–humus micelles (see below); it also means cohesion and plasticity are extremely low, leading to a porous character, which can be good for cultivation. However, in dry or wet periods the looseness means there is nothing to bond it together, so it is easily eroded when exposed.

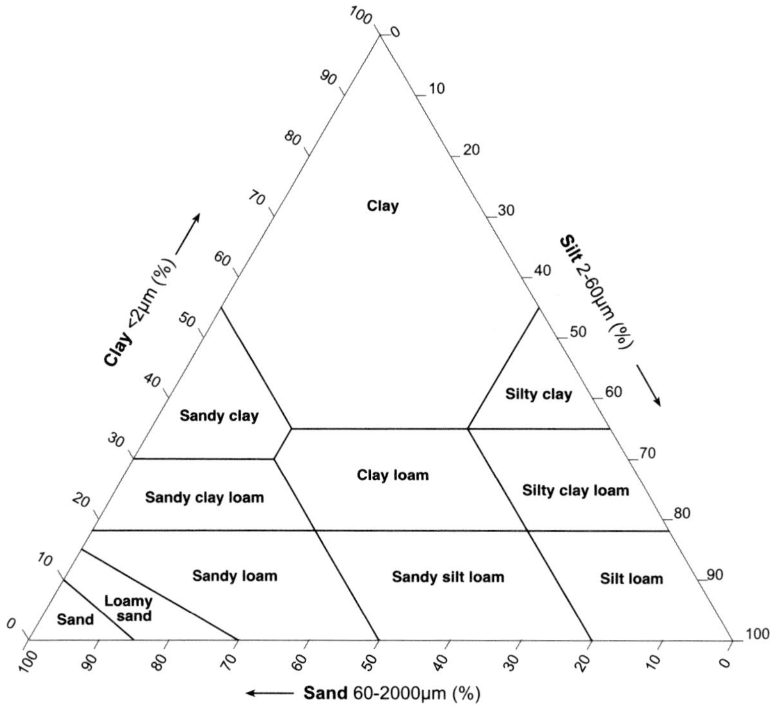

Figure 4.12 Soil Texture Triangle.
Adapted from Hodgson 1973. Soil map @Cranfield University (NSRI) 2011, used with permission.

Structure

Soil structure can be described as soil particles arranging themselves into larger aggregates or lumps. Structure can affect permeability, root penetration, infiltration rates, percolation and erodibility. Certain structural configurations develop in soil. If soil aggregates organise themselves into horizontal layers (a platy structure), this can inhibit downward or upward movement of water, as in gleys. Large agglomerations of soil aggregates create a block structure, which makes root penetration through the soil difficult.

Ideally a crumb structure is sought in the top 15cm, providing a good seed bed for germination and good water transmission. Beneath this a mix of (angular and sub-angular) smaller block-like aggregates, allowing good root penetration. These ideal structures are unlikely to occur in upland soils because of the high proportion of clay they contain. These clays can also be slaked at the surface as aggregates break down (forming a type of 'pea soup'), and a crust when dry, prohibiting seed germination. Farmers have striven by various management regimes to develop a crumb structure in grassland to enhance growth and thus avoid these issues.

The breakdown of clays at the surface or their build-up lower down the soil profile can lead to reduced soil porosity, defined by the size and number of gaps (pores) which occur between the mineral particles and aggregates. The pores are filled by water or air, or a combination of the two, depending on conditions.

Closely related to porosity is bulk density, which is the mass of a unit volume of dry soil. Bulk density is important as a measure of soil's ability to hinder water movement and root penetration. High bulk densities around 2.0g/cm^{-3} indicate poor root penetration and impeded water movement. Soil will become compacted and lead to increased surface run-off. Typically these soils also see the formation of plough pans.

Peat's bulk density is far lower than any other type of soil, at about 0.1 to 0.3 g/cm^{-3}, in contrast to cultivated mineral soils at 0.8 to 1.4 g/cm^{-3}. But peat's porosity is the highest, 0.93 to 0.79 cm^3/cm^{-3} compared to cultivated mineral soils of 0.69 to 0.46 cm^3/cm^{-3}. The combination of these two properties means that peats are not compact, with plenty of air pockets between particles. Though this may be good for plant growth, it can exacerbate the erosive potential of the soil as there is little to hold it together once the water has gone.

The relationship of these physical properties is the subject of soil mechanics, which focuses on aggregate stability, soil strength and soil consistence (Ellis and Mellor 1995). Aggregate stability is the potential of aggregates to remain structurally intact after wetting. If air is locked into the aggregates during wetting, the soil can slake. Excessive water encourages any cementing agents in the soil, such as $CaCO_3$, to dissolve. Soil strength is a combination of texture, bulk density, moisture and organic content. At one extreme a soil can be coarse textured, have a low bulk density, high water content and low organic content – this will lead to low soil strength. At the other end of the spectrum, a strong soil will have a fine texture and high bulk density, with low water and high organic content. Finally, soil consistence refers to the cohesion and adhesion of soil in terms of water content and its effect on the soil's ability to retain its stability under stress. Typically it can be assessed by a change of state of the soil from solid to plastic to liquid. Solid soils are more stable, whereas plastic soils will deform and smear with pressure, as often happens when machinery is introduced too early in the season into wet, clay-rich fields (Strutt 1970).

Water Content

The final physical property of note is water content and availability. This is very much a function of the mineral component of soil. Clay particles come in different forms, some of which have the ability to draw in and hold water inside their chemical structure or between different crystalline layers of clay.

Some of these clays can release water fairly easily back into the soil for plant use; others do not. For example, smectite and vermiculite clays can hold water effectively, but the former retains water longer than the latter (hence the use of vermiculite as a horticultural germination medium). Despite this variation, most mineral soils can hold only one fifth to two fifths of their dry weight in water, whereas organic soils can retain four to five times their weight due to the decomposed plant matter content – another reason why peat is an excellent growing medium.

Soil Chemistry

Soil's other main properties (see Figure 4.11) are the chemical aspects of organic matter content, cation exchange and acidity. In mineral soils, organic matter is concentrated near the surface where various fauna have mixed it in, assisting decomposition. On occasion, organic matter can be found lower down the soil profile as a result of mechanical eluviation typically found in various podzol sub-groups (see above).

Organic matter is an important property even though on average it comprises only 5% of a mineral soil's content. Once decomposed to humus, it can chemically bond with clay minerals to form larger particles known as clay–humus micelles. These greatly improve the structure of a soil and its water- and nutrient-retention capabilities. They also carry a net negative charge that attracts positively charged ions (known as cations) and locks them on (Gerrard 2000). These cations can then be moved in and out of the soil–water solution, providing nutrients for plant growth when required.

Some cations are better able than others to escape the attraction of this negative charge. Cations like hydrogen (H^+), sodium (Na), magnesium (Mg) and calcium (Ca) are easily moved to and from the soil solution. Others, such as iron (Fe), aluminium (Al) and titanium (Ti), are attracted onto micelles and stay attached for much longer, only being released into the soil when the acidity level (pH) falls sharply. Clay–humus micelles are valuable for their high cation-exchange capacity (CEC).

Other useful ions are nitrogen (N), phosphorus (P) and potassium (K). Phosphorus is particularly important for grass growth, helping grasses to out-compete broad-leaf plants. Soils in the UK are naturally low in P, so it is added via fertilisers and dung to a level of 16 to 25 mg/l (P Index of 2). In addition, P is very insoluble and rather immobile, meaning the soil releases it very slowly. P fertiliser that was added fifty years ago can still be releasing into the soil today. Unlike agricultural pasture, semi-natural grasslands have an Index of 0 (0 to 9 mg/l), so low P levels are crucial to controlling sward diversity (RDS 2005). This is a good example of when agriculture and nature conservation are in tension.

Attached cations can be lost from the micelles by leaching (Box XVII). However, the soil will attempt to maintain the concentration of cations by a process called buffering. As the cations are leached, cation exchange takes them off the micelle and into solution in order to maintain previous levels. In exchange, H^+ ions are put onto the micelles. The H^+ ions determine soil acidity, which is measured on a log scale: so pH2 is ten times more acidic than pH3 and pH1 is one hundred times more acidic than pH3. In soils, pH14 represents few H^+ ions and many OH^- (hydroxide) ions, pH1 is many H^+ and few OH^-, and pH7 is a balance between the two.

As a result wetter climates tend to produce more acidic soils, compounded by vegetation inputs and lack of microbial activity, whereas alkalinity is a function of the parent material, vegetation and hydrology. As soils acidify and pH falls below 5.5, Al (Aluminium) ions become the most mobile cation, disattaching themselves from the micelle for cation exchange. This leads to soil podzolisation and ultimately the Al^{3+} can be leached out into the groundwater at toxic levels. As the pH rises above 5.5, other cations become dominant on micelle surfaces. This explains why farmers seek a neutral pH (6 to 7) because at this level micelles contain exchangeable cations of most value to crop growth. This relationship between cation availability and soil pH is shown in Figure 4.13. In contrast, upland semi-natural vegetation occurs across a relatively wide pH continuum as different plants adapt to different nutrient statuses (Miles 1987), another cause of conflict between agriculture and conservation.

As noted above, most upland soils are acidic. Whether this is the result of an acidic parent material, intense leaching processes or the build-up of rain-fed peat, depends on the soil's local environment. All these situations lead to low nutrient status (Figure 4.13). Peat soils are more deficient in potassium and phosphorus than mineral soils, but have more of other nutrients such as Ca^{2+}, Mg^{2+}, K^+, S^- and N^-.

We can see that in the uplands, soil is controlled by the parent material, climate and topography. The relationship between upland soils and biota (people, habitat and animals) is more complex and is a recurring theme of this book. As this discussion and Figure 4.11 show, the physical and chemical properties of soils are closely inter-related. For upland farmers these properties have led to four soil-management challenges to agricultural activity: poaching, poor drainage, infertility and erosion.

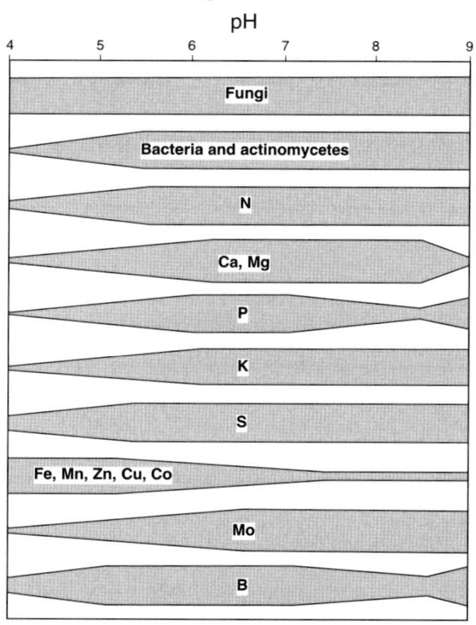

Figure 4.13 Nutrient Availability with Changing Soil pH.
Reproduced from Brady NC (1984) 'The Nature and Properties of Soil' 9th Edition, Figure 6.10 p207 with permission from Prentice Hall.

Challenge 1: poaching

Soils under grass are susceptible to structural damage when the grass is in a vulnerable state. This is likeliest in periods of high rainfall or waterlogging, when soil structure can be weakened by physical and chemical processes, or in soils that have a high silt or clay content. Gleys, hydromorphic podzols and peat soils in the uplands are particularly at risk in these conditions.

Grazing animals can compact and smear the soil simply by their weight and frequent movement over the same ground. Although dairy cows are considered the worst offenders, sheep can have a similar effect if they gather repeatedly in one place (for example, at a feeding trough or repeat location for supplementary feeding). Poaching is a combination of compaction, poor water infiltration, slaking of the soil surface and reduced porosity. It is a particular problem with higher stocking densities, encouraged in upland areas by the Hill Livestock Compensatory Allowance and Sheep Annual Premium (see Chapter 8).

Grass reacts to poaching by increases in shallow rooting and aerial tillering of plants, because roots cannot penetrate the compacted layer. It can be good to encourage tillering, but if the ground remains wet the entire plant can be torn out by a grazing animal and hence grass yield is reduced. The other problem is that poaching leads to grass becoming muddied and then increasingly rejected by livestock as stocking rates rise.

Farmers can lessen poaching by reducing or removing stock as quickly as possible. Frame (1992) suggests a number of other management strategies to support this:

- maintain effective soil-drainage systems
- use tolerant grass species and varieties
- make use of long-term rather than short-term swards
- encourage the growth of dense, vigorous swards
- maintain an adequate supply of grass in a field
- optimise winter grazing management of swards
- use only light stock (young cattle/sheep)
- reduce stock traffic to other fields across the sward

All these solutions are eminently suitable for agricultural land, but putting in drainage and switching to more tolerant grass types are not desirable for nature conservation, because it is usually intolerant species that have high conservation value. Conflict between farming and nature conservation can be reduced by using other options to control poaching.

Challenge 2: poor drainage

In upland areas, iron and plough pans are a particular problem with podzols and gleys, where downward water movement can be inhibited or a high water table exists.

Iron pans develop as part of the process of podzolisation (Box XVI); plough pans result from repeated cultivation and occur at a point just beyond the maximum depth to which the plough share reaches. In the plough zone, soil is churned and broken up, capturing air in the process, developing a well-structured soil that allows effective root penetration. Below the plough zone, soil has not been mixed and aerated and, as one ploughing season follows another, this lower level is smeared and compacted by the plough passing over. Gradually the layer becomes impermeable and eventually all drainage ceases, leading to rush (mainly *Juncus effusus*) infestations, exacerbated by bare patches on the soil created by poaching (Miles 1987). To deal with this, the rushes can be mown and harvested as a crop or drainage can be put in, but rush seeds can remain dormant for decades and, if the drains do not lower the surface water enough, the rushes will re-establish.

Despite these problems, general drainage management is usual on inbye and intake land in uplands. Drains, mole ploughing and subsoiling are options. In the late 17th century, farmers began using tile drains; after the introduction of drainage grants in 1939 and land improvement grants in 1951, 1956 and 1957, extensive areas of pasture were drained and their fertility improved (Evans and Attwood 1961). Tile drains are expected to last 60 to 70 years with good maintenance. They are made from fired clay and come in lengths of 30 to 45cm. They are laid in a v-shaped groove, to prevent the overburden collapsing the drain, and then filled over with soil or gravel (Figure 4.14). These traditional clay tile drains have been replaced by rigid plastic pipes up to 6m long or corrugated flexible plastic pipe, which comes off a spool in lengths up to 150m. Perforated plastic pipes are used for permanent drainage, but they are less effective than traditional clay tile drains because they take less than half the water away (Farr and Henderson 1986).

Where the soil profile is highly impermeable due to a high clay content, the best means to improve the soil is mole drainage (Farr and Henderson 1986), created by pulling a mole plough (shaped like a torpedo) through the soil, forming a channel of 50mm diameter at a depth of 500 to 700mm (Figure 4.14). Ideally, this is done in May or June when the land is drying out. Mole drains last about 6 to 7 years. However, in upland areas mole ploughing is not as effective as more permanent drainage solutions, because the higher precipitation results in more clay being mechanically eluviated and the mole drains simply clog up with the clay rapidly.

A third option is a subsoiler, which works at depths of 375 to 500mm. Unfortunately it is a temporary solution that may need repeating every two to three years. The subsoiler breaks up plough pans beneath the cultivated topsoil. Subsoilers differ from a mole plough in that the tip is flattened and chisel-shaped, producing many fractures in the soil (Figure 4.14). This operation is meant to disrupt any compacted layers in soils with low clay content, so it can be effective in podzols but hopeless in gleys (Farr and Henderson 1986). Davies *et al* (2001) demonstrated the effect of subsoiling on infiltration[3] rates of various soil textures (Table 4.6). Here the clay-rich soils saw little drainage improvement, whereas the loams got substantially raised capability. Subsoiling also helps grass-root penetration with a fourfold increase below 200mm (Table 4.7).

Table 4.6: The Effect of Subsoiling on the Infiltration Rates of Various Soil Textures

soil texture	*infiltration rate (cm/hr)*	
	not subsoiled	*subsoiled*
Clay	0	27
Silt loam	16	27
Loam	111	176
Sandy loam	360	1028

(Source: Davies *et al.* 2001)

Table 4.7: Root Penetration into the Soil

depth (mm)	*percentage of total grass roots*	
	not subsoiled	*subsoiled*
0 to 100	82.4	69.8
100 to 200	13.3	13.1
200 to 300	2.7	8.0
300 to 400	1.0	5.0
400 to 500	0.6	2.4
500 to 600	0.0	1.2
600 to 650	0.0	0.5

(Source: Davies *et al.* 2001)

Given that much upland farmland is of low economic value, it is unusual for farmers to drain their hill land, usually reserving such processes to only inbye and intake. However, in the past, unenclosed moorland has been drained using a system known as 'gripping' or 'gryping' particularly on peat soils (Farr and Henderson 1986). Gripping produces a series of small surface channels, about 40cm deep and semi-circular in section, which form a herringbone pattern along the contours about 20 to 30m apart (Figure 4.14). Depending on the management objectives, grips can be dug to various depths – deeper for peat cutting and shallower for afforestation (Tallis

[3] Infiltration - the way in which water moves down through the soil.

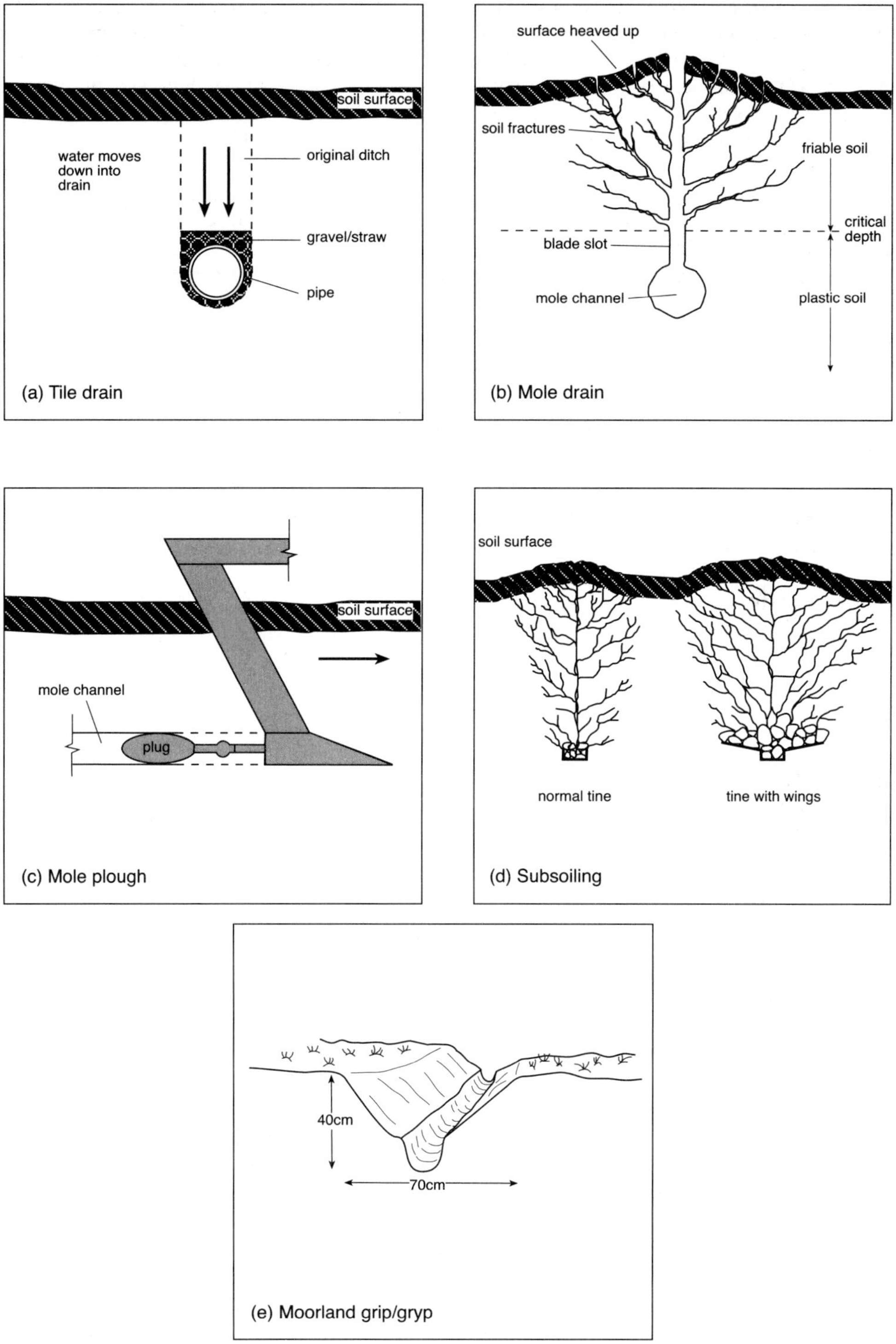

Figure 4.14 Drainage options

1998). Grips deteriorate fairly rapidly because of erosion, deposition or bank collapse caused by frost heave or poaching; consequently the process needs repeating every ten years to ensure effective drainage continues.

Grips were dug extensively on British uplands from the 1950s to 1980s with financial support from MAFF. It was believed they would reduce the wetness of peaty soils and create more productive vegetation, particularly heather, but the effect is localised to within 0.5 to 2m of the ditch (Tallis 1998). Since the 1980s it has been proved that – far from being advantageous for agriculture, nature conservation or water collection – grips actually create problems (Robinson 1985; Stoneman and Brooks 1997; Barrett 1997):

- young birds falling into and drowning in the steep sided grips
- lambs and young sheep falling in and drowning
- loss of hydrological integrity of peat systems
- damage to the bog vegetation
- exacerbated erosion caused by grips overdeepening
- flash flooding as run-off accelerated through the grips
- increased siltation choking upland rivers and reservoirs down-catchment
- subsidence of peat surface
- increased bulk density of surface layers

Grips are now blocked, using peat dams, to help re-establish the natural drainage of the open fell. Over time, grips vegetate over and run-off is halted. Positive effects quickly develop, initially for birds and livestock, and then for the vegetation itself. Medium-term effects of reduced erosion and siltation of rivers have been observed, as well as their long-term value as carbon sinks.

Challenge 3: infertility

Infertility is a result of many of the properties and processes in the soil combining to stop upland soils retaining nutrients necessary for plant growth. As a result, much of the vegetation on these soils is of limited diversity.

From medieval times to the early 20th century, hillsides in northern England and Scotland's Southern Uplands were irrigated with lime-rich water to improve fertility (Heddle and Ogg 1936). Light dressings of nitrogenous fertiliser can increase heather growth, though heavier applications can change floristic composition. Adding NPK at agricultural rates, if accompanied by grazing, encourages purple moor grass. Other options are ploughing and reseeding, usually alongside drainage activities, but the cost of such practices far outweighs the profits, so most upland grazing is left to its own devices.

The species growing in uplands have thus adapted to the impoverished conditions and many are palatable for only short periods of their life cycle. For instance, sedges (*Carex* spp.), heath rush (*Juncus squarrosus*) and cotton grass provide good grazing early in spring, but are unsuitable for the rest of the year (Table 4.8). Young ling is most nutritious in winter. Cattle will eat purple moor grass when it is green, but not once it matures and flowers. Other species are rich in specific minerals: for instance, bilberry is high in calcium, cotton grass in phosphorus. The most valuable species for grazing are bent and fescue grasses as they are nutritious and give the highest yield. Consequently, sheep are known to graze selectively, congregating around these plant communities on open fell. Evans (1977) recorded this in the Peak District, where sheep stocking densities were 1.67 sheep/ha on burnt heather moor and 4.54 sheep/ha on bent–fescue grassland. This was also observed by Rawes and Welch (1969) on summer grazing land at Moor House in

Table 4.8: Seasonal Variations in Grazing on Various Upland Plants

Month	*Jan*	*Feb*	*Mar*	*Apr*	*May*	*June*	*Jul*	*Aug*	*Sept*	*Oct*	*Nov*	*Dec*
Ling	■	■							■	■	■	■
Hare's tail cotton grass	■	■	■	■	■							
Purple moor grass						■	■	■				
Deer grass			■	■	■	■	■	■				
Cloudberry						■	■	■				
Bilberry	■	■	■	■								■
Common cotton grass								■	■	■		
Heath rush	■	■										■
Mat grass		■	■	■	■							
Crowberry									■	■	■	■
Common Bent				■	■	■	■	■	■	■		
Sheeps and Red Fescues				■	■	■	■	■	■	■		

(Source: Brooks and Stoneman 1997, Tallis 1998.)

the northern Pennines, where there were 0.04 sheep/ha on blanket bog and 8.7 sheep/ha on bent–fescue grassland. This uneven distribution is a function of the production potential of vegetation, which is very much dependent on the soil type (Table 4.9).

In some upland areas, more fertile acid brown earth soils exist, supporting the desired bent–fescue vegetation and, with fertiliser application, producing up to 9t/ha dry matter. Unfortunately these soils are often exploited by bracken, which also prefers these conditions and will shade

Table 4.9: Potential Production of Various Upland Soil Types

soil type	*pH range*	*principal species in community that are grazed*	*potential production (t/ha DM)*
Acid Brown Earth	5.3 to 6.0	Common bent grass Red and Sheeps fescue grass Meadow grasses White clover Abundant herbs	1.0 to 3.6 (L) 3.5 to 4.5 (M) 6 to 9 (H)
Podzols	4.0 to 4.5	Mat grass Wavy hair grass Ling Bilberry Bell heather	} 91 to 2 } 1.5 to 3
Peaty Gleys	4.0 to 4.5	Purple moor grass Sedges Rushes	1 to 2.5 7.5 (fertilised)
Peat	3.5 to 4.0	Cotton grass Ling Purple moor grass Sphagnum Deer hair sedge ('deer grass')	} 1 to 3

NB 1 Dairy Cow needs 1t/ha DM for 67 days, using standard Livestock Units as a conversion this is equivalent to 6.67ewes (1 ewe = 0.15 LU)

(Adapted from: Frame, 1992)

out grasses. The only real exception is grassland production from calcareous swards, which is significantly higher (O'Donovan 2001).

Variations in grassland sward composition and seasonal fluctuations in digestibility mean that one heft may contain a wide range of fodder sources, which different livestock species and breeds may graze selectively. Thus, reliance on annual stocking densities in agri-environment initiatives is not helpful, because any fell can really support stock only for limited periods, which depend on its fodder composition. Felton and Marsden (1990) investigated this for heather moorland using the MLURI (Macaulay Land Use Research Institute) productivity model for upland grazing. They found hill grazing was compatible with the maintenance of heather at 1 to 1.5 ewes/ha. If the whole flock was off-wintered, then a rate of up to 2 ewes/ha was feasible when the stock returned.

Further research by Armstrong and Milne (1995) demonstrated that levels of utilisation of each vegetation type in a given area, taking into account seasonal variations in grazing behaviour, was a more effective method of estimating sustainable stocking densities. They also found that doubling stock densities did not double the rate of utilisation (how much is eaten) of grass for fodder, and that little heather was eaten until total grass utilisation reached 70% (i.e. 70% of what could be eaten was eaten). In this way Armstrong and Milne calculated that ling has its highest utilisation rate at 40% in the pioneer stage of its life cycle (0 to 7 years). Once it enters the building stage (7 to 15 years) the rate drops to 10% and beyond this period to 5%. Consequently livestock will seek out young ling at the expense of more mature heather. This places undue pressure on heather moorland, slowing its recovery rate, because pioneer ling is least tolerant to grazing pressure (Felton and Marsden 1990).

Variations in plants' digestibility and palatability suggest that the agricultural potential of uplands can be maximised either by eliminating plant diversity to just a bent–fescue mix or by ensuring that the landscape supports a mosaic of species, to allow livestock to seek subsistence at any time of year. However, as noted already (Box III), reducing diversity to just high-nutrition grasses is not favourable to native breeds, so it seems the only option is a management approach that maintains a mosaic of vegetation types while keeping an eye out for evidence of increased grazing pressure.

Challenge 4: overgrazing and erosion

The final major management issue with upland soils is overgrazing, with the possibility of subsequent erosion. Though it varies from one region to another, overgrazing is probably one of the most sensitive issues in upland agriculture in the UK. Many organisations over the last fifty years have used it as proof of poor grazing management of upland habitats. DEFRA officially define overgrazing as

> Grazing land with livestock in such numbers as adversely to affect the growth, quality and species composition of vegetation (other than vegetation normally grazed to destruction) on that land to a significant degree. (SI 1996 1500 – the Hill Livestock (Compensatory Allowances) Regulation 1996)

Despite being so vague, this definition has been used to introduce overgrazing clauses into livestock support under the LFA Directive, as well as advice and guidance for farmers about how to avoid overgrazing and unsuitable supplementary feeding (DEFRA 2005a).

Overgrazing has three main stages. First, vegetation deteriorates because of pressure from selective grazing: more palatable species are eaten out by stock, leaving behind plants such as mat grass and purple moor grass. Patches of calcareous grassland are particularly susceptible to this differential grazing, because they are made up of the most palatable grasses found in unimproved swards. This is problematic when the stocking density of a heft has reduced, lowering the intraheft competition. This allows drift from one heft to another to seek the tastiest dinner, leaving behind the unpalatable species. Evidence of this activity can be seen on Warcop Fell, part of the Moor House NNR in the North Pennines (Scott 2005).

The second phase of overgrazing is the complete removal of vegetation and exposure of bare soil. Livestock create and maintain bare ground in uplands in four ways: their tracks, grazing pressure, resting places and congregational behaviour. Trampling as stock move about is a particular problem for montane vegetation, where plants grow more slowly than elsewhere and are eaten out, exposing the soil (Britton and Pearce 2004; 2005). Grazing pressure suppresses vegetation to the point where it ceases to grow. It varies from species to species depending on parameters like palatability, digestibility or season (Armstrong and Milne 1995) as well as a plant's morphological and physiological ability to recover from grazing (Crofts and Jefferson 1999). Stock seek shelter on open hillsides, sheep especially. Evans (1977) describes how sheep lean against changes of slope and rub away vegetation to form scars of exposed soil. These crescent-shaped features lead to high levels of erosion. Evans monitored a small catchment in the Peak District for 25 years and demonstrated that, as sheep numbers were curtailed, so sheep scars re-vegetated and disappeared (Evans 2005). Finally, trampling where animals congregate can be exacerbated by repeated supplementary feeding at the same location.

The final phase of overgrazing can see the removal of the bare soil by wind or water erosion (Table 4.10). For uplands, soil fragility means that they are far more susceptible to erosion once soil is exposed. Because of their immaturity, lithomorphic soils and lithosols lack structure, and this is exacerbated by steep slopes. For peats, lack of structure, caused by the lack of mineral content, is magnified by such extreme environmental conditions as greater minimum and maximum temperatures, longer periods of sub-zero temperature and more freeze–thaw cycles breaking up and loosening surface structure (Tallis 1998). However, peats can also suffer from natural erosion, climate change, fires, grypping, air pollution, recreational pressure and mass movements (Bragg and Tallis 2001; Evans 2005; Bayfield 1979; Coleman 1981).

Table 4.10: Forms of Erosion

type	*description*	*notes*
Microerosion	Peat surfaces are covered by a network of shallow channels up to 10cm deep with drier ridges in between	
Erosion complex	Deep gullies (<2.5m) interlink forming upstanding ridges (*haggs*) in between. Peat gradually gets washed away, which eventually can go down to the subsoil or bed rock	Haggs typically colonised by dry heath communities. Slumps can aid in recolonising the complex
Downhill gullies	Parallel channels running downslope usually 20 to 30m apart Sides and bottom can be vegetated slowing erosion	Ridges between are colonised with dry heath
Sheet	Areas of bare peat over $10m^2$ are exposed and the surface is removed en masse	Surfaces dry out from a combination of wind exposure and freeze–thaw cycles. Water and wind can then remove the top in one go
Bog Burst	The whole peat mass on an incline simply shears away from the mineral soil beneath and slides rapidly downhill	Related to heavy rainfall events probably caused by a combination of hydraulic overload in the peat, peat cutting, drainage management or stream erosion at the foot of the area

(Source: MacDonald *et al.* 2001; Warburton *et al.* 2004.)

Reducing erosion on upland grazing is simply a function of management re-adjustment. A common solution is to reduce the stocking rate, as Evans (2005) found in the reduction of sheep scars in a Peak District catchment. Unfortunately de-stocking does not always solve the problem because selective grazing can develop. As the competition between stock reduces, animals have more choice in their grazing selection and thus graze out the most palatable species. Elsewhere, however, the sward composition begins to change in favour of less palatable species and undergrazing develops. A further complication is the loss in economic viability by running a smaller flock; Felton and Marsden (1990) estimated this cost was about £37/ha in 1989. One solution to these issues is active shepherding

Summary

The physical character of uplands fundamentally determines what agricultural activity happens there. The British Isles' location on the Atlantic seaboard plays a major role in the climate experienced by highland areas. Regional and localised climatic conditions mean that grazing and grass-growing seasons vary from one upland to another.

Other physical factors that influence farming are geology and soil variations. The British uplands are geologically diverse, with acidic rocks dominating. As a result, most uplands have soil types that limit agriculture and present management challenges for farmers. Poaching, poor drainage, infertility and overgrazing/erosion severely limit the productive potential of the upland environment. Different soils have more or less severe problems. Thin soils such as lithomorphics and lithosols suffer from erosion, podzols from infertility, gleys from poor drainage. Few upland soils support vegetation with high agricultural productivity and palatability, but it is these limitations that provide the ecological diversity valued by nature conservation interests. We explore this ecological value in the next chapter.

5

A landscape of High Nature Value

The agricultural challenges discussed in Chapter 4 are inextricably linked to soil and vegetation, which in turn limit the productive potential of the uplands. Yet the vegetation produced for agriculture has given the uplands their High Nature Value landscape that is seen as worthy of conservation (Hoogeveen *et al.* 2004).

Farming and conservation

Potter *et al.* (1996) showed that upland farms contained 57% of Britain's environmental resources. Unfortunately this creates conflicts of interest between the needs of farming and those of nature conservation (from here on, simply 'conservation'), because the ecological interest of upland vegetation is removed long before its agricultural viability is exhausted. In other words, the agricultural carrying capacity of the land is greater than its ecological carrying capacity (see Box XXVI). Upland nature conservation therefore becomes the lowest common denominator for land management. This issue was compounded by agricultural intensification during the 20th century, when ecological levels fell on 36% of upland farms, 38% of lowland pastoral farms and 60% of arable farms (Potter *et al.* 1996).

There are many reasons why conservation is important (Table 5.1) and there are also sound economic reasons for farmers to maintain these habitats – for instance, they provide grazing and shelter for livestock, and may attract agricultural grants (see Chapter 8). It is in a farmer's interest not to let such habitats deteriorate. Unfortunately, upland farms in Britain may contain several or all of the habitats shown in Figure 5.1; consequently, conservation management for uplands is complex because different farms contain unique mosaics of vegetation in response to local physical conditions.

In valley bottoms a pattern of meadows, enclosed pastures and woodland has evolved. Above this lies the sub-alpine zone (roughly 240 to 650m asl), a mosaic of dwarf shrub heaths, grasslands and blanket bogs, with some woodland surviving in inaccessible gulleys. In the uppermost areas, those above the tree limit (about 650m), are alpine communities of heath and grassland. This variety of semi-natural habitats could be seen as a constraint on upland farm management; in fact these patterns developed historically as a direct result of resource management, as people systematically adapted their environment to extract what they needed for subsistence or profit.

Thus in the British Isles any upland ecology and landscape is a construction of the needs of upland farmers and other land users (Bonn *et al.* 2009). Following this argument to its logical

Table 5.1: The Importance of Nature Conservation

Ethical reasons	Matter of conscience Moral justification of stopping extinctions Duty of stewardship/responsibility
Aesthetic values	Pleasure of wildlife and wild country Appreciation of landscape Nostalgia and childhood memories Art & poetry
Cultural & scientific values	Intellectual benefit Leisure needs Pollution indicators Baseline information Religious or cultural symbolism
Material benefits	Sustaining natural resources Commercial value Gene pool/ancestral variety maintained Medicinal value
Ecological balance	Natural systems need homeostasis Loss of predators can cause plagues Loss of keystone species can collapse food webs and ecological communities

(Source: Green 1985.)

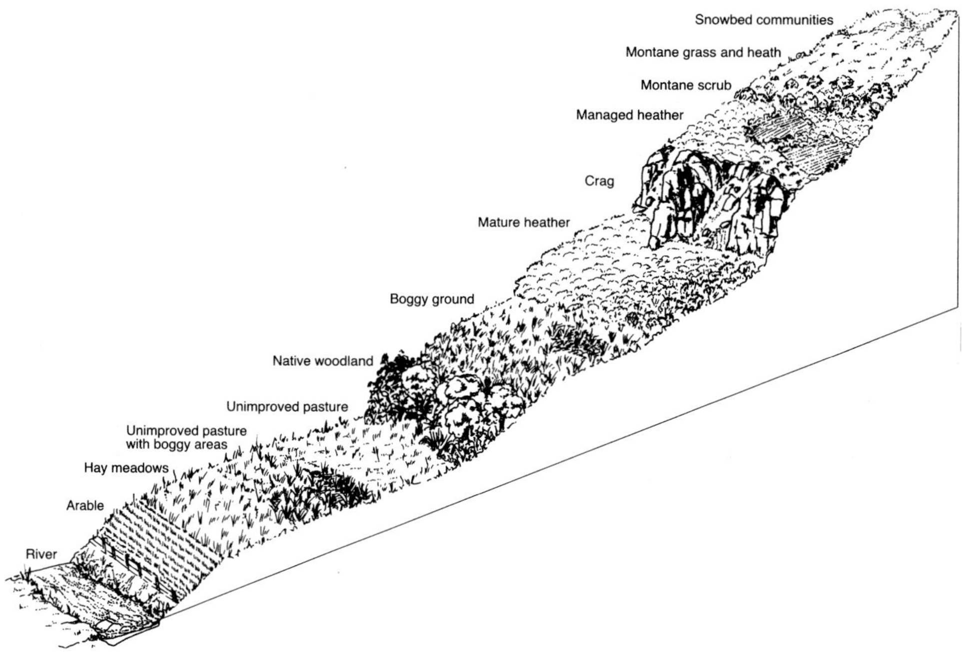

Figure 5.1 The Ecological landscape of Upland Farms.
Adapted with permission from Figure 1 Dodds *et al.* (1996) '*A Management Guide to Birds of Upland Farmland.*' RSPB: Sandy

Box XIX: Favourable Condition

Favourable Condition is a term used as part of the standard approach to monitor nature conservation throughout the UK. The monitoring regime is one of the main roles of the Joint Nature Conservation Committee (JNCC) who represent UK nature conservation internationally. Condition surveys are carried out on SSSIs, ASSIs (Areas of Special Scientific Interest in Northern Ireland), SACs and Ramsar sites (see Box XXI), although as Williams (2006) points out, the same process can be used on any land. A site may have one or several features of interest (species, habitats or earth science) and each is assessed separately as part of the condition survey. An assessment measures various characteristics of each feature against key attribute targets previously identified as crucial to the make up or maintenance of it. For example, with respect to montane moss heaths the following three attributes are evaluated:

- The cover of *Racomitrium* moss
- Mean depth of moss/lichen/dwarf shrub mat
- Grazing impact

Each attribute is set a target standard to achieve or exceed, all of which must be met to classify the habitat as in *favourable condition*. In this montane example the following standards are set for each attribute:

- *Racomitrium* cover must exceed 66% of the ground
- The mean depth must exceed 5cm across the area
- Grazing impacts should be light

Of interest to us is what is considered 'light' grazing.A range of observable phenomena are provided for the surveyor to evaluate which include: no evidence of grazing on any dwarf shrubs; almost no grazing on a range of identified grass species; less than 10% of fine leaved grass species favoured by stock by (eg sheeps fescue); negligible cover of broad leaved grasses (eg. common bent); sporadic appearance and no more of certain herb species; no *Juncus squarrosus*; no uprooted plants and less than 5 clusters of sheep pellets per 100m sq. – these phenomena suggest substantially less than one sheep per hectare. The standards are quite exacting but for montane heath, grazing pressure is one of the key reasons for its disappearance (see main section in this chapter about montane habitats).Whilst this seems relatively straightforward for this habitat, condition surveys for other habitats are far more complex; as is the case for dry dwarf shrub heath.

If the surveyor cannot meet all the key attribute standards, then the feature will be classified as in *unfavourable condition.* Both favourable and unfavourable condition can be recorded in different ways:

1. *Favourable maintained* – the feature has remained favourable since it was last surveyed
2. *Favourable recovered* – the feature has moved from unfavourable since it was last surveyed
3. *Unfavourable recovering* – the features is moving towards favourable condition, probably as a result of some form of change in its management
4. *Unfavourable declining* – the feature is deteriorating in its condition away from that of favourable, again probably the result of some form of management activity
5. *Unfavourable no change* – the feature is neither improving nor deteriorating

A final category also exists, *Destroyed (partially or completely)* which means that the feature is no longer present and there is little or no chance of bringing it back.

Part of the JNCC's responsibility is to report on the national picture of condition every six years.As the scheme was only piloted in 1998, there has so far, been only one six-yearly review for 2005.In 2003, the English Government set up a public service agreement (PSAs[1]) with the then English Nature, which stipulated a target of 95% of SSSIs to reach favourable condition by 2010. Initial evidence showed that the baseline from which they were starting was so low that this target was unachievable by 2010. A period of negotiation ensued, out of which the 95% target remained but would now include both *favourable* and *unfavourable recovering* accounting towards it (DETR 2004).Looking briefly at the six-year 2005 review around 57% of the total possible features where reported on (Williams 2006).From the graph below we can see that generally the UK uplands were

[1] Public Service Agreement – these were developed by the then Government to demonstrate that public bodies were spending public money properly to achieve set objectives

in more favourable condition than the lowlands, but further analysis shows that many upland habitats were in relatively poor shape, particularly montane grasslands and heaths with only 43% in favourable or unfavourable recovering condition.

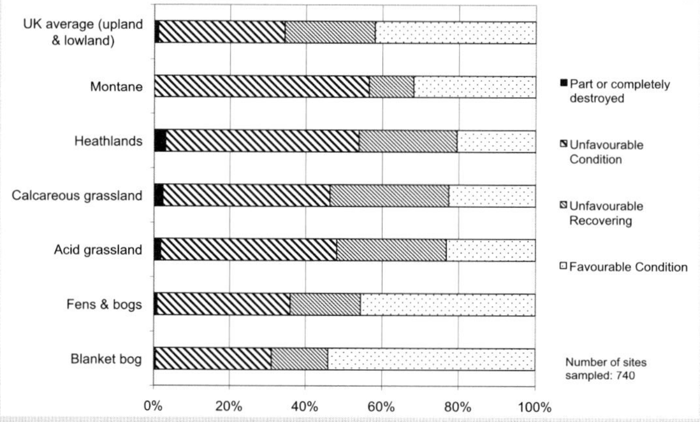

Condition Status of Upland Habitats 2006

(Based on Table 3 *'First 6-year Report'* © Crown Copyright. http://jncc.defra.gov.uk/page-3595)

The arrival of the PSA and the six year review were timely. Concurrently, agri-environment initiatives had just gone through the Mid Term Review (see Chapter 8).These different strands coalesced presenting the opportunity to use the Higher Level Environmental Stewardship Scheme to support the achievement of the PSA by creating ecological buffering on land adjacent to designated nature conservation sites.As a result agri-environment scheme management switched to Natural England from DEFRA, and HLS baseline site assessment guidelines emerged that are very much based on the key attributes of the SSSI condition surveys (Natural England 2008).Unfortunately the level of technical botanical knowledge needed to conduct these baseline surveys is beyond most farmers requiring the employment of ecological consultants (see Chapter 8)

conclusion, clearly these semi-natural habitats can be sustained only if we perpetuate the management practices that produced them. This is what environmental and agri-environmental initiatives of the late 20th and early 21st century have sought to encourage (e.g. Whitby *et al.* 1994). In other words, we <u>need</u> upland farming if we wish to maintain the ecological diversity of the bulk of our upland landscapes.

In the European Union, upland and mountain agriculture is seen as a way to bolster compliance with the 1992 Habitats Directive (Hopkins 1996), since 42 habitats in Annex 1 and eight species from Annex 2 are found only in upland environments. Some habitats, for instance limestone pavements, are very limited geologically; others, such as blanket bog, are extensive. In many cases, favourable condition (see Box XIX) cannot be maintained by designation alone; for example, hay meadows require continued agricultural management (Hughes and Huntley 1988) whereas montane willow scrub is damaged by agricultural activity (Britton *et al.* 2005). Some habitats remain unaffected by farming, for example rocky shelf vegetation. Each habitat interacts differently with agriculture, so farmers and conservationists need to understand each other's fundamental philosophy and management issues if a way forward is to be found.

Upland vegetation

Almost 99% of vegetation in the British Isles is a function of land use, firstly for food, timber and minerals, then for recreation, water and energy. Most vegetation is therefore 'semi-natural' (Tansley 1939) and such a habitat is:

> Composed of native species with structural affinities and community processes corresponding to those of natural vegetation, although the abundance of species, and in some cases the actual species composition, has been changed by man's activities. (Backshall *et al.* 2001 section 2:3)

Along with Table 5.2, Figures 5.2 and 5.3 show the extent of various upland habitats. Following the classification of Haines-Young *et al.* (2003) discussed in Chapter 2, environmental zones 3, 5 and 6 are upland; environmental zones 1, 2 and 4 relate to lowlands.

Lowland Britain has far more improved agricultural land and less of the semi-natural habitats, excluding calcareous and neutral grassland (of which it has 58.3% and 78.8% respectively, both geologically controlled) and broadleaved woodland, its climatic climax community (of which it has 76.9%). In contrast, the upland environmental zones support:

- acid grassland (87.1% of it, or 1.13 million ha)
- coniferous woodland (71.8% or 972,000 ha)
- bog (93.6% or 2.08 million ha)
- dwarf shrub heath (89.4% or 1.33 million ha)
- fen, marsh and swamp (68.1% or 372,000 ha)
- bracken (66.5% or 292,000 ha)
- inland rock (51% or 34,000 ha)
- montane habitat (100% or 49,000 ha)

However, not all this land cover is semi-natural; for example, a large percentage of upland coniferous woodland was planted in the 20th century and bracken has invaded as a result of changes in land use and management (National Forest Inventory website, accessed 06/04/2011).

If the three upland environmental zones are analysed in more detail, the pattern becomes more complex, but some general trends can be identified (Table 5.2):

> Environmental Zone 3 contains much more of the improved agricultural land; this parallels the notion that England has the least extent of Less Favoured Areas. This zone also has the largest proportion of bracken (38.7%, mainly in Wales) and the second largest percentage of acid grassland (35.8%). These are related because overgrazing or 'improvement' of heather moorland has encouraged the spread of bracken and grassland.

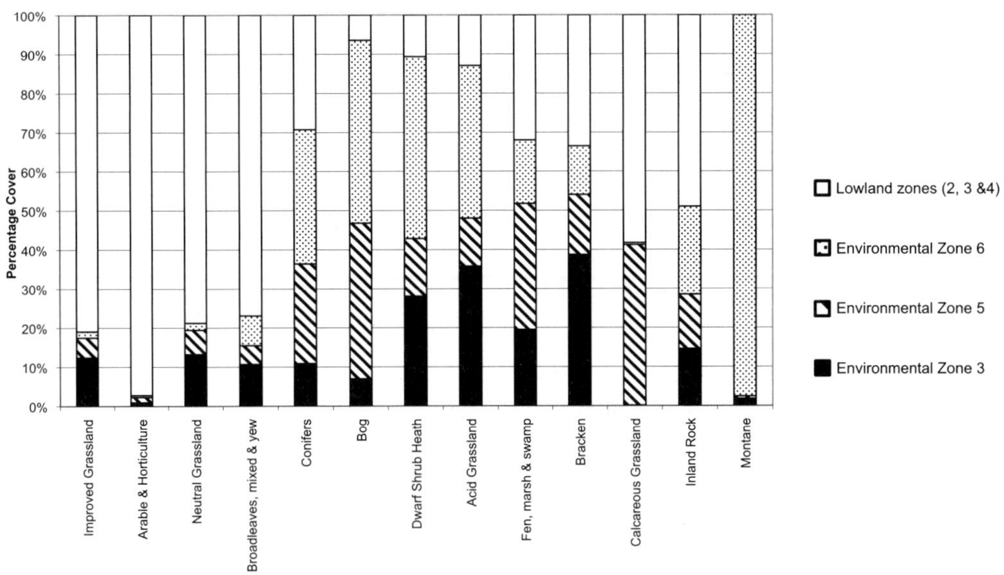

Figure 5.2 Broad Habitat Types in Britain 2000
Adapted from CS2000, www.CS2000.org, accessed on: 06/02/06. Acknowledgement: Countryside Survey data owned by NERC – Centre for Ecology & Hydrology.

Table 5.2: Broad Habitat Types in the Upland Environmental Zones as defined by CS2000

broad habitat	*zone 3*		*zone 5*		*zone 6*	
	000s of ha	*%*	*000s of ha*	*%*	*000s of ha*	*%*
Improved grassland	730	28.2	299	12.0	92	2.9
Arable and horticultural	54	2.1	100	4.0	4	0.1
Neutral grass	81	3.1	38	1.5	11	0.3
Broadleaf, mixed and yew	157	6.1	71	2.8	111	3.5
Conifer woodland	150	5.8	351	14.1	472	14.8
Bog	154	5.9	884	35.4	1037	32.4
Dwarf shrub heath	417	16.1	220	8.8	691	21.6
Acid grassland	464	17.9	159	6.4	506	15.8
Fen, marsh and swamp	107	4.1	177	7.1	89	2.8
Bracken	170	6.6	67	2.7	55	1.7
Calcareous grassland	0	0.0	27	1.1	0	0
Inland rock	8	0.3	8	0.3	13	0.4
Montane	1	0.03	0	0	48	1.5
Standing water and canals	23	0.9	36	1.4	43	1.3
Rivers and streams	7	0.3	8	0.3	7	0.2
Built-up and gardens	28	1.1	30	1.2	8	0.3
Boundary and linear features	39	1.5	23	0.9	11	0.3
Totals	**2590**	**100**	**2498**	**100**	**3198**	**100**

(Source: adapted from Haines-Young *et al.* 2003: 270.)

Acknowledgement: Countryside Survey data owned by NERC – Centre for Ecology & Hydrology.

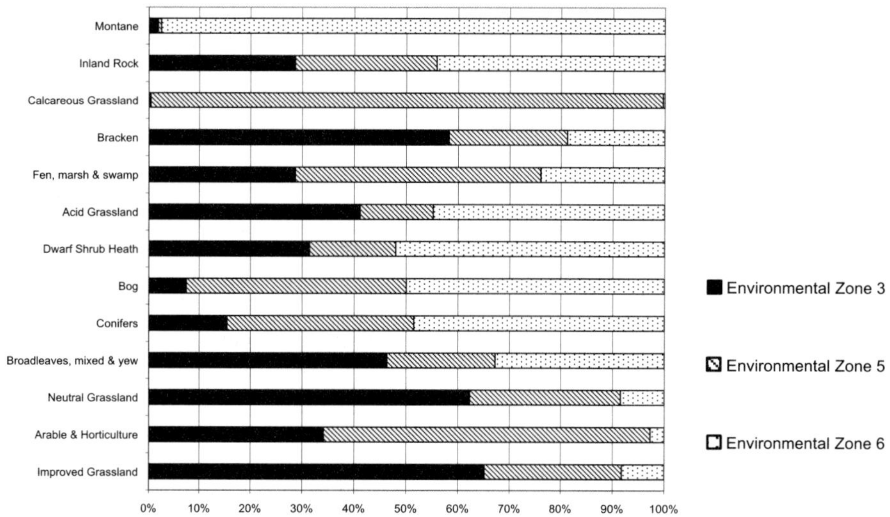

Figure 5.3 Broad Habitat Types by Environmental Zone in Britain 2000
(Adapted from Haines-Young *et al.*, 2003:270). Acknowledgement: Countryside Survey data owned by NERC – Centre for Ecology & Hydrology.

Environmental Zone 6 contains the greatest proportion of 'upland' semi-natural habitats of the three zones, notably bog, dwarf shrub heath and acid grassland, and nearly all the montane habitat for the entire UK (97.6%).

Environmental Zone 5 has the highest combined proportion of wet environments (bog and fen/marsh/swamp) and most of the remaining calcareous grassland outside the lowland.

What these data do not show, however, is the quality of these habitats, something which is important from a conservation point of view.

Quality is a function of diversity and the typicalness of the habitat in comparison to an ideal one (Ratcliffe 1977). For designated and protected land in Britain this diversity is measured against what is known as favourable condition (Box XIX). Upland farming activity can impoverish (decrease species diversity) semi-natural plant communities simply by increasing stocking densities, adding fertilisers or implementing soil improvements of agricultural value. Put simply, any procedure that moves plant communities away from their low-nutrient/low-utilisation status will decrease species diversity. Thus even though the British Isles has 1.3 million hectares of heather moorland, nearly 240,000 ha is in poor ecological condition (JNCC 1999). Rarity is as important as typicalness: some communities of plants and animals may be relatively abundant in the UK, but are internationally rare, such as limestone pavement (inland rock), raised and blanket bogs (bogs) and upland Atlantic oakwood (English Nature 2001).

The remainder of this chapter considers in greater detail a range of habitats which typically occur on upland farms – montane, woodland, dwarf shrub heath, grassland, wetlands and inland rock – looking at their general character, distribution and management issues in relation to agriculture. This review is by no means exhaustive and more detail can be found in other sources such as Backshall *et al.* (2001); Habitat Action Plans (JNCC website at www.jncc.gov.uk) or Rodwell *et al.* (1991, 1992, 1993, 1994, 1995).

Montane habitats

As Chapter 2 demonstrates, the uplands of Britain only just reach into the montane or alpine zone. This contains some of the harshest environments in Britain and biodiversity is limited to species that have adapted in various ways to these conditions (Table 5.3).

The occurrence of montane vegetation is influenced mainly by climate, then regional variations in topography and geology, and finally land use (Thompson and Brown 1992). The unique vegetation on the Cairngorm mountain tops, for example, is controlled by the

Table 5.3: Plant Adaptations to Alpine Environments

strategy	*adaptation*
Life form	Perennial – longer time to build energy reserves or avoid poor years Prostrate – to reduce wind and frost damage Most biomass underground – two- to six-fold that on the surface
Seed dormancy	Long dormancy – germinate in optimal conditions Seeds ripen in a second year of existence
Seedling establishment	Rare Very slow – taking until second year to put out seed leaves
Flowering	Flower buds pre-formed in the year before
Pollination	Insect mainly

(Source: Korner 2003.)

geomorphological effects of wind and the periglacial processes operating there (see Chapter 4). Alpine soils are either lithosols or lithomorphic.

In spite of the harsh environment, plant and animal diversity are surprisingly high in the montane zone. Such diversity allows the exploitation of all possible microniches, makes it likelier that some species will survive if conditions alter slightly and avoids the extinction vortex whereby the fewer the species, the lower the chances of the community surviving (Korner 2003). This diversity is in contrast to the traditional view of ecological succession, that many alpine areas are held in the pioneer stage when biodiversity is lowest.

Characteristics

In the British Isles the middle alpine zone occurs over 900m asl. It is dominated by sedges, rushes and snow-tolerant mosses, in complete contrast to the rest of north-western Europe where large tracts of bilberry exist at these altitudes. The harsher oceanic climate of the British uplands creates plant communities similar to upper alpine environments at higher altitude in continental Europe. Snow lie has a particular role in controlling plant communities (a role which increases with altitude) along with variations in precipitation, shelter, altitude and grazing pressure. According to Drewitt and Manley (1997) there are fifteen types of plant community in the English alpine zone, among them heaths, mires, springs, grasslands, ledge and periglacial terrain. Table 5.4 summarises the characteristics of four montane communities, using the National Vegetation Classification (NVC; see Box XX).

Snow can lie for long periods in sheltered hollows, developing communities like U11, whereas exposed summits develop fell field communities of windswept rocky debris and small patches of tolerant vegetation (U13) (see Table 5.4).

The lower alpine zone occurs between 900 and 600m asl. It is in this zone that shrubs make their first appearance, although many grow in prostrate form (along the ground) to avoid high winds and exposure. There are three typical plant communities collectively referred to as montane heath (Table 5.4). Bilberry is the dominant vegetation, though much of it is confined to rocky ledges in response to grazing pressure.

At the base of the lower alpine zone is the tree line (see Box IV). At this level, what are usually understorey woodland shrubs can begin to form a canopy, notably dwarf birch (*Betula nana*), montane willows (*Salix* spp.), rock whitebeam (*Sorbus rupicola*) and juniper (*Juniperus communis communis)*. European ecologists refer to this vegetation zone as *krumholz* ('twisted wood'). In Britain it is known as montane scrub (Scott 2000).

The willows that grow in the montane scrub are perhaps of most interest. They form the southernmost extension of the arctic-alpine vegetation normally found in Scandinavia (Ratcliffe 1977). There are eight main species, of which the woolly willow (*Salix lanata*) and downy willow (*Salix lapponum*) are distinctly rare (Scott 2005). A survey in 1994 by Wigginton (1999) found only 14 sites where woolly willow grew, with only five viable colonies of between 30 and 1000 individuals. The scrub provides food and shelter for birds like black grouse (*Tetrao tetrix*), redwing (*Turdus iliacus*) and fieldfare (*Turdus pilaris*). Along with montane willow, another dioecious[1] plant grows in the British uplands, though it grows elsewhere too. Juniper, one of only three conifers native to Britain, grows at all altitudes to 1013m asl, though the dwarf sub-species is confined to montane habitats (Plantlife 2004). It is typically found on grassland or moorland (W19, H15 on Table 5.5), but it grows in a range of environments including montane, maritime heath and woodland.

These alpine and tree-line communities support a range of invertebrates that have arctic-alpine affinities, making them nationally rare, like the wolf spider (*Tricca alpigena*) and

[1]. Dioecious – male and female reproductive organs exist on different individual trees.

Box XX: National Vegetation Classification (NVC)

The NVC was devised as a standard way to classify vegetation in Britain (Rodwell 1993). Extensive fieldwork was carried out in the late 1980s and 1990s to identify the structure of plant communities.

A straightforward methodology, using quadrat surveys, was designed so that new surveys could be completed and the data compared to standardised tables, to allow vegetation to be matched against them for classification. Quadrat sizes varied, depending on the type of vegetation under inspection:

- 2 × 2m – short herbs and dwarf shrub heath
- 4 × 4m – open tall herbs/ sub-shrub heaths/ low woodland field layers
- 10 × 10m – dense shrubs
- 50 × 50m – woodland canopy

For each taxon recorded in the quadrat, its abundance (the percentage area covered by the plant in the quadrat) was recorded using the DOMIN scale (these are the bracketed values on the floristic tables):

%	*numbers*	*DOMIN value*
91–100		10
76–90		9
51–75		8
34–50		7
26–33		6
11–25		5
4–10		4
<4	many individuals	3
<4	several individuals	2
<4	few individuals	1

A comprehensive site description was generated for each surveyed site. This recorded a wide range of physical variables, such as slope angle, geology, soil depth, pH and soil type. Five sample quadrats were taken at each site and, using a software package known as VESPAN, the frequency (how often the plant is found from one quadrat to the next, irrespective of its abundance) of their occurrence on the site was averaged:

frequency	*% of sites*	*description*
V	81–100	constant
IV	61–80	constant
III	41–60	common/frequent
II	21–40	occasional
I	1–20	rare

Thus a frequency of V or IV means the plant occurs constantly in any quadrat surveyed in that type of vegetation. From this information a floristic table was created, which recorded the typical combinations of plants, their abundance and frequency (see below; Rodwell *et al.* 1992). It took six volumes to summarise the voluminous data from this project (Rodwell *et al.* 1991a/b, 1992, 1993, 1994, 1995).

	sub-community a	*sub-community b*	*U8*
Carex bigelowii	V (1–10)	V (1– 10)	V (1–10)
Polytrichum alpinum	V (1–6)	V (1–4)	V (1–6)
Cladonia arbuscula	V (1–4)	V (1–4)	V (1–4)
Dicranum fuscescens	V (1–4)	V (6–10)	V (1–10)
Cetraria islandica	IV (1–3)	IV (1–3)	IV (1–3)
Cladonia bellidiflora	IV (1–3)	IV (1–3)	IV (1–3)
Ptilidium ciliare	IV (1–4)	II (1–4)	III (1–4)
Cladonia pyxidata	IV (1–3)	II (1–3)	III (1–3)
Dicranum scoparium	III (1–3)	I (1–3)	II (1–3)
Deschampsia cespitosa	III (1–4)	I (1)	II (1–4)
Cetraria delisei	III (1–3)		II (1–3)
Caldonia coccifera	II (1–3)	I (1)	II (1–3)
Salix herbacea	II (1–6)		I (1–6)
Lophozia alpestre	II (1–3)		I (1–3)
Luzula spicata	II (1–3)		I (1–3)
Drepanocladus uncinatus	II (1–3)		I (1–3)
Hylocomium splendens	I (1–3)		I (1–3)
Racomitrium canescens	I (1–3)		I (1–3)
Juncus trifidus	I (1–3)		I (1–3)
Racomitrium lanuginosum	II (1–3)	V (1–5)	III (1–5)
Festuca ovina/vivpara	I (1–3)	III (1–6)	II (1–6)
Vaccinium myrtillus	I (1–3)	III (1–4)	II (1–4)
Agrostis capillaries	II (1–3)	III (1–6)	II (1–6)
Galium saxatile	I (1)	II (1–4)	II (1–4)
Cladonia rangeferina	I (1)	II (1–3)	I (1–3)
Cornicularia aculeate		II (1–3)	I (1–3)
Cladonia macrophylla		I (1–3)	I (1–3)
Caldonia uncialis	III (1–3)	III (1–3)	III (1–3)
Deschampsia flexuosa	II (1 6)	II (1–4)	II (1–6)
Cladonia gracilis	II (1–3)	II (1–3)	II (1–3)
Pohlia nutans	I (1–3)	I (1–3)	I (1–3)
Pleurozium schreberi	I (1–2)	I (1–4)	I (1–4)
Huperzia selago	I (1–3)	I (1–3)	I (1–3)
Nardus stricta	I (1–3)	I (1–3)	I (1–3)
Racomitrium fasciculrare	I (1–3)	I (1–3)	I (1–3)
Caldonia squamosa	I (1–3)	I (1)	I (1–3)
Caldonia floerkeana	I (1–3)	I (1)	I (1–3)
Saxifraga stellaris	I (1)	I (1)	I (1)
Ochrolechia frigida	I (1–3)	I (1–3)	I (1–3)
Sphagnum compctum	I (1)	I (1)	I (1)
Number of samples	11	16	27
Number of species/sample	17 (11–26)	13 (8–18)	15 (8–26)
Vegetation height (cm)	4 (2–5)	5 (2–12)	5 (2–12)
Vegetation cover (%)	99 (95–100)	98 (90–100)	99 (90–100)
Altitude (m)	1050 (853–1235)	1066 (823–1250)	1059 (823–1250)
Slope (o)	5 (0–15)	5 (0–10)	5 (0–15)

CONSTANTS – species that always occur and define the community code

SUB COMMUNITIES – species listed that occur together with the constants to form a sub-community

ASSOCIATES – species that occur continually in the community but at very low levels

Note: Bracketed values represent the range found in the samples.

The plant communities are organised into 12 main types, each denoted by a prefix:

habitat	*prefix*	*communities*
Woodlands	W	25
Heaths	H	22
Mires	M	38
Calcifugous grasslands & Montane	U	21
Mesotrophic grasslands	MG	13
Calcicolous grasslands	CG	14
Aquatic	A	24
Swamp & Tall herb fens	S	28
Salt marsh	SM	28
Shingle, Strandline & Sand dune	SD	19
Maritime cliff	MC	12
Open habitats	OV	42

Some communities have sub-communities (e.g. U8a and U8b). Here certain species are constant in one community but not in another, though overall the dominant species are the same. In the NVC guides, each community has a description under these standard headings:

- Synonomy – comparison with previous classification systems
- Constant species – species always present in the community
- Rare species – species seldom present in the community
- Physiognomy – a description of what the community looks like
- Sub-communities – descriptions of the plants found in each sub-community
- Habitat – physical parameters (like rainfall, snow cover, altitude, soil) typical of the community
- Zonation and succession – the transition from one community to the next
- Distribution – location in the United Kingdom
- Affinities – more detail on similar vegetation classifications
- Floral table – like the example (U8) here
- Maps of community and sub-community distribution

The uplands contain at least 10 woodland, 15 heath, 30 mire and 27 grassland communities (Fielding and Howarth 1999). For more detail, see Tables 5.4, 5.5, 5.6, 5.8 and 5.11.

The development of the NVC generated much debate in nature conservation and countryside management circles. A particular issue was the evolution of 'recipe book management' whereby anyone knowing the NVC code of a plant community can follow a standard set of management objectives to reach favourable condition. The computer software known as Countryside Management System has adopted this approach. One criticism of this method is that, by standardising management, the number of British plant communities will remain fixed by the number found in the NVC. Another fear is that managers may come to rely on software telling them the answer, which may not be appropriate for the unique management problem or site character they face, a problem Rodwell himself never anticipated. Nonetheless the NVC has much to offer in understanding and simplifying the complex relationships between plant communities and the wider environment.

Table 5.4: Alpine Vegetation: montane heaths and related vegetation

NVC	*dominant species*	*environment*	*examples*	*fauna*	*zone*
U9a	*Juncus trifidus* *Racomitrium -* *heterostichum* rush heath	Exposed plateaux and summits Minimal snow cover due to wind Precipitation 1600mm pa 1000 to 1200m asl lithosols	Central Cairngorms	**Mammals:** sheep deer feral goats mountain hare voles mice shrews foxes weasels stoats **Birds:** ptarmigan dotterel snow bunting golden plover dunlin snowy owl **Invertebrates:** Diptera Lepidoptera	MIDDLE ALPINE
U8	*Carex bigelowii* *Polytrichum* *alpinum* sedge heath	Flat or sloping ground 100+ days of snow cover Precipitation 1600 to 3200mm pa 800 to 1200 m asl stagnopodsols	Cairngorms Ben Lawers		
U12	*Salix herbacea* *Racomitrium* *heterostichum* snow bed	100+ days of snow cover Precipitation 1600 to 3200 mm pa Above 900m asl (in snow beds only) gleyed lithosols	Isle of Mull		
U11	*Polytrichum* *norvegicum -* *Kaieria starkeii* snow bed	100 + days of snow cover Precipitation 1600 to 3200mm pa above 900m asl (in snow beds only) lithosols	Central Highlands east of Ben Nevis		
U7	*Nardu stricta* *Carex bigelowii* grass heath	50 to 1200 days of snow cover Precipitation 1600mm pa 600m to 900 m asl stagnopodsols	Lake District Southern Uplands		MONTANE HEATH LOWER ALPINE
U10	*Carex bigelowii* *Racomitrium*| *langunisom* moss heath	Broad windswept plateaux 500m in E to 900m asl in W lithosols	Lake District Southern Uplands N. Pennines N. Wales		
H19	*Vaccinium myrtillus* *Cladonia arbuscula* *heath*	Sheltered areas < 100 days of snow cover Precipitation 1600mm pa 660 to 1000m asl rankers	Skiddaw Cross Fell cent. & east Highlands		

(Sources: Rodwell *et al.* 1991a, 1991b, 1992; Drewitt and Manley 1997.)

the beetle *Amara alpina*. Montane birds, such as the dotterel *(Eudromias morinellos)*, snow bunting *(Plectrophenax nivalis)* and ptarmigan *(Lapogus mutus)*, are limited to this environment (Gordon *et al.* 1998). Ptarmigan were present in the Lake District until the late 1700s and in the Southern Uplands until the 1800s. Their distribution is now restricted to the Scottish Highlands following habitat deterioration from overgrazing, recreation and predation pressures (Galbraith *et al.* 1988; Thompson and Brown 1992).

According to Countryside Survey 2000, there are about 67,000 ha (3% of the land area) of montane habitat remaining, almost all (66,000 ha) in Zone 6 (in the Scottish Highlands and outliers in the Lake District and Snowdonia). Between 1984 and 1998, the amount of montane vegetation in Zone 3 did not change, but it expanded by nearly 40% in Zone 6 (Haines-Young *et al.* 2003). The best examples of montane snow bed, fell field and montane heath are found in Scotland, around Ben Lawers (Thompson and Brown 1992) and the Cairngorms (Gordon *et al.* 1998). Other notable communities exist on the Carneddau massif in North Wales, the most southerly montane vegetation in the UK (Britton *et al.* 2005). The outliers are nowhere near as pristine as those in Scotland, owing to land-use pressure and access issues (JNCC 1995).

Management Issues

Montane scrub now covers about 1500 ha; once it was 2000 ha (Mackenzie 2000). The expansion of shieling grounds led to rapid overgrazing of these slow-growing habitats, which could not recover quickly enough from the presence of cattle and sheep. The swift increase of red deer in the 20th century threatened to remove them altogether. Juniper has suffered latterly, with roughly one third of sites being lost between 1987 and 1999 (Plantlife 2004). Over 60% of the remaining plants are mature or old, with much of the younger growth dating back to the 1950s myxomatosis outbreak, when grazing pressure collapsed.

Montane vegetation are mainly climatic climax communities and thus should not require management. However, British montane areas have been modified (to a certain extent) by human activity. Grazing by sheep and browsing by deer, heather management, recreation pressure and public access are all challenges faced by these alpine habitats. Particular long-term concerns are the effect of global warming on species adapted to cooler climates and problems of environmental pollution (Grieve 2001). Thus there is some debate as to whether any form of management will enable their recovery (Backshall *et al.* 2001).

The Carneddau plateau in North Wales has especially suffered from a combination of grazing pressure, from deer and sheep, and nitrogenous atmospheric pollution. The latter has retarded the recovery ability of *Racomitrium* from grazing pressure by changing the soil chemistry inappropriately (Britton *et al.* 2005). The horizontal stems of the prostrate ling and *Racomitrium* moss are also trampled as stock graze on adjacent grasses and sedges (Britton and Pearce 2004). Similar problems have been encountered in Glas Maol in Scotland where a successional continuum develops, changing vegetation from fern-dominant to heath-dominant, then to grass–sedge/moss mixes as grazing pressure increases (Welch *et al.* 2005).

Apart from the long-term concerns of climate change and atmospheric pollution, grazing management needs to take account of the naturally slow growth of montane heath species. Carrying capacities are much lower than at lesser altitudes, and sometimes the only solution is to remove stock altogether to allow amelioration and then to re-introduce animals gradually until numbers reach an appropriate level. What that level is, unfortunately, is a matter of trial and error over several growing seasons. Hefting systems do not aid the recovery of vegetation because sheep will tend to congregate in sheltered areas or move onto grassland patches to seek food, so extra walling, fencing or active shepherding may be needed. This can be technically difficult, owing to lack of soil in which to sink fence posts, expensive (if large areas have to be fenced off) and less than straightforward if there are questions of common rights (see Box III). Britton *et al.* (2005) suggest that changes in soil chemistry mean it is better to target sheep reductions on the less degraded montane heath rather than on seriously degraded areas, simply because the soil chemistry needs to improve before grazing reduction will have any effect.

Particular management challenges face willow and juniper populations of the montane scrub (Long and Williams 2006). These dioecious plants have the disadvantage that they need several individuals of each sex to reproduce or else rely on vegetative reproduction (with new aerial growth tips getting rubbed off quickly). Many remnant areas of willow, for example, are simply too small to support a viable sex ratio, which is exacerbated by sheep preferring to eat the male rather than female foliage! For juniper, complete grubbing out – either to improve pasture, or by sheep seeking shelter and nutrition from the plants – has left many populations fragmented, plants defoliated, bark damaged, limbs broken and roots trampled (Clifton *et al.* 1995). Juniper is also placed under stress by rabbits grazing and undermining it, bank erosion, an ageing population (with reduced fruiting ability) and shade intolerance through ecological succession to woodland. All types of montane scrub are also under threat from climate change (Hester and Brooker 2007).

To overcome these issues for montane scrub, Scottish Natural Heritage and the Highlands Birchwood Trust conducted extensive joint research projects (Gilbert 2006) and suggested these initial management techniques to stabilise and expand willow populations:

- Re-introduce missing sexes through plant propagation so that each site has at least 50 individuals of each sex

- Ease grazing/browsing pressure by reducing stocking densities and erecting exclosures (keeping stock and people out)

- Provide information for landowners

These approaches, trialled extensively at Ben Lawers, have successfully created a natural scrub-like landscape with supporting flora and evolving insect populations (Mardon 1997; 2003; 2007).

In contrast, juniper management focuses on encouraging natural regeneration and reducing damage to existing populations (Clifton *et al.* 1997). Prostrate juniper is usually restored by controlled burning/grazing. Common juniper seeds require bare ground, light summer grazing only and the application of shrub guards (Plantlife 2005). If natural regeneration is not possible, then seeds, cuttings or plantlets can be used. New plants or existing populations can be fenced into exclosures, but this can be costly, affect sheep heft management and lead to bird collisions and snow build-up, allowing access to herbivores.

Given its great rarity and struggle for survival, woolly willow and its community seems an ideal candidate for legislative protection under the UK Biodiversity Action Plan (BAP) invoked by the EC Habitats and Species Directive (Box XXI). Woolly willow does indeed have its own Species Action Plan (Joint Nature Conservancy Council 1998), but montane scrub did *not* have a Habitat Action Plan (HAP) until 2008. Scott (2005) suggested this was because the habitat is so rare and small that it simply does not show up on the 'BAP prioritisation radar'. He also postulates that the ecotone nature of the scrub, between alpine and sub-alpine zones, means it gets lost in the compartmentalised BAP system, which deals inadequately with transitional habitats. At the end of 2008, JNCC produced a mountain heaths and willow scrub HAP covering 18 NVC communities, which goes some way to justify Scott's view (www.ukbap.org.uk/PriorityHabitats.aspx – accessed 13/01/2009).

From an agricultural viewpoint there are benefits in maintaining montane vegetation. At a fundamental level, retaining any vegetation in uplands slows the processes of soil erosion, particularly in montane areas where the soils are thinner, more fragile and easily affected by human or animal activity. Any vegetation at this altitude offers more forage for light grazing regimes in the upper reaches of hefts. Montane scrub can also help shade out less desirable forage grasses such as mat grass. Scrub retains snow later into the year, increasing soil moisture and thus encouraging more plant growth than in open grassland. There are also diversification opportunities in visitor experience, cover for shoots and other ecosystem services

Box XXI: Nature conservation legislation and designation

In the UK and EU there are many forms of legislation that protect natural and semi-natural vegetation. Comprehensive reviews of its intricate development can be read in Evans (1992) and Reid (2009). The European Union adopted the Habitats Directive (Directive 92/43/EEC on the conservation of natural habitats and of wild fauna and flora) in 1992, aiming to:

- maintain and restore the extent and quality of rare habitat types,
- ensure rare species survive and maintain their populations and natural range in the long term.

The directive includes a number of annexes that list habitat types and species considered rare, against which each member state has been required to submit an inventory.

Habitat types are classified by a system called CORINE, which does not directly transpose to the NVC. A further complication is that the annexes use an earlier CORINE coding system, later revised. Examples are shown in the table here.

Comparison of CORINE and NVC using various upland examples

original CORINE code	*current CORINE code*	*Annex 1 habitat type*	*NVC code*
31.11	4010	Northern Atlantic wet heaths with *Erica tetralix*	H5, M14, M15, M16
31.2	4030	Dry heaths (all types)	H1–H4, H7–H10, H21, H16, H18, H21
31.4	4060	Alpine and Sub-alpine heaths	H10, H12–H22
31.622	4080	Sub-arctic willow scrub	W20
36.32	6150	Siliceous alpine and boreal grassland (natural)	U7–U12, U14
36.41	6170	Alpine calcareous grasslands (natural)	CG12–CG14
38.3	6520	Mountain hay meadows	MG3
52.1, 52.2	7130	Blanket bog	M1, M15, M17–M20, M25 (all in part)
42.51	91C0	Caledonian Forest	W18 and some of W4, W17, W19

CORINE codes are gradually being replaced by a Europe-wide classification known as EUNIS (European Union Nature Information System). This five-level hierarchy starts with broad habitat types, such as Heathland, Scrub and Tundra (Level 1: code F); this is divided into 11 categories of which Temperate shrub heath (Level 2: F4) contains three classes (Level 3: wet heath F4.1, dry heath, F4.2 and Macronesian heaths F4.3). The 'dry heaths' class contains six groups, of which Atlantic *Erica-Ulex* heaths (Level 4) represents NVC classes H2, H3, H4 and H8 (http://eunis.eea.europa.eu/habitats-code-browser.jsp?expand=#level_F – accessed 19/01/2009).

The habitats and species listed in Annex 1 are to be protected by designation of Special Areas of Conservation (SACs), part of the Natura 2000 network, which also encompasses Special Protection Areas (SPAs) created under the Birds Directive 79/409/EEC in 1979. Member states then identify sites which best represent the range and variety of those habitats and species. Within the UK, SACs and SPAs are drawn from our own pool of legislatively protected sites – Sites of Special Scientific Interest and National Nature Reserves (designated under the National Parks and Access to the Countryside Act 1949, updated by the Wildlife and Countryside Act 1981/1986).

Running parallel to these developments has been the UK Government's response to the Convention on Biodiversity, ratified in 1992 (an output from the Rio Earth Summit). The Biodiversity Action Plan (BAP) lists species and habitats requiring protection and provides management plans for each. There are Species Action Plans (SAPs) for 391 species in the UK. In the uplands there are four priority Habitat Action Plans (HAPs): Blanket Bog; Upland Calcareous Grassland; Upland Heathland; Limestone Pavements. Within this framework local authorities can write their own BAPs, like *A Living Landscape: Biodiversity Action Plan for the Peak District* (www.ukbap.org.uk/lbap.aspx?ID=389 – accessed 20/01/2009). This BAP is a partnership of 36 organisations and identifies a range of targets and actions for seven species and eleven habitats found in the Peak District National Park area (four of which are specifically upland).

We can use the woolly willow to exemplify how these different pieces of legislation and procedure relate to one another. As discussed in this chapter, the woolly willow is a rare dioecious tree which grows exclusively in montane scrub in Britain. Its existence is threatened by grazing pressure and sex imbalance within the fragmented populations that remain. Its rarity and need for protection is recognised by the Wildlife and Countryside Act 1981, Schedule 8 (Plants that are protected) and it is listed in the Red Data Book.[1] The species has its own SAP (www.ukbap.org.uk/UKPlans.aspx?ID=556 – accessed 06/12/2006) and is protected and managed through SSSI management agreements and local BAPs, such as that for Stirling Council (2002). Under the Habitats Directive, the plant community in which the woolly willow grows is recognised as Sub-arctic willow scrub (4080) and is in the process of receiving a HAP.

---

[1] Red Data Books - texts containing lists of all rare and endangered species in the UK published by the Joint Nature Conservancy Council.

(see Chapter 9). On the downside, scrub rejuvenation can make gathering of sheep more difficult and time-consuming, and the cost of exclusion from recovering scrub may not outweigh the benefits of increased hillside shelter.

Woodland habitats

Because of exploitation, management and clearance, original natural woodland is now practically non-existent, confined to inaccessible gulleys and steep slopes, as in Wistman's Wood, Devon. Other woodland is secondary regeneration, which may be ancient (pre-1600) or modern (post-1600). Two types of upland woods have conservation value (Table 5.5):

- deciduous birch, ash and oak woods,
- native pinewoods.

Characteristics

Birch grows on free-draining acid soil that has poor fertility. Mostly found above 250m asl, birch woods cover 15,000 to 25,000 ha, in the north-west Highlands of Scotland, where cool, wet, windy weather prevents other species from dominating the woodland (Forestry Authority 1994a). In western areas, downy birch (*Betula pubescens*) grows as low as 150m asl on the thinnest soils. Moving eastwards, soil conditions improve to acid brown earths and woods become dominated by silver birch (*Betula pendula*). The two species also readily hybridise. These woods support a profusion of rare mosses and liverworts (W11 and W17 in the NVC).

Ash woods (*Fraxinus excelsior*) are found on neutral or alkaline soils in northern and western Britain, where the climate is cool, wet and windy. They are particularly associated with limestone areas such as the Pennines and Mendips. There are two main types, W8 and W9 (Table 5.5). W9 ash woods grow in the more oceanic north-western areas and on and around limestone pavement, whereas W8 has a more southerly distribution and the woodland community is more diverse (Forestry Authority 1994b). W9 woods are thought to be the remnants of secondary woodland resulting from reduced grazing, hence some were originally wood pasture[1]. Upland ash woods cover about 67,000 ha (English Nature 1998).

By far the greatest area of upland deciduous woodland is ancient oak, about 100,000 ha (English Nature 1998). They are classified as W11 or W17, so can be confused with upland birch woods. They are typical of northern and western Britain, and extensive areas can be found in the upland areas of Cornwall, Exmoor, Snowdonia, the Lake District, Argyll and the Western Isles, for instance Eigg.

Most upland oak woods are dominated by sessile oak (*Quercus petraea*); a few, like Wistman's Wood, consist of pedunculate oak (*Quercus robur*). One notable type of wood is western Atlantic oakwood, which is restricted by climate to the western seaboard. This type of wood has survived on areas under-utilised for agriculture because of their inaccessibility and poor grazing potential. Alongside oaks, a wide range of canopy and understorey trees support a unique assemblage of lower plants (ferns, lichens, mosses and liverworts) for which the woods are world renowned (Pankhurst 2007). Historically these woods were exploited for fuel and charcoal by coppicing and used for tanning materials. Many of the trees have been pollarded since the Viking era.

Topographically these three types of deciduous woodland are complementary. Ash prefers the dampness of the lower slopes and valley floors; further upslope, oaks emerge where the soil is more free-draining; finally, birch woods extend onto the moorland where the climate is more inclement (Figure 5.4). Each type can overlap the adjacent zone(s) in terms of a gradual tree line (see Chapter 2); thus ash and oak woods that exploit the microclimates of hillside ghylls may be found at higher-than-normal altitudes.

---

[1] Wood pasture – an area of treed grassland providing shelter and forage at the same time, historically much more extensive than now.

Table 5.5: Sub – Alpine Vegetation: Tree line and Woodlands

<table>
<tr><th>NVC</th><th>dominant species</th><th>environment</th><th>examples</th><th>fauna</th><th>zone</th></tr>
<tr><td>W20</td><td>Salix lapponum
Luzula Sylvatica
scrub</td><td>Dalradian limestone or schist
Steep shelving rock ledges
N- or E-facing slopes or corries
600m asl +
Late-lying snow
Calcic lithosols</td><td>Coire Cheap in Strathspey</td><td rowspan="3">Mammals:
As alpine habitats in Table 5.4

Birds:
black grouse
red grouse
snipe
woodcock

Specialist invertebrates:
sawflies
beetles
juniper carpet moth
pug moth</td><td rowspan="3">MONTANE SCRUB
TREE LINE</td></tr>
<tr><td>W19</td><td>Juniperus communis
spp. communis
Oxalis acetosella
wood</td><td>Open landscapes
240 to 1030m asl
Heavy rainfall
Free-draining, nutrient-poor lithomorphic soils</td><td>Yorkshire Dales
Cumbria Fells and Dales
North Pennines
Snowdonia
Scottish Highlands</td></tr>
<tr><td>H15</td><td>Calluna vulgaris
Juniperus communis
ssp communis
heath</td><td>Shady slopes
Snowblown areas
330 to 875m asl
Cool maritime climate</td><td>north-west Highlands
Scottish western Isles</td></tr>
<tr><td>W8</td><td>Fraxinus excelsior
Acer campestre
Mercurialis perennis
woodland
ash–maple–dog's mercury wood</td><td>Limestone, shale and clays – drift deposits
Precipitation 750 to 1000mm pa
rendzinas or calcic lithosols</td><td>Mendips
Wales
S. and N. Pennines
North York Moors</td><td rowspan="4">Mammals:
long-tailed field mice
badgers
polecats
pine martens
red squirrels
red & roe deer

Birds:
buzzard
red kite
greater spotted woodpecker
jay
pied flycatcher
wood warbler
tree pipit
redstart

Invertebrates:
butterflies
moths</td><td rowspan="5">WOODLANDS</td></tr>
<tr><td>W9</td><td>Fraxinus excelsior
Sorbus aucuparia
Mercurialis perennis
woodland
ash–rowan–dog's mercury wood</td><td>Carbonate-rich sedimentary rocks
Summers cool and days cloudy
Negligible snow lie
Precipitation >1200mm pa
rankers or rendzinas, depending on rainfall</td><td>Yorkshire Dales
Durham Dales
North York Moors
Skye
Moine Thrust</td></tr>
<tr><td>W11</td><td>Quercus petraea
Betula pubescens
sessile oak–downy birch–wood sorrel</td><td>180 to 450m asl
Cooloceanic climate
Precipitation >1200mm pa, 3000mm in extreme W
Typically unenclosed, allowing grazing
brown earth or brown podsolic earth</td><td>Lake District
North Wales
N. Pennines
NW Scottish seaboard</td></tr>
<tr><td>W17</td><td>Quercus petraea
Betula pubescens
dicranum moss
sessile oak–downy birch</td><td>Humid oceanic climate
Damp shady ravines up to 600m asl
Precipitation >1600mm pa
Typically unenclosed, allowing grazing
humic rankers and podsols</td><td>NW Scottish seaboard
Wales
Lake District
cent. Highlands
Pennines
Dartmoor</td></tr>
<tr><td>W18</td><td>Pinus sylvestris
Hyocomium spendens wood
Scots pine</td><td>Found as high as 600m asl
Precipitation 1600 to 3000mm pa
Deer and sheep readily graze
Podsols</td><td>cent. and NW Highlands of Scotland
NW Cairngorms</td><td>Mammals:
red squirrel
pine marten
wildcat
Birds:
capercaillie
crossbill
crested tit</td></tr>
</table>

(Source: Rodwell *et al.* 1991, Forestry Commission 2003abc, Plantlife 2004, 2005).

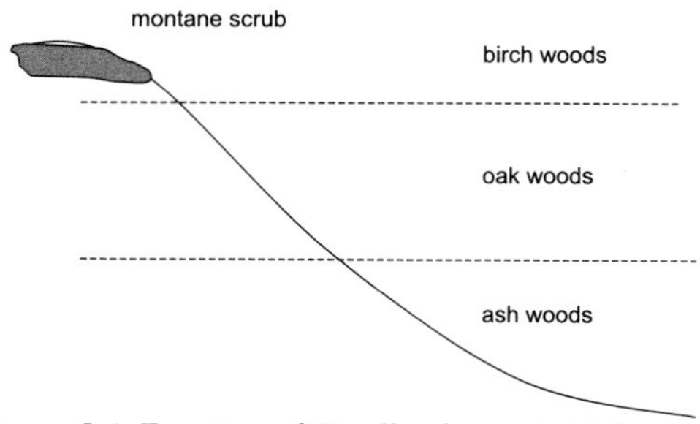

Figure 5.4 Zonation of Woodland types in Uplands

Known in the British Isles as Caledonian pine forest or native pinewood (W18 in Table 5.5), this woodland is ecologically limited to Scotland. It grows on podsolic soils in cooler parts of the Highlands to about 650m asl, though natural regeneration of the dwarf version has been found as high as 850m. The NVC recognises four sub-types as a result of climatic variation from west to east (Rodwell and Cooper 1995, cited in Mason *et al.* 2004).

Such woods are 95% Scots Pine (*Pinus sylvestris*) and Birch (*Betula* spp.), with a fairly impoverished understorey (31 species in all) of acid-tolerant bilberry (blaeberry) and ling. Mosses and liverworts are extensive, increasing in diversity from east to west as conditions become more oceanic. Scots pine is a pioneer species, requiring high levels of light and being intolerant of shade. These woodlands therefore tend to be made up of stands of even height, but not necessarily of similar age (Mason *et al.* 2004) though poor regeneration, caused by shade and grazing, can lead to even-aged trees (Gimingham 2002; Forestry Commission 1994d). They cover about 16,000 ha, mainly in the Highlands and Cairngorm fringes (Figure 5.5); it is thought they once covered 1.6m ha (Forestry Commission 1994d). Over half of the remaining wood occurs in six large blocks at Abernethy, Rothiemurchus, Glen Affric, Glen Tanar, Glen Strathfarrar and Mar.

Native pinewood also grows on ombrogenous bogs (ones supplied by rain-water only), but the high water table and poor nutrients stunt the pines to a height of 1m. Fragmentation of pinewoods has removed the gradation into other types of upland wood where soils become more fertile, with a trend towards W17, or where elevation increases, merging into juniper woodland of type W19 (Mason *et al.* 2004). Important fauna found in these woods include birds like the capercaillie *(Tetras rogallos)*, black grouse *(Tetrao tetrix)*, scottish crossbill *(Loxia curviostra)*, crested tit *(Parus cristatus)*, wryneck *(Jynx torquilla)* and spotted flycatcher *(Muscicapa striata)* and mammals, notably the red squirrel *(Sciurus vulgaris)* and pine marten *(Martes martes)*, though neither is restricted to pine forest. Over 60 species of Red Data book invertebrates are found in W18, benefiting from the amount of dead or rotting wood and the light level on the forest floor (Mason *et al.* 2004).

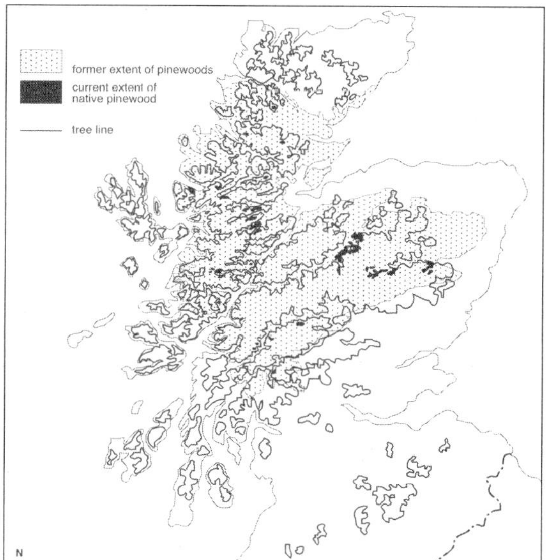

Figure 5.5 Extent of Native Pinewoods in Scotland.
Reproduced from Atherden M (1992: 66) *Upland Britain: a natural history* with kind permission of Manchester University Press.

Large upland forests were first planted in the early 1920s, when the Forestry Commission bought estates from landowners needing to pay death duties (Pringle 1994). The aim was commercial timber production and thus coniferous species were chosen that could provide rapid returns for little outlay on poor-quality land. Unfortunately much was agricultural rough grazing, whose loss caused resentment in the farming population (Balfour 1942; Pringle 1994; Moore-Colyer 1998).

Sitka spruce, Norway spruce and larch were typical choices, but Scots pine was also planted. The ground was often deep-ploughed before planting to improve drainage and the trees were planted in grid formations across many fells. Private landowners also planted new coniferous woodland to raise income from land deemed uneconomic otherwise. Governments have offered afforestation grants, initially to reduce agricultural surpluses, but latterly to enable farmers to diversify their income streams. The Woodland Grant Scheme, the Farm Woodland Premium in England, Better Woodlands for Wales and Scottish Forestry Grants have latterly supported expansion of broadleaved woods instead (Starr 2005).

Management Issues

The conservation value of upland woods is undeniable and easy to defend, but it is harder to justify their maintenance for farming, even though light grazing is important for their perpetuation. Historically, these woods were part of the wider subsistence economy (Holl and Smith 2007; Chapter 3). In some parts of the British uplands, woods still hold some importance in farm-management systems (e.g. Pollock *et al.* 2005), particularly by giving livestock some winter shelter and extra browsing, but often upland woodland is perceived as a complication to farm management. Even their value as shelter is less now that fells are destocked in winter, encouraged by agri-environment grants and conservation agreements. On the plus side, the Higher Level Environmental Stewardship Scheme will pay for the restoration, maintenance and creation of ghyll woods and blocks of other woodland under 1 ha in total (DEFRA 2008). Larger woodlands (over 3 ha) can receive financial grant aid from the Forestry Commission.

Upland birch woods have not traditionally had positive management, and instead have been exploited for wood fuel. Early clearances, up to about 1750, were replaced by bilberry-dominated heath (Dodgshon and Olsson 2006). Thereafter the remaining areas were used for shelter and grazing by sheep and deer. In the 20th century, birch woods were harvested for chipboard and paper pulp, and even underplanted with conifers up to 1985. Birch woods now suffer from heavy grazing by sheep and browsing by deer, causing stunted growth, and from poorly managed muirburn (heather burning). Both problems have halted natural regeneration, making the woods homogeneous in age. To reinvigorate birch woods, temporary exclusion of sheep and deer is needed (Pollock *et al.* 2005), but birch woods do naturally expand and contract because it is a pioneer species, so it is important to monitor the wood as it spreads beyond the fence, which should be moveable. The changing extent of birch woods makes them difficult to manage, in cost and grazing, so some form of wood pasture may be more effective as long as stocking densities are low.

Oak woods also are used as shelter for deer and sheep, which graze out understorey vegetation and prevent regeneration by nibbling the growth tips of seedlings. Many oak woods have at some time been coppiced, a practice gradually abandoned between the 1890s and 1950s, making woods even-aged and structurally less diverse. The perpetuation of pollarding is a must, for without it regrowth eventually becomes too heavy, toppling the tree in high winds and killing it. Other management problems have been caused by the introduction of *Rhododendron* and beech (*Fagus sylvatica*), invasion by bracken and underplanting with exotic conifers (Truscott *et al.* 2004).

The upland ash woods too have been exposed to grazing pressure, mainly from sheep, which remove the ground flora, replaced by herb-rich grass swards, and curb natural regeneration (Dodgshon and Olsson 2006). Coppice management has been abandoned (Box XXII). For upland oak and ash woods, the solution is fairly straightforward: keep out stock with some form of boundary (fence, drystone wall or hedge depending on local custom), which needs regular

Box XXII: Pollarding and Coppicing

Woodland management takes many forms, but two techniques have been used for hundreds, if not thousands, of years to encourage trees to re-sprout and produce more timber.

With pollarding, tree limbs are pruned back to the main trunk about 2m above ground level. The timber can then be periodically harvested for such purposes as fuel, fencing, wattle mears, spars and hurdles.

It was used to minimise grazing damage to trees by deer, sheep or cattle. For this purpose it has all but vanished and now is usually maintained as a landscape aesethetic or to extend the life of a tree and protect the species that symbiotically live with it. Oak trees in upland Atlantic oak wood in Borrowdale, in the Lake District, are managed for this purpose using pollarding (Pankhurst 2007).

(a) Coppice (b) Pollard

Coppicing is when growth is cut back to ground level. The tree re-sprouts with multiple stems, which thicken and lengthen year on year to create 'underwood'. Certain species coppice better than others, particularly hazel, sweet chestnut *(Castanea sativa)*, oak, ash and birch. The stems are harvested at different ages, providing a range of wood thicknesses, each for a different purpose. Thinner poles (1–2cm) can be used to make woven fences, 2–4cm stems make stakes for hedgelaying and 10cm+ timber is used for building, construction or charcoal production.

The coppice woodland is divided into a number of sections (coupes or haggs) which are coppiced in rotation, thus ensuring there is at least one part of the wood able to provide a certain girth of timber every year. On occasion individual trees are allowed to reach maturity to provide large baulks of timber for bigger construction jobs. Historically, livestock were permitted to graze the field layer (grasses and herbs) after four to seven years from cutting. Most farms and estates, secular or otherwise, operated coppice management in at least part of their woodland, particularly in the medieval period (Stamper 1992; Bond 2004). Over time coppice wood was replaced by other materials and the need for coppice management fell into disuse. Fuller and Warren (1993) suggest, however, that large-scale coppicing came late to western and northern Britain (circa 1700) as a response to the need for oak bark for tanning and charcoal for iron smelting; but again these uses faded as industrial processes changed.

The rotational management regime of a coppice woodland gives structural diversity, which creates many microhabitats and niches with a variety of light levels exploited by different species, which can move to a nearby similar habitat in the woodland when their home is harvested. With increasing interest in nature conservation from the late 19th century, the ecological value of coppicing was recognised and it was re-instated in many abandoned woods to encourage diversity to return. Any material that is harvested as a by-product offers the possibility of economic revenue.

Pollarding and coppicing can be funded by the Higher Level Environmental Stewardship scheme if they form part of the farm management system. Grants under the English Woodland Grant Scheme (www.forestry.gov.uk/forestry/infd-6dfk2u – accessed 21/01/2009) and Better Woodlands for Wales (www.forestry.gov.uk/forestry/infd-5z8jqk – accessed 21/01/2009) became in 2010 part of the Rural Development Programme (2007–13). Scotland uses a slightly different woodland management process by creating rural development contracts revolving around various biodiversity priorities.

maintenance because livestock will naturally seek the shelter of woodland. However, low-density grazing and browsing is a natural feature of these woods and helps to maintain diversity in structure and composition. Experiments by Truscott *et al.* (2004) using brash to protect regeneration instead of fencing were only partly successful because the brash eventually inhibited seedling growth, probably through shading and/or water and nutrient deficiency.

Reinstatement of coppicing is more difficult. For oak woods, a 5- to 10-year rotation can be introduced only if old trees are avoided; these re-sprout weakly, providing easy targets for browsing animals. For ash woods, coppicing can be reintroduced only if the wood has been managed in this way within the last 50 years and then only if new shoots are protected from browsing animals.

Protected by their remoteness, the native pinewoods escaped exploitation until about 1700; but peaked during the Napoleonic Wars, when timber was extracted and heather and bilberry heath replaced the woods (Dodgshon and Olsson 2006). In the 1840s timber prices fell and landowners ceased extraction, switching to more profitable deer stalking and later to sheep rearing, both of which suppressed natural regeneration (Gimingham 2002). Felling became popular again in the first half of the 20th century. Native woods had to compete with other land uses such as afforestation with commercial conifers, muirburn for sheep, clearance for housing, sawmills, powerline erection and flooding for reservoirs (Mackenzie 2002). All these changes have fragmented the woodland into small pieces that struggle to maintain themselves.

The remaining stands of native pine are especially at risk from sheep grazing and deer browsing because many are unenclosed and used by stock for winter shelter; this risk is greatest in the first 20 years of a tree's life. Also, red grouse suppress any regeneration on moorlands by eating out the growth tips of seedlings and halting ecological succession to woodland.

Simply excluding stock would seem to be the answer, but it is important to encourage thicket-like areas in the woodland and allow some light grazing – this increases floral diversity. Perimeter fencing is sometimes recommended, but the fragmented nature of many areas makes this operation expensive (Forestry Authority 2003c). Though sheep can be controlled in these ways, deer are a bigger problem because they can cross fence lines, unless these are over 2m high. Otherwise, deer can only be controlled by a managed cull, which requires the agreement of adjacent landowners, or else the remaining deer simply migrate onto the culled area.

Moorland

The archetypal upland landscape, heather moor, is in fact three habitats forming a mosaic: dwarf shrub heath, grassland and bog. Moorland – a function of historical woodland clearance and the arrested development of ecological succession – forms the major example of a plagioclimax community in sub-alpine environments of the UK. The three habitats are discussed separately and then considered collectively with agricultural management.

Dwarf shrub heath: characteristics

Dwarf shrub heath actually occurs in all environmental zones of the British Isles – alpine, sub-alpine and lowland. We considered alpine heath under montane environments (above) so here the focus is on sub-alpine heath. Most is plagioclimax dating from 5000BP and it was the result of woodland clearance for grazing and climatic deterioration (see Chapter 3). Since then it has been continually managed for sheep grazing, building and bedding materials, or grouse for sport shooting. Some sub-alpine heaths are subclimax, a result of local environmental factors arresting succession, but these are much rarer.

It is estimated there are about 1.33 million hectares of upland dwarf shrub heath, with 52% found in Environmental Zone 6, the Scottish uplands (Haines-Young *et al.* 2003), and about 31% (417,000 ha) in upland England and Wales. However, between 1984 and 1998 the area of dwarf shrub heath contracted massively in zones 5 and 6 (by 25,000 and 57,000 ha), whereas in zone 3 it expanded by 39,000 ha (Haines-Young *et al.* 2003).

Various ericoid species dominate dwarf shrub heath, notably ling, cross-leaved heath and bell heather. Ling and bell heather occur in drier conditions alongside bilberry, crowberry (Empetrum nigrum), cowberry (Vaccinium Vitas-videa), bearberry (Arcto-Straphylois Urva urst) and gorse, collectively known as dry heath (Table 5.6). Cross-leaved heath prefers damper conditions, developing a plant assemblage that includes *Sphagnum* moss, purple moor grass, heath rush and deer grass. This combination, known as wet heath, tends to occur on more peat-rich soils on blanket mires and raised bogs; it is discussed in detail under wetlands (see below). However, it is not unusual to have a mosaic of dry and wet heath together juxtaposed with acid semi-natural grassland.

Drewitt and Manley (1997) identified seven NVC sub-types of wet heath and eight of dry heath in sub-alpine areas; those of greatest distribution and use in upland agriculture are shown in Table 5.6. The various sub-types are associated with different fauna and/or are endemic, and as a result are internationally important (e.g. H4, H8, M16 and M17; Thompson *et al.* 1995). Many of them are restricted to the seaboard of north-western Europe, notably Ling-dominated heath (Figure 5.6).

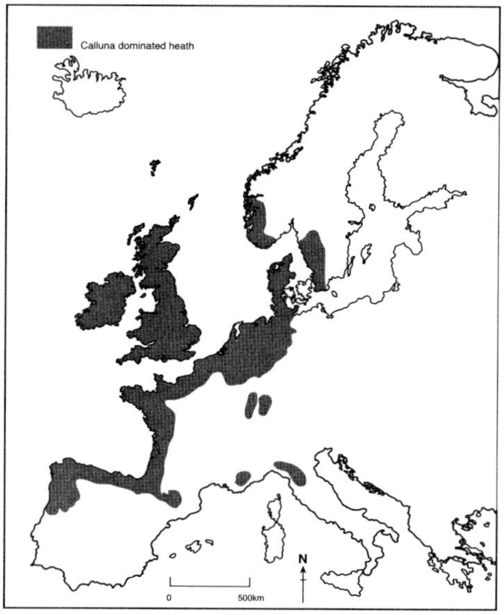

Figure 5.6 Distribution of Sub Alpine Heaths in Europe.
Based on materials first published by Scottish Natural Heritage in: Hester A, Thompson DBA, & Usher MB (1995) *'Heaths & moors: a cultural landscape'* p207 with permission from SNH.

Although floral diversity is low on dry heaths (only 2% of all British vascular plants), Usher and Thompson (1993) showed that faunal diversity is disproportionately large, including 15% of carabid beetles and 20% of spider species. Individual counts have found about a hundred species of insect on Ling, whereas plants like mat grass and purple moor grass support only about ten apiece (Yallop *et al.* 2005). Among birds, lapwing *(Vanellus vanellus)*, golden plover, curlew *(Numenius arquata)*, hen harrier *(Circus cyaneus)* and endemic red grouse *(Lapogus lapogus)* inhabit dry heath. The mountain hare *(Lepus timidus)* is also at home here, since much of its diet (50%) is Ling, along with cotton grass and bilberry. Historically, its range was forced back into the Scottish uplands through moorland management. Attempts were made to reintroduce it, most of which failed except in the Peak District (Mallon *et al.* 2003). A positive correlation exists between grouse distribution and hares because of their restricted diets; thus they are rarely found where sheep graze.

Dwarf shrub heath (DSH) occurs between 200 and 700m asl where precipitation exceeds 1000mm p.a. (Thompson *et al.* 1995); its range can extend down to sea level or up to 900m asl (MacDonald *et al.* 2001), at which altitude the plants can be prostrate and wind clipped (see earlier). DSH develops on mineral soils, from shallow rankers to acidic podsols. Typically, the uppermost soil horizons have well-developed, peaty, humic layers about 10cm thick. Peat is rarely deeper than 50cm underneath (MacDonald *et al.* 2001). Most of the nutrients are in the litter layer, but the plants are deficient in many respects; upland animals grazing on DSH can suffer from mineral deficiencies, notably of copper and boron. Ericoid species also help to increase the acidity of their own environment, as Grubb *et al.* (1969) found. Soil underneath the canopy of these shrubs is more acidic than other soil immediately adjacent.

Table 5.6: Sub-Alpine Vegetation: Dry Heath

NVC	*dominant species*	*environment*	*examples*	*fauna*	*zone*
H21	*Calluna vulgaris* *Vaccinium myrtillus* *Sphagnum capillifolium* heath	Mean annual maximum temps <22°C 200 to 600m asl on steep NW- or E-facing slopes of 25°–45° Precipitation 1600 to 3200mm pa, 220+ wet days Free-draining lithosols with bryophytic surface layer easily damaged by burning/light grazing	NW Highlands Lake District	black grouse ring ouzel various cranefly spp. various ground beetles	SUBALPINE SUBCLIMAX
H16	*Calluna vulgaris* *Arctostaphylos uva-ursi* heath	Mean summer temperatures 21–23°C Precipitation 1000mm pa, 140–180 wet days Snow lie >40 days pa at 250 to 600m asl Brown earths, brown podsolic and humo-ferric podsols Maintained by burning and exposure to grazing	Cairngorms	red grouse black grouse golden plover twite curlew merlin hen harrier northern eggar moth heath fritillary butterfly mountain hare red deer	SUBALPINE PLAGIOCLIMAX
H12	*Calluna vulgaris* *Vaccinium myrtillus* heath	At 200 to 600m asl on brown soils or podsols Mean maximum annual temps <24°C Precipitation 1000 to 1600mm pa, 160–180 wet days Snow lie <40 days per year Maintained by burning for grouse moors	cent. and NE Highlands Southern Uplands Northern Pennines cent. and E. Wales		
H10	*Calluna vulgaris* *Erica cinerea* heath	Mean annual maximum temps <22°C but dropping below freezing From sea level to 400m asl with wind exposure in places Base-poor humic rankers, brown earths and podsols (pH3.5–6) Some grazing and burning, some subclimax where exposed	Shetland & Orkney Southern Uplands		
H9	*Calluna vulgaris* *Deschampsia flexuosa* heath	Cooler climate than for H4/H8 but still above freezing 800 to 1000mm pa, <160 wet days Base-poor soils (pH3–4): rankers, iron pan and stagnopodsols Regular burning and grazing	Peak District North York Moors	golden plover curlew	
H8	*Calluna vulgaris* *Ulex gallii* heath	Mean annual maximum temps 25°C, winters mild (over 1°C) 1200mm pa on upland margins, >180 wet days affected by salt spray from sea Free-draining rankers, brown earths and brown podsolic soils Grazing is the main land management; burning is accidental	Dartmoor Exmoor Pembrokeshire Glyders	ring ouzel red grouse common lizard emperor moth fox moth	
H4	*Ulex gallii* *Agrostis curtisii* heath	Fewer than 40 days' frost pa Mean annual max temps 27°C 1000mm pa, 140 wet days up to 500m asl Acid brown earths and podsolic soils with impeded drainage Managed by burning, grazing, gorse cutting and turf paring	Dartmoor Exmoor	nightjar Dartford warbler tree pipit grayling butterfly green hairstreak butterfly Kugelann's ground beetle	

(Source: Rodwell *et al.* 1992; Dodds *et al.* 1996; Backshall *et al.* 2000.)

The ericoids that comprise DSH have a four-stage life cycle. Gimingham (1975) describes how ericoid plants take 25 to 40 years to pass through this cycle (Figure 5.7). As the plants mature, stems physiologically change from being soft and green to brown and woody. The plants' digestibility and nutritional value also deteriorates as it ages, which influences sheep grazing behaviour (see Chapter 4). To maintain the nutritional value and palatability of DSH, burning can be used as a management technique (Box XXIII). Each phase of the life cycle is important in maintaining faunal diversity, notably of birds (Thompson *et al.* 1995).

Dwarf Shrub Heath: Management Issues

Historically, upland DSH was the domain of extensive cattle grazing, and later sheep, with extraction of peat as fuel and ling for bedding or thatching. This was a low-input/low-output economy suited to the habitat, not in the least controlled by lower human population pressure. During the 19th century, the emphasis changed with DSH put to sheep (for wool and meat) and grouse (for recreational, latterly profitable, shooting). Both grouse and sheep require all stages of the heather life cycle. Sheep prefer young and building heather for forage, whereas grouse like building and mature heather for forage and cover respectively. Regular rotational burning spread as a standard management technique (Box XXIII). Furthermore, from the 17th century onwards

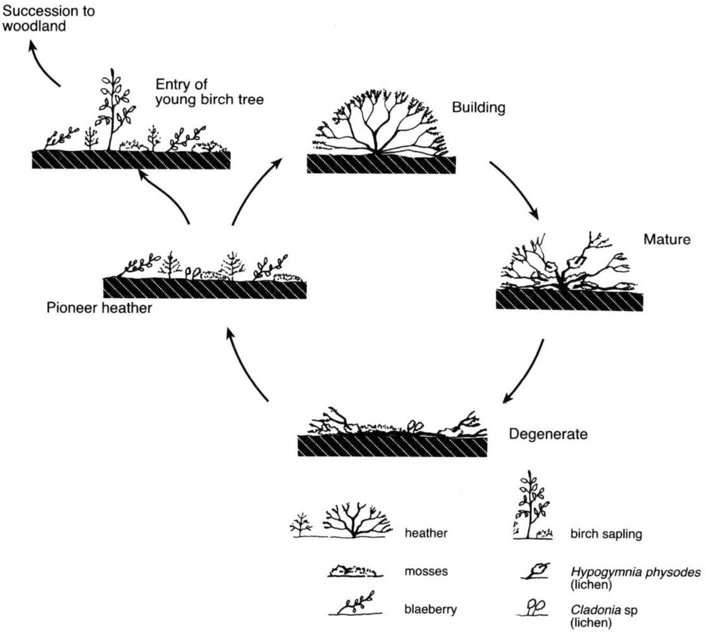

Figure 5.7 Life Cycle of Ericoid Species.
Adapted from Gimingham (1975:42) *Heathland Ecology* Oliver & Boyd: Sevenoaks.

Box XXIII: Burning as a management technique

Known variously as swaling, paring or muirburn, fire has been used to manage upland vegetation since mesolithic times. Apart from clearing woodland for temporary or permanent agriculture (see Chapter 3), burning has been used for just as long as a tool to encourage regrowth of vegetation.

Dwarf shrub heath (DSH), acid grassland and blanket mire can all be burnt, but it is not always necessary or wise. DSH above 600m asl in the montane zone has probably never been managed in this way, and nor is it necessary to burn heather for it to go through its lifecycle, since the plant will layer its stems naturally (MacDonald 1996). Cutting is also a legitimate and effective technique for grassland and heather management, but more problematic for blanket mire because of slope angles, sinking machines and the need for specialist contractors (see later in Chapter 5).

Grassland has traditionally been burnt in the management system used on Exmoor, the Southern Uplands and the north-west Highlands of Scotland (Miles 1987). The big issue is how to deal with re-sprouting purple moor grass, which will rapidly become dominant if not controlled. Cattle performed this function in the past, when the upland livestock economy preferred them to sheep. It has become a management issue now simply because sheep do not like to eat it. Gorse too has been controlled in the past using fire, but similarly it re-sprouts more vigorously if burnt. It is therefore important to consider post-burn management options as part of the overall strategy, or to employ burn alternatives. Furthermore, there are species which do not re-sprout when subjected to fire, such as juniper and Scots pine, so areas containing these should be managed differently.

As a management tool, rotational DSH burns are largely confined to Britain and Ireland, except for a few small areas of western Norway and the adjoining islands (Thompson *et al.* 1995). It is burnt to speed up the life cycle (about 25 years rather than 40), to avoid degenerate stands that could encourage the encroachment of bracken or other forms of ecological succession, to create mosaics which offer greater biodiversity and economic exploitation or to inhibit the possibility of accidental fire spreading unchecked. Burning also removes litter around plants, which cutting cannot do (Tucker 2005). This is beneficial as seeds will not germinate in litter, or those that do are starved of water and eventually die.

Surveys of Ling moorland have found poor burns are the second biggest cause of unfavourable condition (Drewitt and Manley 1997). So the use of burning as a conservation management tool has created much debate. Particularly poor burns can reduce the carrying capacity of heather for stock, increasing the risk of overgrazing as it recovers. There is risk of setting fire to the underlying peat if the burn is not properly controlled, causing destruction of rhizomes, rootstocks and seed banks (Tucker 2005). The temporary loss of ground cover in the recovering pioneer community favours plants that can quickly take advantage of post-fire regeneration, such as Purple moor grass and Mat grass. Being unpalatable they come to dominate the sward as other species are grazed out first, leaving behind an acid grassland. There is also evidence that burns trigger peat erosion (e.g. Tallis 1998).

In favour of the technique is the way it helps maintain the heather's vigour earlier in the life cycle when its nutritional value is greatest. This is of particular use in grouse rearing and sheep profitability (Davies 2005). Burning allows conservationists to create mosaics of different aged and structured *Calluna*, which increases the moorland's biodiversity dramatically (Usher and Thompson 1983). Regular burning also helps reduce the risk of high-temperature wild fires; these do the most damage.

Important variables are the characteristics and behaviour of fire itself. Tucker (2005) shows that frequency of fires, their temperature, ground-to-surface intensity,[1] duration and size are all significant, and are controlled by angle of slope, windspeed, soil-moisture levels and state of the Ling. Poor burns derive from high temperatures and ground-to-surface intensity typically kill many moorland species, and any factor that encourages these characteristics should be limited if possible. Burning of heather, gorse, bracken and bilberry is controlled under the Heather and Grass etc. (Burning) Regulations 1986, which state that:

- the burning season is from 1 October to 15 April in the uplands,
- managers must give written notice of intent from 24 hours to 7 days before the burn starts,
- fires may not be lit between sunset and sunrise,
- sufficient persons and equipment must be in place to control the burn.

The optimal time to burn heather is before it reaches 12 years old (or 20 to 30cm high). After this, heather will not regenerate so well, only about 50% of plants survive and the moor becomes over-run by bracken or purple moor grass. Other guidelines to ensure good burns include: choosing a closed canopy with no grasses; avoiding peat bogs, scree, steep slopes and wet heath; avoiding all land over 600m asl; limiting burns to manageable areas of about 0.5ha in long strips 20 to 30m wide; having one person per 5m of fire front; avoiding anything more than steady gentle breezes (about Force 3); and operating an effective communication system between fire wardens and the fire brigade.

[1] The surface zone is where vegetation grows and heat gets trapped; the ground is where the soil or rock begins.

heather moorland was converted to grassland in an attempt to raise stocking rates (Stevenson and Thompson 1993). The decline was regional, with some areas such as western Wales and Dumfries and Galloway losing Ling cover much earlier than other locations (see Chapter 3).

The superior profitability of sheep farming over cattle has encouraged overstocking of upland DSH; much of it has become overgrazed, damaged by inappropriate management caused by poor burning regimes and selective grazing or lack of management creating overly mature (tall, leggy or collapsed to prostrate) plants, the corollary of which is unfavourable condition (Box XIX). Bardgett *et al.* (1995) found the greatest damage in south-west England (39% of total cover) and least damage in the north-west (19%). In addition, grazing had stopped the sprouting of pioneer plants. Again south-west England recorded the worst suppression (44%) and the North York Moors (20%) the least. This was probably because northern areas put more emphasis on grouse rather than sheep, compared to the south-west. The pattern is not straightforward, though, because individual massifs suffered differently; for example, over 60% of the Long Mynd in Shropshire had extreme damage from overgrazing (Felton and Marsden 1990).

Sheep grazing of heather can even lead to its complete replacement by other habitats. Estimates of 27% complete loss are given for England and Wales between 1947 and 1980 (NCC 1984) and 18% in Scotland (SNH 1994). Regional variations exist here too; for example, Cumbria lost 36% of DSH between 1940 and 1980 (Felton and Marsden 1990). Thompson *et al.* (1995) modelled this loss of heather moorland and its subsequent switch to grasslands (Figure 5.8). However, such loss is usually the indirect result of a poor burning regime or overgrazing, rather than a deliberate policy to create more grassland on open fells, which was typically the case in the 19th century.

Pastoral farming is not the only cause of DSH deterioration in British uplands. For thousands of years peat has been extracted to provide fuel for domestic and industrial purposes. Many common rights relate to the cutting of peat turves on moorland (Aitchison *et al.* 2000), some dating back over 700 years, and a few are still utilised. The extent of peat extraction has been grossly underestimated and many moorlands show obvious signs of peat removal dating back to the Iron Age. Rotherham *et al.* (2004) have calculated that a single farm in Farndale, North York Moors, would have used 4 ha of peat (1 metre deep) in a 300-year period

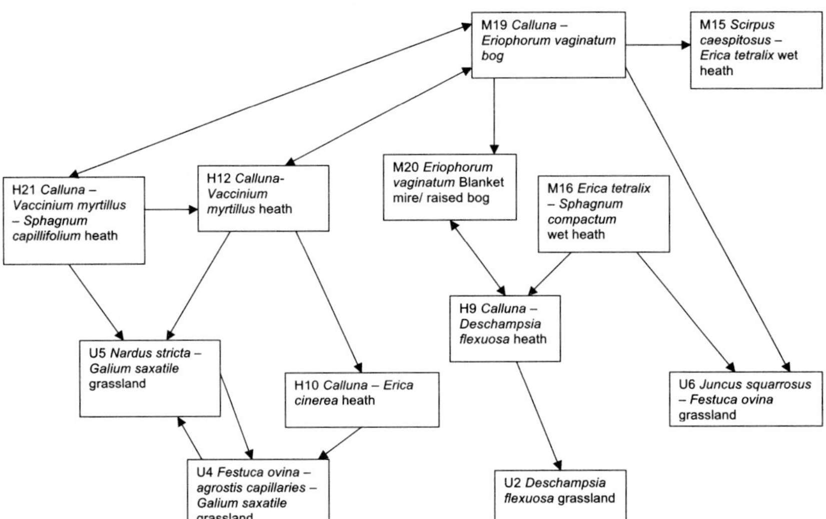

Figure 5.8 Replacement of Heather Moorland Communities.
Reprinted from *Biological Conservation* Volume 71, Bardgett RD., Marsden JH. & Howard DC 'The Extent and Condition of Heather in Moorland in the Uplands of England and Wales,' 1995:170, with permission from Elsevier.

DSH has also been affected by the Forestry Act 1919 and the planting, particularly during the 1920s and 1930s, of many thousands of hectares of upland with exotic conifers. Accounts of how forestry 'migrated uphill' can be read elsewhere (Pringle 1994), but as a result forestry became a direct competitor with the existing semi-natural vegetation and also with the upland farmers (De La Warr 1944; Stapledon 1951; Zuckerman 1955).

Undergrazing is a third problem for the quality of dwarf shrub heath. Destocking of grazing animals occurred widely in England and Wales during the agricultural depression (see Chapter 3). As a result, unwanted species – usually fast-growing competitors with little value for nature conservation or stock nutrition – were no longer suppressed. Particularly problematic are purple moor grass, gorse and bracken (Box XV). These species grow faster than ericoids and thus limit or even halt the growth of DSH, particularly if heather is in the pioneer phase.

Purple moor grass is problematic after a burn because it is unpalatable to sheep, so it quickly comes to dominate the sward, leading to acid grassland. As a deciduous plant it produces large amounts of litter which, if eliminated by burning, simply encourages more to grow. If it is left unchecked the whole hillside can gradually scrub up as a result of ecological succession, removing all traces of the ericoid communities. The expansion of gorse and bracken is partly due to changes in resource exploitation. Historically these two plants were highly prized and collected, gorse as fuel and bracken for bedding (Rotherham *et al.* 2004), neither of which now takes place.

Dwarf shrub heath therefore suffers from two main issues: deterioration of quality and total loss. To avoid or overcome these, a range of management techniques is employed. Deterioration due to overgrazing has been the subject of research and practical experiment (eg Pakeman *et al.* 2003; Milligan *et al.* 2004). Results recommend:

- lower stocking rates to 1.0–1.5 ewes/ha,
- reduce or eliminate overwintering on fells,
- actively shepherd to keep grazing uniform,
- reinstate appropriate burning regimes,
- control unwanted species, especially deer, Purple moor grass and bracken,
- cut heather where burning cannot be carried out.

To prevent total loss, many similar strategies can be used. A fell may seem devoid of ericoid species, yet often resprouts a couple of years after stock are removed (Johnston 2004). Clearance of competitive species was employed on Crompton Moor, near Oldham, where acid grassland dominated by Purple moor grass was removed using a flail to break up tussocks and the root structure without causing erosion; the land was then re-seeded with Ling (Roworth and Morris 1998). Purple moor grass can also be removed by the application of herbicides such as glyphosate (Anderson 2002). Milligan *et al.* (2004) tried a number of approaches including herbicides, various cutting regimes and Ling brash addition. They found that a multiple cutting regime was the most effective when linked to appropriate sheep grazing. However, they did state that, though this integrated system worked for the case-study site in North Yorkshire, it might not always work elsewhere and land managers must adapt the approach to the seed bank character and floral composition. None of these experiments used cattle (notably White Park) to control Purple moor grass, which is possible, though cattle are now far fewer in the uplands, as agricultural policy favours sheep (see Chapter 8).

Heather can be reinstated where plantation forestry once stood. The latest technique is to use a 'three-in-one' Universal Forestry Mulcher, which pulverises brash, grinds stumps and prepares a seed bed in a couple of passes (Bacon 1999). The only drawback, as with all machines, is the maximum slope on which it can operate – a problem in uplands for many large-scale tasks.

Although all these techniques to reinstate or improve DSH do work, the issue for the farmer is cost. In a grazing economy they will provide little return and thus the cost of such work is simply unprofitable. There are really only four ways to protect these habitats on upland farmland:

- A full management agreement with a nature-conservation quango using statutory powers to designate land as some form of nature reserve (Box XXI).
- Financial support via agri-environmental initiatives like Environmentally Sensitive Area agreements or Environmental Stewardship (see Chapter 8); however, the economic and environmental validity are matters of conjecture.
- Cut and bale the heather for commercial uses, such as seed source for moorland restoration, thatching, garden furniture, soil erosion control or in bale boilers for heating systems (Wall 1993).
- Compromise the heather-management regime and develop a grouse moor.

For this last option, grouse need older heather for nesting areas and younger heather as a food source, yet the heather requires burning to help eliminate parasitic threadworms and perpetuate all stages of the ericoid life cycle (Atherden 1992). Grouse also need the moorland compartmentalised into small units of different ages, which helps with the territorial requirements of the birds, thus allowing as high a density as possible of grouse on a single moorland block (Felton and Marsden 1990). Hence grouse requirements differ substantially from those of sheep, wildlife and landscape amenity value (Table 5.7).

Table 5.7: Heather Requirements of Different Land Uses on a Moorland

use	*% moorland which is heather units*	*% of heather cover in the heather units*	*burn area required (ha)*	*burn frequency (years)*	*% not burned*
Sheep	50	50	15–20	7–10	none
Wildlife	50–75	50–75	2–6	10–15	10–50
Grouse	75–85	75–90	1–3	10–15	< 10

(Source: Felton & Marsden, 1995)

Developing a multifunctional DSH is complex. A management regime is needed, but it may have to be a compromise if there are multiple objectives for a particular piece of moorland. These are a few basic guidelines for the upland farmer:

- grazing utilisation should be less than 40% of this year's growth,
- heath is least sensitive from December to March,
- building heather is the most grazing-tolerant,
- heather is most palatable for sheep in autumn and winter,
- heather should be burnt when it reaches 30cm in the building phase.

Put simply, stock in summer to the level that grass on the fellside can support; stock in winter to a level below 40% of the season's growth; then develop a controlled burning regime to fit in with multiple uses of the fell.

Grassland habitats: characteristics

Upland grassland can be enclosed to form fields, as inbye and intake, or unenclosed. It can be divided into roughly two main agricultural types: first, pasture, which is grassland that is grazed directly by livestock; and second, meadow, where the grass is grown, cut and then fed as fodder to livestock elsewhere. In contrast, from the point of view of nature conservation, grassland occurs in three semi-natural states: acidic (calcifugous), neutral (mesotrophic) and alkaline (calcareous). All these are unimproved from an agricultural viewpoint: there has been no intervention to raise their production potential (Figure 5.9). Acid and alkaline grasslands are usually pasture. They are ecologically valued for their wide diversity of flora and associated fauna, many of which form communities not seen other than in the British uplands.

About 20 sub-types of acid grassland are recognised by the NVC (Rodwell *et al.* 1992; see Table 5.8). Many are temporarily or permanently waterlogged and thus are technically mires, dominated by sedges and rushes (Backshall *et al.* 2001). Acid grasslands are typical of base-poor geologies and acidic soils such as podsols and peats with a pH below 5. They fall into two main groups:

1) those growing on drier acid brown earths, the sward dominated by sheep's fescue and common bent – includes U1 to U4 in England (Drewitt and Manley 1997);

2) those where the soil exhibits gleying and is more peaty, and the vegetation is mainly mat grass and heath rush (*Juncus squarrorus*), typical of U5 and U6 plant communities (Drewitt and Manley 1997; Yallop *et al.* 2005).

Acid grassland is the most extensive type, covering 1.13m ha in the uplands and able to replace dwarf shrub heath that has been overgrazed (Figure 5.8). It has had drainage put in, particularly in enclosed valley-bottom fields as farmers have sought to increase their inbye land. In 1984 it covered about 5% more land than today, but this disguises regional differences: Environmental Zones 3 and 5 have seen declines since 1984 (to 51,000 and 26,000 ha), whereas Zone 6 has had an increase of 31, 000 ha (Haines-Young *et al.* 2003). At local level, change can be stark: the Culm area of Devon and Cornwall has only 8% of the purple moor grass and rush pasture it had in 1900 (JNCC 1995), hence its inclusion in the Countryside Stewardship scheme as a priority reclamation project.

Upland alkaline (calcareous) grasslands are limited to thin soils with pH between 6.5 and 8.5, which are found on Carboniferous limestone, the alkaline Borrowdale volcanics of Cumbria, the Dalradian schists and the machair of north-west Scotland. Despite this limitation, 14

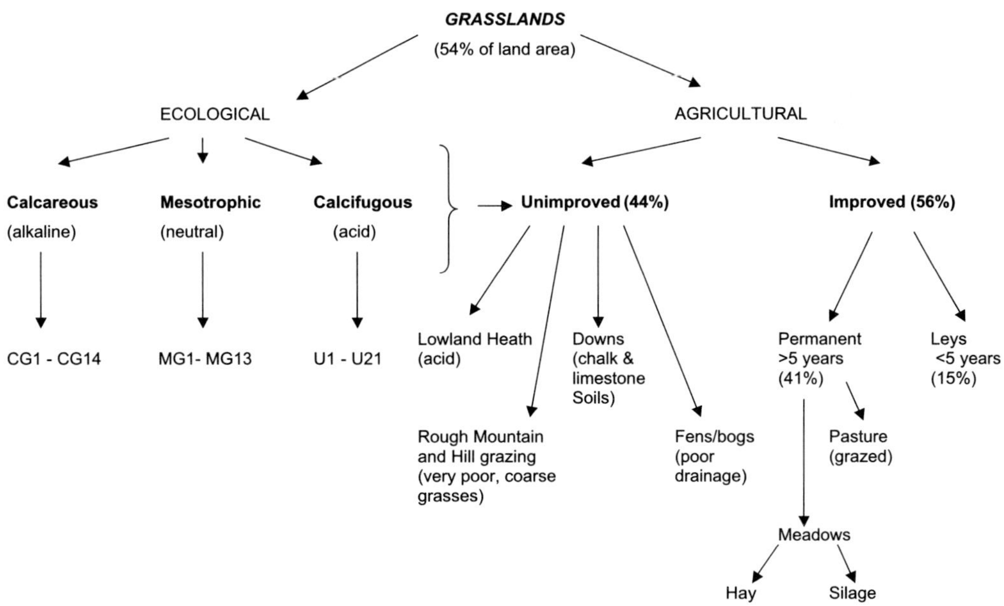

Figure 5.9 A Comparison between Ecological and Agricultural Terms for Grassland

Table 5.8 Sub-Alpine Vegetation: Grasslands

NVC Community	*Dominant species*	*Environmental Conditions*	*Distribution*	*Associated Fauna*	*Zonation*	
U1	*Festuca ovina – Agrostis capillaris – Rumex acetosella* Grassland	South facing rocky slopes of granite Rankers and podzols Precipitation <1000mm and <140 wet days pa Maintained by grazing	Long Mynd Stiperstones SW Peninsula	**Mammals:** Fox Badger Field voles Shrews Bats Roe deer **Birds:** Merlin Curlew Golden Plover Black Grouse Redshank Twite Short Eared Owl Lapwing **Butterflies:** Heath Fritillary Marsh Fritillary Narrow bordered bee hawk **Reptiles:** Slow worm Adder	Upland Fringe	SUB MONTANE / SUB ALPINE
U2	*Deschampsia flexuosa* Grassland (*Vaccinium* sub community in uplands)	Millstone grit, granite and Devonian sandstone Base poor humic ferric podzols Precipitation >1000mm and 140 to 160 + wet days pa 300 to 500m asl Maintained by grazing	S Pennines Dartmoor S Wales			
U3	*Agrostis curtisii* Grassland (gorse sub community in uplands)	Sandstone and granite Brown podzolic soils and podzols sensu stricta Precipitation 800 to 1600mm and <140 wet days pa Perpetuated by grazing and burning	Bodmin Moor Exmoor Quantocks			
U4	*Festuca ovina – Agrostis capillaris – Galium saxatile* Grassland	Many igneous and metamorphic geologies Brown earths and brown podzolic soils 150 to 500m, occasionally up to 800m asl Precipitation 800 to 1200mm and 140 to 180 wet days pa Perpetuated by grazing as a plagioclimax, good quality	SW Peninsula Wales Welsh Marches N England Grampians S Uplands Lower and middle slopes above the intake		Lower and middle slopes above the intake	
U5	*Nardus stricta – Galium saxatile* Grassland	Parent material typically shale Stagnopodzols and stagnogleys Precipitation 1200mm> and 160+ wet days pa 300 to 700m asl Perpetuated by grazing, through extensive as poor quality	SW Peninsula Wales Pennines Scotland			

Table continues over

Table 5.8 – *continued*

NVC Community	*Dominant species*	*Environmental Conditions*	*Distribution*	*Associated Fauna*	*Zonation*	
U6	See Table 5.11					
MG3	*Anthroxanthum odoratum – Geranium sylvaticum* Grassland	Superficial glacial drift deposits with brown soils, some gleys Precipitation 900 to 1200 mm, 180 to 200 wet days and <50 days of snow pa 200 to 400m asl Perpetuated through cutting for hay	N Pennine Dales Yorkshire Dales SW Cumbria	**Mammals**: As above **Birds:** Skylark, Corncrake **Butterflies:** High Brown Fritillary, Pearl Bordered Fritillary, Dark Green Fritillary, Duke of Burgundy	Dale Heads	SUB MONTANE / SUB ALPINE
MG10	*Holcus lanatus – Juncus effusus* Rush Pasture	Brown earths, brown calcareous earths and gleying Perpetually moist soils Maintained by low level grazing and abandoned land	All upland margins		Middle slopes	
CG2	*Festuca ovina – Avenula pratensis* Grassland	Calcareous geologies supporting rendzinas, pH7 to 8 Precipitation <1000mm and <160 wet days pa Plants adapted to drought conditions Maintained via a balance of sheep and rabbit grazing	North York Moors Durham uplands Mendips Derbyshire dales	**Butterflies:** Northern Brown Argus, Small Blue		
CG9	*Sesleria albicans- Galium sternei* Grassland	Restricted to Carboniferous limestone Rendzinas pH6 to7.5 Precipitation >1000mm and 160+ days pa 250 to 500m asl	N Pennines W Yorkshire Dales		Montane fringe	
CG14	*Dryas octopetala – Silene acaulis* Ledge Community	Calcareous schists and limestone crags Calcareouse lithosols, pH 7 Precipitation 2400mm> and 160+ days pa Only where grazing is non existent	Scottish Highlands N Wales Glyders Lake District Mull & Rum Orkney		Ledges	

(Sources: Rodwell *et al.*, 1992; Backshall *et al.*, 2001)

sub-types have been identified by the NVC (Rodwell *et al.* 1994), of which three dominate the English uplands (CG2, CG9, CG14; see Table 5.9). Sward composition is extremely diverse, a few typical species being listed in Table 5.9. Forb (flower) counts can be as high as 80 species within a square metre. The agricultural potential of these grasslands has targeted them for improvement and overgrazing, reducing the diversity to just a few grazing-tolerant species, as sheep have 'mown' the sward down to a few centimetres high. They are mainly found in Environmental Zone 5, where 27,000 ha were recorded in 1998, a decline of 5000 ha since 1984 (Haines-Young *et al.* 2003).

Table 5.9: Indicator Species of Semi-natural Grasslands

acidic	*neutral*	*calcareous*
Grasses		
common bent (d)	meadow foxtail	meadow oat grass
velvet bent (d)	sweet vernal grass	quaking grass
sheeps fescue (d)	false oat grass	blue moor grass
mat grass (w)	crested dogs tail	
purple moor grass (w)	Yorkshire fog	
deer grass (w)	cocks foot	
Herbs/ forbs (some examples)		
heath bedstraw (d)	orchid spp. (d)	birds foot trefoil
tormentil (d)	common knapweed (d)	salad burnet
bilberry (w)	ragged robin (w)	thyme
heath rush (w)	lady's mantle (d)	bistort
	meadowsweet (w)	field gentian

Note: (w)=wetter soils, e.g. podsols; (d)=drier soils, e.g. acid brown earths.

Neutral (mesotrophic) grasslands in the uplands cover substantially less land than acid grasslands, about 130,000 ha in 1998, mostly in Environmental Zone 3 (accounting for 60% of the total) – NVC MG10 – on intake land in England (Drewitt and Manley 1997). Since 1984 zones 5 and 6 have lost some 42,000 ha of neutral grassland, though zone 3 has gained 37,000 ha (Haines-Young *et al.* 2003). This could be a result of upland farmers seeking economies of scale and deciding to abandon semi-improved intake land, which has then reverted to unimproved grassland. National schemes such as EA and CSS have assisted here.

A much rarer mesotrophic grassland type is MG3, the upland hay meadow. This is where traditional agricultural management developed inbye land for hay production (Hughes and Huntley 1988), leading to flower-rich grass swards, these being cut and dried towards the end of August. They are confined to English northern uplands in the valleys of the Pennines and south-east Cumbria. Their area now is about 2,024 ha, most of which is protected through SSSI (Site of Special Scientific Interest) status (English Nature 2008). They are managed by a combination of grazing and cutting (Figure 1.11), making the best use of the fodder either from direct grazing or supplementary feed in winter.

Grassland: management issues

All semi-natural grasslands in the sub-alpine zone are a product of management by people to some degree and thus it is important to maintain their agricultural use by traditional methods and appropriate stocking densities that do not eliminate their conservation value. Many of the plants adapted to living on semi-natural grassland require a low-nutrient soil, light grazing and no disturbance of the sward by periodic ploughing (see Duffey *et al.* 1974 and chapters 3 and 4 here).

The main challenge in managing semi-natural grassland is improvement. Maintaining its low-nutrient status provides minimal nutritional support for stock, but biodiversity interest for conservationists. The largest changes have come from attempts to raise the dry matter content and palatability of the grass, particularly on intake and inbye land, where improvements have most economic benefit. A well-known example of this intensification is the decline of northern hay meadows (MG3) and replacement by silaging (Figure 5.10), slowed only by hay-meadow prescriptions in ESA and CSS agreements (see Chapter 8). Silage requires the complete improvement of the field by drainage, reseeding and fertiliser treatment. The meadow is re-seeded with one or two nutritious species such as perennial ryegrass (*Lolium perenne*) or timothy (*Phleum pratense*), resulting in high-nutrient, high-productivity grassland with low or negligible diversity. In the Pennines this has caused sharp declines in upland breeding wader and passerine populations, such as lapwings, skylarks *(Alauda arvensis)*, reed buntings *(Embrizia schoeniclus)* and snipe *(Gallinago gallinago)* (Fuller *et al.* 2002).

Another issue with unimproved conservation grassland is overgrazing. Acid grasslands now are a product of overgrazing, either of dwarf shrub heath (see above and Figure 5.7) or of the grasslands themselves. In the latter case, livestock either selectively graze out the more palatable species, leaving grasses, or eliminate the vegetation completely so that bare soil is exposed and then eroded. Calcareous grasslands in particular suffer from differential grazing, because they are made up of the most palatable grasses in unimproved swards (see Chapter 4).

Grasslands can also be undergrazed. Calcareous grassland can be selectively grazed out on a fell, while the stock ignore less palatable vegetation such as that in the contiguous acid grassland or blanket bog. The undergrazed sward can become coarse and rank, dominated by grasses growing unchecked by grazing pressure. Build up of grass litter develops as the material, not

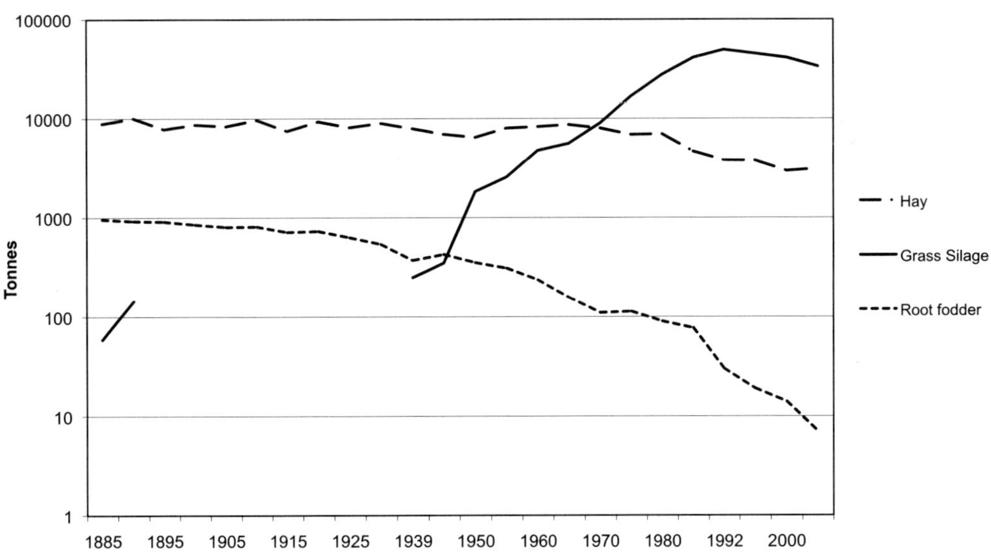

NB Note log scale

Figure 5.10 Hay and Silage Production in England 1890 to 2005.
Adapted from: DEFRA 2008, Brassley 1999 and Marks & Brittan 1986.

eaten, simply dies and creates mats across the soil. Certain species of rare plant also seem to require grazing to flourish. A study by Miller *et al.* (1994) investigated the effect of grazing/no grazing on alpine gentian (*Gentiana nivalis*) at Ben Lawers in Scotland. They found that lack of grazing inhibited growth of gentian plants, which suffered from over-competition from grass species and lack of bare ground created by trampling. There is also evidence from central Europe that the breakdown of transhumance systems has meant sheep no longer replenish floral diversity in isolated calcareous grasslands through fleece transmission of seeds (Poschlod and Wallis DeVries 2002). Worse still, as these plants grow older their palatability decreases further and stock ignore them, preferring more appetising fare (Armstrong and Milne 1995). Enclosed calcareous grassland can also be undergrazed. On the Carboniferous limestone of the Burren in Ireland, undergrazing has led to decreased biodiversity and an increase of scrub. Dunford and Feehan (2001) traced how the decline in winter grazing (winterage) had allowed hazel and blackthorn to spread across grasslands. Once the shrubs reach 1.5m high, they become impervious to grazing pressure and clearance is uneconomic.

For nature conservation, improvement, overgrazing and undergrazing can all lead to complete loss of conservation grasslands, their replacement by silage fields and a grazing regime unbalanced between sheep and cattle, reducing sward diversity. Economics and politics ostensibly drive land improvement (see chapters 6 and 8), but changes of livestock – from cattle to sheep, from native to commercial breeds – also play a part (see chapters 1 and 3). If farmers revert to native breeds, these challenges can be avoided as long as a ready market can be found for the products (see Chapter 6). At the same time ecological variation can be maintained, particularly if agri-environment grants continue to support low-intensity traditional farming systems (see Chapter 8).

Economics and politics influence overgrazing and undergrazing. The British Government saw overgrazing of uplands by sheep as a function of the Less Favoured Areas Directive, expressed through the Hill Livestock Compensatory Allowance and Sheep Annual Premiums of the 1970s, 1980s and early 1990s (see Chapter 8). Undergrazing is even more of an economic and social issue, because abandonment of hefts and entire farms continues (see chapters 6 and 7). Solving these grassland management issues is therefore inextricably bound up with economic and political policy, requiring a change in direction. Agenda 2000 and the Rural Development Regulation (see Chapter 9) have shown the way, but what matters is whether policy can be articulated on the ground in a way that benefits ecologies and farmers.

Appropriate conservation management for grasslands is to reduce their nutrient status to allow plant diversity to flourish. This requires a low-intensity, low-input, low-output farming system, maintaining the grassland by the farming practices that created it. These beneficial practices include cutting and/or light grazing, burning and some cultivation.

The first task is to cease silaging and reduce nutrient status by desisting from fertiliser use. Over time the sward will respond, with low-nutrient species returning – many will have seeds still resident in the seed bank. The forb component of the grassland may come back naturally or need re-introduction by turf transplant or plug planting. A more drastic option is to strip the topsoil off or plough up the sward, and then sow a grassland/forb mix to create an 'instant' unimproved grassland (Table 5.10).

Both approaches work, but the second can be somewhat disconcerting: it has been found that certain wildflowers in the sowing mix can become dominant, such as a flush of ox-eye daisies (*Chrysanthemum leucanthemum*) early on in the process. Given time, other species will begin to compete, restoring a more natural-looking balance of flowers. There is also the real problem that some species need other organisms to be in the soil before they germinate and thrive. An example is yellow rattle (*Rhinanthus minor*), which requires a mycorrhiza (symbiosis with a fungus) that can take 20 years to develop.

Table 5.10: Commercially Available Upland Hay Meadow Seed Mixes

wildflower species name	*wildflower common name*	*% by weight*		*grass species name*	*grass common name*	*% by weight*
Centaurea nigra	common knapweed	3.2		*Agrostis capillaries*	common bent	10
Conopodium majus	pignut	0.2		*Anthroxanthum odoratum*	sweet vernal grass	20
Geranium sylvaticum	wood cranesbill	0.2		*Festuca rubra*	red fescue	30
Lathyrus pratensis	meadow vetchling	1.5		*Dactylis glomerata*	cocksfoot	10
Lotus corniculatus	birdsfoot trefoil	1.5		*Poa trivialis*	rough meadow grass	20
Plantago lanceolata	robwort plaintain	3.2				
Primula veris	cowslip	1				
Ranunculus acris	meadow buttercup	3.2				
Rhinan minor	yellow rattle	2.5				
Rumex acetosa	common sorrel	2.5				
Sanguisorba officinalis	great burnet	0.1				
Succisa pratensis	devils bit scabious	0.1				
Trifolium repens	white clover	0.2				
Vicia sativa	common vetch	0.5				
Viola riviniana	common dog violet	0.1				
Total % by weight		**20**		**Total % by weight**		**80**

*Note: based on MG3 of the NVC, seeds are usually sourced to ensure local provenance for the meadow being re-created.

(Source: wwww.scotiaseeds.com accessed: 11/04/2011).

The second phase is to introduce an appropriate maintenance regime using combinations of cutting, grazing and/or burning. Cutting occurs where grazing cannot be used, usually because of proximity to built-up areas, access problems or lack of grazing animals. It works by mimicking the height of sward produced through the eating process (Figure 1.11).

The second maintenance option is light grazing, which is the main method of choice, particularly for large areas where cutting is too time-consuming and costly. When setting up a grazing regime, several factors need to be considered, which include:

- management objectives,
- type of stock to be used,
- forage available – its seasonality and palatability,
- grazing regime – temporary, seasonal or permanent,
- stocking densities.

Of these, the decision on stocking density – to avoid under- or overgrazing – is probably the most difficult. These two extremes are complex because they depend on a huge number of variables, such as soil, climate, geology, breed requirements, vegetation grazing tolerance, sward structure and diversity, seasonal palatability and management regime.

Much of research has focused on identifying optimum stocking rates using empirical and mathematical modelling (e.g. Armstrong 1993). Empirical work has been used to increase the economic efficiency of hill pastures, as at the Cahn Hill Improvement in Wales between the 1930s and 1970s, and to de-intensify them to protect nature conservation value (Griffith 1951; Duffey *et al.* 1974). Though all these exercises provide useful data on possible stocking densities, the uniqueness of every upland means that stocking density should be tailored to that environment; identifying optimum rates is often a process of trial and error. In fact, historically this is how fells were managed before economics and politics distorted the markets and agri-environment schemes set standardised stocking rates.

A 'suck it and see' approach is all very well, but it can take several years to identify the optimum; even then, if the weather is unusual or climate change is apparent, the optimum rate may no longer be optimal (see Chapter 4). By then, damage may have been done or change may not be rapid enough for the funding agencies. To this end, agri-environment initiatives that help fund many low-intensity grazing regimes in the British uplands are based on a combination of empirical research and mathematical modelling, setting required stocking rates within specific time frames. These can be problematic for farming if cutting times are too late for the crop or too early ecologically; this was found in the Pennines ESA management prescriptions (Saunders 1994). Farmers will also need to re-jig their management system to compensate for de-intensification on some parts of the farm.

Another grazing option is to change the livestock to breeds that selectively graze out the species not required. Certain breeds of native cattle, such as White Park, actually prefer to eat Purple moor grass and Mat grass (see Box III). However, the destructive impact of cattle hooves also needs careful management, by adopting some form of stock rotation. Another advantage of cattle is that they control bracken, not by eating it, but simply by trampling and when they settle to ruminate. This breed, if carefully marketed, can provide an additional premium at sale times and is among the management options in the High Level Stewardship Scheme (Natural England 2008).

A third, less used option is to burn grassland. Traditionally known as swaling[2] or 'pare and burn', burning effectively removes dead litter and encourages young nutritious shoots, and was known to raise the nutrient status, particularly in phosphates (Duffey *et al.* 1974). Tucker (2003) notes that little else is known of its effects, though its value is double-edged: for conservation, higher nutrient status is the opposite of what is required.

Supplementary techniques include rolling and chain harrowing, managing pests, weed control, appropriate drainage and scrub management. The decision to manage a grassland for nature conservation cannot be taken lightly, however; altering or stopping management regimes can lead to grasslands changing from one sub-type to another, or in the worst case initiating ecological succession (as in the Burren – see above).

Bogs and mires: characterisitcs

Many wetland environments exist in the uplands, from saturated ground to open water (Drewitt and Manley 1997). 'Mire' is often used as a generic term by ecologists to describe wetlands (see Chapter 4 and Table 4.5), which all have particular conservation value; they are priority habitats for protection under EU and UK legislation (JNCC 1998). Some are natural, rather than semi-natural: for example, the accumulation of peat in waterlogged conditions over thousands of years created the mires of the Flow Country in Caithness (Atherden 1992).

---

[2]. Editor's note: the form 'swealing' was used in Lancashire in the 1920s to mean deliberately setting fire to grass.

[3]. Water efflux is the explusion of water from a wetland body, in contrast to influx.

Important controls on mire evolution are the inputs of organic matter and water. The organic material comes from the slow decomposition of plant matter to form peat. For true mires and bog pools, various *Sphagnum* mosses are a fundamental component. *Sphagnum* has no true roots: it absorbs water directly through its leaflets, relying on precipitation. It can hold four to five times its own weight in water and reacts poorly to acidic deposition from the atmosphere, despite its preference for acidic environments. A number of species of *Sphagnum* may develop across the microtopography of a mire, some preferring the slightly drier hummocks (e.g. *Sphagnum rubellum*), others staying in or around hollows that lie below the water table (e.g. *Sphagnum cuspidatum*). In contrast, wet meadows tend to develop on stagnogleys (see Chapter 4) and are found along upland fringes.

The water component balance (input/output) is known as the water flux. Inputs are precipitation and drainage waters from surrounding land; outputs include lateral seepage, evapotranspiration and water efflux.[3] Different mires have different water fluxes, as Bragg (2002) shows (Figure 5.11). In the uplands, blanket mire is the most extensive type of wetland, covering over 2.2m hectares, roughly 8% of the British Isles (Tallis 1998). These habitats have had a chequered history: data from Countryside Survey 2000 (between 1984 and 1998) show that 'bogs' (including wet heaths and mires) have declined in all upland environmental zones by about 104,000 ha, yet 'fens, marsh and swamp' have increased by 56,000 ha (Haines-Young *et al.* 2003).

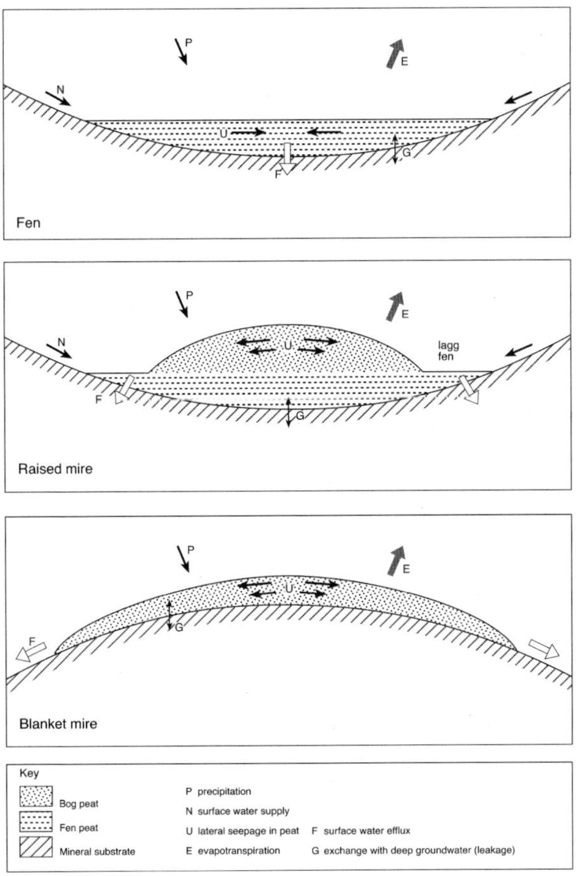

Figure 5.11 Water Fluxes in different Mire Types.
Reprinted from *The Science of the Total Environment* Vol 294, Bragg OM 'Hydrology of peat-forming wetlands in Scotland.' p111 to 129, 2002. With permission from Elsevier.

Blanket mire occurs from sea level to 1070m asl (Ratcliffe 1964). These extensive peat deposits carpet flat ground and gently slopes in areas of high precipitation and high humidity in northern and western Britain. It needs precise conditions to develop:

- annual precipitation over 1200mm,
- at least 160 wet days per year,
- a rainfall surplus of 200mm from April to September,
- annual mean temperature in the warmest month no more than 15°C,
- impermeable underlying geology,
- slope angle under 30° (Moore 1993; Backshall *et al.* 2001).

It supports a limited range of acid-loving vegetation due to its ombrogenous nature. The combination of organic matter and water flux lead to a distinctive two-layer structure. Uppermost lies the acrotelm, the most recently dead plant matter, still decomposing, which is spongy with low bulk density. It fluctuates from being waterlogged to aerated as the watertable changes. Beneath this, somewhere between 6 and 50cm down, lies a denser, permanently waterlogged layer called the catotelm. The acrotelm is the zone of water transmission, helping to protect the blanket mire from erosion (Tallis 1998).

Typical plant communities found on mires can be generically termed wet heath and bog pools. They vary considerably due to geography and altitude. The largest areas are dominated by the NVC communities of M17, M18 and M19 (Table 5.11). Where local conditions make the peat much wetter, M1, M2 and M3 have evolved; at the edges of the blanket peat where deposits are thinner, M15 and M16 are found (Tallis 1998); these cover about 400,000 ha. From west to east in Britain, natural blanket mires follow a general M17–M18–M19 trend; M17 occurs up to 500m asl, with M19 typically above this level up to about 1000m asl. Characteristic plant species include cross-leaved heath, *Sphagnum* mosses, sedges (*Carex* spp.), cotton grasses, purple moor grass and deer grass (*Scirpus cespitosus*). Interesting insectivorous plants also grow here, such as sundews (*Drosera* spp.) and butterworts (*Pinguicula* spp.). Management activities like grazing, burning and drainage divert the vegetation away from these eight communities, towards others such as M20, M25, H9, U6 and H12 (Shaw *et al.* 1997 and Yeo 1997, cited in Tallis 1998). These cover some 900,000 ha.

Blanket mire plant communities support a range of fauna, among which birds and invertebrates have received most attention. Few birds rely solely on blanket mire and the adjacent dwarf shrub heath (DSH) for their existence, except for red grouse, greenshank *(Tringa nebularea)* and dunlin *(Calidris alpiria)*. Water birds like the arctic skua and red-throated diver prefer the open pools and undrained bog (Bibby 1988). Compared to the adjacent drier DSH, the invertebrates in blanket mires include more species associated with the cooler climate of Scandinavia and the subarctic (Coulson 1992, cited in Tallis 1998). Two groups are important: enchytraeid worms, to break down plant matter ready for peat formation, and tipulids (craneflies) as food for a range of moorland birds.

Wet heath plant communities can also be found where the peat is thinner, usually at the edges of raised or blanket mires. They eventually shade off into dry heaths if drier conditions prevail. There is often a mineral component beneath, usually making the soils form some sort of stagnopodsol or stagnogley. For wet heaths much of the organic matter is derived from decomposition of cross leaned heath, which can tolerate a lower water table. Other typical plants include deer grass, purple moor grass, bog asphodel and heath spotted orchid. At higher altitudes, plants such as cloudberry *(Rubus chamaemorus)*, bear berry and dwarf cornel *(Cornus suecica)* appear (Atherden 1992). Seven wet heath communities are classified under the NVC, the more notable of which are shown in Table 5.11.

Tallis (1998) estimated that erosion affects 350,000 ha of blanket peat in the UK, with more erosion at higher altitude (above 400m asl in the east, 450/500m in the west). Peat accumulates

Table 5.11: Sub-Alpine Vegetation: Bogs and Mires

NVC Community	*Dominant Species*	*Environmental Conditions*	*Distribution*	*Notable Fauna*	*Zonation*
M17	*Scirpus caespitosus – Eriophorum vaginatum* blanket bog	Controlled by oceanicity and hydrology of location Ombrogenous peat 2 to 4m deep at pH4 or less Precipitation >1200 mm and 160+ wet days pa <500m asl on slopes up to 25°C Highly stagnant watertable	Scottish Highlands Dartmoor Bodmin moor Lake District	Golden Plover Red Throated Diver Dotterel Butterflies: Large Heath Various ground beetles Various money spiders	SUB SLPINE SUBCLIMAX
M18	*Erica tetralix – Sphagnum papillosum* raised and blanket mire	Controlled by oceanicity and surface relief Ombrogenous peat which raises above watertable in lowland areas Precipitation 800 to 1200mm for 140 to 160 wet days pa <500m asl developing into a domed profile Highly stagnant watertable	Borders SW Scotland Solway Plain		
M19	*Calluna vulgaris – Eriophorum vaginatum* blanket bog	Ombrogenous peat <2m deep at pH 4 or less Precipitation > 1200mm and 160+ wet days pa 300 to 850m asl and 0 to 10° undulating ground Harsh winters with Feb <0°C Burning, grazing and drainage have shaped this community	Southern Uplands Lake District Pennines Scottish Highlands Cheviots		
M20	*Eriophorum vaginatum* blanket and raised mire	Ombrogenous peat Precipitation 1200 to 1600mm and 160 to 200 wet days pa 500 to 700m asl Perpetuated by grazing and burning Severe peat erosion is typical	Peak District	Various specialised beetles	SUB APLINE PLAGIOCLIMAXES
M25	*Molinia caerulea – Potentilla erecta* mire	Confined to low moorland stream sides and open slopes Shales and gritstone parent material Acid/ neutral peats and stagnopodzols Range pH 4 to 5.5 influencing sub communities Perpetuated by grazing, burning and drainage	Wales Southern Scotland SW England		
U6	*Juncus squarrosus – Festuca ovina* wet grassland	Underlain by superficial deposits eg glacial/ periglacial Stagnohumic gleys and humic stagnopodzols pH 3.5 to 5 Precipitation 1000mm +and 160+ wet days pa 400 to 800m asl Perpetuated by grazing	Breadalbane Lake District Pennines Wales S Uplands	*Coleophora alticolella* moth	

(Source: Adapted from Rodwell *et al.*, 1994; Backshall *et al.*, 2000) very slowly, about 4 to 10cm per century (Chambers 1984), so a deposit that took centuries to develop can be destroyed in a few years by exploitation. The peat in a blanket mire has potential for pure conservation, recreation, agriculture, afforestation and extraction (Tallis 1998). The first management task is to identify whether the peat mass is still accumulating, damaged, eroding or modified, and thus ascertain its conservation potential and the possibility of restoration (Table 5.12).

Table 5.12: Blanket Mire Activity

terminology	*process*	*definition*
Accumulating	Active	The continued growth of Sphagnum which adds to the peat mass, or the presence of NVC communities M15, M17, M28 or M19
Damaged	Degrading	Bare peat exposed as a result of some form of land use
Eroding	Degrading	Bare peat is being eroded by wind and water
Modified	Degrading	Blanket mires which exhibit NVC communities other than M15, M17, M18 or M19

(Source: Tallis 1998)

Bogs and Mires: management issues

The most significant threats to mires come from land uses that dry out or remove the peat completely. Extraction for fuel or horticultural purposes and afforestation present the greatest problems. Tallis (1998) estimated peat was being cut on 100,000 ha in the UK and 119,000 ha in Ireland and afforestation was affecting 208,000 ha in the United Kingdom and 140,000 ha in the Irish Republic. The extent of past peat extraction for domestic use in the uplands was long overlooked. Ardron *et al.* (1997) surveyed moorland in the south Peak District and estimated that 34 million cubic metres had been extracted, compared to the 30 million cu.m. that formed the Norfolk Broads. They also estimated that moorland gripping for drainage (see Chapter 4) had affected 500,000 ha in Scotland alone.

A second group of land uses degrade peat and vegetation; these include agricultural grazing, burning management and recreational pressure. Much of the grazing impact is similar to that found on dry heath (see above), with selective grazing and marked seasonal exploitation. Tallis (1998) suggested that overgrazing was responsible for the most damage to blanket mires, citing a litany of upland areas where the natural communities have been replaced by M25 (species-poor Purple moor grass mire); some areas had been overgrazed for much longer than others, depending when sheep-based upland agriculture became dominant in that area. The effect has been magnified by the abandonment of transhumance, complete by the late 1700s, and the huge, rapid rise in sheep numbers in LFAs of 87% between 1957 and 1981 (Sydes and Miller 1988). Productivity of vegetation on wet heaths (15%) is much lower than the 40% found on dry heaths (Grant and Armstrong 1993), so sustainable stocking densities on these plant communities are extremely low: less than 0.15 LU/ha (Backshall *et al.* 2001). However, various studies have found few estates stocking at this level (e.g. Hudson 1992). In some cases stocking rates were as high as 0.37 LU/ha, two and a half times the recommended level, leading to exposure of the underlying peat and subsequent erosion (Anderson and Radford 1994). These grazing effects can change plant communities from one type to another, as Thompson *et al.* (1995) modelled (Figure 5.12). Compounding the effect of grazing is livestock trampling (see Chapter 4).

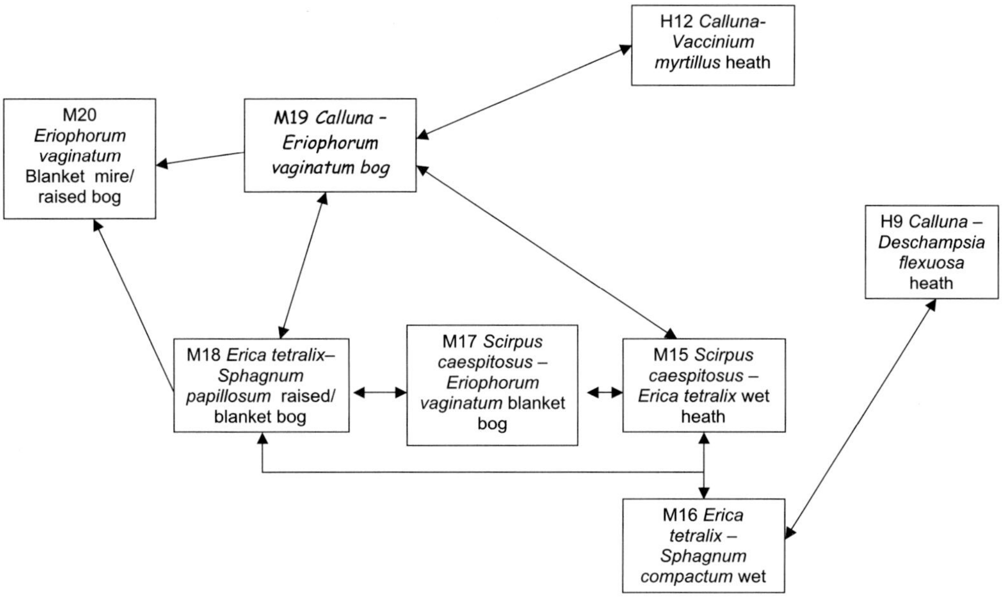

Figure 5.12 Replacement of Wet Heath Moorland Communities.
Reprinted from *Biological Conservation* Volume 71, Bardgett RD., Marsden JH. & Howard DC 'The Extent and Condition of Heather in Moorland in the Uplands of England and Wales,' 1995:170, with permission from Elsevier.

As burning is mainly a tool of grouse management on the more economically viable DSH, it is hard to understand why it would be used on wet heaths, especially since it is challenging to use. This is because the ericoid species do not enter the degenerate phase, as they do on DSH; instead, the woody stems are subsumed into the accumulating peat from below. Thus any burning can remove not only the surface vegetation but also the top layers of peat, especially if dry weather or a low water table has dried out the acrotelm (Kuhry 1994, cited in Yallop *et al.* 2005). Burning also alters the vegetation composition and renders the surface more susceptible to erosion by exposing the acrotelm. However, Mallik *et al.* (1984) showed that soil pores clog up and initiate peat development (see Chapter 3).

There is evidence that atmospheric pollution, particularly nitrogen oxide and acid rain, is affecting the conservation value of blanket bogs. *Sphagnum* and Ling are negatively affected as these pollutants increase. Some areas are more susceptible than others: the Pennines, Wales, the Lake District, Galloway and western Scotland are all regions where acid soils and underlying geologies lower the capacity of the soil to buffer the effects of acid precipitation (Woodin 1988; Gorham 1998). A particular problem is occult precipitation, where moisture from low cloud condenses straight onto the vegetation on hillsides.

The management issues of blanket mires are twofold: to avoid complete destruction and to avoid degradation or restore degraded areas. How these issues are dealt with is very much a function of the aims for the blanket mire. For instance, conversion to forestry may be seen as positive management by foresters, but negative by nature conservationists. As the intention of this book is to investigate the relationship between agriculture and nature conservation in uplands, it will be assumed that any loss of blanket mire in a quantitative or qualitative way is undesirable, since blanket mire provides some grazing potential and masses of biodiversity. It is also increasingly valued as a carbon sink (see Chapter 9).

Degradation is a trickier issue, because it requires strict control of any management techniques that reduce biodiversity or initiate damage, erosion or modification of NVC communities (if they form the aim). A management plan should therefore aim to:

- control grazing to appropriate stocking densities,
- follow controlled burning practices,
- retain the water table,
- revegetate bare peat,
- prescribe variations in line with individual species requirements.

As noted above, the ideal stocking density for blanket mires is below 0.15 LU/ha. However,

> Published data on acceptable stocking levels rarely distinguish between different breeds of sheep or between different levels of NPP [net primary production] (Tallis 1998: 94)

Hulme and Birnie (1997), investigating the 0.15 LU/ha stocking density, found that native Shetland sheep can stock at this level, but Scottish Blackface need to be stocked at only 0.06 LU/ha on blanket mire. Stocking densities need to remain economically viable for a farm business but, as Anderson *et al.* (1997) have calculated for the Peak District, loss of vegetation can lead to losses of up to £80/hectare annually from sheep grazing. Using this figure, some basic accounting calculations can be made for a blanket mire and the effect of reducing its stocking density, allowing for the breed.

Burning is not recommended for blanket mire at all if nature conservation is the objective. However, no burning over many seasons can encourage the build-up of dead leaf litter, which would normally create a fire hazard in long dry periods. It may therefore be wise to carry out a controlled burn roughly every 15 years. Retention of the water table is addressed by blocking grips with any of a variety of materials (Stoneman and Brookes 1997). Sluices can be used to control water flow – especially useful if separate units of hydrological management are needed to give different water levels in different parts of the mire, to serve the varied ecological requirements of various species and perhaps integrate more fully with grazing requirements.

Revegetation of bare peat is more complex and drawn out, but many large projects have investigated the practicalities of this, notably in the Peak District, North York Moors and Yorkshire Dales (Harvey 1993; Pearce-Higgins and Yalden 1997). All these projects tend to recommend the same tasks:

1. remove all stock for at least five years,
2. install dams and sluices in any grips,
3. replace the original vegetation if lost – seeds, turf transplants or potted plants,
4. stabilise peat with a nursery grass crop (dying out from competition in a few years),
5. extremely mobile areas of peat to be stabilised using geotextiles (Box XXIV).

Such exercises are expensive. For example, revegetating 4.5 ha of Holme Moss in the southern Pennines cost £27,000 using hydroseeding[4] (Anderson 1997, cited in Tallis 1998) and it cost £25,000 to restore one kilometre of Pennine Way path which had devoured the surrounding blanket mire (Tallis 1997). There is no reason for a farmer even to consider such costs, though they might put animals to graze the mire once conservation work was complete. Stocking density limits would be severely limited on the mire, only gradually rising in the light of experience, so short-term compensation might be offered in a positive management agreement (Chapter 8) or

[4] Hydroseeding sows seeds in a liquid spray, which can also contain pesticides and fertilisers if necessary.

Box XXIV: Geotextiles

Geotextiles can be defined as:

> Flexible fabrics of natural or man made materials used on or in the soil to perform a variety of functions. (Blunt 1993: 8)

They were first used in road building and other civil engineering schemes to aid stabilisation of soil and vegetation. They are now common in engineering applications where permanence is important and in bioengineering, where the textile is used alongside vegetation.

Geotextiles are particularly useful under paths which receive heavy use (such as cycleways), where soils are fragile (such as peat or lithosols; see Chapter 4), and beside paths where vegetation is easily damaged, exposing soil to erosion. There are three types:

- artificial – thin woven nets
- artificial – thick 3-D mats
- natural – jute or coir mats

Artificial textiles have two main purposes. First they act as separators – for instance in path trays – to encourage water to stay where it should be, away from the path. These look like soft plastic weave; the best-known brand is Terram™, though Covertex™ is also popular. The second purpose of artificial geotextiles is structural reinforcement. This type is much thicker and is usually made from polythene derivatives. Wyrtex™ is a good example – it comes as plastic-coated wire nets or geogrids. These are effective in bank stabilisation, where rubble and debris need holding back to allow soil particles to build up to create a growing medium. Both thin and thick artificial geotextiles can be hydroseeded afterwards. They are permanent and expensive.

The other main type of geotextile is made from natural materials, typically jute and coir (coconut). According to Agate (1995), natural geotextiles have particular benefits in being:

1. biodegradable,
2. cheap compared to artificial products,
3. mouldable to the landscape,
4. able to create microclimates for plant growth,
5. suited to allowing soil capture,
6. a source of insulation and shade,
7. able to adsorb water (jute mesh can adsorb 660% of its dry weight; coir mesh can adsorb 300% of its dry weight), and
8. wildlife-friendly – animals don't get trapped!

These materials come in rolls, which are simply rolled out across the hillside and pegged down, weighted with stones or dug under the soil. The main disadvantage of natural geotextiles is actually their biodegradeable nature. In the uplands Agate (1995) notes that jute lasts only about a year, which is not long enough for vegetation to establish. It also breaks up and flaps around, which is aesthetically unpleasing and potentially a danger to wildlife.. In contrast, coir matting lasts for up to 10 years in the uplands and is better for slow plant establishment, as with Ling.

The general verdict by most managers is that natural geotextiles are good for small, steep slopes that are difficult to re-vegetate and where trampling can be excluded. If the slope angle is over 25°, these mats are more effective than hydroseeding, but more expensive to use. It is important to note that geotextiles can only succeed in vegetation re-colonisation and soil stabilisation if other factors, such as trampling, grazing and drainage, are also dealt with.

some form of carbon grant (Chapter 9). Finally, some researchers believe the erosion of blanket mires into hagg complexes is part of the natural evolution of these landscapes (Evans and Warburton, 2010), though this could be an instance where overgrazing has triggered a sensitive response (Gordon *et al.* 2001; see Chapter 4).

Upland moorland mosaics

Dwarf shrub heath, grassland and mires all develop in similar environments on open fells. They are often found next to one another, their distribution controlled by variations in the local geology, geomorphology and hydrology. The effect is a mosaic of three habitats which can easily cover a whole heft, if not several.

The agricultural management challenge in moorland mosaics is how to ensure that use of the fell does not unduly degrade one or all parts of the mix. This is particularly difficult with grouse and sheep, which seek different proportions of the vegetation types (Table 5.7). Furthermore, the needs of birds vary from species to species, with black grouse requiring moorland plants as adults, and ants and sawflies as chicks; whereas golden plovers eat only insects, bar the occasional earthworm (Dodds *et al.* 1996). As already stated, decisions on a particular site are a function of the management objectives. Most moorland is used for some form of economic return, with nature conservation as a secondary objective, so it needs to be profitable to continue that management. For example, Felton and Marsden (1990) calculated the loss in gross margins as stocking rates are decreased in dry heather moorland (Table 5.13).

Table 5.13: Costs of Reducing Stocking Rates for farmers on DSH in 1988

down to 1 ewe/ha		*down to 1.5 ewes/ha*	
current level	*gross margin change*	*current level*	*gross margin change*
4.0	46.20	4	31.80
3.5	46.20	3.5	31.80
3.0	41.85	3.0	27.50
2.5	35.90	2.5	21.50
2.0	26.70	2.0	12.34
1.5	14.40	n/a	n/a

(Source: Felton & Marsden, 1995)

Of course Table 5.13 shows 1988 prices (date of survey), but the general trend is that, as expected, the higher the original stocking density, the greater the loss in gross margin. Nor is abandoning the grazing an option, because undergrazing delivers as many problems as overgrazing. Only occasionally is nature conservation the primary objective, as at the RSPB reserve at Geltsdale where the site is specifically managed for birds, using grazing to achieve this, making the grazing value a financial by-product.

The other issue on moorlands is the appropriate stocking density. The three main habitats making up the mosaic have been shown to have different carrying capacities, depending on plant species composition, seasonal palatability, life-cycle stage and even the type and breed of livestock. Identification of an optimum level seems difficult, particularly when experimental research produces contradictory answers. For example, Hope *et al.* (1996) found that reducing grazing did not necessarily lead to DSH recovery, whereas Pakeman *et al.* (2003) and Milligan *et al.* (2004) found that it did.

Three approaches can be adopted to resolve this. Firstly, using the lowest common denominator as the controlling stocking density, namely 0.15 LU/ha. Secondly, calculate the area of each habitat on the site, then scale up and add to give an overall figure. Finally, use the 'suck it and see' system and adjust the density year on year until some form of equilibrium is reached. All three options have their drawbacks and benefits.

The first option, stocking at 0.15 LU/ha, is the conservative choice: it will alleviate overgrazing pressure instantly and allow suppressed vegetation to flourish (as at Wasdale in the western Lake District: Johnston *et al.* 2005). However it may create differential grazing, as happens when calcareous grassland is present in patches (see above), and it can cause interheft drift of stock if not everyone using the common has reduced to the same level. It could be financially crippling unless grant aided and might lead to undergrazing. Next, the proportional approach usually gives a figure much closer to the current stocking density, which is more manageable for the farm business, though it will probably lead to selective grazing behaviour across the mosaic, undermining the management objectives. The final option, 'suck it and see', allows gradual adjustment of numbers, which again can help the business adjust, but can be a seesaw financially with year-on-year changes. It does allow for gradual recovery to meet nature conservation objectives, but it can be a slow process. This, together with active shepherding to keep stock on-heft, is probably how our ancestors dealt with the problem.

All these options are equally valid in practice. Nonetheless, there have been strenuous efforts to develop predictive tools using computer software, feeding in data for identified variables to calculate a stocking rate. The Hill Grazing Management Model (HGMM), for example, takes stocking rates, heather/grass cover and vegetation utilisation by stock, then uses multiple regression[5] to predict appropriate stocking rates (Armstrong *et al.* 1997a/b; Palmer *et al.* 2004). MLURI[6] are at the forefront of these developments, working on upland grazing systems for Scotland; their latest model is HILLPLAN. However, as with all models, a great deal of time and effort is spent on empirical trials and gathering field data to ensure that all parts of the grazing system on a hillside are fully understood.

Inland rock habitats

Uplands contain three main bare-rock communities: ledges, crags and scree slopes, and limestone pavement. Vegetation is usually sparse and patchy, as there is virtually no soil. In 1998, surveyed inland rock habitats covered 29,000 ha, 45% in Environmental Zone 6, 55% split between zones 3 and 5 (Haines-Young *et al.* 2003). This was 8,000 ha more than in 1984, probably owing to better surveying; these habitats cannot really be created anew.

There are two main types of plant community on inland rock:

- plants using the rock to shelter from climatic extremes or grazing,
- species which are poor competitors, but can stand extremely infertile conditions.

Ledges, crags and scree are important as they support a range of vascular plants, lichens and bryophytes which are internationally scarce. They provide refuge for species that are particularly intolerant of high grazing pressure, though plants on ledges, crags and scree can expand their distribution outwards if grazing is kept in check. Plants in these habitats are called chasmophytic; many of them are ferns, such as brittle bladder fern (*Cystopteris fragilis*) and maidenhair spleenwort (*Asplenium trichomanes*), forming NVC community OV39 or OV40 on alkaline geologies. Woodrush (*Luzula sylvatica*) dominates on acidic geologies to form U16 (Drewitt and Manley 1997).

---

[5] Multiple regression is a statistical technique to account for possible multiple factors affecting the issue in question.
[6] MLURI - MacCaulay Land Use Research Institute now the James Hutton Instittue.

The other main type of inland rock is limestone pavement, a rare habitat in Europe: over 40% is found in Britain (Drewitt and Manley 1997), with most of the rest in Ireland. They are of international importance for the plant communities that develop on, in and around them. Pavements are formed by the action of ice scouring burden off limestone beds of Carboniferous, Dalradian or Cambrian age. Loss of the overburden exposes them to water erosion through carbonation. The water penetrates natural weaknesses in the limestone, which gradually widen to form vertical fissures known as grykes, separating large blocks, or clints (Figure 5.13). The surface of clints can be pitted with various runnel-like features (karren) as well as pot holes and solution hollows.

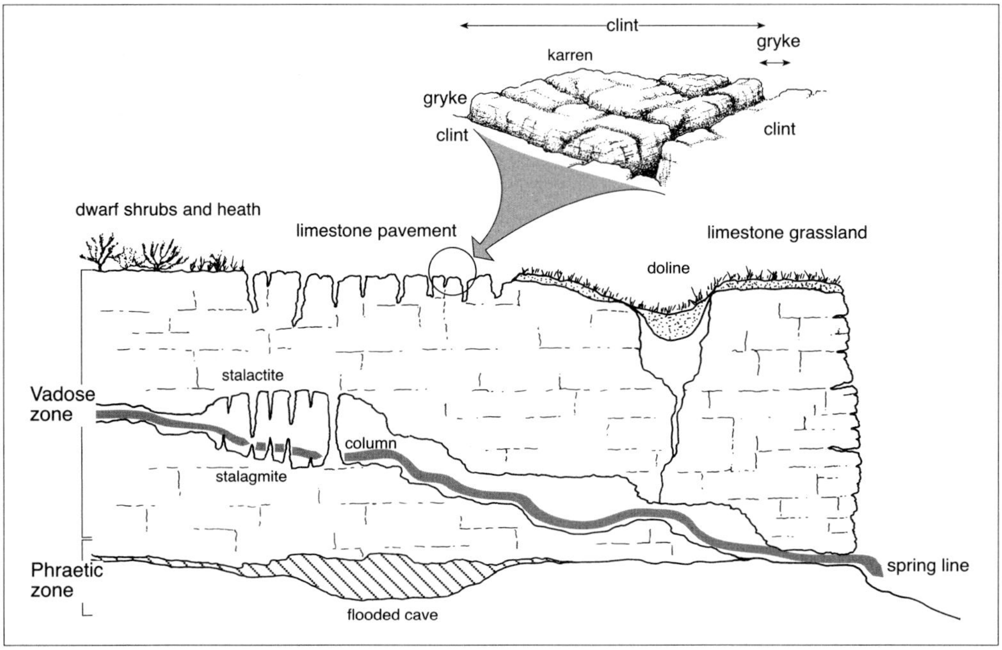

Figure 5.13 Limestone Pavements and related landforms

The grykes, karren, pot holes and hollows provide microniches where certain plant species thrive. The grykes are particularly important because most pavement flora live in their humid, shaded microclimates, similar to upland ashwoods (W8 or W9). Over 300 plant species are recorded on the pavement habitat, six of which cannot exist anywhere else (Ward and Evans 1976). The flora support a range of invertebrates, particularly snails and butterflies, though many are not unique to this habitat (Webb and Glading 1998). In the UK, pavements cover about 3,000 ha in north-west England, Wales and Scotland. Limestone pavement occurs in three main situations:

- in pasture, e.g. Asby Scar, Cumbria;
- in woodland clearings, e.g. Whitbarrow, Cumbria;
- beneath closed-canopy woods, e.g. Gaitbarrow, Cumbria.

Each type has different geomorphological features and flora. A survey by Ward and Evans (1976) found the best pavements were at Hutton Roof and Whitbarrow in Cumbria, around Morecambe Bay in Lancashire and on the flanks of Ingleborough in North Yorkshire.

The very nature of screes and ledges protects plants from the herbivory of most grazing animals. That is why their flora remain so rare. Screes are particularly immune as sheep cannot navigate the uneven, leg-breaking debris. Ledges can be invaded if grazing pressure forces stock along them, though more agile rabbits and deer are likelier to destroy these habitats. This has been a problem for montane willows (see above). Some pavements are more accessible than others, depending on their type. Many have been damaged by illegal extraction because unfortunately their water-worn surfaces are highly sought after for rock gardens (Ward and Evans 1976).

Limestone pavements can suffer agricultural damage from grazing, stone removal for walling, application of surface treatments (pesticides and/or fertilisers) and stock feeding, particularly where pavement is on common land. Stock feeding can cause poaching and localised nutrient enrichment, which can percolate down into the grykes. This can create a difficult management issue, because the solution is usually to fence off the pavement and adjacent grassland so it can recover. However, before fencing can be erected on common land, all commoners must agree and so must the Secretary of State for the Environment. Once a pavement has been damaged by rock removal, little can be done. All one can hope is that limestone grassland will eventually develop as a thin skin over the surface.

For pavements that can be saved from destruction, it is grazing management that is important. A grazing density of no more than 0.15 LU/hectare is recommended by Natural England (Backshall *et al.* 2001). This leads to:

- increasing species populations and diversity,
- more grazing-intolerant species,
- more species which favour grykes and clint tops,
- greater physical structure in plant communities,
- the provision of invertebrate shelter,
- the spread of pavement-fringe habitats like limestone grassland or dry dwarf shrub.

Cattle are usually preferred because they will not venture onto the pavement, unlike sheep, but farming practices in upland areas usually mean sheep rather than cattle are stocked. Supplementary feeding should be avoided if at all possible; otherwise, sheep need to be fed away from the pavement area, preferably in an area of low conservation value to contain the poaching to one place.

Summary

As the Countryside Survey 2000 showed, the uplands contain much of the UK's semi-natural and near-natural vegetation, which supports faunal biodiversity. This struggles to adapt when its environment is degraded or lost through human activity. Many of these habitats have been exploited for subsistence since the arrival of humans 5,000 years ago.

Since 1900, exploitation has increased and diversified to include afforestation, peat extraction, water collection, recreation and capitalist agriculture. However, the impact of these land uses has not been uniform, neither by habitat nor by upland massif. An audit by Drewitt and Manley (1997), for example, investigated habitat condition in 18 natural areas above the fell wall in upland England (Table 5.14). The higher the score, the more NVC communities exist in the area, indicating a richer ecology. The Cumbria Fells and Dales are the ecologically richest upland, with fine examples of dwarf shrub heath, montane, wetland, acid and calcareous grassland, and inland rock. At the lower end of the spectrum are the Oswestry Uplands, which contain only calcareous grasslands and wetland.

Table 5.14: Habitat Diversity in Upland England Natural Areas

category	*number of NVC communities*	*Natural Areas*
Outstanding	10+	Cumbria Fells & Dales Dartmoor North Pennines Yorkshire Dales
Considerable	7 to 9	Border Uplands Bowland Fells Exmoor & the Quantocks North York Moors South Pennines Staffordshire Uplands
Notable	5 to 6	Bodmin South West Peak Black Mountains & Golden Valley
Some	1 to 4	Central Marches Dark Peak Shropshire Hills
Negligible	0	White Peak Oswestry Uplands

(Adapted from: Drewitt & Manley, 1997)

More detailed fieldwork by Drewitt and Manley (1997) explored the issues inhibiting favourable conditions in those habitats that remain in each of these natural areas (Table 5.15). Unfortunately, most management problems stem from farming and in fact every habitat in every natural area is affected by some form of agriculture. The harmful operations most extensively recorded are inappropriate grazing and overgrazing. Lack of management has resulted in scrub invasion, bracken encroachment and undergrazing. Poor burning has affected communities with ericoid species, though Drewitt and Manley do not clarify whether this is a result of farming or grouse keeping.

Table 5.15: Land Management Issues in Upland England Natural Areas

natural area	*broad habitat*	*key issues*
Black Mountains & Golden Valley	Dwarf Shrub Heath Wetlands Acid Grasslands Calcareous grasslands	G/BU/F/BR/H G/F AI AI
Bodmin Moor	Dwarf Shrub Heath Wetlands Acid Grassland	BU/O/V/H BU/O/V G
Border Uplands	Montane Dwarf Shrub Heath Wetlands Acid Grasslands Calcareous grasslands Mesotrophic grasslands Scrub/upland wood O/P/AD/R/AC	BU/O/RC/M G/O/BU/D/AF O O O O/RE

Table continued over

Table 5.15: *continued*

natural area	*broad habitat*	*key issues*
Central Marches	Dwarf Shrub Heath	G/BR/S/R/V
	Wetlands	G
	Acid Grassland	AI/O/BR/AF
Cumbria Fells & Dales	Montane	O/R/P
	Dwarf Shrub Heath	O/AF/BU/SF/R/BR/W
	Wetlands	PC/S/D/E/O
	Acid Grasslands	S/BR/O/U/AI
	Calcareous grasslands	S/O/U/AI
	Inland rock	L
	Scrub/ upland wood	O
Dark Peak	Dwarf Shrub Heath	R/F/ER/O/BR/BU/FR
	Wetlands	D/BU/FR/O/F/ER
	Acid Grasslands	D/AI/G/BR/S/AC/R
Dartmoor	Dwarf Shrub Heath	O/BU/RC
	Wetlands	BU/G/O/ER/WA
	Acid Grasslands	BR/O/D
Exmoor & Quantocks	Dwarf Shrub Heath	U/BU/RC/BR/SF/V
	Wetlands	BU/D
	Acid Grasslands	G/RC
Forest of Bowland	Dwarf Shrub Heath	O/R/AC/BU/RC/AI
	Wetlands	O/R/AC/BU/RC/D/P
	Acid Grasslands	AI
North York Moors	Dwarf Shrub Heath	AC/R/BU/BR/SF/SH/RG/GR
	Wetlands	BU/D/FR
	Acid Grasslands	AI/AF/S/T
North Pennines	Montane	O/R/AD
	Dwarf Shrub Heath	O/BU/R/SF/BR/RG/W
	Wetlands	O/D/BU/R/SF/SP/W
	Acid Grasslands	O/R/SF/RG/BR
	Calcareous grasslands	O/R/SF/RG
	Mesotrophic grasslands	O/SF
	Scrub/ upland wood	O/RG
Oswestry Uplands	Wetlands	G/M/FR/RC
	Calcareous grasslands	G/M/Q/AI
Shropshire Hills	Dwarf Shrub Heath	G/BU/SF/FR/BR/RC
	Wetlands	G/SF/FR/RC
	Acid Grasslands	G/RC
	Calcareous grasslands	G/RC
South Pennines	Dwarf Shrub Heath	G/BU/F/AC/D
	Wetlands	AC/R/D/W/BU
	Acid Grasslands	O/BR/AI/AC/R
South West Peak	Dwarf Shrub Heath	BR/O/BU
	Wetlands	O/D/BU/AI
	Acid Grassland	AI

Table continued over

Table 5.15: *continued*

natural area	*broad habitat*	*key issues*
Staffordshire Uplands	Dwarf Shrub Heath Wetlands Acid Grasslands	G/BU/RC/S/BR/FR G/S G/ST/S/BR
White Peak	Dwarf Shrub Heath Acid Grasslands Calcareous grasslands	O/AI/BR/FR M/BR/U/S/E/R M/U/S/E/R
Yorkshire Dales	Montane Dwarf Shrub Heath Wetlands Acid Grasslands Calcareous grasslands Mesotrophic grasslands Scrub/upland wood	O/P O/BU/D/AI O/D/P O/AI L/Q/O/AI AI RE

Key: AC access; AD acidification; AF afforestation; AI agricultural improvement; BR bracken encroachment; BU inappropriate burning; D drainage; E eutrophication; ER erosion; F accidental fires; FR habitat fragmentation; G grazing issue; GR grouse management; L rock removal; M lack of management; O overgrazing; P pollution; PC peat cutting; Q quarrying; R recreation; RC reclamation; RE poor regeneration; RG woodland regeneration; S scrub encroachment; SF stock feeding; SP Sphagnum harvesting; SH poor shepherding; U undergrazing; V off-road vehicles; W wind; WA water catchment.

(Source: adapted from Drewitt and Manley 1997.)

Montane habitats suffer most from overgrazing, pollution and recreation. Dwarf shrub heath is most affected by bracken encroachment, inappropriate burning and overgrazing. Both the latter adversely affect wetlands, which also suffer from inappropriate drainage. Grasslands endure agricultural improvement and overgrazing, with more specific issues of bracken encroachment on acidic sites and stock feeding on mesotrophic grassland habitats. Scrub and upland woods experience overgrazing and thus poor regeneration.

Whatever the cause of deterioration or loss of habitat, there are practical steps that can improve things. Often it is a matter of discontinuing inappropriate operations, by exclusion or extensification, or re-introducing management techniques now considered redundant. These are things all land managers can do, but it is another matter whether they can afford to do so. For farmers on the physical margins of cultivation, to go back to a less intensive, low-fertility system would be economic suicide without financial support. If they are to embrace successfully the low-input/low-output systems that helped create the multiplicity of upland ecological habitats, they need to be presented with options that render the farm business economically viable. To understand why this can be such a challenge, the next chapter addresses the economics of upland agriculture.

6

The economics of upland agriculture

In Chapter 5 we saw that upland agriculture has been responsible for, and continues to produce, a range of high-quality natural environments which are valued by society. Chapters 4 and 5 also showed that certain agricultural operations can impact negatively on upland biodiversity. The key is to maintain an appropriate balance between agriculture and conservation, but in such a way that neither is undermined by the other. In order for this inter-relationship to work, upland farming needs to remain economically viable at a fundamental level. How this can be sustained is the focus of this chapter.

The broad economic picture

Many upland and mountainous areas are on the periphery of current economic development. Earlier in history, uplands played a more central role in the economy, first in livestock management and then in mining and water-powered industries, until changes in technology, transport and international trade pushed this economic core downhill, leaving agriculture behind. The resultant marginality was exacerbated by upland agriculture's location on the physical margins of cultivation (Chapter 4). Then 19th- and 20th-century afforestation policy in the UK intensified this marginalisation by converting the better rough grazing land to tree plantations (chapters 3 and 5).

Although the new industries of recreation and tourism have gone some way to stabilising UK upland economies, the close relationship of these areas with agriculture can never be ignored. For example, in 2001, when most of the countryside closed down as foot-and-mouth disease swept Britain, tourism lost about £3.2 billion (Curry 2008). Farmers who had attempted to diversify their farm business into non-farming occupations were doubly affected. Where there is little tourism, the dominant occupations remain agriculture and forestry, as in Dumfries & Galloway and the Northern Pennines. Consequently anything that reduces the effectiveness of agricultural industry in these areas can have an even greater effect on the wider community.

At farm level, the main factor influencing enterprise choice is the physical environment, as we saw in Chapter 3, but economic processes and structures also play a major part in controlling the viability of a business through the laws of supply and demand. In a capitalist economy, the supply of agricultural commodities depends on satisfying the demand of a market[1] (Ilbery 1985). The relationship between supply and demand is one of three kinds:

[1] In contrast to a subsistence economy, which simply provides food to the populace; this has not been the case in Britain since about 1650.

- With constant supply, the higher the demand, the higher the price. For example, the supply of paintings by Turner is a limited resource, so people bid at auctions against one another to own a desired work, thus pushing up the price.
- With constant demand, the higher the supply, the lower the price. Even if everyone needs a computer, once everyone has one any more production means prices fall in an attempt to get people to buy a replacement.
- In a perfect market, prices oscillate around an equilibrium where supply and demand meet.

Elasticity

Related to supply and demand is the concept of elasticity, which can be defined as a price change that brings about a more-than-proportional change in supply or demand. Inelasticity occurs when a price change brings about a less-than-proportional change in supply or demand.

An agricultural example may help to explain this. The demand for most staple agricultural produce (wheat, potatoes, milk, rice) is inelastic, simply because consumers require only a certain amount of such goods. For farmers, the law of supply and demand means the more they produce, the more they end up overproducing, as people simply will not buy any more. The market response is that prices start to fall as there is a glut. To overcome this, the farmer increases production, aiming to increase income but deflating prices even more. This is the fundamental reason why governments intervene in agriculture, thus controlling the market place and ensuring prices are kept as high as possible for the producer, even if this is an artificial situation.

For upland farmers, inelasticity in the market for meat, wool and milk is compounded by the physical limitations to expanding their enterprise choice. Upland farm businesses have specialised in a market from which they struggle to diversify, and thus are forced to intensify to avoid financial loss if costs increase. This situation is also distorted by government intervention, which has undesirable side effects such as overproduction, overgrazing, habitat deterioration and erosion (see chapters 3, 5 and 8). The combined effects of inelasticity, restricted enterprise choice and rising production costs have a seriously deleterious effect on farm incomes.

Income

This set of circumstances places upland agriculture at the economic margins of cultivation. Because production costs closely match prices obtained from the sale of goods, upland farmers make small or negligible profits annually. This is borne out by the small gross and net margins produced by any livestock business, in contrast to other UK farm types (Table 6.1a; Box XXV). Furthermore, net farm incomes (often called 'the farmer's reward' and roughly equivalent to take-home pay) can be tiny on livestock farms, between £3,700 and £5,800 including subsidies (Table 6.1b).

Table 6.1a: Gross and Net Margins of UK Farms 2002

*farm type**	*gross/ha (£)*	*farm size (ha)*	*gross margin per farm (£)*
Cereal	542	153	82,926
Mixed	773	177	136,821
Sheep & Cows:			
Scotland	87	370	32,041
Wales	289	133	38,437
England	212	188	39,856
Dairy	908	86	78,088

(Adapted from Chadwick, 2003)

Box XXV: Approaches to measuring farm size

No two farms are alike in terms of land, labour and capital; thus generalisations about agricultural patterns and trends need some form of objective measure to allow statistical comparison between enterprises or farm units. The most straightforward method is to calculate areas of land used for various enterprises or the total farm area using hectares or, historically, acres. Another technique is to use the standardised Livestock Unit system. Here animals are proportionally weighted against one another, based on how much they eat in a day. One livestock unit is equivalent to one dairy cow, from which all others are derived.

	livestock unit
Dairy cows	1.00
Dairy bulls	0.65
Beef cows	0.75
Beef bulls	0.65
Barley beef	0.47
Horses	0.80

	livestock unit
Ewes (light weight)	0.06
Ewes (medium weight)	0.08
Ewes (heavy weight)	0.11
Rams	0.08
Lambs	0.04

Source: Beaton *et al.* 2007.

Livestock Units (LUs) are useful for calculating stocking densities grazing different types of vegetation, for conservation purposes (see Chapter 5). Unfortunately native breeds do not fit this pattern as they are generally smaller and require less nutrition. Examples of native breed LUs include (Crofts and Jefferson 1999):

- Hebridean ewes (0.06 to 0.10 LU),
- Cheviot ewes (0.08 to 0.13 LUs),
- Dexter cows (0.47 to 0.64 LUs),
- Galloway cows (0.73 to 1.00 LUs).

An alternative measure to farm size or LUs is to calculate labour requirements. A popular system historically was to use Standard Man Days (Tarrant 1974; Crofts and Jefferson 1999). Standard Man Days (SMDs) allowed all types of farm production to be reduced to their standard labour requirements, where 1 SMD equals 8 hours of manual work:

type of stock	*SMD per head per year*
Suckler Cows	1.4
Bull	3.5
Bulling heifers	1.0
In calf heifers	1.4
Beef stores	1.5
Breeding ewes	0.5
Rams	0.5
Shearling ewes	0.3

Field type	*SMD per ha*
Silage (2 cuts)	4
Hay (1 cut)	2.5
Pasture	0.5

Source: Crofts and Jefferson 1999.

Although this looks like a good idea, it does assume that all farmers are equally efficient. People who are learning how to complete a task take much longer, and need more supervision, than those who have performed them for years (Mansfield *et al.* 2005). The calculations also need to be updated regularly to take in changes in economies of scale, capital investment, grants and labour-saving technological improvements. Standard Labour Requirements (SLRs) are a newer concept. An SLR is roughly equivalent to 95 ha of cereals, 50 dairy cows or 400 ewes (FBS 2006). These are representative hourly labour requirements by head or hectare under typical conditions for enterprises of average size and performance (DEFRA 2008). Those farms that equate to less than 1 SLR (or a person year) are considered to be part-time businesses.

Another way to measure farm size is to use financial information. The two most common methods are net and gross margins, which use the costs of production (variable and fixed) and the financial output from a farm. Usually farm business incomes are analysed by gross margins, as a way of protecting the 'salary' information we would not expect others to know about ourselves.

Farm economics: fundamental terminology

term	*definition*
Output	returns (sales) + used or consumed resources with no cash returns + subsidies
Variable costs	material added to farm as inputs e.g. fertilisers, contractors, casual labour
Gross margin	GM = output – variable costs
Fixed costs	labour, fuel, repairs, depreciation, rent, machinery
Net farm income	= gross margin – fixed costs
Total income from farming	= (receipts from farm enterprise sales + subsidies) – (variable costs + fixed costs + interest)

As the table above shows, net farm income (NFI) is the difference between gross margins and fixed costs – in other words, the family's reward for farming, excluding money from grant payments, and therefore is loosely equivalent to take-home pay at the end of the year. It allows a level comparison to be made between tenanted farms and owner-occupied farms by taking into account the imaginary cost of rent for farmers who own their land (DEFRA 2004). DEFRA no longer consider NFI to be the most appropriate measure of farm income because:

- it assumes most UK farms are tenancies when increasingly they are not,
- NFI does not include all on-farm activities (i.e. all diversification),
- the definition excludes financial benefits that go to partners,
- as a consequence of the above NFI does not represent the total income a farm business accrues.

The current favoured approach to measuring the economic size of a farm business is to use total income from farming (TIFF), which represents the business profits and income to farmers, partners and directors of the farm business (DEFRA 2004). The definition is therefore much broader than net farm income, and takes in, for instance, payments made to businesses from the foot-and-mouth cull in 2001 and income from diversification. TIFF therefore is calculated thus:

TIFF = (Gross output + other subsidies) – (total intermediate – total consumption)
at basic prices less taxes) consumption) of fixed capital)

The final technique to calculate farm income is to use European Size Units (ESUs). These are designed to allow comparison of data across the whole European Union, where farms vary immensely from 2-hectare olive farms in Greece or intensive pig breeding units in Denmark to pre-industrial Romanian peasant farms. ESUs use standard gross margins, where 1 ESU = 1200 SGMs. An ESU is roughly equivalent to 1.3 ha of cereals, 1 dairy cow or 25 ewes, and 8 ESUs are considered to be the threshold dividing full- from part-time farms (http://statistics.defra.gov.uk/esg/asd/fbs/sub/min_size.htm accessed 6/9/07).

The challenge with all these measures is that they do not equate directly to one another. For instance, 1 SLR is equivalent to 50 dairy cows. Using ESUs this would be still be 50 cows, but way above the 8 cows needed to be considered full-time under the ESU system. On the other hand, the SLR–ESU comparison does work with ewes. It means that statistical comparisons between data sets, using different measures, need to be treated with care.

6.1b: Gross and Net Margins of UK Farms 2002

*farm type**	*net farm income (per ha)*	*farm size(ha)*	*net farm income (£)*
Cereal	40	153	6120
Mixed	104	177	18,408
Sheep & Cows:			
Scotland	10.2	370	3767
Wales	28	133	3724
England	31	188	5828
Dairy	162	86	13,932

* Two thirds of gross margin comes from this enterprise

(Adapted from: Chadwick 2003)

Whichever measure is used, net farm income, gross margin or total income from farming (TIFF), the cost of production changes year on year due to changes in inflation and prices at market, so the income of farm businesses fluctuates. Since 1999/2000 the average income of UK farms has risen, but is still less than in 1995/6, when it reached a thirty-year high of £31,900. Cereal farms are still the most profitable, followed by dairy farms. Cattle and sheep businesses in LFAs have seen an increase in their net farm incomes (on average, six times what they were in 1999/2000) but they are still small in comparison to other farm types. Regional trends exist, with English LFA cattle and sheep farms doing three times as well as those in Northern Ireland (Table 6.2). This is believed to be due to the old headage payment system and higher prices for finished/store cattle and lambs (DEFRA 2004).

Economic controls on farm inputs

Why upland farm incomes are so limited can be explained through a range of established economic principles, particularly:

- economic man,
- economic margins of cultivation,
- factors of production,
- economies of scale.

Table 6.2: Net Farm Incomes 1999 to 2004

£ thousands/farm	*1999/00*	*2000/01*	*2001/02*	*2002/03*	*2003/4*
England:					
Dairy	9.4	14.0	30.7	18.3	20.5
Cattle & sheep (LFA)	5.3	5.8	7.9	18.0	19.0
Cattle & Sheep (non-LFA)	0.5	–0.4	–	5.8	4.5
Cereals	11.7	6.2	4.1	12.1	32.0
Wales:					
Dairy	14.2	12.3	29.6	18.6	18.5
Cattle & sheep (LFA)	3.1	3.7	1.7	12.5	17.0
Cattle & Sheep (non-LFA)	0.6	0.7	2.2	9.1	8.5
Scotland:					
Dairy	1.9	13.6	33.1	5.1	9.0
Cattle & sheep (LFA)	2.7	5.6	5.8	13.3	16.5
Cereals	10.6	0.7	–0.6	–1.6	17.5
Northern Ireland:					
Dairy	7.0	14.6	17.5	6.5	13.0
Cattle & sheep (LFA)	-1.6	0.7	4.0	4.6	6.0
UNITED KINGDOM:	**9.0**	**13.9**	**27.9**	**14.9**	**17.5**
Dairy	**2.2**	**3.9**	**4.5**	**12.0**	**14.5**
Cattle & sheep (LFA)	**0.1**	**-0.1**	**0.8**	**5.8**	**5.5**
Cattle & Sheep (non-LFA)	**11.5**	**5.4**	**3.3**	**10.0**	**30.0**
Cereals	**5.7**	**7.8**	**11.0**	**12.2**	**20.5**

(Adapted from: Defra 2004a, p. 23, Table 2.8)

Economic man

According to Johnston *et al.* (1985: 96), economic man can be defined thus:

> Given perfect knowledge and a perfect ability to utilise such information in a rational fashion, economic man maximises returns and minimises costs.

A farmer would study their options, calculate the costs and profits, and choose the optimal enterprise for each piece of land they managed. This concept has been used extensively in conjunction with the idea of economic rent[2] to theoretically model agricultural land use (Morgan and Munton 1971). In other words, s/he considers all possible land uses and enterprises for a piece of land and then chooses that which will yield the biggest surplus and hence provide the largest profit.

Such a concept assumes that people are completely rational, have access to all relevant knowledge and use it to its maximum potential. This extreme view that farmers are optimising in their choices has now been rejected and replaced by satisficing behaviour (Simon 1957), where farmers do not operate with perfect knowledge and understanding of all variables affecting their farming. Simon said that, because the world was so complex and individuals never have perfect knowledge, satisficing behaviour was more likely. Wolpert (1964) examined the effect of imperfect knowledge by calculating the actual and potential distributions of agricultural labour productivity in central Sweden. This principle was also considered by Pred (1967), who devised a theoretical model known as a behavioural matrix (Figure 6.1). Pred proposed that changes in land use were the result of a combination of two factors:

- quality and quantity of information,
- ability to use the information.

[2] Ricardo (1817) put forward the idea of economic rent as "The net income accruing to an agricultural producer over and above the net income he would get, if he were producing at the margin" (Johnston *et al.* 1985: 96).

Figure 6.1 employs three farmers, A, B and C, to illustrate how this idea influences agricultural land use. A farmer's position in the matrix depends on his/her location on the information and ability axes (Ilbery 1985). Factors like farm size, education, experience and aspiration would influence their position. According to Pred, residing in the bottom right-hand corner of the matrix would be economic man (Farmer A) with complete knowledge of the business (farm), how to farm it and complete ability to do this. In contrast is Farmer B, at the top left-hand corner of the matrix, who has no knowledge of the land or how to farm and no ability to use any of the information he/she already has. Most business people, Pred argued, were somewhere between these two extremes, with differing levels of knowledge and abilities to utilise it (Farmer C being an example).

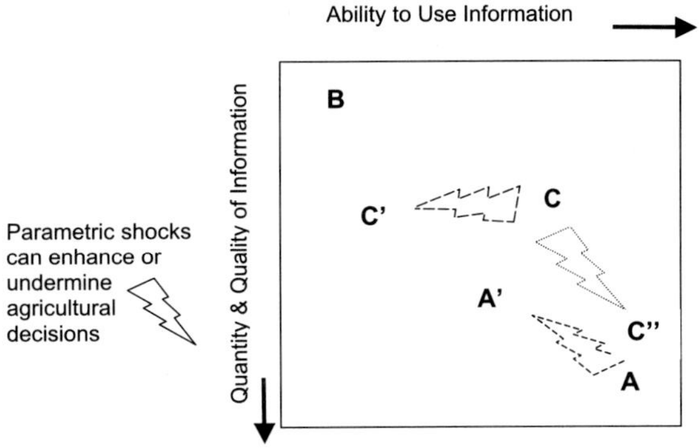

Figure 6.1 Pred's Behaviourial Matrix
Adapted from Ilbery BW, 1984.

Pred built into his matrix the recognition that decision makers (farmers) could improve their knowledge base and thus move around in the matrix in a never ending quest to reach Farmer A's exalted location. Pred also recognised that people could slide back towards Farmer B's position, often as a result of changes in the world beyond the farm, destabilising the basis on which the farm is managed and the farmer's knowledge of a particular enterprise or other business. These events were termed parametric shocks by Johnston (1979). Unfortunately Pred's behavioural matrix is theoretical and generally impossible to test in reality because it would require a complete knowledge of the situation by the researcher as well! Attempts were made by Selby (1980, 1987) researching field afforestation and sawmills in Finland. In both instances he believed he had proved that these business people operated at a satisficing level rather than optimising, and that there was a relationship between quality and quantity of information and the ability to employ it.

Pred's ideas do give us an insight into the effect of external forces on agricultural land-use patterns and farming in upland areas. Examples include the recent change of livestock subsidies from headage to hectarage payments and the arrival of the Upland Entry-Level Scheme (see chapters 8 and 9). For upland and hill farmers some of these parametric shocks are more profound than others, for instance the concept of ecosystem services and the impacts of climate change (see Chapter 9). Pred's views on information access and ability to use it are also interesting, in showing that we need to consider individuals as part of the economic viability of a farm (as explained in Chapter 7).

Economic margins of cultivation

A second economic concept usefully employed, and one related to physical margins of cultivation, is that of economic margin of cultivation. This is the point at which production costs equal income made; thus it usually lies within the physical margin of cultivation, so that income exceeds costs (Figure 4.2). In some economies, the economic margin has been extended beyond the physical margins to allow food production where it normally would not occur, as in Saudi Arabia where irrigation systems bring green to the desert. Conversely in other areas, if not managed sensitively, overextending the economic margin can lead to environmental deterioration because the land is working beyond its biophysical and ecological carrying capacity (Box XXVI). Symptoms of this in upland farming systems include changes in vegetation, overgrazing and soil erosion (as noted in chapters 4 and 5).

Box XXVI: Carrying capacity

Carrying capacity is concept that allows us to evaluate a resource's ability to absorb the impacts of use without deteriorating. Originally the idea was developed by Hawden and Palmer (1922; cited in Mitchell 1989) for livestock management on the US prairies, corresponding to "the number of stock which a range can support without injury to the range". A more encompassing definition is that proposed by Patmore (1983):

> Carrying capacity is reached when the demand for a resource is just matched by the supply of that resource.

The concept is now used in a number of disciplines to explore resource exploitation. Biophysical carrying capacity is the point when biological or physical phenomena are damaged beyond natural recovery. For example, vegetation can be damaged by pedestrian traffic or overgrazing, or soil erosion occurs (Morgan 1995). Perceptual carrying capacity is exceeded when people feel a place is overcrowded (Mitchell 1989).

It is quite possible for a resource to have different levels of carrying capacity by type. Often perceptual carrying capacities are lower than biophysical ones, on footpaths for example. Another characteristic is that a particular type can vary from individual to individual; the classic example being where we compare the views of casual visitors to a site with those of wilderness walkers. The former will have a much higher perceptual carrying capacity of how many people a site can hold than the latter. Furthermore, different plant species or habitats have different carrying capacities.

The concept has received a great deal of attention in scientific literature in an attempt to use it as a resource-management tool. However, it presents many problems, summarised by Lindberg *et al.* (1997):

- Definitions often provide little guidance for practical implementation, e.g. how many sheep are too many?
- There is a perception that carrying capacity is scientifically objective when it is clearly subjective, i.e. we all see the carrying capacity of an area differently;
- It tends to focus on use levels or numbers of visitors, yet management options relate to condition of resources;
- It relies on the state of interaction between physical and biotic environments, but trying to quantify something that constantly changes is impossible;
- Humans are limitless in their ingenuity and will find solutions to exceeding carrying capacity;
- Carrying capacity will vary over time as societies' perceptions and demands on a resource change;
- The number of variables affecting the carrying capacity can be vast, leading us to overlook important controlling factors.

Grazing pressure is a typical phenomenon that exhibits many facets of this critique, complicated by the various objectives of management for an area. Usually agricultural carrying capacity is greater than the nature-conservation capacity, which characteristically focuses on species that are intolerant of grazing. Within this are the variable capacities of different habitats, as we saw in Chapter 5, which makes it hard to calculate appropriate stocking densities. Complicating it further is the shift towards ecosystem services in uplands, which in some cases (such as maintaining carbon sinks) may not require maintaining the biodiversity value of an open fell, but simply stabilising vegetation. Critical review of carrying capacity as a management tool has led to the development of other concepts, such as 'limits of acceptable change' (Manning 2007); this has yet to be embraced fully in the UK and Europe. 'Favourable condition', used by JNCC for habitats, is an example (Backshall *et al.* 2001).

A related concept gaining attention is that of resilience. This idea originated in the discipline of ecology, specifically for population studies, where Holling described it as:

> the persistence of relationships within a system and is a measure of the ability of these systems to absorb change of state variable, driving variables, and parameters, and still persist. (Holling 1973: 17)

A better definition is:

> the capacity of a system to absorb disturbance and re-organize while undergoing change so as to still retain essentially the same function, structure, identity and feedbacks. (Folke 2006: 259)

Ecological or environmental resilience refers to the way ecosystems respond to human intervention either positively or negatively. Most of the literature has focused on issues such as biodiversity impoverishment or other resource degradation. There is also increasing discussion about the role of resilience in social systems. Adger (2000) regards social resilience as part of the framework within which communities and people adapt to environmental change. Relevant to upland agriculture, ecological and social resilience – often termed socio ecological systems (SES) – are linked, where certain societies depend on ecological resources for their economic activity. He also suggests that societies and related institutions (defined as habitualised behaviour, rules and norms) that are dependent in this way may be less resilient.

Other concepts which have become linked to resilience are vulnerability – the capacity of a phenomenon to resist or recover from the impact of a natural hazard, and adaptation – the ability of systems to offset potential damage, particularly of climate change (Gallopin 2006). Resilience is therefore not only about the capacity to buffer or be robust to disturbance (see Chapter 4), but refers to opportunities to exploit adaptive capacity inherent in systems. Adger (2000) considers social vulnerability to be a corollary of the response of individuals and/or communities to the impacts of environmental change, manifesting itself as disrupted livelihoods and loss of security – symptoms which have arguably appeared in upland farming communities since the 1980s. It is therefore important to understand the nature of social resilience, which can be a mechanism to resist vulnerability and consequently allow individuals and communities to be more adaptable as their environment changes. We need

> to develop governance systems that make it possible to relate to environmental assets in a fashion that secures their capacity to support societal development for a long time into the future. (Folke 2006: 253)

We will return to these themes in Chapters 8 and 9.

There are yet other situations where economic margins can retreat substantially away from physical margins. The classic study by Varjo (1979) investigated why there had been a southwards shift in the northern limit of cereal production in Finland in the 1960s compared to earlier in the century. Varjo was able to discount deteriorating climate as records showed no change in temperature or precipitation. Instead, he demonstrated how the rise in production costs had outweighed the rise in prices that farmers received for cereals, thus forcing southerly retreat of cultivation to compensate for declining profitability.

There has been much debate over this economic 'marginality' for upland and hill farmers. Often expressed as a combination of physical and economic circumstances, this type of marginality exists only because it is perceived as existing in relation to the rest of the agricultural economy. Of course, if we limit our investigations to farm incomes and physical limitations of uplands, this marginality is not just a matter of perception and often manifests itself in rural poverty. Fundamentally, this is the essence of the hill farming problem that emerged in the 1930s and 1940s; it was explained away by a poor combination of the factors of production, which we consider below (Evans and Attwood 1961). This character of upland farming as lagging behind the rest of agriculture has lasted to the present day, and more and more subsidy has been directed at the upland sector to offset lower incomes, in line with classic agricultural policy objectives (see Chapter 8).

Marginality was not seen as an issue prior to the Agricultural Depression, because the emerging upland farming system worked with the environmental limitations of the valley–fell system (Winchester 2000: 9):

> the lower slopes carried perhaps the greatest variety of resources: here the margin of cultivation ebbed and flowed; comparatively good grazing was available for livestock; woodland often survived in patches, and vegetation such as bracken and gorse could be harvested.

Indeed this type of environmental determinism is shown in the development of regional specialisation throughout England in the late 17th century (Thirsk 1987). Dodgshon and Olsson (2006) examined the carefully restrictive management practices of Scottish clans before 1750, which avoided over-exploitation in contrast to the later, unrealistic improvements of Clearance landlords (see Chapter 3). The current re-alignment of EU policy away from production, in parallel with increasing recognition of the value of upland farms in providing ecosystem services, further suggests that economic marginality may no longer haunt the sector (more on this in Chapter 9).

Factors of production

Originally these factors were land, labour and capital; Tarrant suggested adding management as a fourth, for

> It is clear that agriculture is much more that the sum of physical properties of land. It is a combination of physical capabilities and man's reaction to them and his ability to overcome them. (Tarrant 1974: 120)

The importance of management is demonstrated by Pred's theoretical work and farmers' ability to move toward a position of economic man. For upland farming, Evans and Attwood (1961) believed that these factors of production were responsible for its poor economic state in the mid-20th century. Later, Rieder and Wyder (1997) argued that these factors also control the absolute and comparative advantage of goods and products of uplands.

Absolute advantage is when the cost of production is always lower in uplands than lowlands for specific goods, for example the production of High Nature Value. In contrast, livestock production has comparative advantage; it is possible to rear stock, but costs are more than in the lowlands. Thus the farmer must seek to reduce the cost of production to remain competitive with lowland farmers, usually achieved by lowering wages. However, if wages have a nationally set minimum (as in the UK) the only option an upland farmer has is to reduce labour completely (see below).

Land

As a factor of production, land can vary in two main ways: by quantity and by quality. In terms of quantity, land in the UK is generally a static commodity in relation to stock[3] and individual farm size. The total stock of land can change if upland agricultural potential deteriorates through mis-management or diversion to other uses, which reduces the total farmed area available. Particular upland challenges are complete abandonment or reduced grazing of hefts. Both have taken place all over Cumbria since 2000 (Burton *et al*. 2005), despite the Farm Practices Survey (DEFRA 2009a) suggesting otherwise. Generally the FPS showed that 81% of farmers had maintained grazing at similar levels for the previous four years and 88% would continue to do so over the next two, despite changing agri-environmental policies (see Chapter 9).

More positively, land can also be improved and brought into production through liming, drainage or re-seeding operations, and this has happened regularly throughout history (see Chapter 3). Contemporary patterns of change suggest that 60% of farmers continue to maintain drainage on enclosed rough grazing, and 40% plan to reduce artificial fertiliser inputs and 8% lime inputs (DEFRA 2009a). The FPS also demonstrated that 85% of farmers did not intend to change the land area of their holding in the next two years. There were some regional variations, notably in the uplands of south-west England, where there seems to be a more fluid land sales and rental market (28% of farmers foresaw this happening).

Without improvement, land quality varies greatly because of differing combinations of environmental conditions (soil, climate and altitude). In England and Wales, land is graded using

[3] Stock – a resource management term to describe the total quantity of minerals and land available on Earth, including those consumed, theoretically recoverable or recyclable. Flow resources are those that are naturally renewed within a human lifespan, e.g. water, plant and animal life (Rees 1990).

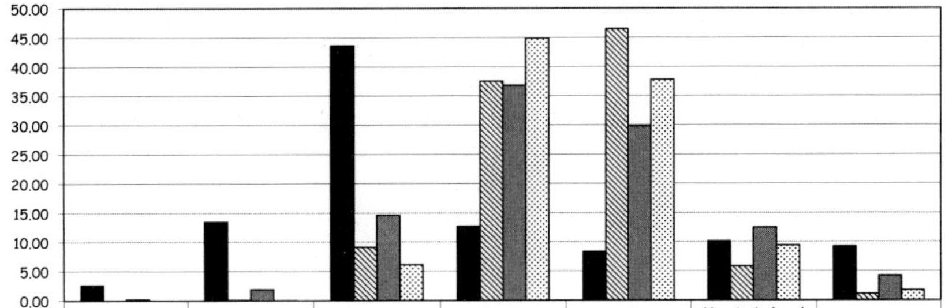

	Grade 1	Grade 2	Grade 3	Grade 4	Grade 5	Non-Agricultural Land	Urban Land
■ England	2.60	13.50	43.60	12.70	8.30	10.10	9.20
▧ LFA England	0.00	0.10	9.10	37.50	46.50	5.80	1.10
▩ Wales	0.20	1.90	14.60	36.80	29.80	12.50	4.20
□ LFA Wales	0	0.04	6.11	44.91	37.76	9.38	1.72

% total land area

Sources: Defra, 2007; Welsh Assembly Government, 2009

Figure 6.2 Agricultural Land Classification for England & Wales
Based on data taken from MAFF, 1983.

the Agricultural Land Classification (ALC) from 1 to 5. Grade 1 land has few limitations for agriculture; Grade 5 has a large range of cultivation problems such as flooding, poor nutrient status, erosion, thinness and so on (see Box XXVII). Nationally we have little Grade 1 or 2 land, about 2.8% of the total land area (MAFF 1983), most of it (58% Grade 1 and 41% Grade 2) in the lowlands of eastern England. So, until recently, planning laws protected this land from any change away from agricultural use (Cullingworth and Nadin 2006). The biggest part of England and Wales, about 40%, is Grade 3 land, whereas most English and Welsh LFAs are classified as Grade 4 or 5. Figure 6.2 also shows a greater disparity between LFA and non-LFA land in England than in Wales, a simple function of Wales's physical geography. As seen in Box XXVII, Scotland has a slightly different system for classifying the quality of agricultural land from 1 down to 7. Despite this, Scottish LFA land still represents the poorest, with over 94% graded Class 4.1 or lower (Wright *et al.* 2006).

Although MAFF (1999) do not recommend use of the ALC for financial valuation of agricultural land, on average, Grade 1 has a tendency to command the highest price and Grade 5 the lowest. This can, of course, be distorted by proximity to urban areas, enterprise flexibility, development speculation, crises in agriculture, availability and management ability. Curiously, after the foot-and-mouth epidemic of 2001, as some farmers withdrew from farming, the glut of land did not precipitate a decline in financial value, but quite the reverse. Land in parts of Cumbria, for instance, escalated in value with Grade 2 land being sold for over-inflated prices such as £7,101/ha (Westmorland Herald, 12/4/03). It could well be that the exceptional supply of something usually unavailable encouraged a temporary price war among farmers with compensation money to spend – a good example of the laws of supply and demand in operation.

The other important characteristics of land price are that it varies by region and over time. Quarterly reviews by the Royal Institute of Chartered Surveyors (RICS) record price fluctuations by region and farm type. At the end of 2006 the average price of land was £8,164 per hectare, with pasture land averaging £7,878 (RICS 2006). In 2004 the average price was about £5,900/ha, with pasture at £4,900, increases of 38% and 61% respectively. RICS analysts suggested the rising prices were caused by a combination of factors:

- Higher commodity prices for sales
- Investment in land by foreign EU farmers (notably Irish and Danish)
- Increased purchases by 'lifestyle' buyers
- Lack of supply of land in general

Box XXVII: The Agricultural Land Classification

The Agricultural Land Classification (ALC) was introduced and first mapped in 1966 in England and Wales. It was designed to assess the quality of land for growing food, based on an analysis of the long-term physical limitations of land for agriculture. The grade is affected by:

- climate – temperature, precipitation, aspect, exposure and frost risk,
- site – gradient, microrelief and flood risk,
- soil – texture, structure, depth, stoniness and chemical properties that cannot be corrected.

All these factors are discussed in some detail in Chapter 4. By combining them, the soil wetness and droughtiness can be calculated. Wetness controls the ease of mechanical operations and grazing management; droughtiness affects yields. The two in combination influence the types of enterprise and their yields, the consistency of the yield and the cost of production (MAFF 1999). The higher the rating of the ALC, the more flexible the land use is, requiring lower inputs to provide the highest yields.

There are 5 grades, with 1 being the best and 5 the worst. Originally Grade 3 was one class, but as over 40% of England and Wales is in this class, it was re-surveyed in 1976 and subdivided into 3a, 3b and 3c (MAFF 1999). In 1988 the climatic criteria were further developed, which led to 3c being subsumed into Grade 3b. Details of the exact criteria of each class can be found in MAFF (1988), but it is instructive here to understand the descriptions of grades 4 and 5, which dominate the English and Welsh uplands (Figure 6.2):

> **Grade 4** – land with severe limitations due to adverse soil, relief or climate, or a combination of these. Adverse soil characteristics include unsuitable texture and structure, wetness, shallow depth, stoniness or low water holding capacity. Relief and climatic restrictions may include steep slopes, short growing season, high rainfall or exposure. For example, land over 183m (600ft) which has over 1270mm (50 in) annual rainfall or land with a high proportion of steep slopes (between 1 in 5 and 1 in 3) will generally not be graded above 4.
>
> **Grade 5** – land with very severe limitations due to adverse soil, relief or climate, or a combination of these. The main limitations include very steep slopes, excessive rainfall and exposure, poor to very poor drainage, shallow depth of soil, excessive stoniness, low water holding capacity and severe plant nutrient deficiencies or toxicities. Land over 305m (1000ft) which has more than 1525mm (60in) rainfall or land with a high proportion of very steep slopes (greater than 1 in 3) will generally not be graded above 5. (MAFF 1983: 23)

As a result Grade 4 land is generally used for grass production and occasionally oats, barley or forage crops. Grade 5 land is restricted to grass or rough grazing.

In Scotland, a slightly different approach was adopted and first used in 1981. The Land Capability for Agriculture (LCA) system classified land by its biophysical limitations (climate, soil properties, wetness, erosion risk, slope) to describe its agricultural potential (Bibby *et al.* 1991). There are seven classes ranging from 1, where the land is capable of producing a wide range of crops, to 7, where the land is of limited agricultural value. Classes 3, 4, 5 and 6 are subdivided further; 5, 6 and 7 are roughly synonymous with upland areas. The figure below shows the LCA for Scottish LFA and non-LFA land:

Land Capability for Agriculture Classes in Scotland in and outside Less Favoured Areas, 2006.

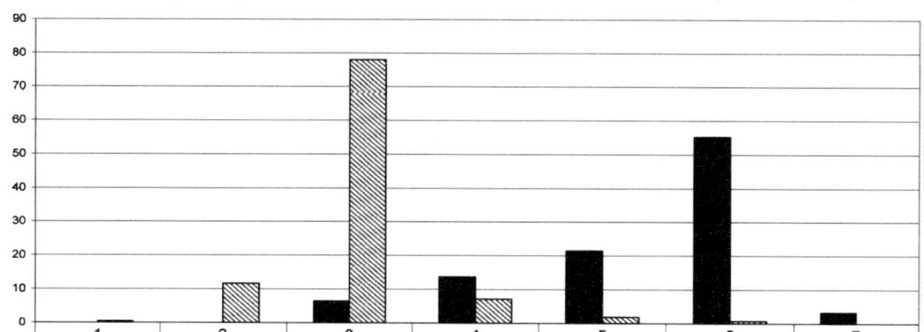

	1	2	3	4	5	6	7
■ LFA	0.0	0.0	6.4	13.7	21.4	55.2	3.3
▧ non LFA	0.4	11.6	78.1	7.1	1.9	0.8	0.0

(Source: Wright *et al.*, 2006; p9)

Regionally, in 2006 pasture land was most expensive in the North West (averaging £9,884/ha in 2006), followed by Wales and the West Midlands. The cheapest land could be purchased in Scotland at about £4,400/ha.

Finally, land varies by tenancy. Historically, most land was owner occupied by the overlord to whom people were bonded to work. With the move to a more capitalist economy in the 17th century, entrepreneurial farmers could rent land through a cash tenancy from local landowners, giving them a greater say in how land was farmed. In 1891, 85% of land was under cash tenancy, with the other 15% owner-occupied. This state of affairs was not favourable to hill farming, as Moore-Colyer (1986) showed for Wales (Chapter 3), where tenanting farmers saw no point in improving land which they might not rent again the following year; this compounded the hill-farming problem.

Owner occupation has gradually increased at the expense of cash tenancy. In 1970, 57% of agricultural land in Great Britain was owner-occupied; in 2004 it was 67% (Marks and Britton 1989, DEFRA 2005). Among upland farmers, in 2009 about 55% were owner occupiers in England, and only 17% were just tenants (DEFRA 2009a). There are regional trends once again, notably the North York Moors with fewer owner-occupied farms and more tenancies, whereas the Welsh Borders are skewed the other way (Figure 6.3a).

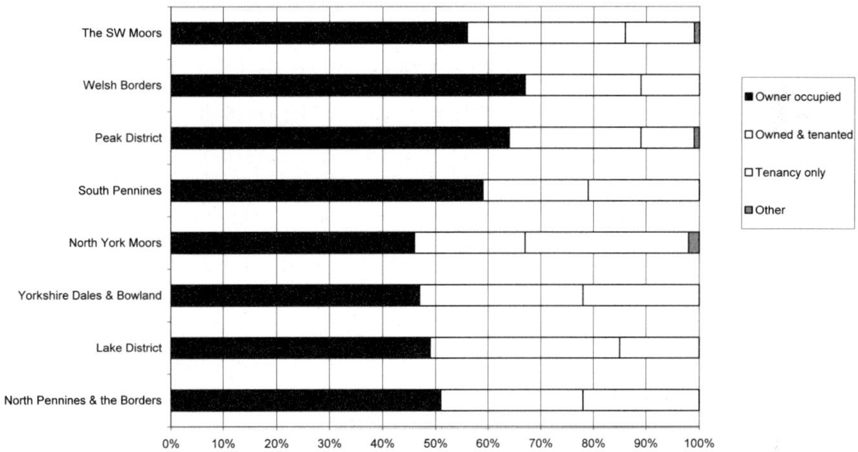

Figure 6.3a Farm Tenancy Uplands of England 2009

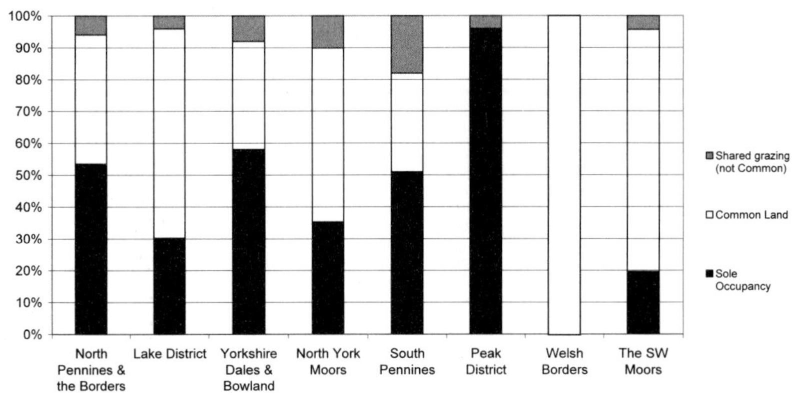

Figure 6.3b Occupancy Types of Moorland area: Upland Farmers in England 2009
Based on data adapted from DEFRA, 2009a.

A third tenancy arrangement is communal ownership. In some countries this amounts to state ownership and control, as in the former Soviet Union and currently in the kibbutz system of Israel. For upland farmers in England, common ownership exists in about 50% of open fell land (DEFRA 2009a). As we saw in Box II, English upland farmers have rights of common to graze stock, but the land is owned by another. Figure 6.3b shows that above the fell wall ownership has a variable pattern. For instance, 65% of open fell farmland in the Lake District is common, but none is in the Peak District. Irish uplands have a similar commonage system covering 400,000 ha, over 21% of the land. Some variations in ownership exist: first, the farms around the commonage may jointly own the land; second, the rights may not be attached to a certain area of the land but held collaboratively across the entire unit (di Falco and Rensburg 2006).

Labour

The second factor of production is labour. How much labour is required on a farm depends on many variables. These are generally of two kinds:

- The needs of the farm business – type of farm system, enterprise mix, land quality;
- The character of the labour itself – cost, availability and output.

Every enterprise requires variable amounts of labour, which can be standardised for purposes of comparison. Dairy cows are quite demanding of labour requirements for each animal, because they require milking (at least) twice a day. On the other hand, some arable crops are even more challenging, for example sugar beet, because the developing beets need to be regularly hoed to remove weeds.

Based on the labour requirements in Table 6.3, a hill farmer with a flock of 500 ewes would need to work a 37-hour week for 54 weeks each year, which is impossible. Of course, these generalisations do not take into account the enterprise mix on a specific unit, the flock or herd composition, other day-to-day estate activities, running farm diversification, transport of livestock, gathering, hay/silage production or complying with agri-environment initiatives, let alone 'having a life'.

Table 6.3: Comparison of Labour Times various Enterprises

enterprise type	*hours p.a./animal or hectare*
Dairy Cows	
cowshed	65
parlour	34
Suckler cows	
single suckler	20
Sheep	
breeding ewe	4
other	2
Arable	
cereals	20
oilseed	20
sugar beet	90

(Source: Chadwick 2003.)

As noted in Box XXV, an SLR is roughly equivalent to a full-time farm business, which means that most farms in the UK are considered part-time operations (Figure 6.4a), requiring less than one person a year to operate them. If we examine the SLR patterns in more detail, there is little difference between LFAs and non-LFAs in the UK, apart from Northern Ireland, where there seem to be fewer farms classified as part-time (<1 SLR) both inside and outside the LFA. This may be partly due to the greater preponderance of middle sized farms (20 to 50ha; Figure 6.4b), which are now rare in mainland Britain, the result of adopting various economies of scale (see below). Whichever way we look at these data, most UK farms are no longer classified as full-time.

A particular issue for upland farming businesses is the cost of labour. Some farm posts are more skilled than others, and wages reflect this. General farm workers in 2002 typically earned

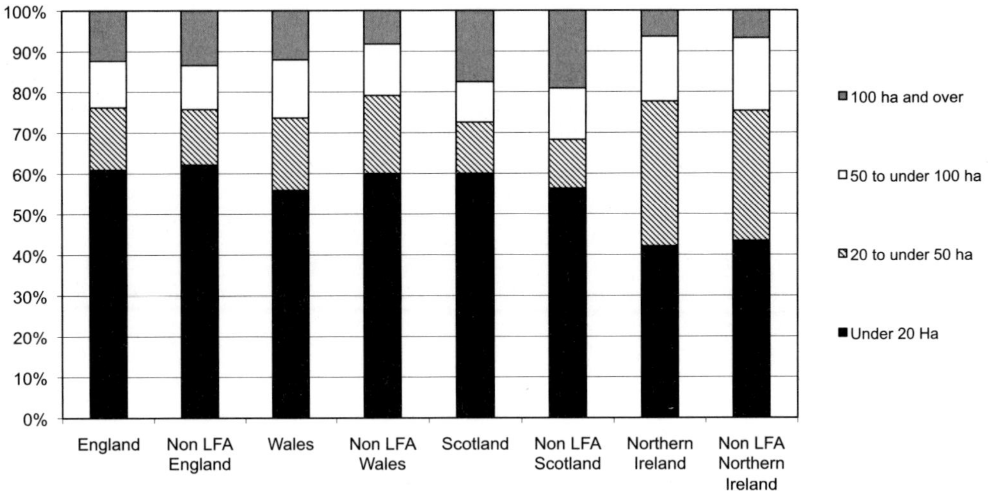

Figure 6.4a LFA and Non LFA Farm Holding Sizes by hectarage

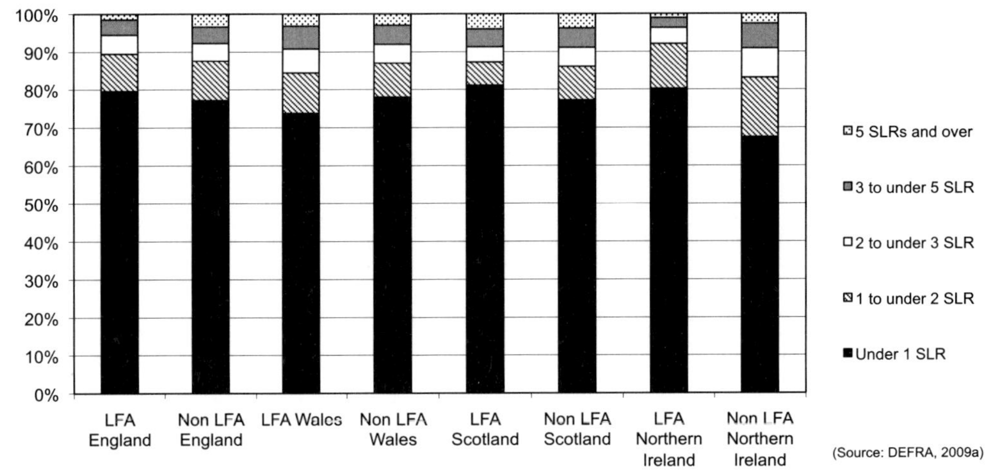

Figure 6.4b LFA and Non LFA Farm Holding Sizes by SLR
Based on data adapted from DEFRA, 2009b.

£16,400 a year, whereas shepherds earned £18,280 and dairy stockmen £23,690 – all figures including National Insurance contributions and insurance (Chadwick 2003). Historically, the cost of labour was controlled by demand or lack thereof; for example, after the Black Death in the 14th century (when there was a shortage of labour) and during the 19th-century agricultural depression (when there was too much; see Chapter 3). Statutory control of farm wages began in the 1920s when the Agricultural Wages Board was set up to ensure a minimum wage for farm employees (dissolved in 2011). Only recently has such a concept spread to the rest of the population in line with EU regulation.

For upland farmers, labour is the greatest component of variable costs with which they can dispense when times get hard. Lobley *et al.* (2004) found that 73% of upland farms in their survey had no employed labour, compared to only 53% of lowland farms. Instead the farmer works longer hours and members of the farm family are dragooned into helping at no extra cost. Overlying this is the general declining trend in farm labour as a result of a series of push and pull factors. In 1861, 2 million people worked on the land; by 2004 the figure was close to

367,000 (Marks and Britton 1989; DEFRA 2009, accessed 12/11/09). Mechanisation of farm tasks, low wages, long hours, poor standard of living, better pay and services in towns and a set working week have all played a part. For upland farmers the situation is compounded because few tasks can be mechanised. Quadbikes may generally have replaced horses or feet, but much open fell remains inaccessible to vehicles, keeping labour needs to the same levels as the pre-mechanisation era.

Unfortunately, economic hardship has still forced many upland farm businesses to shed labour, even if this affects practical operation of the unit. Tables 6.4 and 6.5 show the changes in farm labour over the 1990s: nationally, in the two LFA zones and in the seven upland English national parks. The overall pattern is one of continued decline (about 16%), proportionally more of it outside the LFAs, with a loss of nearly a third of the farm labour force in Wales. Within LFAs England has suffered the most (loss of 18%, 75,000 jobs) and Scotland the least, with a slight gain (+2%).

Table 6.4: Changing Agricultural Labour Force in LFAs 1991 to 2000

	% SDA	*% DA*	*% LFA total*	*% non-LFA*	***average of all***
England	–12.7	–28.6	**–18.4**	–17.6	**–17.7**
Wales	–14.9	–1.3	**–9.0**	–29.5	**–16.1**
*Scotland**	+2.5	–3.3	**+1.7**	–7.2	**–5.6**
Northern Ireland	–6.6	–13.9	**–10.2**	–12.3	**–10.9**
United Kingdom*	–7.6	–14.3	**–9.9**	–15.8	**–15.6**

*1993 to 2000 only.
Note: labour force covers farmers, partners, directors, wives and husbands who work on the farm, salaried managers, regular employees, seasonal and casual workers.

(Source: DEFRA 2002.)

Table 6.5: Changing Agricultural Labour Force in English Upland National Parks 1990 to 2000

national park	*total farm labour 1990*	*total farm labour 2000*	*% change 1990 to 2000*	*% change in labour not including farmers*
Dartmoor	2,005	2,082	+1.0	–24.6
Exmoor	1,423	1,387	–2.5	–24.3
Lake District	3,056	2,952	–3.4	18.9
Northumberland	543	526	–3.1	–19.7
North York Moors	2,605	2,640	+1.0	–21.2
Peak District	3,659	3,567	–2.5	–29.9
Yorkshire Dales	2,397	2,271	–5.3	–24.0

Note: labour force covers farmers, partners, directors, wives and husbands who work on the farm, salaried managers, regular employees, seasonal and casual workers.

(Source: DEFRA website, accessed 12/11/09.)

These general trends hide the detailed pattern, whereby Disadvantaged Areas (DAs) have suffered more than the Severely Disadvantaged Area (SDA), which equates to the true upland core (see Chapter 2). Again the pattern is regional, with only Wales losing more in the SDA than the DA. Slow declines in LFAs can be explained, to some extent, by the fact that hill and upland farmers had already shed much of their labour force to reduce costs and were working as one-person businesses before 1993. The variation between English DA and SDA change is probably explained by the shift away from dairy and beef production in the upland margins, where labour demands were greater than for sheep found in the SDA. Though all these changes are small numerically, they do have a disproportionate effect on upland communities with lower population and fewer alternative employment opportunities.

Considering the English upland national parks, we can see that over the same period total labour change varied. The Yorkshire Dales saw the largest overall reduction (-5.3%), whereas Dartmoor and the North York Moors saw marginal increases (Table 6.5). However, these data obscure a huge reduction in farm labour beyond the main farmer of each holding. The final column of Table 6.6b demonstrates a massive decline in all six parks of 18 to 30% employing this type of extra labour. These losses are probably due to attempts to reduce variable costs or a movement of wives, partners and children off-farm completely, seeking alternative employment to bolster total farm income – in what combination merits further investigation, but the overall effect is the same: longer hours, more complex management and increased personal isolation for the farmer left behind on the farm.

Capital

In many core agricultural geography texts (e.g. Morgan and Munton 1971, Tarrant 1974, Ilbery 1985) the decline in labour is related to increased use of capital, the third factor of production. Capital enters a farm in three main ways: a farmer's own capital, government subsidies and grants, or loan from a financial institution. The lack of absolute advantage of livestock production in mountainous areas and reliance on comparative advantage means that few farmers have large savings (Rieder and Wyder 1997). With consistently low net margins there is often little available to re-invest in the farm business. Individual farmers can therefore:

- carry on business as usual,
- take out bank loans to improve certain parts of the business, and/or
- rely on government support.

Farmers who take the first option expose themselves to the vagaries of the market. The cost–price squeeze (production cost goes up and sale prices come down) has an enormous effect on business viability. As we saw above, upland farmers can only reduce costs of production by losing labour, but there comes a time when there is no more labour to shed and the outcome is economic ruin and social deprivation.

Choosing the second option can be risky as it relies on a market and good business acumen. Typically loans are used to erect off-wintering barns, improve milking parlour systems, purchase machinery or support farm diversification. In their Farm Practices Survey, DEFRA found that 52% of upland farmers do not borrow and, of the other 48%, some 42% were finding it harder to borrow. However, 54% of upland farms were debt free, though 8% worried about the cost of debts (DEFRA 2009a).

Under the third option, government subsidies provide financial aid to encourage production, whereas grants are used to improve farm infrastructure or diversify into other activities (often called restructuring). Both of these mechanisms have been used widely to support agriculture in Britain in different ways (see chapters 3 and 8). Specific support for upland farmers started in 1940 with the Hill Sheep Subsidy Scheme. Since then upland farm businesses have benefited from a wide range of targeted and non-targeted subsidies and grants. Though upland farmers seem happy to take production subsidies, they seem less willing to take grants to improve the efficiency of farming operations. The reasons for this are explored in the next chapter.

Financial institutions, in the past, have included investment funds, pension bodies and development portfolio operators. There has also been the injection of money from up- and down-stream of the farm (Figure 6.5). Munton (1977) showed how in the 1970s the focus of this capital tended to be in certain regions of the country, such as East Anglia and Lincolnshire. For upland farm businesses, Munton's study showed that capital penetration from financial institutions did not play a big role and in some areas, such as the Cambrian Mountains of central Wales and the Yorkshire upland massifs, none at all.

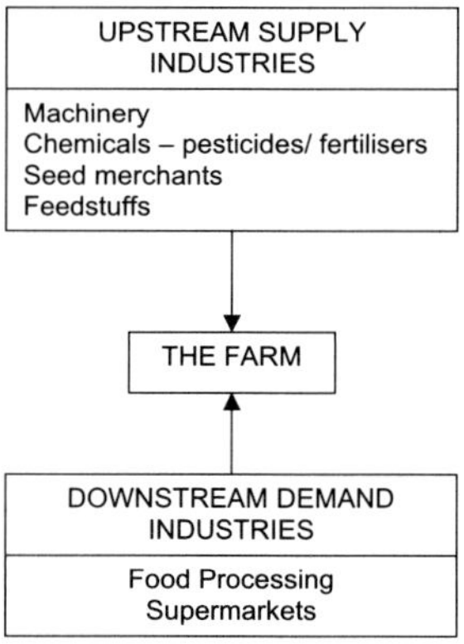

Figure 6.5 Vertical Integration into Farms

Another example of capital penetration is the frozen pea industry. Dalton (1971) showed how downstream industries were controlling the quality of the farm product through various contractual obligations. Though these types of activity were rare in uplands when Dalton was reporting, the situation has now changed to some extent; financial control is starting to come from supermarket chains downstream (Robinson 2004). Upland farmers are signing up to meat contracts with certain supermarkets, whose contracts ensure control of supply between farmer and supermarket, freeing the latter from the burden of ownership. However, some contracts could increasingly be construed as including an element of cynical profiteering as supermarkets play on animal welfare and health scares to raise customer fears and then allay them through branding and the use of words like 'local', 'quality', 'homegrown' and 'hillbred', emphasised by farmer photographs on labelling.

The factors of production play a pivotal role in the operation and survival of all farm types. Upland farm businesses are perhaps most affected because the land quality is so poor, and labour and capital so limited. To make the best of the situation, farmers can manipulate their resource base by applying what are known as economies of scale.

Economies of scale

Farmers can improve the economic efficiency of their farms by employing economies of scale in two ways: internal or external. First, internal economies of scale can be used to re-organise inputs within the farm business, which helps reduce costs of production. This can be achieved by restructuring the farm through modernisation, combining:

- concentration into larger farm units,
- specialisation of enterprises,
- intensification of output/ha of land.

Between the 1950s and 1970s the trend on upland farms was towards fewer, larger units (MacEwan and Sinclair 1983). Then between 1992 and 2006 this pattern reversed, as Figure 6.4b shows, with farms under 20ha accounting for over 60% of units (IEEP 2003, DEFRA 2009b), though the dominance of small farms in uplands is not universal: Northern Ireland retains a more equal mix with nearly 35% of farms 20 to 50ha in size. Patterns varied in 2006 in English national parks, with small farms dominating Dartmoor (69% under 50ha) in contrast to Northumberland (30% under 50ha) in 2006 (Figure 1.10). There are, however, several points of caution:

1. All these statistics cover directly owned or tenanted land, but not common land (notably the Agricultural Census). Unfortunately this distorts the reality of the managed farm area.

Table 6.6 shows this on a smaller geographical scale, since a much lower proportion of land is common in the Pennines than in the Cumbrian massifs. Furthermore, Lake District farms are substantially larger with higher proportions of intake and common land to reflect their greater physical marginality.

2. In some instances farms can go beyond an optimum size when diseconomies begin to occur (Ilbery 1985). For upland farms this is a definite reality particularly with respect to declining labour, where practical management of large fell areas becomes increasingly unviable. So the labour shortage may be a factor in the swing to smaller farms. There has also been a general trend for middle-sized farms to disappear.

3. Farm size based on area provides useful data about changing patterns of farming systems, but not about economic viability. In this respect other measures of farm size could be more valid. Amog those developed or appearing in official returns and research literature (Box XXVII) are:

    - labour – standard man days (SMDs), standard labour requirements (SLRs)
    - income – net margins, gross margin, total income from farming (TIFF), European size units (ESU)

Table 6.6: Farm Sizes in Three Cumbrian Massifs

massif	*average size of inbye (ha)*	*average size of intake (ha)*	*average total farm size (ha)*	*average size of open fell common land (ha)**
Pennines	131	46	176	254
Orton-Howgills	107	15	122	303
Lake District	91	161	252	604

*Many farmers could only cite stint agreements (maximum head of livestock allowed) and did not give the actual hectarage

(Source: Burton *et al.* 2005, unpublished data).

A second internal economy of scale is enterprise specialisation: simplifying or reducing the number of enterprises on a farm. By doing this, a farmer can reduce production costs by eliminating enterprise-specific machines, seeds, livestock and/or skilled labour. As we have already seen, upland farmers have few options in this regard, the choice being some form of livestock production or dairying on the upland fringes. In turn this also constrains diversification options (see below).

The third process of modernisation, intensification, can be achieved in various ways. Mechanisation has played a role in farm enlargement, replacing manual labour with machines. Most of the possible changes occurred on UK upland farms from the 1920s onwards with the introduction of tractors, automated milking and fodder-harvesting machines. As noted earlier, quad bikes have replaced horses and footwork on the fells, but there are still many areas of upland, more specifically hill farms, that can only be reached on foot. Gathering requires large amounts of labour to draw sheep together over many hectares and corral them into an enclosed area ready for driving down to the main farm. Without enough manpower, as the 'net' is closed on the flock, sheep simply slip through and scatter, and the process has to begin again.

Farmers can also intensify production by raising stock yields through:

- Breeding programmes
- Artificial supplementary feeding
- Switching from native to commercial breeds

} improvement of carcass weight

- Switching from hay to silage
- Replacing heaths with grassland
- Re-seeding grasslands to raise DMC
- Draining wet land to increase grass crop
- Liming

} increase of stocking densities

Many of these operations are costly and so over the years government grants have been offered to encourage their adoption (see Chapter 8). The effect has been to increase production per unit of land from 0.0625 LUs of pre-Clearance Scotland to the extreme of 1.31 LUs in the mid-20th-century Peak District, a factor of 20 (chapters 3 and 4). Unfortunately, this has led to impoverishment of much of the HNV of the uplands (see Chapter 5) and in some cases to overproduction of farm products (see Chapter 9).

Farmers can also improve their farm's efficiency by employing external economies of scale. These are mechanisms that develop relations between the farm business and the rest of the economy – for example, support businesses that specialise in the needs of certain farm enterprises, such as quad bike maintenance or feed merchant. Other options include capital penetration through vertical integration, discussed above, or upland farmers adopting Short Food Supply Chains rather than conventional ones (see below).

Responses to contracting profit margins

At the beginning of this chapter, we saw how upland farming businesses typically suffer from low gross and net margins. By applying various economic theories and concepts we can see that these types of business are often constrained by a unique combination of margins of cultivation, factors of production and economies of scale, producing small profits.

Farmers can respond to this situation by restructuring the business in various ways. On upland farms the traditional solutions have been to shed labour or intensify production, both of which have undesirable secondary consequences, but these are only two of a series of possibilities, which Lobley *et al.* (2002, 2004) call the restructuring spectrum (Table 6.7). Alternatives that require more drastic restructuring and greater perceived risk are, however, less likely to be chosen by upland farming businesses (Lobley *et al.* 2004). Let us consider why this is.

One particular restructuring option that has received much attention in both academic and professional literature is pluriactivity (on- and off-farm diversification), by which farmers can avoid the cost–price squeeze and continue farming (e.g. Ilbery 1988, 1991; Robinson 2004; MAFF 1994; CRR 2002). It allows farm incomes to stabilise, makes more effective use of farm resources and creates business opportunities that respond to changes in the family life-cycle (Robinson 2004). The effect is to spread the risk of financial ruin across multiple business strands. Lobley *et al.* (2004) found that 70% of farmers who had restructured by shedding labour or intensifying production relied on their farm income for business viability, in contrast to only 53% of farmers who had diversified.

Unfortunately, opting to diversify is often harder for upland than lowland farmers. Non-economic factors, such as personal circumstances, are crucial considerations for farmers (see Chapter 7) along with a range of economic issues: the marketing mix, physical requirements, farm system flexibility, labour requirements, capital and running costs, grant availability, legal constraints and training needs.

Table 6.7: The Restructuring Spectrum

Static Businesses	No change in practices
Minor change	Marginal changes to business, such as limited investment or input variance
Traditional restructuring	Resources are re-deployed within farm businesses, e.g. specialisation, farm enlargement
Agricultural integrators	Resources are re-deployed within wider farm business sphere, e.g. agricultural consultancy or upstream business
On-farm diversifiers	Resources re-deployed within the farm-based business, e.g. tourist accommodation
Off-farm diversifiers	Labour and possibly capital into off-farm businesses or employment
Capital consumers	Agricultural assets and resources liquidated to provide income
Leavers	Exit from agricultural activity with or without a successor

(Source: Lobley *et al.* 2002, 2004)

The marketing mix

When farmers decide to diversify, they need to take into account the 'four Ps' – product, place, promotion and price. Product choice is complex: many variables need to be considered, like shelf life, competition from other outlets producing the same goods and public desirability. The range of enterprises and resources available to create a product on an upland farm is by far the most limited of any farm type in mid-latitude areas, being derived solely from meat, milk or wool. However, there is a surprisingly large range of possibilities, such as felt, knitting wool, insulation, cheese, yogurt, ice cream, organic, rare breed or premium meat, and processed meat products (sausages, pies, burgers), to name but a few.

Place, or location, is also a challenge for upland and hill farms because remoteness means the product usually has to go to the market rather than the market come to the farm. Transport links and distribution networks are generally more problematic in LFAs than elsewhere (McNally 2001; Ilbery *et al.* 2004). In these situations costs of production will rise, as processing may happen elsewhere, as well as sales. However, for some forms of diversification, remote location is everything. Walford (2001) investigated the occurrence of tourist accommodation around popular tourist destinations in England (national parks, AONBs, heritage coasts). Many uplands – Bodmin Moor, Dartmoor, Lake District, Snowdonia – contained very little accommodation. Others had a build-up of farm accommodation tourism just beyond a number of designated upland – Exmoor, Dartmoor, Snowdonia and the Mendip Hills. The Peak District stood out as having significant clustering within the National Park boundary. So some upland farms have greater potential than others for farm diversification, depending on whether they are within or outside a particular upland massif. Why this pattern varies probably depends such factors as competition, market availability and possibly planning constraints (see below).

Promotion needs vary from product to product, but generally include a range of personal communication skills. Being able to 'talk up your product' to a sales outlet, selling concepts and ideas, advertising and being 'pushy' are not common among upland farmers (see Chapter 7). Often this side of the farm business shows a gender divide, with farmers' wives or partners fronting it. Price depends on knowing the competition, the market into which you are selling, your overheads and cost of production. Some products need consistent quality and production volume to command a reasonable price. Reputation is everything, as Ilbery and Maye (2006) found in Northumberland and the Scottish Borders.

Physical requirements

Every enterprise requires a different resource base. Upland and hill farms have really only three main resources to exploit: land, buildings and livestock. However, many ideas generated under the marketing mix are not feasible when the farm's physical resources are audited. Internal capital may not be available because of the small gross margins, so loans and grants become central to farm business diversification (see chapters 7, 8 and 9). Labour is very limited. Thus a typical physical resource audit could resemble Table 6.8.

Table 6.8: Physical Resource Audit

physical resources		*positive*				*negative*				
		Close to utilities and services	Often already redundant	Lots available	Close to hereditament	Planning permission issues	May affect AES applications	Distant from utilities	Can disrupt farm management system	Inappropriate terrain or costly to manipulate
Buildings	Main Farm	✓				✗			✗	
	Outlying		✓			✗		✗		
	Building materials		✓	✓			✗			
Land	Inbye	✓			✓	✗	✗		✗	
	Intake			✓	✓	✗	✗	✗	✗	✗
	Open fell			✓		✗	✗	✗	✗	✗
Livestock	Sheep			✓					✗	
	Cattle			✓					✗	

Farm system flexibility

Any change to activity on a farm will affect the operation of other enterprises. Loss of land and buildings can reduce the stocking capacity of the unit and thus careful analysis of cash flow and revenue is an important aspect of developing a new farm enterprise.

For many upland farmers, a reduction in their main 'traditional' enterprise is a huge psychological change. Not only does it require a different skills set, but it removes certainty from day-to-day business into uncharted waters. The labour may simply not be available. Questions need to be asked: whether there is enough inbye to feed the remaining stock, enough barn space to store supplementary feed or enough of each land type to accommodate the rotation of stock throughout the year. The same questions are becoming an issue in the adoption of agri-environment schemes, which often require lower stocking densities (see Chapter 8), with the result that stock is spread more thinly on hefts, making gathering more time-consuming or even impossible. It is interesting to note that 36% of upland farmers responding to the Farm Practices Survey (DEFRA 2009a) felt that their farm was not suited to on-farm diversification and 60% had no scope for off-farm diversification.

Capital and running costs

These again very much depend on the enterprise a farm business decides to develop. For upland and hill farmers this is often the crux of the matter. With low capital availability (unless some land is sold), small gross margins and a naturally conservative outlook based on risk aversion, few will take on a bank loan to set up a new enterprise (DEFRA 2009a). Coupled with the fact that most diversification enterprises fail in their first year (another feature not picked up in the Agricultural Census, which underestimates the level of diversification), this limits the upland farmer's choice in what they can develop. Examples of possible upland diversification options are shown in Table 6.9.

Table 6.9: Costs of Two Diversification Examples

	cheese production	*wind power (5Kw size)*
Capital costs	£20 to £100K	£30K
Running costs	£1.25 per Kg	2-3% of capital
Returns	£4.25 to £15/kg from 1000L of milk (140kg cheese)	Per year: Feed in Tariff £3360 Export to Grid £186 Saved Electricity bill £780

Feed in Tariff - grant paid per Kw/hr by current government. (Source: Verdant Engineering, 2011)

Returns

Typically the returns from diversification enterprises are higher than those of traditional farm enterprises. However, the risks are often higher, so farmers are encouraged to complete feasibility studies with help from organisations such as the NFU, Farm Business Link or LEADER+ (see Chapter 7). Returns are very much a function of local competition, as few upland areas can support similar diversification activities in close proximity, as markets are remote and spheres of influence enormous in order to hit certain population levels.

One such example exists in a well-known Lake District valley. Here a farm business has diversified into organic fell-bred meat in boxes. Originally this precluded neighbouring farms from diversifying in a similar way since the market would soon be saturated and the premium for a kilo of lamb would be undermined. However, by simply waiting until the sphere of influence of the first business extended beyond local need, a new similar enterprise was able to slot in behind. This typically happens at farmers' markets when a producer has developed a client base, become known, moved on (or upmarket) and their stall is taken by a new producer.

The relationship between capital, running costs and returns is obviously crucial for any business success. For example, McInerney and Turner (1991) calculated operating costs for different diversification enterprises. They found that diversification based on crops or animals had the greatest overheads and some of the lowest net margins. In contrast, accommodation had the smallest overheads and agricultural trading the largest net margin. This had changed somewhat by 1997, when McNally (2001) found that retail operations were the most profitable form of diversification.

Constraints

A particular limiting factor for diversification is the labour requirement for enterprises already operating on the farm. Evans and Ilbery (1993) considered that arable and livestock farmers, for instance, were more likely to diversify than those farm types requiring high levels of labour, such as dairying, pigs or poultry. The slow strangulation of the upland farm labour force undermines this thesis, reflected in the later DEFRA survey (CRE 2002 – see Figure 6.6). Instead there seems to have been more of an activation of the farm spouse to operate the B & B, for example, around

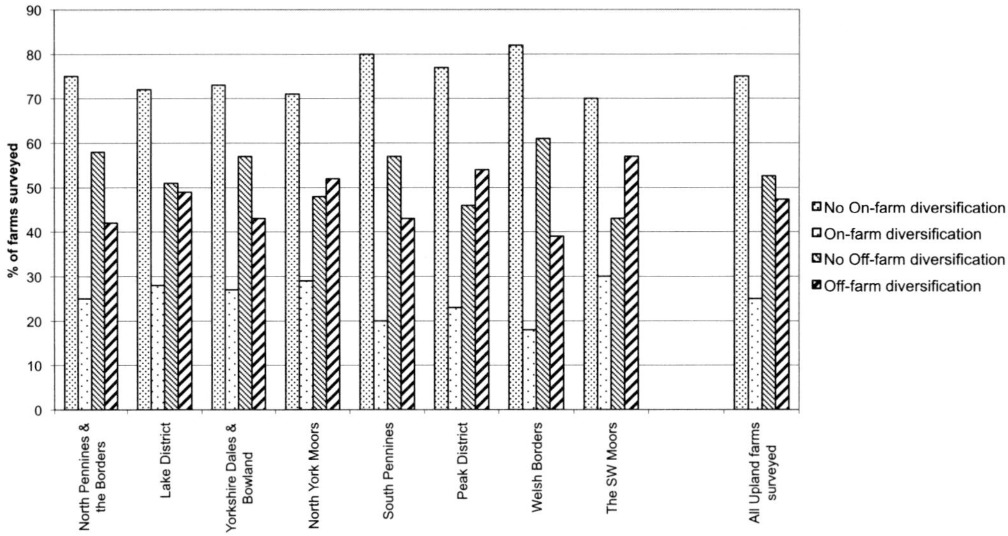

Figure 6.6 Pluriactivity on Upland Farms
Based on data adapted from DEFRA, 2009a.

their domestic chores, child rearing and book-keeping roles. As the children mature, some of these roles are shed allowing time to develop more complex forms of on-farm diversification.

As well as capital constraints, another typical issue is the navigation of planning permission. Using land or buildings for operations that are not classed as agriculture (the production of food) requires planning permission from local councils in the UK (somewhat relaxed in 1997). Conversion of buildings into accommodation or processing plants and provision of camping or caravan sites are two common examples of this. Other constraints can emerge in any diversification enterprise that handles foodstuffs, where food hygiene certification is needed, which leads us on to training.

Training needs

These may be specific to the choice of activity, such as the ability to use a different machine or feed regimes for 'novel' livestock. Typically, they are more related to marketing skills for product promotion as noted above. For upland farmers training is a particular issue, as many currently in the sector left school at 14 with no formal qualifications and thus find it difficult to re-integrate into a learning culture (see Chapter 7).

Grants

Over the years there have been a number of grant schemes available to farmers to enable them to diversify. Grants provide for the completion of feasibility studies, writing of business plans, purchase of equipment, conversion of premises, planting up new crops (e.g. woods), and processing and marketing of new products. One of the first schemes offering support was the Farm Diversification Grant, introduced by MAFF in 1988. It was not that successful, with only 2,000 adopters (Ilbery and Steill 1991).

More recent grants have been routed through the EU Rural Development Programme (RDP) 2000–06 (see Chapter 9). One of these was the Processing and Marketing Grant (PMG), designed to improve the agricultural processing and marketing of farm products in England (MAFF 2000), provided the total cost of the project was between £70,000 and £1.2m. It offered capital grants of 30% of the cost to farmers who needed new or renovated buildings, or new equipment in order to become more competitive and responsive to consumer demand, particularly where they were processing raw materials. The other 70% had to come from the individual farmer (at least 45%) or a private body (up to 25%). The scheme's mid-term evaluation was not favourable, with only 237 projects in England receiving funding (Elliott *et al.* 2003b).

The overall effect of these eight factors is that the diversification options for upland farmers and their business can be extremely limited. We can see this manifested in four ways: temporal, geographical, level of diversification and type of activity chosen.

- Temporal change – between 1989 and 2002, farm diversification in England deepened by about 8% (CRR 2002), suggesting that more farm businesses sought supplementary income as time passed. By 2007, 41% of English LFA livestock farms had some form of diversification compared to the national average of 51%, a figure dominated by cereal farms, of which about 70% had diversified (DEFRA 2009b). In contrast, the FPS (DEFRA 2009a) recorded that 23% of upland farmers had never contemplated diversification as an option for their businesses.

- Geographical variation – farms in Northern England are less diversified than elsewhere, with only 14% of farm businesses involved, compared to 26% in the West of England (DEFRA 2004). By 2009, the situation had changed again, with about 48% of upland farm businesses having diversified in some way; but there are substantial regional disparities, with the South West Moors most diversified and the Welsh Borders least (Figure 6.7; DEFRA 2009a).

- Level of diversification – areas outside the LFAs and national parks are more diversified, with an average of at least two types of activity, whereas inside these areas the figure is below two (Table 6.10). Figure 6.7 also shows a marked imbalance between on-farm and off-farm

Table 6.10: Diversification in LFAs and National Parks 2002

	LFA	*non-LFA*	*national park*	*non-national park*	*all*
Number of enterprises (% of diversified holdings)					
One	51	41	54	42	43
Two	31	26	28	27	27
Three	12	17	8	17	16
Four +	6	15	11	14	14
Average per holding	***1.8***	***2.2***	***1.8***	***2.1***	***2.1***
Enterprise type (% of diversified holdings)					
Agricultural services	30	36	24	36	35
Trading enterprises	31	33	20	33	32
Accommodation & catering	31	23	36	23	24
Equine enterprises	9	25	19	24	23
Recreation and leisure	19	24	31	23	23
Crop-based	17	23	15	23	22
Livestock-based	22	15	17	16	16
Miscellaneous	18	37	20	35	35

(Source: CRR 2002, p. 55.)

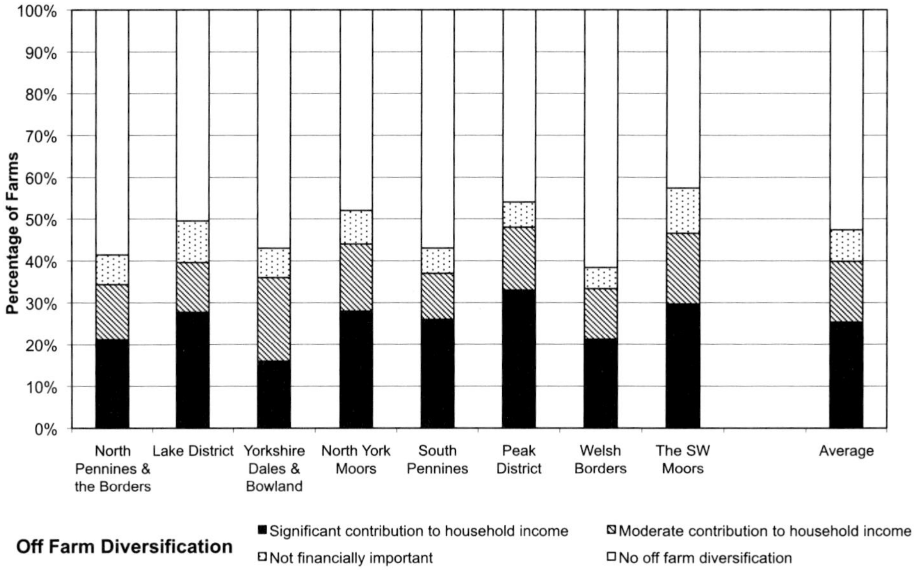

Figure 6.7a The Financial Contribution of Off Farm Diversification to English Upland Farms 2009

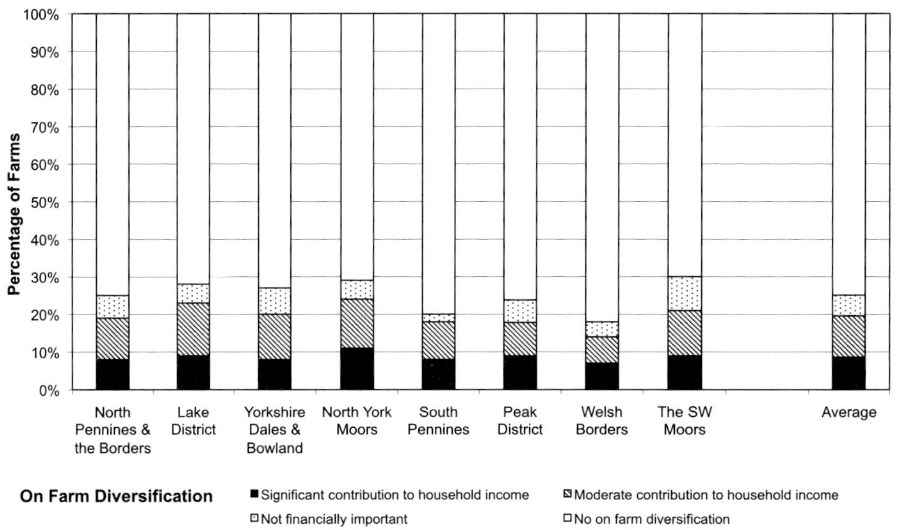

Figure 6.7b The Financial Contribution of On Farm Diversification to English Upland Farms 2009
Based on data adapted from DEFRA, 2009a.

diversification, the latter being more popular on upland farms (25% against 48%). This is most likely to be a function of one person managing the farm unit and their partner seeking off-farm work to supplement farm income. Usually, but not always, this means the male partner on-farm and the female partner off-farm.

• Type of diversification – in England the most popular form of diversification recorded in 2007/8 was letting buildings for non-agricultural purposes, occurring on 37% of all farms (DEFRA 2009). There was a great disparity between crop (48%) and livestock farms (28%), with LFA farms exploiting the buildings least, at 21%. Earlier studies by CRR (2002) and Lobley *et al.* (2004) showed a bias towards accommodation and catering operations, but away from active recreational pursuits in LFAs (Table 6.10).

The character of pluriactivity

The situation that has emerged is one of a paradox: those in most need of increased income seem unable to obtain it through farm diversification. For the minority who can, pluriactivity can be divided into three main types:

- On-farm agricultural diversification, e.g. organic production, biofuels or novel crops, agri-environment initiatives;
- On-farm non-agricultural diversification, e.g. tourism, forestry, processing, direct marketing;
- Off-farm other gainful activities (OGAs), e.g. other employment or independent enterprises.

On-farm agricultural diversification

The most common type of diversification considered by a farmer is to expand into alternative agricultural enterprises. Farmers view it as the most straightforward and, perhaps, least risky as it is an extension of current farming practice. Examples of on-farm agricultural diversification are given in Table 6.11. Many of these are not options in uplands, due to physical constraints as discussed above, with the exception of organic farming and farm wood exploitation.

Table 6.11: On farm Agricultural Diversification Options

Unconventional crops	linseed, teasels, evening primrose, borage, triticale, fennel, durum wheat, vineyards, biofuels, spelt
Unconventional livestock	fish, deer, goats, horses, llaminids, rare breeds
Organic farming	range of goods
Farm woodland	energy forestry, amenity, recreation, timber, biofuels

(Source: Robinson 2004, p. 135)

Organic farming is often cited as a possible agricultural diversification on-farm for upland farmers, on the grounds that the low-input/low-output system they currently operate is almost organic anyway. Organic farming is defined by Marland as:

> A system that attempts to provide a balanced environment in which the maintenance of soil fertility and the control of pests and diseases are achieved by the enhancement of natural (biological) processes and ecological interactions, with only moderate inputs of energy and resources, whilst maintaining optimum productivity. (Marland 1989: 24)

The consumer demand for organic food has increased over the last 40 years for a range of reasons including environmental concerns, food scares, animal health and welfare, pollution reduction and increasing biodiversity (Shepherd *et al.* 2003). Despite these, organic farming in the UK accounts for only 4% of the land area (741,144 ha) and 2% of all UK food sales (DEFRA 2003; Elliott *et al.* 2003a). In Less Favoured Areas a minuscule 31,549 ha of cattle and sheep farms are organic (although the census figures are incomplete as they do not record the amount of common grazing used by organically produced stock).

It seems perplexing why so few upland farmers have embarked on an organic production route, given that their artificial inputs onto the land are close to the official standards required by UKROFS (UK Register of Organic Food Standards) and other UK regulators like The Soil Association. The main incentive for farmers is that organic produce commands a premium over conventionally produced food. For example, organic beef can be sold at a 24% premium, organic lamb 14% and organic milk 11% (Elliott *et al.* 2003a), with the caveat that it does fluctuate seasonally and annually. Also, the cost of production is less than for conventionally produced meat (e.g. £541/ha versus £643/ha for sheep; Elliott *et al.* 2003) mainly due to reduced variable costs.

Unfortunately there are a number of disadvantages to organic production in the uplands which may explain the low adoption rates. First, other products which cannot be grown in the uplands command greater premiums, for example, poultry at 119% and cereal at 100% mark-up. Second, Elliott *et al.* (2003a) have calculated that on average about 10% more labour is needed than for conventional farming, something that would be hard to achieve in a sector where labour is contracting. Another issue is the poor conversion and maintenance rates offered for unimproved grassland under the various organic grants available in the UK since 1994. Both the Organic Aid Scheme (1994 to 1999) and its replacement the Organic Farming Scheme (1999 to 2006) provided £5/ha/yr for conversion, in contrast to prime arable land that received £90/ha/yr (MAFF 1997; Elliott *et al.* 2003a). On top of this farmers need to allocate more land because of lower stocking rates than conventional farming. For example sheep need to be stocked at 1.25 LU/ha for organic production compared to 1.6 LU/ha; and 78.3 ha compared to 61.2 ha in total (Elliott *et al.* 2003a). Such figures leave upland farmers with a difficult decision whether to convert or not. One way of securing high enough profits to warrant the conversion is to increase the end sale-value somehow (usually referred to as 'adding value'). Another option is to top up organic management using the agri-environment Organic Entry Level Stewardship Scheme. This adds another £8/ha for LFA land over 15ha in size (RDS 2005d; Natural England 2008d).

A second main on-farm agricultural option is farm woodland and other related crops. Into this bracket we can include biofuels, which can be used for:

- heat and electricity – short rotation coppice, *miscanthus*, canary grass, straw or forest material,
- transport fuels – cereals, oilseeds, sugar and fodder beet.

For upland farmers it is possible to make greater profits from short rotation coppice than sheep, as Heaton *et al.* (1999) calculated for upland Wales. Under the 2007–13 RDP, in Scotland 50% grants can be obtained to plant up land for biofuels in LFAs up to a maximum of £1,000 per hectare (Table 6.12).

Farm woodlands have a history of general under-exploitation in capitalist agriculture. Many farmers are loath to invest in woods because there is such a long period between initial investment and final return. This is one reason why afforestation grants have long support periods and generous financing. From the mid-1980s many EU forestry schemes targeted agricultural land to reduce food overproduction. In uplands, once again, the incentive to take land out of production was lower and adoption rates consequently small (Ilbery and Kidd 1992). Encouragement of woodland planting has continued into the second decade of the 21st century, but the general aims have shifted towards public benefit and biodiversity. Most related maintenance grants available from the Forestry Commission are paid through Axis 2 of the new Rural Development Regulation (see Chapter 9); thus they are designed with its goals in mind and paid on the basis of agricultural income foregone (Table 6.12). The Higher Level Scheme (see Chapter 8) now includes options for woodland creation, maintenance and restoration, directed as before to amenity and environmental benefit (Natural England 2009c; Forestry Commission 2008).

Few grants encourage farm woodland diversification, with the notable exception of Scotland's Improving Economic Value of Forestry. Farm wood income can be drawn from Axis 1 money of the RDP under measures 122 and 123 (see Chapter 9). In Wales, the organisation *Coed Cymru* (lit. 'Woods Wales') encourages farm woodland diversification to improve business incomes, particularly for exploiting hardwoods (Table 6.13). Firewood and charcoal are two obvious

Table 6.12: Woodland Grants available in the UK 2009

grant	*main purpose*	*rates*
Scotland (Rural Development Contracts)		
Short Rotation Coppice	For biofuels & wildlife	50% of costs in LFAs
Improving EconomicValue of Forestry	Pruning, thinning, squirrel control, natural regeneration	50% of costs
Woodland Creation	Planting of new woods for amenity and biodiversity benefits	£630–1961/ha (80% of costs in LFA)
Sustainable Management of Forests	To maintain forest resource in a sustainable state	£101–229/ha
Woodland Improvement Grant	To improve the management of a woodland	Various rates from £2/m for fencing to £1375/ha woodland clearance
Management of Ancient Wood Pasture	To restore wood pasture systems	Capital grants: Inbye land £105/ha/yr, Rough grazing £50/ha/yr
England (Axis 2 and some Axis 1)		
Woodland Creation Grant (WCG)	Creation – public access and biodiversity	£1000/ha + £1000/ha special priorities
Farm Woodland Payment (FWP)	Creation – public access and biodiversity linked to Single Payment System	£60/ha/yr in SDA, £200–300/ha/yr outside SDA
Woodland Planning (WPG)	Management for existing woodlands – assessment and writing a plan	<100ha £20/ha , >100ha £10/ha
Woodland Assessment (WAG)	Management for existing woodlands – to assess ecological/ landscape/ historical value	From £5.60/ha minimum payment £300
Woodland Regeneration (WRG)	Management for existing woodlands – to change species composition	£360–1760/ha
Woodland Improvement (WIG)	Management for existing woodlands – capital investments needed	50 or 80% costs
Woodland Management (WMG)	Management for existing woodlands	£30/ha/yr
Wales (Better Woodlands for Wales)		
Management plan preparation	Preparing woodland management plans	Set payments by task
Replanting		£500–1500/ha
Woodland Improvement		£500–1500/ha

(Sources: Forestry Commission Website, 12/04/11; Better Woods for Wales, 12/04/11)

Table 6.13: Market for Farm Timber

	Veneer	Sports goods	Tool handles	Furniture	Joinery	Flooring	Kitchen ware	Turnery	Structural	Construction	Boat Building	Telegraph poles	Sleepers	Fencing
Hardwoods														
Ash	✗	✗	✗	✗	✗	✗								
Beech	✗		✗	✗	✗	✗	✗							
Birch				✗		✗		✗						
Oak	✗			✗	✗	✗			✗					✗
Sycamore	✗			✗	✗	✗	✗	✗						
Alder				✗	✗	✗								
Cherry	✗			✗	✗									
Sweet Chestnut	✗			✗	✗	✗								✗
Softwoods														
Douglas Fir										✗				✗
Larch										✗	✗			✗
Pine												✗	✗	✗
Spruce										✗				✗

Note: other species with distinct uses are lime, walnut, yew, apple, pear, hawthorn, plum, damson, poplar.

(Source: Coed Cymru 2005.)

products with ready markets irrespective of species, though some timber is more suitable than others. They are also seasonally compatible, so there can be income from woodland all year round (Coed Cymru 2005).

On-farm non-agricultural diversification

Organic farming and woodland exploitation may not be possible for many upland farmers, but there are other options for on-farm diversification related to changing consumer demands. Trends in how food is sold to the consumer have been driven by various food scares (salmonella in eggs, BSE and FMD), alongside mistrust of conventional food production, animal welfare issues, and ethical and environmental concerns. As a result, alternative food systems have emerged.

Instead of putting food into the conventional retail sector (supermarkets), many farmers are exploiting what Ilbery and Maye (2006) call 'individual retail perspectives'. These systems of sale typically minimise the number of stages between producer and consumer (Figure 6.8), so that goods pass through only one or two 'pairs of hands'. Farm produce is sold unadulterated but semi-processed, as with meat or milk, or else it is processed to add value, as in cheese, butter or pies. These goods are then sold through Short Food Supply Chains (SFSCs), of which there are three recognisable types:

- face-to-face, in farmers' markets and on-farm shops;
- spatial proximate, where producers sell to local retailers in the region;
- spatial extended, where producers sell to consumers outside the region.

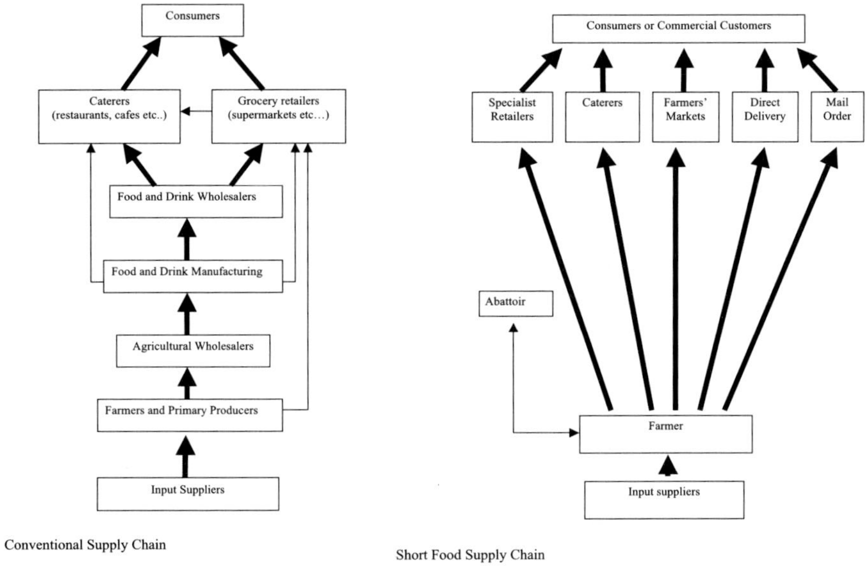

Figure 6.8 Comparing Food Supply Chains
Adapted from Ilbery BW & Maye D, 2005.

SFSCs are perceived as opportunities for the lagging rural regions (LRRs) of Europe to improve their economic and social structure. Upland areas are classic LRRs with remoteness, poor infrastructure, low population density, limited employment opportunities and poor development capacity (Ilbery *et al.* 2004). On the other hand, consumers often make inaccurate inferences about the quality, localness, social embeddedness (connections or associations between product and place) and sustainability of SFSC products. Studies have shown that quality can vary (Ilbery and Maye 2006; see below), social embeddedness may be stretched (Ilbery and Kneafsey 1998, 2000) and 'localness' and sustainability are all a matter of perception (Ilbery and Maye 2005).

Farmers' markets are an example of SFSCs that do provide opportunities for upland farmers to diversify. In the UK they are a recent revival of an old phenomenon, the first being in Bath in 1997. Much has been written about their character, function, distribution and role (e.g. Ilbery *et al.* 1999, Holloway and Kneafsey 2000), but less about their effectiveness for upland farmers, who face issues that lowland farmers are less likely to encounter. For example, lowland studies have focused on Bath, Ashford and Stratford upon Avon, but all these farmers' markets are in urban areas with large pools of customers and good turnover. These factors encourage competition and keep prices down to a point where people can buy as part of the week's shopping, rather than the occasional treat. In contrast, farmers' markets in small towns and villages in an upland area like south-east Cumbria have been successful despite the small local population (Table 6.14). In 2007 there were 14 farmers markets in Cumbria (the second largest county in England with the smallest and most dispersed population). In such places, 'localness' can be a somewhat stretched concept with some stallholders travelling over 50 miles to sell their wares. Sales are also made at a premium, which puts the goods more into the luxury category than part of the week's shopping.

Some advantages do exist for farmers' markets in upland areas. Firstly, if they are in or near HNV landscapes, national parks or areas of outstanding natural beauty, this will attract consumers. People are more likely to buy premium products as a treat while on holiday or short visit. Secondly, since many market locations in Cumbria have few permanent shops, shopowners benefit from an influx of market shoppers on the day. Thirdly, quality is vital but it will pay off because stallholders need to attract and maintain customers where the client base is restricted.

Table 6.14: Farmers Markets in South East Cumbria 2009

	Orton	*Brough*
Settlement population	594	680
Nearest arterial road	M6 (7 miles)	A66 (0.5 miles)
Permanent food shops	2	1
First market	2000	2002
Number of stalls	40	12
Farm produce	10	7
Non-farm produce	30	5
% stalls from livestock farms	25	58
Frequency	Once/month	Once/month

Location is particularly advantageous in branding farm goods, by linking quality to the area of origin. This helps to link the product and producer to the area's landscape, culture and heritage, which in turn can allow both farmers and retailers to ask premium prices for produce (Kuznesof *et al.* 1997). Quality thus becomes an important sales angle for farm businesses, but is fraught with subjectivity. Ilbery and Kneafsey (1998) demonstrated that 'quality' could be shown in a number of ways:

- gaining certification (objective),
- requiring specification (objective),
- identifying geographical or historical association (subjective),
- generating perceptions of special quality (subjective).

In practice, these aspects blend to give the consumer some form of quality assurance, which is promoted by different parts of the agro-food chain, such as a supermarket; a breed association or a processor. Individual farmers can develop their quality brands in direct contact with consumers and middlemen; groups of farmers can also develop quality assurance by using EU regulations.

These also help to protect traditional foodstuffs from cheap imitations or misrepresentation. Under EU Regulation 2081/92 'on the protection of geographical indications and designations of origin for agricultural products and foodstuffs' and Council Regulation (EC) No. 510/2006 'on the protection of geographical indications and designations of origin for agricultural products and foodstuffs', products can attach the following labels:

- Protected Designation of Origin (PDO) for foodstuffs produced, processed and prepared in a given place using recognised know-how;
- Protected Geographical Indication (PGI), where the geographical link must occur in at least one of the stages of production, processing or preparation;
- Traditional Speciality Guaranteed (TSG), which does not refer to the origin but highlights the traditional character, either in the composition or the means of production.

Ilbery and Kneafsey (1998) viewed these types of appellation as an important exploitation mechanism to help marginal agricultural areas develop their economies. Applications are made to the European Commission on behalf of producers by DEFRA in the UK. Accreditation of PDO or PGI is particularly stringent. It must comply with the following (Ilbery and Kneafsey 2000):

- a product name must be used, with a geographical descriptor linked to a specific area, e.g. Buxton Blue Cheese;

- methods of production have to be specified;
- historical evidence must link the product to the specific area;
- inspection must show the above conditions are met.

Consequently an agricultural region needs to find what makes it distinctive from other areas, ensure that it is authentic and be able to promote the 'place' effectively. Beaufoy (2007) was able to show how PDO foodstuffs commanded higher prices and thus helped to maintain farming systems that might otherwise have disappeared. At the same time it is important to realise that labelling under these regulations is not linked to HNV, nor to the quality of environmental conditions of farm production.

There are currently 14 PDOs, 20 PGIs and one TSG in the UK (Food from Britain website, accessed 18/08/08). The majority are cheese (37%), followed by fresh meat (20%) and cider (17%). In upland agriculture, seven cheeses and seven meat food names are protected. The cheeses are location-specific and the five PDOs include, for example, Swaledale cheese. The meat foodstuffs include Orkney beef and lamb (PDOs), Shetland lamb (PDO), Scotch beef and lamb (PGIs) and Welsh beef and lamb (PGIs). Among notable foodstuffs seeking protection at the time of writing are Lakeland Herdwick Lamb (PDO) and Yorkshire Wensleydale Cheese (PDO), which will directly aid upland farming businesses, also Traditional Cumberland Sausage (PGI) which will have an indirect impact for farm businesses.

Although this sounds positive for upland farmers, there are particular problems with slaughtering arrangements in some areas. Under PDO criteria livestock must be slaughtered within the designated area, but many abattoirs have closed down as a result of uneconomic legislative demands, and those which remain are at full capacity. For the Rough Fell lamb group in south-west Cumbria this has stymied achieving PDO status for the foreseeable future (Mansfield 2008).

Off-farm diversification

The final option for adding new sources of income is off-farm employment. Other Gainful Activities (OGAs) are an interesting concept as part of upland agriculture. Historically this is how the farming system tended to work, with farming being a more part-time occupation linked to other jobs such as mining. Crofting is an extant example of this, though it is a development from the pre-1800 Highland farming system (Warren 2002).

It could be argued that the current definition of a full-time farm demonstrates that many farms in uplands are pluriactive businesses anyway (Box XXV). Whether this has arisen by necessity or by the simple application of the 'SLR' definition is partly academic; the fact that many farms are now considered part-time operations (77% in LFAs) suggests off-farm employment should be an important part of maintaining the upland sector. Yet, as we have seen, diversification statistics show that LFAs and national parks have the lowest off-farm employment rates of any farm group (Table 6.10, Figures 6.5 and 6.7). The corollary is that many upland farm businesses are part-time, but the labour requirement is full-time.

Opting for OGA is often a combination of pull factors off the farm (more opportunities) and push factors (financial condition of agriculture). It would seem logical that the lower the farm's income, the more likely OGA would be an option, but the pattern of response is more complex than this. Edmond and Crabtree (1994) found in their Scottish study that farms with the lowest gross margins were not responding to this push factor and embarking on OGA. Instead, they found that there were regional patterns, with crofting areas seeing the most OGA where it was traditional to have multiple employment streams already. The push factor of low gross margins was also not operating in a study of hill farmers in Cumbria (unpublished data, Burton *et al.* 2005). For upland farming this lack of OGA is probably a function of the small labour pool these systems now rely upon.

Diversification as an option for hill farms is a 'Catch 22' situation. On the one hand, for the farm business to survive, new income streams are needed to reduce the reliance on food production. For upland farmers these opportunities are enhanced by their location in HNV environments. The only problem is that, to survive, many have adopted a traditional restructuring approach which has in effect boxed them into a corner with regard to labour. Without slack in the labour system, there is no time to develop diversification opportunities. The changing policy environment in Europe and the world is not helping, now that European farmers are required to shift part of their efforts from production into non-agricultural agendas (see Chapters 8 and 9). It is difficult to adopt the new diversifications of environmental management and rural development when there is no financial slack left in a particular farm system.

Summary

In this chapter we have investigated the economics of upland farms. These businesses produce some of the lowest gross margins of any farm type in the UK and historically this has led to a range of government support mechanisms to lift these households out of poverty. Unfortunately, the very subsidies and grants designed to improve farm incomes have led to a suite of less desirable environmental changes because the economic margins of cultivation have been pushed closer to, or even beyond, the physical boundaries.

As the cost–price squeeze has continued, uplands farmers have sought to reduce overheads by limiting their outgoings. Many have taken a traditional restructuring approach by reducing labour or intensifying production. Neither choice has had the desired effect. Instead, by over-exploiting the HNV environment of uplands or removing divertable labour, both have inherently harmed farm businesses' potential to diversify. A minority of farms have embraced diversification as a way of maintaining the business and farm household. What drives farmers to respond to economic crises in these different ways is the subject of the next chapter, as the social dimension of upland agriculture is explored.

7

The social dimension

Historically, upland areas of Britain exhibited economic superiority over their lowland counterparts as they were richly endowed with land, minerals and water resources which underpinned agricultural and industrial activity (see Chapter 3). Over time the uplands' comparative advantage in minerals, then water power and finally agriculture dwindled. As international trade and industrialisation grew, the focus of activity switched to the more accessible lowlands, leaving many upland areas on the periphery of economic activity.

Decline

With stagnating economies, uplands have become mainly agrarian, with some renaissance supported by recreation and tourism exploiting HNV landscapes. A declining agricultural sector and few other job opportunities act as push factors, forcing out-migration of the indigenous population. In turn lower population levels weaken social service provision, which itself becomes a push factor, amplifying emigration; thus the economy spirals gradually downwards. This cycle of decline is typical of many peripheral areas like the English national parks, as described by MacEwan and MacEwan (1972; Figure 7.1). It is a situation well understood by social, rural and agricultural geographers and historians (e.g. Howkins 2003, Moseley 1979, Newby 1985).

In response, socially responsible governments seek to halt rural depopulation from these areas by developing policies and strategies that create:

> well balanced demographic structures, appropriate real incomes with sufficient purchasing power, secure working and living conditions, social institutions, and finally, a sense of value in the daily life of the population – a life without war, oppression, and need, and also a viable local culture and language. (Reider and Wyder 1997: 89)

The EU has supported lagging regions with regional development policies and sector-specific schemes (for example, the Less Favoured Areas Directive 268/72), but the effectiveness of such policies has been questioned remorselessly, simply because peripheral areas like uplands continue to decline economically and socially, while also damaging the very environment they rely upon for their economy (see chapters 5 and 8). This set of circumstances suggests that the situation is more complex than just providing economic support to upland agriculture, a view now more readily accepted with the EU's Rural Development Regulations (RDRs) (2000–2013), whereby economy, society and environment are perceived as a three-legged stool on which all-encompassing rural development rests (see chapters 8 and 9).

There are two flaws in the RDR approach. First, a cursory inspection of both iterations of the RDR shows a tendency to rely on better formulated, tried, tested and more objective economic and environmental initiatives in the hope that the third subjective social 'leg' will spontaneously

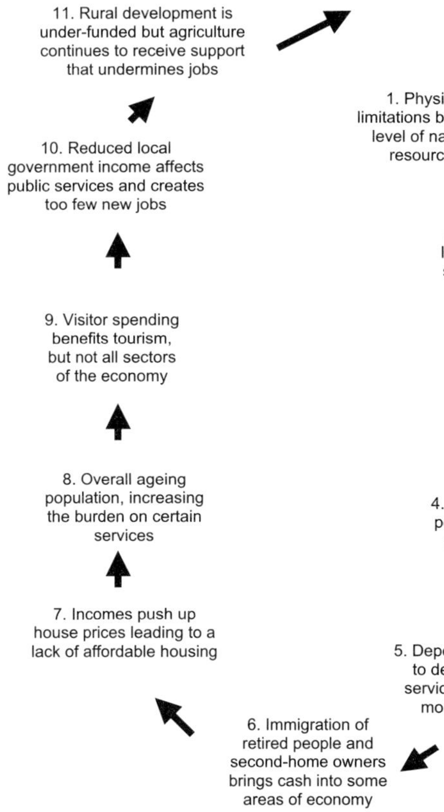

Figure 7.1 Declining Rural Populations
Adapted from McEwan M & McEwan A, 1987.

improve as a result, not unlike the earlier, less-sophisticated LFA directive (see Chapter 8). Second, RDR philosophy is to move away from food production into environmental management and general rural development, neither of which is core to the psyche of farmers or the farming community. The challenge is to develop effective social policy which can take on board the complexity of processes in upland farming communities, between farmers and individually, while identifying and maintaining those social processes and structures that satisfy EU policy and sustain farming communities.

One concept that may help to identify and maintain relevant social processes and structures is social resilience: the capacity of individuals and/or communities to absorb change, based on inherent characteristics of that society (Box XXVI). Adger (2000) suggests we can look at this in two ways. First, social systems depend on the environment: society's resilience is linked to ecosystem resilience. In other words, if the ecosystem falters, then society's capacity to adapt will falter too. Chapters 1 to 5 have shown how upland agriculture relies on ecological systems and what can happen following poorly judged management. Social resilience can also be seen in the ability of the community's social institutions to maintain social organisation. This type of social resilience in upland farming communities is typified by local knowledge systems (LKSs), meaning the way indigenous peoples develop management systems for their environment by trial and error (Mitchell 2002). Even if resource degradation does occur occasionally, the local population learn to avoid repetition because their survival depends on it; thus over time a LKS becomes more and more attuned to its environment.

LKSs consist of natural, human and social capital. Natural (environmental) capital relates to any stock of natural assets that provides goods and services year after year. There are three types of natural capital: non-renewable, renewable and capacity to absorb unwanted by-products. Human capital is the knowledge, skills, tradition, practices and motivation that guide a person's everyday activities. This will vary from individual to individual and over time as a person moves through their lifecycle (Kristensen and Thenail 2004). Social capital refers to "features of social organization, such as trust, norms, and networks that can improve the efficiency of society by facilitating co-ordinated actions" (Putnam 1993: 167). An important feature of social capital is that it cannot be created instantly; it takes time to develop, as people learn to respect and trust each other (Roseland 2000). Natural, human and social capitals are interrelated, so if one changes the other two do too. For example, social capital can be eroded by neglect and not only can the collective suffer, but individuals can lose their own human capital as a result (Ostrom 1993, quoted in Roseland 2000). The key to sustainable development is to ensure that change in one does not negatively affect the others. The decoupling of EU agricultural policy from food production, moving it towards environmental management and rural development, has increased natural capital (see chapters 5 and 9). Unfortunately, this has undermined human and social capital as upland farming has responded with economic contraction. Thus any solution to the 'hill farming problem' requires equitable support of natural, human and social capital.

The upland farming local knowledge system contains all three elements of capital (Figure 7.2). In comparison to other farming systems, the upland LKS is unique in certain structures and processes:

- the farmer's role in deciding how the farm management system manages land use (inbye, intake, heft) and livestock production in line with the natural capital (human capital);
- the heft, and the process of hefting, which relies on co-operative working (social capital);
- localised skills and knowledge of the natural environment (fell knowledge) being transmitted orally and visually rather than written down, hence the need for succession (human capital);
- the farming community and neighbouring farms being needed for hefting and gathering processes as well as creating a sense of place and tradition among everyone (human and social capital);
- the pivotal role upland farming plays in stratification of the livestock industry (social and natural capital);
- the production of HNV landscapes as a result of the relationship between farm management techniques and upland environment (natural capital).

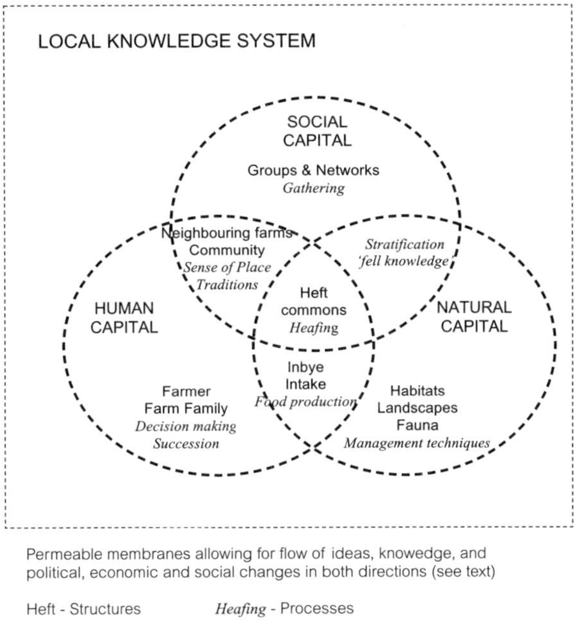

Figure 7.2 The Upland Farming Local Knowledge System

Chapter 5 dealt extensively with the character of natural capital on upland farms. The purpose of this chapter is two-fold: to explore the human and social capital of upland farming local knowledge systems; and to use this information to identify which social processes and factors are essential for sustaining these communities, enabling them to feed into more effective tripartite policy development.

Human capital on upland farms

The human capital of an upland farm consists of the farmer and his/her family, plus any additional labour, which is increasingly rare on upland farms due to economic constraints. Labour is usually the first variable cut to reduce the cost of production. On many upland farms the human capital is therefore limited to the farmer and their family (if any).

The farmer's human capital

As we saw in Chapter 6, classic economic theory assumes people are economically rational and consequently make the best decisions for their business based on perfect knowledge. A farmer would therefore maximise their profit margins. Such theoretical ideas met increased resistance from researchers in the 1960s because there is a gap between optimising and satisficing behaviour, representing the human element in farming. A farmer's consideration of risk and uncertainty will be influenced by personal factors like

- health,
- wealth,
- values,
- goals,
- motives,

as well as the broader concerns of

- technological change,
- market structures,
- institutional arrangements (such as government policy),
- physical factors.

To this list Tarrant (1974) added farm family status and the diffusion of information. Burton (2004a) argued that the role of society is important also, because society places behavioural norms on people which most of us are bound to obey. Mitchell (1984) considered an individual's perception to be governed by both past experiences and present outlook, and conditioned by values, moods, social circumstances and expectations. Consequently no two individuals receiving the same stimulus will have similar world images. Put simply, farmers operate in an environment constructed by their own perceptions and those of other people. Many of these views can be drawn together to form a concept known as the Decision Making Process (Figure 7.3), originally suggested by Found (1970). This process is pivotal in explaining how and why a farmer decides what to do, or not to do, in their farming activities. Ultimately it shapes agricultural land-use patterns.

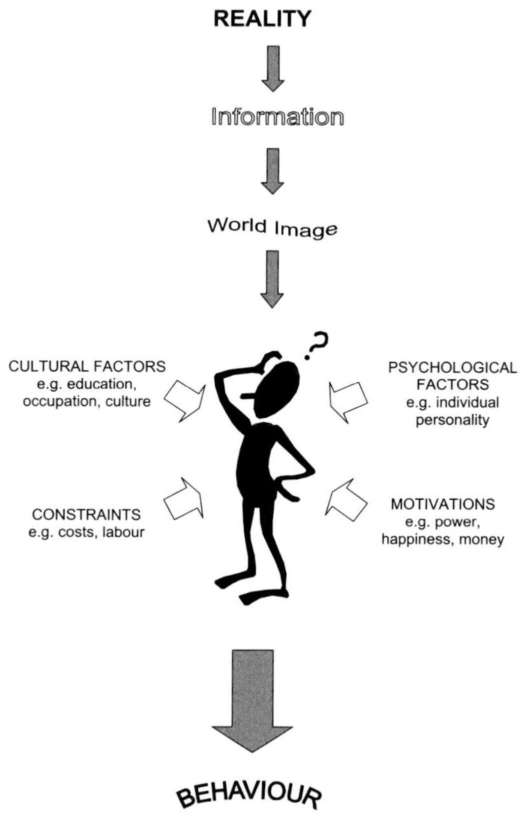

Figure 7.3 The Decision Making Process

Another important concept for the decision making process is the relationship between attitudes and behaviour. An attitude can be described as 'someone's view or opinion about something', whereas behaviour is an action. Attitudes were originally believed to translate directly into behaviour (Ajzen and Fishbein 1975), but in reality what we say and what we do can be two different things (Bunting and Guelke 1979). This dichotomy, which has been observed among farmers in relation to nature conservation, is either because:

- attitudes have little bearing on subsequent behaviour, or
- farmers and conservationists differ about what is environmental behaviour.

Carr and Tait (1990) found the latter occurred among Bedfordshire farmers, where weeds and vermin in the farmers' eyes were flora and wildlife to the conservationists. MacEachern (1992) found that livestock farmers in Upper Yorkshire Dales also differed on the nature of countryside and its conservation. It needed to be agriculturally rather than environmentally sustainable because it was their business. For farming, as discussed in chapters 4 and 5, conservation issues revolve around practices that inhibit production rather than negatively affect the environment, for example stocking densities and overgrazing. Thus any grant schemes proposed by conservationists to manage the countryside may be seen as ludicrous and incompatible with farming. Since that research was carried out around 1990, agricultural reality has become less production-orientated and more environmental. However, empirical studies suggest that farmers still cling to their production-focused attitudes (e.g. Burgess *et al.* 2000, Burton 2004b, Mansfield and Martin 2003), despite their enrolment into the agri-environmental schemes on offer (see Chapter 8). This implies that these schemes are used by farmers to support farm incomes and not to improve the environment as the government would wish (e.g. Hughes 1994, Mansfield 2008). Some evidence does suggest that a positive attitude to conservation led to more rapid adoption of an ESA agreement in Scotland than among farmers who were less positive, but it did not the overall probability of joining (Wynn *et al.* 2001).

So attitudes do not necessarily translate into behaviour. Azjen and Fishbein (1980) refined their approach, suggesting that attitudes have three components: cognitive, affective and behavioural (Figure 7.4). Burton (2004a) expanded this idea by recognising other social influences on attitudes and behaviour, such as habit, moral obligation, social or subjective norms and identity (of self, region and nation). To this we can add sense of place for upland farmers (see below).

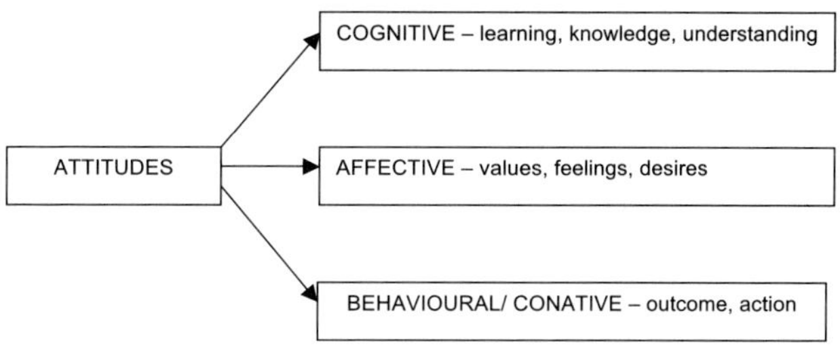

Figure 7.4 Tri-Component View of Attitudes
Adapted from Ajzen I & Fishbein M, 1980.

Cognitive aspects of attitude

The cognitive component of any attitude is the learning of skills and knowledge, which form the cornerstone of any farmer's ability to farm and/or run a business. Whether the business is financially viable depends on a combination of access to (quantity and quality of) information and the ability to use it, as Pred's work showed in Chapter 6 (Figure 6.1). Quality and quantity of information can be increased in three main ways:

1. informally through practical experience – personal or handed on by other farmers, at demonstrations and special events, from TV and radio programmes or increasingly the Internet;
2. formally – via the farming press, government material and services or agricultural education.
3. Trial and error.

How informal knowledge or information is spread to and adopted by farmers has received much attention in the academic literature. Hagerstrand (1953) assumed information would spread among farmers through face-to-face contact, but in a random spatial manner as a result of non-economic and non-rational behaviour. A central theme of his model was that the farmers furthest apart would have the lowest probability of contact. He tested his theory using the diffusion of information on, and adoption of, a Swedish farm subsidy to improve cattle pastures. He found that simulated and actual patterns of diffusion were remarkably similar. It is important to remember that Hagerstrand was writing in the 1950s; ideas now spread without face-to-face contact. Distance is no longer an issue.

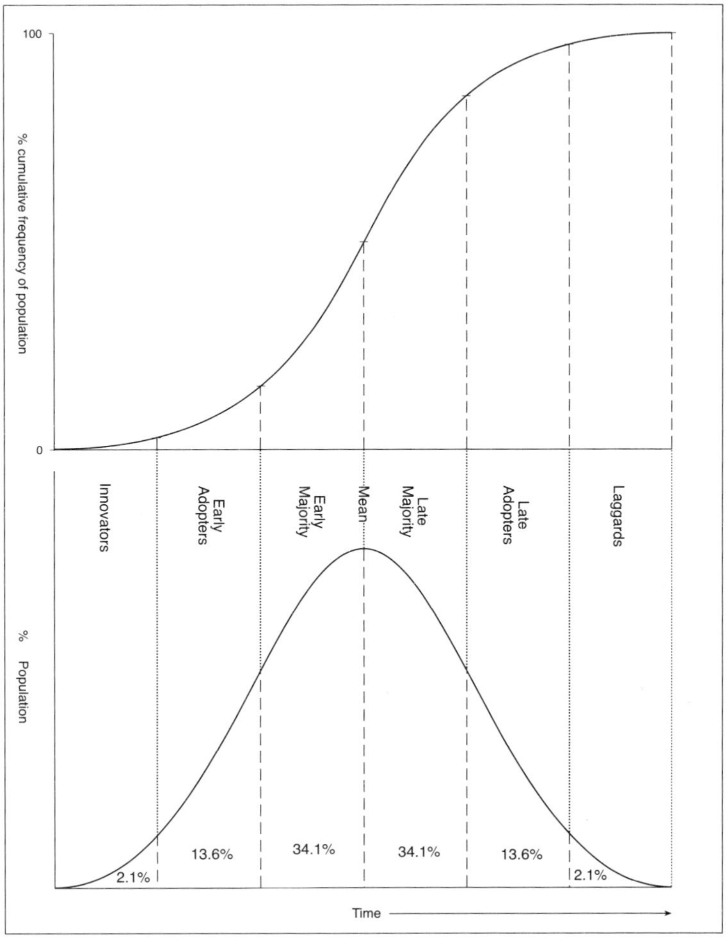

Figure 7.5 Diffusion of Innovation
Adapted from Ilbery BW, 1984.

Further work by Hagerstrand (1953) and Rogers (1962) focused on diffusion rates of ideas and innovations. The rate of diffusion of an idea takes the form of an S-curve in a population: an idea is adopted slowly at first by a few, followed atan accelerated rate of adoption by the majority. As adoption nears saturation, the rate decelerates (Figure 7.5). Empirical research on rates of innovation has shown that some 2.5% of farmers are Innovators. Then come the Early Adopters (about 13.5% of farmers), who quickly see the benefits of an innovation. They in turn are followed by the Early and Late Majority, then the Late Adopters and finally the Laggards (about 2.5%), the farmers who resist change and innovation and may never adopt a 'new' idea. The whole pattern can be translated into the classic mathematical statistical shape – the normal curve.

The case of laggards is interesting, because they are not necessarily unwilling to adopt an idea on personal grounds, but simply cannot, usually due to lack of capital, resources or information (Healey and Ilbery 1990). Even as early as the 1950s, lack of capital played an important role in non-adoption by upland farmers of the Livestock Improvement Scheme. In a study by Ellison *et al.* (1955) 30% of those surveyed cited this as the main reason why they did not apply for a grant. Wynn *et al.* (2001) found that farmers in Scotland failed to join the various ESA schemes as a result of limited awareness and lack of information. A review by Natural England (2008) into the narrow uptake of options in the entry-level Environmental Stewardship Scheme found that limited scheme knowledge and lack of environmental skills to recognise farm potentials were responsible. This latter, in contrast, was addressed in Wales in the *Tir Cymen* agri-environment scheme, with training days made a mandatory part of grant allocation (Welsh

Assembly 1999). These conclusions resonate with Pred's theoretical work, which also suggested that the ability to use information was important for farmers.

The five phases of adoption were later found by Jones (1963; 1975 cited in Ilbery 1984) to have distinctive characteristics in terms of geography, economic features, demography and social parameters (Table 7.1). Jones' original study found that more educated farmers sought and used new information to a greater degree than less educated. Older farmers were more conservative and rarely sought new information or changed what they were used to. Though useful as indicators, these sorts of typology are fraught with problems; Ilbery (1984) gives a good summary critique. Furthermore, these characteristics were identified empirically many years ago, when there was much more emphasis on production in agriculture and headage payments were central to upland farming systems. Since the 1990s, things have changed: financial hardship affects most of the adoption phases, the farm population is ageing, capitalist agriculture is over-specialised and the computer revolution has had a huge impact on access to information. As a result, later studies have identified other factors responsible for adoption or non-adoption by farmers (e.g. Brotherton 1991, Morris and Potter 1995, Potter and Lobley 1992, Battershill and Gilg 1997):

Table 7.1: Farm Characteristics and Diffusion of Innovation

adopter class	*personal characteristics*	*values and social relationships*	*communication behaviour*
Innovators	Highest social status Largest & most specialised operations Young & well educated Non-farm environment experience	**Venturesome** Willing to accept risk Provide leadership to community Cosmopolitan	Close contact with scientific resources Interaction with other innovators High use of impersonal channels of information
Early Adopters	High social status Large & specialised operations	**Respected** A role model in community with some influence Provides opinion leadership	Greatest contact with local agents of change (e.g. advisory services) Competent use of mass media
Early Majority	Above average social status Average sized operations	**Deliberate** Willing to consider new ideas only after peers adopt Some opinion leadership	Considerable contact with change agents and early adopters Receive mass media
Late Majority	Below average social status Small operations and little specialisation Relatively low income	**Sceptical** Pressure from peers needed before adoption occurs Little opinion leadership	Interaction with peers who are mainly early or late majority Less use of mass media
Laggards	Lowest social status Smallest operations and little specialisation Lowest income Often the oldest	**Traditional** Oriented towards past Avoids risks Little if any opinion leadership Almost isolated socially	Neighbours, friends and relatives with similar values are main information source Suspicious of change agents

(Source: Jones 1975, cited in Ilbery 1984.)

- geography – location, remoteness, land quality, physical restrictions (these were investigated in Chapter 4);
- economic features – farm attributes like income, size, type, labour requirements (these were considered in Chapter 6);
- demography – farmer characteristics such as age, educational background, goals, values and aspects of the farm family like marital status, succession, dependants and stage in life cycle (these are discussed in this chapter);
- social parameters – tradition, culture and heritage (also considered here).

Individual farmers' age and educational status are often cited as challenges for the upland sector (e.g. DEFRA 2000, IEEP 2004b). Farmers in the uplands are perceived as generally older than lowland farmers, possibly explaining their reluctance to adopt new ideas (DEFRA 2000). Data issued by the Farm Business Survey (2007) demonstrate that, in fact, the age profile of LFA farmers is no different from that of the general farming population. Between 1990 and 2005 the average age changed from 55 to 58 years, with over 75% of farmers over 45. There are regional differences: the Welsh Borders and Peak District have more ageing farming populations; the Yorkshire Dales and Bowland have much younger age profiles (Figure 7.6).

Though an ageing population does not bode well for the industry as a whole, the real issue for upland farming communities is that the job is physically more demanding. Much of the time people have to rely on personal fitness, rather than mechanical devices, such as tractors or quad bikes, as a consequence of the physical environment. These comments from hill farmers over 40 are typical:

> It gets harder every year. I used to be able to get up the fell behind the farm in 13 minutes; it now takes me an hour.
>
> I'm getting overweight because I don't walk enough. I have to work off the farm as well and therefore I don't get enough exercise.
>
> I don't go up now, only for gathering times. My son goes on his quad bike a lot. (unpublished data, Social Capital in Hill Farming project 2004)

And:

> Hill farming is a young person's game: once you get over 40, you're knackered. (Mansfield and Martin 2003)

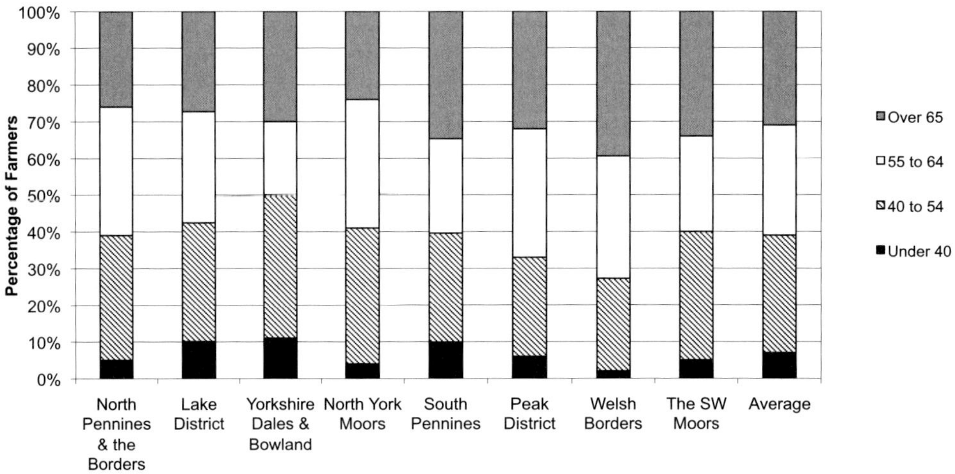

Figure 7.6 Age Profile of Farmers in English Uplands 2009
Based on data adapted from DEFRA, 2009a.

With respect to agricultural education, the 2005 UK survey showed that most farmers (77%) relied on practical experience only (DEFRA 2006; Figure 7.7). Beyond that, 11% had what is called basic training – that is, formal training of less than two years – and 12% full agricultural training, meaning further and higher education courses in agricultural college or university. The most formally educated were cereal farmers: 24% had at least two years' training. For LFA sheep and cattle farmers, the level of agricultural training was below the national average in 2005, with 82% of them relying on practical experience only, 9% on basic and 9% on full training (pers. comm. Robin Karfoot 29/02/08). This was actually the least educated profile of all farm types recorded. In their defence, formal agricultural training only really took off after 1945 (Gasson 1998) and became generally available in the 1970s; thus anyone born before 1960 is unlikely to have had even one year in college. Therein lies the problem for a sector whose average age is 58. The trend shows improvement year on year: in 1990 'practical experience only' accounted for 86% of LFA farmers.

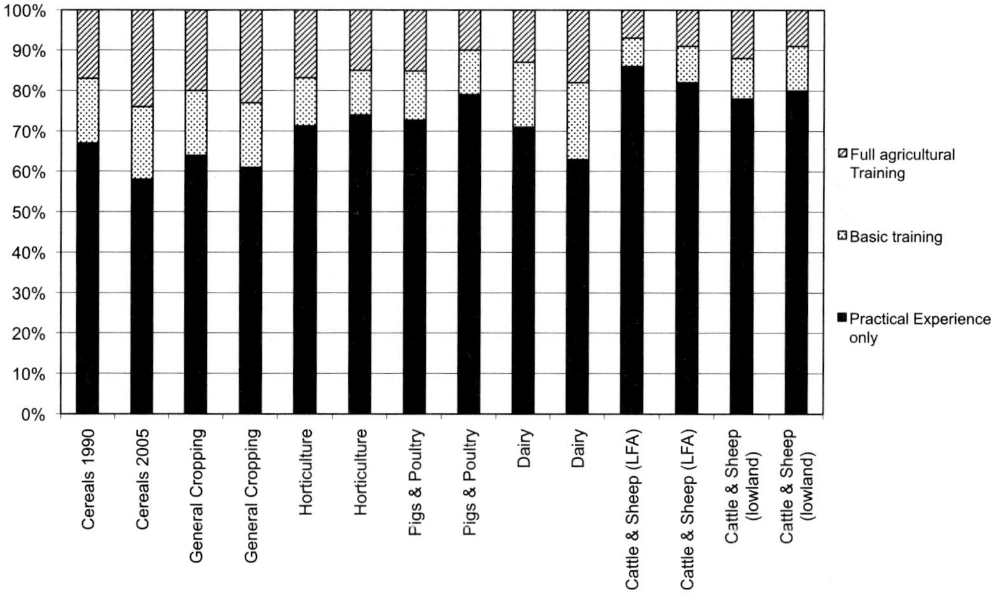

Figure 7.7 Changing Education Levels of English Farmers between 1990 and 2005
Based on data provided with permission from DEFRA provided by R Karfoot, 29/02/08.

The lack of formal training sounds problematic, but there is scant evidence that farm businesses run by those with no or little formal agricultural education are less profitable (Gasson 1998). However, there is evidence that farmers with more education will access more information readily and take more risk. Perhaps a more important question should be: are they getting the right information as part of their education? Historically, many further education programmes in agriculture had no content on conserving biodiversity. Even in 2000 the environmental content of such programmes was fairly limited, often focused on subjects like 'hedge management', 'how to apply for agri-environment grants' and 'how to diversify your business' – topics seen in terms of finance rather than conservation. Battershill and Gilg (1997) found this too with younger farmers (trained post-1980) seeing conservation as a business opportunity, particularly the value of environmentally-friendly food marketing, influenced by college curricula. At the same time young people seem more interested in farming tasks than diversification issues (Mansfield and Martin 2008).

Affective aspects of attitude

The second component of attitude that Ajzen and Fishbein (1980) identified was the affective: values, feelings and desires. These personal attributes derive from three main sources: the farmer themself, the farm family and the world beyond the farm gate (which can include other farmers, the wider local community, stakeholders in land management and the non-farming public).

Individual farmers values

Individual farmers' values were a focus of academic literature in the 1970s and 1980s (e.g. Gasson 1973, Ilbery 1983). Gasson suggested farmers have four main values:

- intrinsic – farming is valued as an activity in its own right;
- expressive – farming is valued as a means of self-expression or personal fulfilment;
- instrumental – farming is viewed as a means of obtaining an income and security;
- social – farming is carried out for the sake of interpersonal relationships.

Within these four are a number of other values that help to explain why farmers farm and hence what can influence their decision making (Table 7.2).

Table 7.2: Values and Motivations of Farmers

Instrumental
Make the maximum amount of money
Safeguard the business
Make a satisfactory income
Provide congenial work conditions
Expressive
Meet the challenge
Feel pride in ownership
Be creative & individual
Gain self-respect for doing a worthwhile job
Exercise special abilities and aptitudes
Intrinsic
Enjoy the work tasks
Gives me control in various situations
Be independent
Work outdoors in a healthy environment
Social
Continue the family tradition
Work with other family members
Belong to the farming community
Gain respect and prestige

(Source: adapted from Gasson 1974, Ilbery 1985 and Gillmor 1986.)

Attempts have been made to analyse differences in these values among distinct farmer groups, using an objective technique called Point Score Analysis (Box XXVIII). Ilbery (1978) examined Worcestershire hop farmers and found intrinsic values were the most important, as did Gillmor (1986) for farmers in Leitrim and Meath in Ireland. In cereal-growing areas Mansfield (1994) found that instrumental values dominated, but there were variations with farm size and farm income. Since then the technique has fallen out of use to some degree (Burton 2004). In uplands the only 'values' study formed part of a research project to investigate social capital (Burton *et al.* 2004a). Data were collected from a small farmer sample for three Cumbrian massifs (Lake District, Howgill–Orton Fells and Pennines). Table 7.3 and Figure 7.8 show that, on average, instrumental values were the most important, followed by intrinsic types. The least emphasis was placed on external social values, though this result was severely skewed by a more isolationist view among Lake District farmers (30.7% importance compared to Pennines at 76.7% and Orton-Howgills at 69.4%). This is somewhat confirmed by the Farm Practices Survey (DEFRA 2009a) which showed 27% of Lake District farmers perceived maintenance of the environment as making upland farm management more difficult, compared to the average response of 18%. This could be due to the influence of national park status and the greater proliferation of SSSIs and NNRs.

Box XXVIII: Point Score Analysis

Point Score Analysis is a quantitative technique for analysing decision making in agriculture (Ilbery 1977). It enables the researcher to measure the relative importance of various opinions and values – social, economic, political and physical – that influence farmers' decisions. A list of randomly ordered factors is presented to the interviewee, who is then asked to judge whether factors are relevant to their farming situation. For each statement that is relevant, the farmer is asked to assign a level of importance on a graduated scale:

- Very Important
- Important
- Unimportant
- Irrelevant

Points are assigned to each level of importance on the scale, in this case 3, 2, 1 and 0. The main advantages of the technique are that it allows some quantification of subjective data, hence statistical analysis is possible, and it allows comparison between different groups, for example upland and lowland farmers. Burton (2004) has expressed concern that this *Likert* scale is being used inappropriately – there should be five categories – but social research has shown that, given a middle category, people often tick this to avoid committing themselves. An even number of choices forces the respondent to be definite about 'less' or 'more'.

The importance of each factor is measured by summing the point scores given by each farmer for that factor. Each factor has a possible maximum score and can be examined using three indices (Van der Vliet 1992, quoted in Ilbery 1977):

- *Percentage Applicability* (%A) – the number of cases where the factor applied, expressed as a percentage of the total number of cases investigated: this shows the applicability of the factor to this group of farmers;
- *Percentage Total* (%T) – the total score as a percentage of the maximum possible score for all cases investigated: this statistic recognises the level of relevance of a value to the farmers;
- *Percentage Importance* (%I) – the total score as a percentage of the maximum possible score for those where the factor was relevant: this indicates the importance of a value in terms of its relevance and applicability.

The three indices can be used to rank the individual values in Table 8.1 to identify variations between groups of farmers and the four main values: intrinsic, expressive, social, instrumental. The process thus allows researchers to add their own values to suit the nature of the work. For instance, unpublished data from Burton *et al.* (2004) contained a number of social capital values, given the focus of the project:

- Maintain good relations with visitors to the area.
- Take responsibility for the countryside.
- Upkeep drystone walls.
- Maintain good relations with my neighbours.
- Maintain good relations with the locals.

Importance-Performance Analysis is a similar tool, used in visitor satisfaction surveys in tourism management (Hall and McArthur 1998). A list of phenomena is presented to survey respondents (for example, visitor centre, car parking, nature trail, toilets) who use a *Likert* scale to grade the importance of each phenomenon for their visit experience and then in terms of performance. In this case, the scale runs from 0 'irrelevant' and 1 'not satisfied' to 5 'completely satisfied'. All scores for all respondents are totalled and averaged to calculate overall 'importance' and 'performance' scores. These figures are plotted as two axes on a graph known as an action grid (as below).

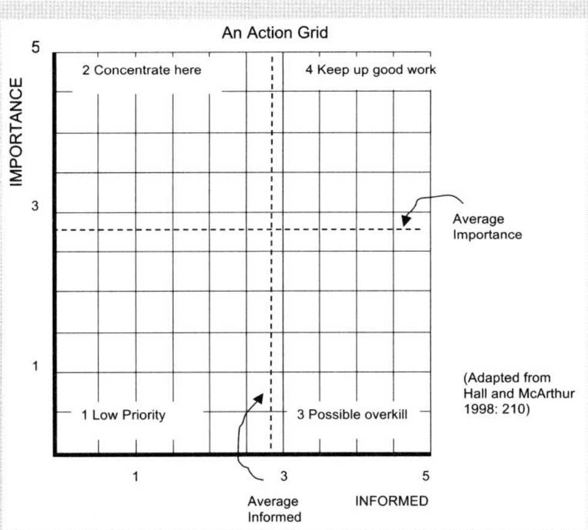

After each attribute's average scores for importance and performance are plotted as graph co-ordinates, the data will fall into one of four quadrants:

- Low importance – little informed (1)
- High importance – little informed (2)
- Low importance – well informed (3)
- High importance – well informed (4)

These four scenarios help visitor service managers to see which attributes are important but need improved performance (concentrate effort here), those which already perform well (keep up the good work; management is sufficient), those that visitors perceive as of little importance (low priority for management) and finally services considered unnecessary (possible overkill). It is of course crucial in using IPA to consider the different views of visitors and service providers of what is important and what is not, because their objectives could be vastly different.

Such a technique is of interest here as it can be applied to farmers, exploring awareness of new ideas, services and grants. The IPA axes can be, for instance, 'how important do you think X is for your farm business' paired with 'how well informed do you feel about X?'. This has been applied to a small survey of ecosystem services in the Bassenthwaite catchment of the Lake District (Mansfield 2011).

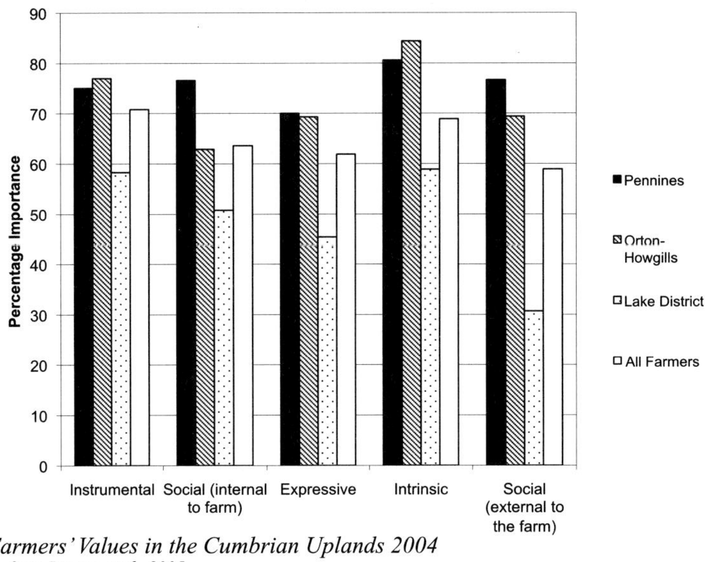

Figure 7.8 Farmers' Values in the Cumbrian Uplands 2004
Unpublished data from Burton *et al.*, 2005.

Table 7.3: Individual Ranked Point Scores for 3 Upland Massif Farmers

value class		*Pennines*	*Orton-Howgills*	*Lake District*	*average*
Instrumental	**%A**	100.0	93.8	66.7	88.9
	%T	75.0	72.2	38.9	62.0
	%I	75.0	77.0	58.3	70.8
Social (internal)	**%A**	99.4	97.2	95.0	97.4
	%T	76.1	61.1	48.3	61.9
	%I	76.6	62.9	50.8	63.6
Expressive	**%A**	100.0	99.4	89.4	96.3
	%T	70.0	68.9	40.6	59.8
	%I	70.0	69.3	45.4	61.9
Intrinsic	**%A**	100	97.9	54.2	95.6
	%T	80.6	82.6	31.9	65.0
	%I	80.6	84.4	58.9	68.9
Social (external)	**%A**	100.0	99.4	85.0	94.4
	%T	76.7	69.0	26.1	56.9
	%I	76.7	69.4	30.7	60.3

Note: %A = % applicability; %T = % total; %I = % importance (see Box XXVIII).

(Source: Burton *et al.* 2004, unpublished data)

More detailed analysis of individual values in the small upland farm survey showed that for the whole sample the three highest ranked values were the ability to have control in a variety of situations, independence and belonging to the farming community (Table 7.4). The pattern was not consistent across the three massifs: different intrinsic values were ranked first in each. The three lowest ranked values were (lowest first): maintaining good relations with visitors, then (these three equal 20th) maintaining good relations with locals, upkeep of walls and gaining self-respect for doing a worthwhile job. The highest ranked scores suggest that the upland farmers of this survey value the inherent nature of farming above all else, but financial concerns matter more to Orton-Howgill and Lake District farmers than to those on the Pennines. The response to continuing the family tradition is considered in more detail below.

Whilst a huge variation in individual human capital exists amongst the farming community, we can make some tentative conclusions about their characteristics. They are driven by agricultural objectives, age around 58, have minimal formal education and are motivated by instrumental and intrinsic values. However, few people live and work in complete isolation and therefore the influence of the farm family human capital merits consideration.

The farm family and succession

Another affective control on farmers is the farm family. The family will influence attitudes and decision making, depending on the farmer's marital status, level of dependence, succession and stage in life cycle. Once again, all these issues have been considered in depth in the research literature, particularly Whatmore (1991) on the role of women in the farm household and the traditional farming patriarchy (see Robinson 2005 for an overview).

Table 7.4: Farmer Values by Class of 3 upland Massifs

value	*Pennines*	*Orton-Howgills*	*Lake District*	*all farmers*
Instrumental				
Making the maximum income	13	14	2	4
Making satisfactory income	5	1	21	13
Safeguarding the business	10	3	15	9
Providing congenial working conditions	19	19	7	16
Social (internal to farm)				
Gaining recognition, prestige as a farmer	21	14	4	11
Belonging to the farming community	6	8	8	3
Continuing the family tradition	12	14	11	16
Working with other family members	16	22	10	18
Maintaining good relations with workers	2	23	6	8
Expressive				
Feeling pride of ownership	6	3	20	13
Gaining self-respect for doing a worthwhile job	17	13	18	20
Exercising special abilities and aptitudes	14	18	3	7
Chance to be creative and original	19	14	8	15
Meeting a challenge, achieving an objective, personal growth	22	11	4	9
Intrinsic				
Enjoyment of work tasks	6	1	18	11
Preference for a healthy, outdoor farming life	11	3	23	19
Independence	1	3	13	2
Control in a variety of situations	15	8	1	1
Social (external to farm)				
Maintain good relations with visitors to the area	17	19	17	23
Take responsibility for the countryside	4	8	13	5
Upkeep drystone walls	9	11	22	20
Maintain good relations with my neighbours	2	7	15	5
Maintain good relations with the locals	22	19	11	20

(Source: unpublished data, Burton *et al.* 2004.)

There is copious evidence to demonstrate that in many farming families a division of labour occurs between the sexes. The role of the farmer (generally but not always male) is to focus on enterprise management and production aspects of the business. The farmer's spouse or partner (generally but not always female) brings up the children, manages the farm accounts, often develops and runs any diversification on farm and may also hold down part-time or full-time off-farm employment (Gasson 1992). On top of this they usually lend a hand at busy times of year, harvesting, lambing, dipping or the myriad of other farm tasks that occur.

The child stage of the family life cycle has been shown to make farmers more cautious in farm business changes. The cost of bringing up a family adds an extra burden on the farm household income. Work by Nalson in the Yorkshire Moors (1968, cited in Potter and Lobley 1996) suggested that each stage in the family life cycle puts different strains on a farm business, in turn creating periods of labour excess and undersupply, alongside changing ambitions of the farmer, the corollary being a pattern of "expansion, consolidation and retrenchment". Other factors have since been suggested that may complicate this simple cycle, notably that of succession.

According to Marsden *et al.* (1992), succession differs from inheritance: succession is a gradual transfer of management control from one generation to another, whereas inheritance is the transfer of legal ownership of possession. Potter and Lobley (1996) argue that the existence of a successor to the business encourages farmers to expand and restructure the farm, ready to hand it on in a fit economic state. In a countrywide survey they were able to demonstrate that many farmers in this position not only acquired more land but also increased the intensity of their production. Conversely they went on to conclude that:

> Farmers lacking successors and/or heirs are most likely to be disengaging or withdrawing from full time agriculture and extensification is most in evidence on farms being managed by elderly operators who lack the incentive and/or opportunity to continue developing the farm in retirement. (Potter and Lobley 1996: 189)

The challenge of succession is a real one for upland farmers and is often cited as one of the most difficult issues faced by the sector, both by observers and by the farmers themselves (e.g. Midmore *et al.* 1998; Wilson *et al.* 1998; Lobley *et al.* 2002; IEEP 2003c; NW Hill Farming Gathering, Jan. 2009). Once again there are regional variations, with the South-West Moors and North York Moors most at risk, the Welsh Borders and South Pennines the least (FPS 2009). In Wales the only way to get into upland farming is through inheritance, where older farmers tend to be more optimistic than younger ones (Midmore *et al.* 1998). A particular issue is lack of interest by farmers' children. They are also discouraged by poor wages (in comparison to other types of agriculture and other economic sectors), long and irregular working hours and being tied to the farm for livestock management.

Cumbrian upland farmers are particularly concerned about the challenge of succession (Burton *et al.* 2004). A third of all farmers surveyed had no successor and 14% were unsure if they had one or not (these tended to be farmers with young children). Among the three Cumbrian uplands investigated, the succession situation was worst in the Lake District where 50% of farms had no successor (Figure 7.9), though the national FPS (DEFRA 2009a) recorded 28% for the same area. The fact that continuing the family tradition was ranked – by farmers perceived as coming from a strong farming tradition and even farmers who saw themselves in that way – only 16th overall (over halfway down their list of values) does not bode well. It could be that the problem has become so acute and insurmountable to many that they have simply given this up as motivationally important. Views of their children, however, contradict this negative attitude, as Burton *et al.* (2004) discovered. They seem far more positive about continuing to

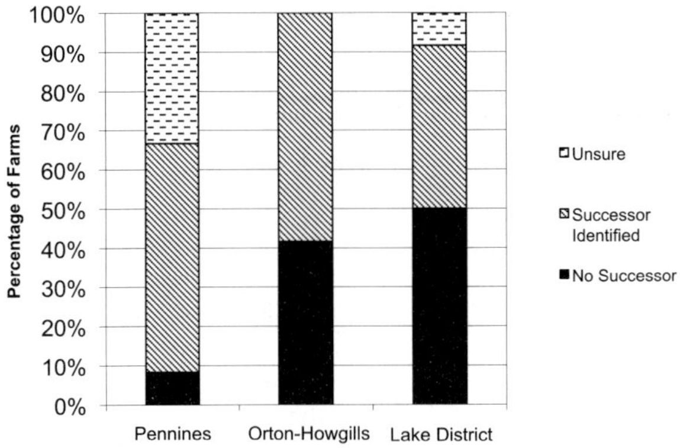

Figure 7.9 Farm Succession in the Uplands
Unpublished data from Burton *et al.*, 2005.

farm, despite attempts by their parents to discourage them. Of course, this could reflect the optimism of youth as against the experience of the previous generation!

The issue of lack of succession in the Cumbrian uplands motivated the local LEADER+ project (Box XXX) to devise and invest in a pilot apprenticeship scheme to encourage young people into hill farming (Mansfield *et al.* 2004). The Fell Farming Traineeship Scheme (FFTS) was designed

> To tackle the issue of providing the effective transmission of the culture (skills, knowledge and understanding) of hill farming in a situation where the labour market is falling.

Young people aged from 16 and 30 with little opportunity to enter the sector, and who would therefore be lost to other occupations, were given the opportunity to train on hill farms. The hope was that these young people would remain in the sector and take over farms as the older generation retired. Every trainee was placed with a cluster of farms each of which had a different enterprise mix and diversification elements. Each 'farmer ring' had a lead farmer who acted as a mentor, allowing the trainee to move around the farms to an agreed pattern of attendance. During their time on the farm, the trainee worked with the farmer and any other staff to learn stock tasks, estate and machinery skills in context. At various points the trainee was taken off the farm and put through a range of competence-based certificated courses or training days.

Whether the scheme was effective or not can be considered from the point of view of the trainee, farmer and wider community. For the trainees it gave them the opportunity to start work in hill farming, something which only one of them would have been able to do otherwise. It broadened their knowledge and skills base in farm management, but to a much lesser extent in what can be termed added value (diversification, agri-environment, public goods). This preoccupation with 'farming' is not unusual among agricultural trainees in general (Mansfield and Martin 2003). For the farmers, three benefits stood out. First, the value of trained, experienced people was apparent, because hill farming relies on local knowledge systems. Second, the increase in available labour made gathering easier on fells. Also the shared labour around the farm ring compensated for low profit margins, which did not allow full-time employment. Third, it promoted (among the farmers) knowledge transfer of good practice, using the trainees as a conduit. It was acknowledged this did not occur normally. For the wider community there was extra income directly through the trainees using local services and accommodation. The main indirect benefit was to lay foundations for hill farming in the future and the recreational, environmental and added value this would bring to the area.

The FFTS is interesting because it crystallises a number of points in solving the infamous hill farming problem. First, the LEADER approach requires bottom-up planning; so local people are empowered to solve local problems and thus they tend to invest more. For upland farming communities this is important because they retain their independence from the state, which they value so much, creating community resilience. Second, the land-based stakeholder organisations often expected too much of the trainees, but it helped the various agencies to understand better the training needs of young people in a difficult economic sector. Third, local fell management knowledge was able to be transferred effectively to the next generation. Finally, and perhaps most importantly, it gave the farming community hope for the survival of the hill farming tradition.

Influences Beyond the farm gate

The third element shaping a farmer's attitudes and behaviour comes from influences beyond the farm gate. Often people are affected by what others think about them – what Burton (2004b) calls social and subjective norms. Social norms are behaviours that are approved of by respected others; subjective norms refer to the individual's perception of how others have viewed certain behaviour (this relates to the psychological concept known as 'Johari's windows'). Opinion is divided whether farmers are influenced by their neighbours. Battershill and Gilg (1997) found this did not happen, whereas Burton considered this made little sense in an economic sector "known for its conservative nature and which is heavily imbued with status symbols" (Burton 2004b: 203). Good examples of these include livestock conformation[1] and prizes at agricultural shows, affirmations of self-worth, achievement and community standing among their peers.

Related to these norms are the concepts of habit, moral obligation, self-identity and sense of place (Burton 2004b; Mansfield, in press). For upland farmers, self-identity is reinforced by their role in hill farming tradition regionally and nationally as well as their own sense of place. Indeed, 60% of farmers interviewed in the (DEFRA 2009a) saw it as "very important" and only 3% as unimportant. Tradition, as a concept, is difficult to grasp. Hervieu-Leger (1996, quoted in Bessiere 1998: 26) suggests that it is:

> The combination of representations, concepts, theoretical and practical know-how, behaviour, attitudes etc, that a group or a society accepts to ensure the continuity between the past and present.

It is particularly important for upland farmers and their families because of the unique system that relies heavily on co-operative management of commonly held resources: the open fell and hefting (see Chapters 1 and 3). Erosion of the hefting system has led to many agricultural and environmental management issues (discussed in depth in Chapters 4 and 5), but it also leads to deterioration of tradition. Demographic decline and an ageing population have left fewer people to pass on the traditional skills. There has been, in effect, a gradual collapse of a local knowledge system for livestock management and related infrastructure, also practices like walling, fencing, vegetation management and hefting, which in turn underpin HNV landscapes, so valued by the public (Hoogeveen *et al.* 2004).

Closely related to tradition is sense of place, which can be described as the distinctiveness of a place for people, whether the local population or transients (short-term residents or visitors). Typically the concept includes natural and man-made features; heritage, culture and traditions; and produce and industries developed from that place. The idea of place has received extensive attention from geographers (see the review by Castree 2004), who see structures, processes and networks as important. In other disciplines the psychological experience of place is significant (Hay 1998). It can be argued that upland farming exhibits both facets of sense of place: the tangible geographical idea of place and the intangible concept of lived experience (Graham *et al.* 2009). Tangible features such as land types, walls and livestock breeds exist, to which can be added intangible unique processes, such as hefting, gathering and shepherds' meets, derived from practicalities of the farming system. Consequently it is reinforced by the relationship between farmer, farm family and wider community.

The importance of upland farming tradition therefore extends beyond the farm gate and thus becomes an issue for society in general. It was in this context that the European Union developed the Less Favoured Areas Directive in 1972 (Wathern 1992) and not, as often portrayed, as an economic policy to support farm incomes in disadvantaged farming areas (see Chapter 8). This approach has pushed more money into the system as income, but it does not on a practical level help to rebuild the social fabric of the hill farming community, namely its human and social capital. In a way it addresses the symptoms and not the cause of the situation. If it had addressed the decline of marginal agriculture, the industry would not still be contracting; it would rather be expanding or, at the very least, static. Whether the newer Rural Development Regulations have been any more successful is considered in chapters 8 and 9.

Tradition as a concept is not, however, neutral as Burton (2004b) suggests. The case can be argued that some traditions, such as bear baiting, are best left to the annals of history. Objectively,

---

[1] Livestock conformation is the distribution of muscles and fat on an animal carcass; muscle is graded into five classes: E (best), U, R, O and P (worst); fat cover is classed 1 (leanest) to 5 (fattest). The two ratings are combined and used to determine market value, with E2 and U2 commanding premium prices and greatest demand for export (EBLEX 2008).

we could argue hill farming is an anachronism simply because it is no longer economically viable. However, as Bessiere (1998) demonstrates, communities can use tradition as a starting point to revive themselves by encouraging local people to promote and transmit skills and expertise. He cites the case of *Le Haut Plateau de l'Aubrac* in France where the community promoted local heritage and gastronomy to help revitalise a flagging economy. Including the tradition of cheese production by local shepherds helped to save the distinctive Aubrac cow from extinction. Projects like this, which draw on the traditional products of wool and meat, are now well-established methods of revitalising marginalised farming communities. They work particularly well for upland farming systems where environmental conditions severely limit enterprise mix or more conventional forms of farm diversification.

In upland Cumbria, a sense of place and local tradition are used effectively by the Hill Lamb and Mutton Marketing Scheme (HLMMS), part of the Cumbria Hill Sheep Initiative. The HLMMS developed a number of projects based on the three local breeds of sheep (Figure 1.9). Meat cutting plants, visitor centres, butter production, light lamb,[2] meat marketing co-operatives, PDO and PGI labelling, wool for craft spinning, carpet, wall insulation and felt making were all supported by LEADER+ (Mansfield 2008).

Although such schemes can be immensely effective regionally, the greater challenge is to extend their philosophies to distant populations who may never visit the uplands and see farming as over-subsidised. To meet this, the affective nature of a farmer's attitudes needs to move not just beyond the farm gate but down the road to the town, where most western Europeans live. The taxpayer needs to reciprocate by recognising the extended role of upland farming and its contribution to society in human, natural and social capital. It is here that problems start to emerge. It is easy to understand the need to produce food, but it is harder to grasp why we must continue to support agriculture in the West, when we regularly overproduce and inflict environmental damage on the landscape (see chapters 5 and 6). With the EU Commission now decoupling farming from food production, it gets harder still because to the person on the street, the EU is supporting farming to not produce food. As Lowe *et al.* comment:

> The social justification both of modulation and of the various measures under Article 33 of the RDR is not so much agricultural survival as the provision of broader environmental public goods for a society that places particular value upon them. Similarly, farming's long-term role is that of developing and responding to particular market opportunities resulting from shifting social demands on the countryside (quality food, regional food chains, farm tourism and countryside management). (Lowe *et al.* 2002: 15–16)

We need, therefore, to ensure that the public recognise that agriculture provides more than just food (see Box XXIX). This is much harder than it sounds because there is a genuine lack of understanding of the indirect benefits agriculture brings to society. Often referred to as public goods or non-market-led outputs, these indirect benefits are hard to justify because it is difficult to place a financial value upon them, as Table 7.5 shows. It is of note that most of the 'new' products developed by the Cumbrian Hill Sheep Initiative, discussed above, are market-led outputs rather than non-market activities. Furthermore, we could argue that these products hold more resonance for upland farming communities and thc public bccause they are traditionally based by-products that have recognisable monetary and practical applications.

We have seen that farmers often do not perceive environmental outputs as 'real farming', but they have a good grasp of the importance of the other three non-market outputs, supposedly because these more clearly affect their business (health and safety) and way of life (culture/ tradition and farming community). For the public, the value of non-market goods in health and safety and the environment is pretty much resolved. Food scares such as Salmonella in eggs, livestock diseases like BSE and FMD, and inappropriate transport of live animals for slaughter have adequately demonstrated the benefits of health and safety in the livestock industries, and

[2] Young, unweaned lambs are considered to provide sweeter, more succulent meat (light lamb) than sheep that have fed on grass.

Table 7.5: Market and Non-Market Outputs from Agriculture

market outputs	*non-market outputs*	
Traditional Economy • Food quality • Food quantity • 'Cheap' food • Employment • Farm incomes • Rural multiplier • Export income • Rural tourism	**Environmental** • Landscape • Habitats • Biodiversity • Water quality • Soil conservation • Carbon sinks	**Health & Safety** • Animal health & welfare • Prevention of animal to human infection • Food safety • Managed genetic modification • Biosecurity
New Economy • Recreation/access • Organics & other niches • Controlled appellation & regional origin	**Cultural/traditional** • Landscape • Traditional industry • Rural character • Local food • Recreation • Prevention of urban sprawl	**Social** • Local employment and economy • 'Vibrant' communities • Tourism • Social cohesion • Educational resource

(Adapted from: McVittie *et al.* 2005)

Box XXIX: Public goods

All environments contain elements that cannot be valued in the economic or monetary sense. These are referred to as intangibles or non-market goods. Typical examples include landscape, habitat, biodiversity, traditional skills and vibrant communities. A particular type of intangible or non-market good is a public good, which we might define as something that is difficult to charge for, but benefits lots of people.

Public goods have two main features. First, they are non-rivalrous, which means that using them does not diminish them. Second, they are non-excludable, meaning people cannot be stopped from taking advantage of them, regardless of whether they have paid for them or not, and thus there is little incentive to safeguard them. It is important to understand that public goods are not commons (Box III). Commons, though non-excludable, are rivalrous: that is, they can deteriorate with overuse (Hardin 1968). Typical commons include the oceans and Antarctica. Anyone managing an environmental resource must ensure that it does not move from being a public good to being a common, which can deteriorate.

	non-excludable	*excludable*
Non-Rivalrous	Public good	
Rivalrous	Commons	Common property resources

Because we cannot easily place monetary value on public goods (and commons), they are open to abuse or seen as subjective items. Governments often refer to this as market failure and use it as an excuse to intervene in the management of public goods and commons (Vogler 2000). One way that land can be abused is through conflict of use. In a capitalist society (one driven by the need to accrue money), being unable to calculate the economic value of something often leads to people ignoring a public good's value. If public goods are to be fully valued by the public and governments, we must find ways to calculate an economic value for an intangible phenomenon. Here we consider just one way: the *Contingent Valuation Method* (CVM).

CVM is a stated preference technique (Willis and Garrod 1993). People say how they will behave in a certain hypothetical situation and this is then compared to how they actually behave (known as their revealed preference). Typically, a sample of people is asked to provide a figure for their willingness to pay (WTP) for a certain public good, for example, to maintain a drystone wall or heather moorland. White and Lovett (1999) used this technique to establish how much people in the UK would pay for national park landscapes calculating £119 per household per annum; given that there are about 25 million households in the UK, this means the public are willing to pay £2.975 billion per annum, rather more than the current budget of national parks. In the discipline of agricultural geography, the technique has been used extensively to calculate the public's WTP for environmental goods in environmentally sensitive areas (e.g. Willis and Garrod 1994; Bullock and Kay 1997).

The technique requires a questionnaire to gather the data. The researcher sets up a plausible scenario. They then ask how much a person will:

- pay for a positive change or to avoid a negative change (*willingness to pay*)

or

- accept as compensation for the perceived negative change (*willingness to accept*: WTA).

The researcher then chooses how they will elicit the monetary value from the respondent. There are a number of ways to do this: a bidding game, payment cards, open-ended or dichotomous choice. A sample of people is then interviewed and their WTP or WTA is calculated from their responses. Of course, as with all techniques there are pros and cons. On the plus side it is a well-established, accepted technique that can be used for market and non-market goods and services, and it is relatively simple to operate. Unfortunately, a critique by Ventatachalam (2004)identifies a number of issues:

1. bias (error and inaccuracy) – usually the difference between what people say they will do and what they actually do;
2. sampling bias;
3. problems with questionnaire design – the order of questions may cause one answer to colour another;
4. embedding – is the good valued on its own or part of a larger package?

Given these issues, other techniques are now favoured, such as *Choice Experiment*, as used by Cumulus *et al.* (2005) to assess various policy scenarios to replace the HFA. A number of attributes are identified at different levels which, when combined as a package, offer different outcomes for a landscape. For example, the choice might be heather moorland and bog at –12%, –2% and +5% change in cover along with cultural heritage set at rapid decline, no change and much better conservation levels. Hanley *et al.* (2007) investigated paying for public goods in English SDAs in four disparate areas. They found, for example, that people were willing to pay more in the Midlands than the North West for heather moorland conservation. They note it would be unwise to base agri-environment payments on this finding; farmers would be right to cry 'unfair'. It could, of course, be a function of the availability of heather moorland, which is far more depleted in the Midlands and possibly valued more there due to greater population density.

surveys show that people think farmers should be paid to produce environmental goods rather than food (see Hall *et al.* 2004: Table 2). Unfortunately many people do not understand what exactly farmers do produce in a landscape. Attempts to rectify this have used the Contingent Valuation (CVM) technique (Box XXIX) as McVittie *et al.* (2005) did to explore the public's preferences among public goods in upland areas. They summarised the qualities of uplands under three main attributes:

- upland landscape – scenic views, traditional buildings, peace and tranquillity, wildlife;
- traditional farm management – family farms, farming skills;
- community culture – local culture and social networks.

The attributes and qualities, in pairs, were then graded by respondents on a sliding scale, as shown in Table 7.6.

Table 7.6: Example of Grading Qualities and Attributes of Upland Landscapes

	much more important			*equally important*			*much more important*			
Peace and tranquillity		X								Wild plants, birds and animals
Upland landscapes					X					Community culture

Thus if a respondent thought 'Peace and tranquillity' moderately more important than 'Wild plants, birds & animals' then they would place an X as shown

(Adapted from McVittie *et al.* 2005: 17).

The attributes (and the qualities separately) were weighted according to their relative importance for five groups of people (rural, remote rural, urban, Manchester and Cumbria). The weightings were then statistically tested to see if significant differences occurred between the five sample groups. The analysis demonstrated that the Manchester sample had a significant preference for the 'upland landscape' quality, in contrast to their Cumbrian counterparts who preferred 'traditional farm management' and 'community culture'. McVittie *et al.* then calculated the respondents' willingness to pay (WTP) for upland goods, deriving a figure of £1.15 billion for the whole UK (£47 per household), a lot more than the £153m spent on LFA support in 2004 (DEFRA 2005). They concluded:

> Geographical qualities of the uplands (scenic views and wildlife) are considered to be more important to people living outwith upland areas. This suggests a disassociation between upland areas and the role of farming in providing public goods. (McVittie *et al.* 2005: 25)

The results of this project, when combined with the investigation into social capital among Cumbrian farmers by Burton *et al.* (2004), led to the conclusion that farmers perceived environmental public goods as less significant than farm-based public goods (Table 7.7). This corroborated earlier work of Carr and Tait (1990), showing that upland farmers still focused on production. The FPS (DEFRA 2009a) found the same perceptions: 52% of upland farmers saw environmental payments as a top-up to the farm business; only 29% considered them as important as production. Furthermore, upland farmers continued to view production-based issues as the most important for their farms rather than non-market activities (Figure 7.10).

Table 7.7: A Comparison on Farmers' Perceptions of the Benefits of Upland Farming with that of the Public of Cumbria and Manchester

qualities	*attributes*	*Cumbria farmers*	*Cumbria public*	*Manchester public*
Traditional Farm Management	Traditional farming skills	1	5	6
	Small family farms	2	4	8
Community Culture	Strong local culture	3	2	5
Upland Landscape	Traditional buildings and stone walls	4	6	7
	Wildlife	5	1	1
	Community culture	6	3	3
	Scenic views	7	8	4
	Peace and tranquillity	8	7	2

(Source: Burton *et al.* 2004).

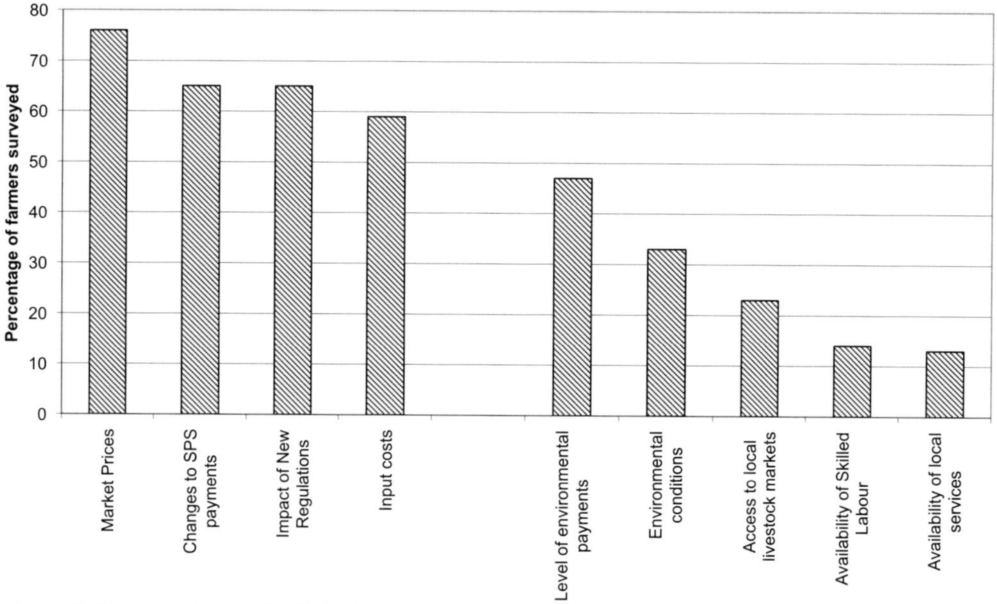

Figure 7.10 Future Issues for Upland Farmers in England
Based on data from DEFRA 2009a

As the work by McVittie *et al.* (2004) and Burton *et al.* (2004) shows, many people are still unaware that the social, traditional and cultural non-market goods produced by farmers are integral to the package. In a way the public are saying they want the effect 'but not the cause, thanks very much'. The argument here is that you cannot have one without the other, so society needs to carry on supporting upland farming systems that are valuable ecologically, but marginal economically. This of course sits badly with the farmers themselves: they want to produce food; it's 'what they are good at'. They do not want to be glorified park keepers, because that undercuts their self-worth, motivation and professionalism. Unfortunately for them, the CAP no longer emphasises production (see Chapter 8).

Social capital on upland farms

The second fundamental feature of a local knowledge system is social capital. According to Pretty and Ward (2001) it has four typical elements:

- relations of trust,
- reciprocity and exchange,
- common rules and norms,
- connectedness, networks and group.

These elements are particularly evident in upland agricultural systems (Table 7.8). Hefting is among the most important processes and demonstrates relations of trust between farmers, who aim to ensure that stock remains on-heft, limiting grazing pressure on neighbouring hefts. The process of gathering also requires trust, to make sure all stock are brought off the fell in a co-ordinated manner.

Gathering provides opportunities for reciprocity and exchange of stock at the traditional Shepherds' Meet, where everyone brings stock that are not theirs to be returned to the owner. This form of specific exchange is unfortunately less common nowadays, mainly because of collapse in the labour force. Furthermore, the quantity of sheep coming off the fell mis-gathered has spiralled upwards as sheep stray further from their own heft. Where once a Land-Rover was

Table 7.8: Social Capital in Upland Farming

element	*tradition*	*characteristics*
Relations of trust	Hefting Gathers	Shared grazing areas require managers to not overstock or let stock drift, affecting others' resource base. Increasingly undermined by destocking and heft abandonment
Reciprocity and exchange	Shepherds' meet Machinery Labour Other equipment	These types of return grow less due to time cost and the increased number of stock drifting further as grazing pressure is reduced
Common rules and norms	Heft management Common land	Ensuring stock are appropriately dipped and wormed to avoid spread of disease. General animal welfare. Maintaining stock on heft.
Connectedness, networks and groups	Breed associations Commons associations Farmers markets Social networks Livestock markets	How people communicate to achieve common goals varies in different ways within upland farming communities

sufficient to return stray stock, now lorries are needed for the long journey from valley to valley. Upland farming also demonstrates diffuse exchange of machinery, equipment and occasional labour (Burton *et al.* 2004).

Heft operation also requires common rules and norms to be followed in the process of active shepherding. Keeping stock healthy on all hefts which share common land is crucial. Without physical boundaries, straying sheep can carry disease to neighbouring flocks, so common grazing rules require farmers to maintain animal health and welfare with regular checks, veterinary inspections and annual dipping. Commoners rely heavily on one another to comply, because a single commoner not complying could have dire consequences for all.

This explains the concern about Foot-and-Mouth Disease entering a hefted flock while over-wintering off fell and on its return bringing the disease onto the common, leading to the entire common's stock being culled. Compounding this loss would be the loss of the flocks' inter-generational hereditary knowledge. There are only two ways to reinstate a heft in this situation: active shepherding is one and the other is fencing, which detracts from landscape aesthetics and raises legal issues related to fencing commons. Opinions differ as to how long it takes to re-establish a heft, perhaps simply because each heft is geographically unique in the way the land lends itself to corralling the sheep to the heft. The time taken for re-hefting is also a function of the shepherd's ability – few hill farmers have the skills to do this because the hefted flock goes with a farm when it changes hands and feware ever in a position to need re-hefting skills. There are exceptions to this, for instance on the North York Moors, in the Wildlife Enhancement Scheme (Box XXXV), and as part of the 2010 Upland Entry Level Scheme (see Chapter 9).

The final and probably most important feature of social capital for upland farming is that of connectedness, networks and groups: the ways people, farmers in this case, communicate and work with each other to achieve common goals. The effect can be one- or two-way, preferably the latter as this strengthens the social capital that develops, as well as creating and maintaining the intangible lived experiences of sense of place. Pretty and Ward (2001) identify five types of connectedness:

- local connections,
- local-to-local connections,
- local-to-external connections,
- external-to-external connections,
- external connections.

Local connections are usually the strongest in an upland farming community because they occur daily, examples being gathering, commons committees and breed associations. At the next level there are local-to-local connections, which extend beyond the farming community as shown in Figure 7.11. This is an interesting aspect as it demonstrates positive reinforcement of sense of place through the creation and maintenance of the network, but also temporal shifts of social capital, bolstering or degrading sense of place. In upland communities there is evidence that the Church is playing less of a role than formerly, being replaced by new networks and groups, such as farmers' markets (Burton *et al.* 2004). Fewer and fewer farmers take stock to market; instead the lorry turns up, the sheep are loaded and those not sold returned at the end of the day. If anything, the social capital of farmers' wives has grown with rising farm diversification and the need for off-farm employment (e.g. Gasson 1993).

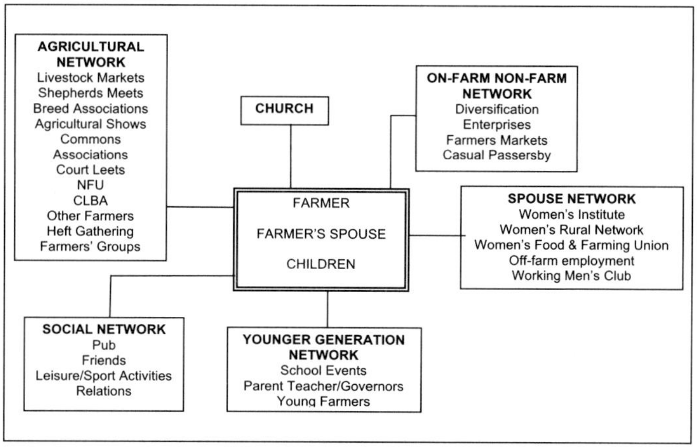

Figure 7.11 Social Capital: Local to Local Connections

There are also instances where local-to-local connections are improving. Historically each common has had its own committee or association, which sorted out disputes and issues related to management of the land. These bodies are beginning to be replaced by commoners' federations, loose affiliations of like-minded associations with similar aims and objectives, which use consensus to manage commons. Aided by the Commons Registration Act 2005, such federations, associations and committees can form statutory Commons Councils as umbrella organisations (FCC 2009 website). These have the ability to manage agricultural activities, rights and vegetation on common land. Though some commoners may be apprehensive about power, control and rule setting, such councils have the potential to enlarge social capital at the next level of connectedness, local to external, moving beyond farming and rural communities in the local area to similar communities elsewhere, even to develop connections with non-farming agencies. This is an important step for farming communities as it allows for the development of 'translocal justice' where similar places far apart can work towards a common agenda and a single political voice (Harvey, cited in Castree 2004).

Another way of looking at these social capital networks is to consider how people move around Arnstein's 'ladder of citizen participation' (Arnstein 1969). Arnstein envisaged that citizens would engage with the wider world at different levels when a decision was needed on a topic of concern (Figure 7.12). As people climb the ladder, their level of involvement with and control of decision making increases. It can be argued that, as people and organisations climb

degree of power sharing			*nature of involvement*
Degrees of citizen power	8	***Citizen control***	
	7	***Delegated Power***	Citizens are given management power for selected or all parts of a programme
	6	***Partnership***	Trade-offs are negotiated
Degrees of tokenism	5	***Placation***	Advice is received from citizens but not acted upon
	4	***Consultation***	Citizens are heard but not acted upon
	3	***Informing***	Citizen rights or options are identified
Non-participation	2	***Therapy***	Power holders educate or cure citizens
	1	***Manipulation***	Rubber-stamping committees

Figure 7.12 Arnstein's Ladder of Citizen Participation
Adapted from Arnstein, 1969.

the ladder, so their social capital increases and thus their sense of place. Each stage is more external looking, building capacity and becoming more effective as a collective.

Many upland farmers, towards the end of the 20th century, felt manipulated by agendas outside their control, placing them on rung 1 of Arnstein's ladder. This was particularly the case for those who suffered most from the cost–price squeeze, labour reductions and poor diversification opportunities. Pretty and Ward (2001) call this Reactive Dependence, a situation where individuals or groups react to a threat or crisis by relying on external agencies for support and typically fear change. In contrast, we could argue that Commons Councils and federations sit above rung 5, depending on their lobbying powers. Pretty and Ward term this Awareness Interdependence, where the group can shape its own reality and begin to resist external influences and threats by joining with other groups to create local identity and forms of translocal justice. Between these two extremes is Realisation Independence, the phase when new capabilities in individuals and groups begin to emerge. Members invest more time in the 'project' and begin to look outwards to develop advantage for the group as a whole – in effect, the beginnings of awareness of the use of sense of place for rural regeneration. It is here that the LEADER (*Liaisons Entre Actions de Development de l'Economie Rurale*) approach comes to the fore (Box XXX).

The next set of connections suggested by Pretty and Ward (2001) is termed external-to-external. For the upland farming community this is, perhaps, the least desired aspect of social capital because it can be perceived as interference by those who know nothing about upland farming including academics! However, it is at this level that effective governance operates and policies are formulated (Roseland 2000), and it is, therefore, a necessary evil. The important thing is to convert this type of social capital into that of external-to-local and thus feed developments up and down the ladder in a form of adaptive management. It is the perceived lack of consultation and resultant disenfranchisement that makes the hill farming community feel most vulnerable. Thus any initiatives which reinvigorate or build new social capital are fundamental to the survival of the sector.

We can now see that these varied connections and the resultant networks develop both spatially and temporally. Therefore, nurturing social capital, both spatial (sense of place) and temporal (the transition from reactive dependence in stage 1 to awareness interdependence at stage 3), becomes central to sustaining marginal farming communities. Pretty and Ward considered the character of the temporal shift, identifying facets which influence the transition, two of which are directly derived from social capital: internal norms and trust (with external

Box XXX: The LEADER approach

LEADER (*Liaisons Entre Actions de Developpement de l'Economie Rurale*) is an EU initiative which uses a bottom-up approach to deliver rural development. Local people are empowered to solve their own challenges, putting them on rung 6 of Arnstein's Ladder. Since its inception in 1991, it has gone through various iterations.

project	*operational period*	*funding*	*UK geographical spread*	*lessons learnt*
LEADER I	1991 to 1993	1.5m Ecus	North Tamar Devon West Cornwall	Have defined outputs Have a monitoring system Recognise time taken to form effective local action groups (LAGs)
LEADER II	1994 to 1999	70% from ERDF 15% from ESF 15% from EAGGF 32m Ecus	21 LAGs in six regional areas	Simplify administration Develop a network from day one Wider community participation in LAGs Bottom-up capacity building High-level promotion Better links to other schemes
LEADER+	2000 to 2006	70% EAGGF 30% DEFRA 75.5m Ecus	25 LAGs in all rural areas	Depends on principles
New LEADER	2007 to 2013		Patchy	

(Source: DEFRA 2001)

The philosophy of LEADER operates through the seven specificities:

- *Area* – operating within a recognised distinct geographical area;
- *Partnership* – a Local Action Group (LAG) comprising voluntary, business and statutory bodies; the latter not dominating.
- *Bottom-up planning* – using local community representation to select which projects receive funding;
- *Innovation* – to explore experimental forms of rural development that might otherwise struggle to receive financial support;
- *Networking* – working across sectors of rural development, rather than vertically (within, say, just agriculture or nature conservation);
- *Co-operation* – working with other LAGs and LEADER areas to build critical mass;
- *Easy access to finance* – by streamlining paperwork, making the application process accessible to all;

Once a plan is accepted, a LAG is formed from local community representatives (private and public sector, charities and individuals) and it then has access to a block grant to be allocated to local projects (Ray 1998). LAGs are a good example of local-to-external connectedness with their mixed membership spreading beyond the farming community. The first three LEADER programmes operated outside the mainstream support system of the EU, allowing them to support ideas that would not receive funding from other grant initiatives. However, it did mean that funding was much lower than for main rural development policies; for example, the England RDP received 996.1m euros, but LEADER+ had only 5.3% of this budget.

LEADER I was a pilot and thus there was a steep learning curve. LEADER II was dogged by late funding, so the capacity-building element of LAG creation was greatly protracted, impairing the effectiveness of many UK schemes, particularly in England and Northern Ireland (Shucksmith 2000). These earlier iterations helped to shape the LEADER+ programme (see Figure below).

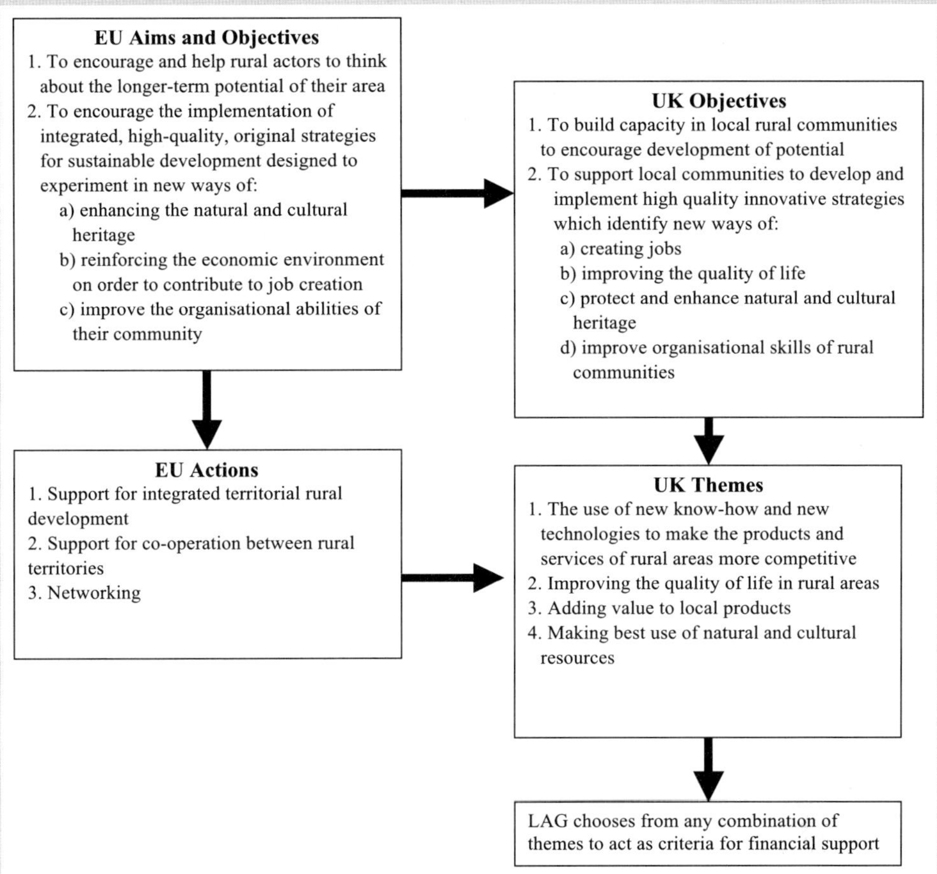

It is important to remember that, while officially it is a scheme designed to tap social capital through increased innovation and capacity building, many of the outputs are economic and environmental, spread across the entire rural economy. The types of project supported through UK LEADER+ were far ranging, including: the re-introduction of the great bustard on Salisbury Plain, the conservation of old apple orchards, provision of childcare, cyber cafés, celebrations of local artists and farm diversification (UK LEADER+ website, accessed: 24/07/07).

The EU LEADER+ programme has become so successful that it is now the main approach to rural development under Axis 4 of the European Agriculture Fund for Rural Development (see Chapter 9). However, there is some tension about the funding allocation, which pushes the approach down from rung 6 to 5 on Arnstein's Ladder. This hints at a criticism often levelled at Arnstein's idea, that it is still a top-down mechanism because those in power still make the final decisions. We can only hope that the new structure is allowed to apply the LEADER philosophy fully in all member states.

linkages and networking). Other key facets include world views and sense-making (considered earlier in this chapter), technologies and improvements (discussed in Chapter 6) and group life span. The combined force of these five factors drives the temporal development of social capital and the move towards awareness interdependence.

Figure 7.13 conceptualises this transition for three upland farming groups – commoners' federations, farmers' markets and fell gathering. As the social groups move from one stage to the next, they become more outward-looking, capacity-building and effective. Whether the transition is staged (1 to 2, 2 to 3) or incremental (groups proceeding at different rates for different facets) is a matter of debate, though it seems to incrementally increase for commoners' federations and farmers' markets.

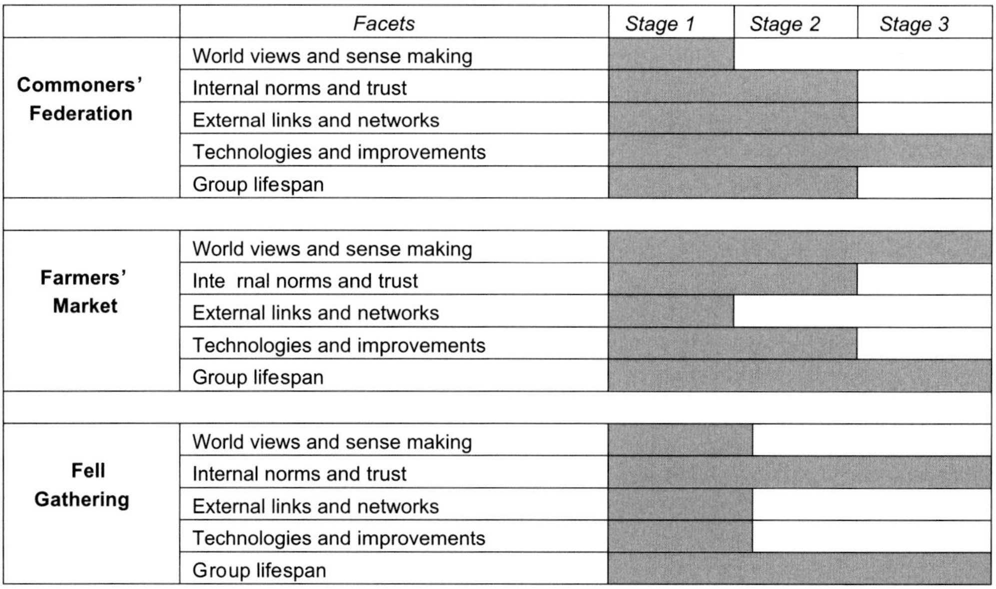

World views and sense making: views of change; attitude and values
Internal norms and trust: rules and norms; recognition of group value; sharing ethic
External links and networks: horizontal; vertical; external facilitators
Technologies and improvements: environmental aspects; source of technologies; capacity to experiment
Group life span: reason for being; resilience; variability

Figure 7.13 Conceptualising the changing facets of social capital with upland farming groups
Based on the ideas of Pretty & Ward 2001.

Unfortunately, we can also consider the erosion of social capital working in the same way but in reverse, with incremental decline or even collapse possible. Burton *et al.* (2004) identified this decline occurring for fell gathering, but also for hay making, shearing rings, local community participation and visits to livestock sales. Given the incremental nature of the various facets of social capital development and disinvestment, it begs the question whether some facets and processes are more important than others in retaining the overall integrity of social capital. The necessity to gather continues even though there are fewer and fewer people to do the job. There comes a point where there is no longer a critical mass to sustain the process; this has happened on the North York Moors plateau where there are only six farmers left operating the hefting system (pers. comm., R. Pickering, July 2007). It could be that some upland farming structures and processes are vital – their loss on a massif will lead to complete collapse of the upland farming system – whereas others are not crucial and their loss would have little impact. The crux is the ability to identify which are keystones (to borrow a term from ecology) and which are expendable. Policy can then focus on supporting the social keystones, making it more cost-effective, with better results.

Based on our findings so far, we could argue that the following structures and processes are central to a sustainable upland farming system:

- hefting, heft management, gathers and related trust,
- inbye, intake and fell land with hefts,
- local knowledge systems,
- social networks and connections,
- sense of place.

But are they keystones? Whether they are or not depends on management objectives for an upland area. If agricultural sustainability is core, then yes, these structures and processes are keystones to perpetuating the system. If agriculture is not the primary policy goal, then we may be able to abandon some of these. For example, if the aim is to maintain or improve the conservation value of uplands, this could be achieved with a lower farming population by a ranching approach or by employing management techniques other than grazing. Using fells for game management demonstrates this: land does not need grazing to maintain a game moor; it can just as well be burnt (see Chapter 5), making heft management almost irrelevant. If we want to maintain the conservation value of the moor by grazing, then one or two farmers can ranch the land (IEEP 2003b). On the other hand, we could argue that maintaining the local knowledge system is important because it allows us to understand the behaviour of grazing livestock and their ability to over- or under-graze a particular hillside (see Chapter 4). The corollary is that it is all a matter of our policy and management objectives whether upland farming and its community survive in an area or not. If this is the case, perhaps policy needs to approach ecosystem management massif by massif, driven by structures like the Water Framework Directive (see Chapter 2) and/or agency policy like that of Natural England (2009b). Of course, this relies on long term planning (30 to 40 years), not five-year government tenure which do not fit ecological cycles.

Capital and cultural landscapes

It is evident that a strong relationship exists between human and social capital in upland farming systems. People are central to both, allowing the continuation of this local knowledge system, which in turn affects the natural capital of uplands, the importance of which became evident in chapters 3, 4 and 5. The overall effect is to produce a cultural landscape – a result of the relationship between humans and their environment (Robertson and Richards 2003; Figure 2.9). Attributes of cultural landscapes show the importance of many human and social capital characteristics discussed here: land use, sense of place, tradition, heritage, public goods, personal memories and associations (Swanwick 2004).

Their importance to society in general is acknowledged at all levels through the identification of World Heritage Sites and latterly by the European Landscape Convention (see Box XXXI), strengthened by initiatives like National Character Areas in England and Natural Heritage Zones in Scotland, which form the basis of more and more policy and planning decisions (see chapters 2 and 9). We therefore appear to have a duty to sustain a farming system that maintains one of our most iconic cultural landscapes. Indeed, the Lake District of Cumbria is now a candidate for inscription on the World Heritage Site cultural landscapes list. The role of upland farming in forming this landscape is central to the Statement of Outstanding Universal Value (Lake District World Heritage Project; accessed 02/02/11):

Summary

The distinctive farming landscape of the Lake District is of outstanding universal value because its terrain inspired fundamental and worldwide changes in the way humans view, value and conserve landscape. A fusion of mountains, valleys and lakes, each with its own specific character, it is one of the world's most beautiful areas and the birthplace of what landscape means to the modern world.

It continues:

Overview

A compact, glaciated upland landscape of radiating U-shaped valleys – many of which contain long narrow lakes – the Lake District is home to England's highest mountain and its deepest lake. Rocky mountain tops, open fell pasture and heather-covered slopes contrast with the native woodland, exotic plantations and stone-walled fields. The dominant land use is upland pastoral farming, often based on the local sheep breed, the Herdwick. This distinctive farming system reached a peak of prosperity in the 17th and 18th centuries when a tradition of independent farming emerged. The resulting landscape has a distinctive aesthetic unity marked by contrasts in detail: mountains, moors, lakes, woods, streams, fields, stone walls, farms, villages and small scale industry, which reflect its slow and gradual development since the 12th century.

Box XXXI: World Heritage Sites and the European Landscape Convention

World Heritage Sites are internationally designated under the Convention Concerning the Protection of the World Cultural and Natural Heritage (UNESCO 1972), which aims to protect sites that have outstanding universal value in cultural and natural heritage. Within the terms of Article 1 of the convention, cultural heritage is defined as:

Monuments – architectural works, works of monumental sculpture and painting, elements of structures of an archaeological nature, inscriptions, cave dwellings and combinations of features, which are of outstanding universal value from the point of view of history, art or science

Groups of buildings – groups of separate or connected buildings which, because of their architecture, their homogeneity or their place in the landscape, are of outstanding universal value from the point of view of history, art or science

Sites – works of man or combined works of nature or of man, and areas including archaeological sites which are of outstanding universal value from the historic, aesthetic, ethnological or anthropological points of view

Cultural landscapes are, therefore, the 'combined works of nature and man' which show how society has evolved over time within physical, social and economic constraints or opportunities (UNESCO 1998). The guidance provided by UNESCO goes on to explain that cultural landscapes must:

Be clearly defined in terms of a geo-cultural region

Reflect specific techniques of sustainable land use which are determined by the physical landscape in which they reside

From this, three distinct types of cultural landscape can be identified:

- landscapes designed and created intentionally by man, e.g. Kew Gardens in London;
- organically evolved landscapes forming as a result of social, economic, administrative and/or religious factors, which develop further in response to and in association with the natural environment, e.g. rice terraces in the Philippines;
- associative cultural landscapes which evolve in response to powerful religious, artistic or cultural associations with the natural environment, e.g. Uluru in Australia.

By demonstrating that it matches these criteria and one of the types, a location can be inscribed on the World Heritage List.

Related to the World Heritage Convention is the European Landscape Convention (ELC), signed in 2000, with the UK joining in 2006. Europe contains many landscapes important to its people, which form a common heritage and shared responsibility (Phillips and Clarke 2004). The ELC defines landscape as: "An area, as perceived by people, whose character is the result of the action and interaction of natural and/or human factors" (Council of Europe 2000). The main aim of the convention is to encourage signatory states to develop landscape policies that protect, manage and create landscapes, and establish procedures for the general public and other stakeholders to participate in policy creation and implementation. Article 5 outlines the main activities (ELC, ETS No. 176, http://conventions.coe.int/Treaty/en/Treaties/Html/176.htm, accessed 19/1/11):

a. to recognise landscapes in law as an essential component of people's surroundings, an expression of the diversity of their shared cultural and natural heritage, and a foundation of their identity;

b. to establish and implement landscape policies aimed at landscape protection, management and planning through the adoption of the specific measures set out in Article 6;

c. to establish procedures for the participation of the general public, local and regional authorities, and other parties with an interest in the definition and implementation of the landscape policies mentioned in paragraph b above;

d. to integrate landscape into its regional and town planning policies and in its cultural, environmental, agricultural, social and economic policies, as well as in any other policies with possible direct or indirect impact on landscape.

The convention continues with Article 6, which focuses on awareness-raising, training and education, identification and assessment of landscapes for inclusion, development of landscape quality objectives and implementation of a system. In England for example, responsibility rests with Natural England, who developed an action plan for 2008–11 with three aims: to review the Landscape Character assessment system, to implement new ideas like the Regional Landscapes and Geodiversity partnerships, and to develop new work.

Although inscription on the list recognises the unique cultural landscape of an upland area and the forces that have produced it, this does not secure the region and its communities from deterioration. That requires some form of institutional intervention, typically achieved by instigating sectoral agricultural policy (see Chapter 8) or rural development work (see Box XXXII in Chapter 8). Unfortunately in both cases the approach so far has been exogenous top-down planning, creating tensions with (and within) farming communities and businesses, with a lack of real policy recognition that culture can help sustain rural areas. Selman (2004) agrees that government intervention is not enough to promote cultural landscape; there also needs to be a continued community input. In other words, top-down needs to be integrated with bottom-up rural regeneration (see Box XXXII).

The value of galvanising community involvement in nature conservation in urban areas, empowering local people to take control of their own environment, has long been recognised (e.g. Emery 1986). In European agriculture, this was not really the case until about 2010, probably because top-down institutional policy support has been the norm since the 1930s. Two symptoms of this have been the widening gap between those who formulate policy and those expected to put it into practice, and the over-reliance on outside 'experts' to drive policy, overlooking the value of local knowledge systems (Stenseke 2008). It is here that peripheral schemes such as LEADER have begun to make headway, allowing communities to slide up Arnstein's Ladder, empowering them to develop their own social resilience to changing situations, taking back control of their own community's destiny. The corollary is that the most effective way forward may be to develop partnerships between governments and communities, drawing on the best of both approaches to sustain upland farming while leaving behind the petty bureaucracies of political leviathans, which foster resentment on both sides.

Box XXXII: Rural development

Woods suggests that governments get involved in rural development for four main reasons:

1. The notion of social well-being and equity between all parts of the population in a state, particularly with respect to public services and stimulation of new economic sectors;
2. Support for the accrual of capital for businesses through such activities as support for infrastructure or limiting risk;
3. A sustainable 'stewardship' role with respect to rural resources;
4. Spatial population control by limiting push–pull factors of out-migration to spread the geographical demand of public services. (Woods 2005)

These four allow for cross-pollination between sectoral, regional and rural development, providing a more integrated approach to resolving economic and social issues in the countryside. Traditionally rural development takes a top-down approach which sees large, state-led infrastructure projects. This methodology is sometimes termed 'exogenous' because it depends on external investment. It is often criticised for causing outflow of money from a region and for its democratic deficit.

To promote rural development, the UK adopted a state-based agency approach from the early 20th century. Robinson (1972, quoted in Pacione 1984) suggests that a rural development agency (RDA) should exhibit a number of characteristics:

- operate at a regional scale,
- economic, social and physical planning should be integrated,
- be innovative in outlook,
- have powers of intervention and its own budget,
- be continuous to provide long-term planning,
- be indigenous to the region to avoid alienation.

In the UK there has been an emphasis on policies using agriculture to solve rural depopulation and deprivation, stemming from the Scott Report of 1942 and the post-war Labour government (Tichelar 2004). This was in direct contrast to the more mixed-economy approach of our European neighbours. The fundamental flaw in UK policy was that, though agricultural policy could re-structure farming in different ways (modernising, diversifying or greening), it could not address all aspects of farming community support.

In 1988 the EU announced a shift away from sectoral economic support towards targeting specific geographical regions suffering from socio-economic disadvantage. The European Commission became committed to ensuring that rural development policy must be multi-sectoral, multi-disciplinary, bottom-up and based on participation. Under the new structural funds, rural areas could apply to be designated as one of the following:

- Objective 1 – lagging areas with a per capita GDP less than 75% of the EU average,
- Objective 5b – fragile rural economies dominated by agriculture and in need of rural development assistance,
- Objective 6 – northern parts of Finland and Sweden.

Between 1989 and 1993 the first Objective 5b programmes were rolled out. In the UK they covered Devon and Cornwall, rural Wales, Dumfries and Galloway, and the Scottish Highlands and Islands. From 1994 until 1999 the scheme expanded to cover eleven areas with a rural population of 3 million and £700m funding (see map). Each area submitted a Single Programming Document, which outlined the main economic and social problems faced and how the funds would be spent. Priorities by region varied but common themes were: economic development and business (50% of funds), tourism (20% of funds) and community support (Ward and McNicholas 1998).

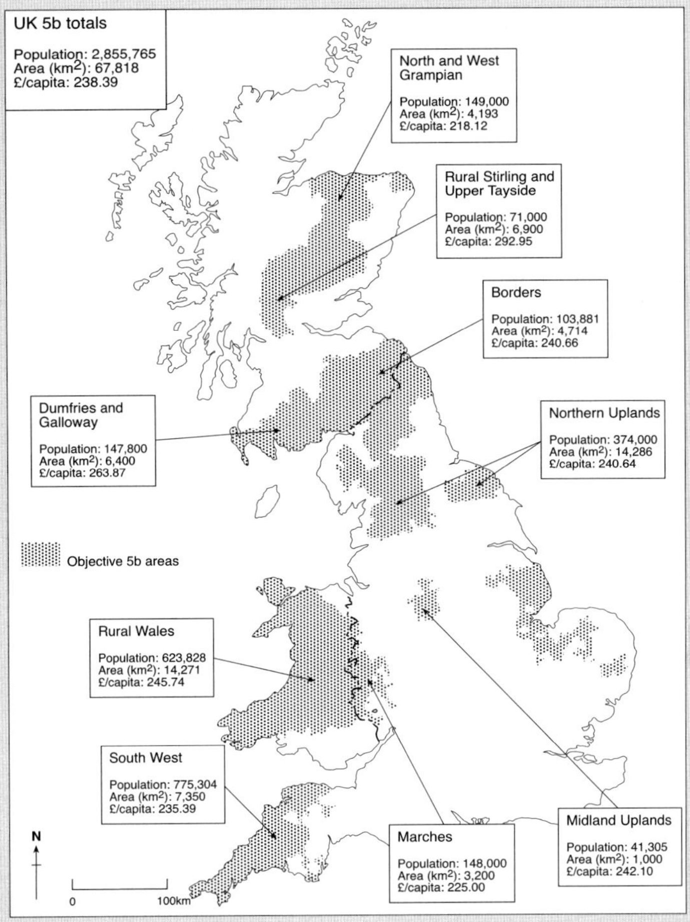

Objective 5b areas

Reproduced by permission of Ordnance Survey on behalf of HMSO.

An example of an initiative that received support from Objective 5b is the Northern Uplands Moorland Regeneration Project (NUMRP). It was managed by The Moorlands Association, which used the £1.5m funding to develop sustainable upland farming and grouse moor systems in the Northern Uplands Objective 5b area. The projects included capital items, gamekeeper training, heather re-seeding, bracken control, erosion management, grouse disease control and red meat marketing. The direct impacts of NUMRP were: improved condition of 86,790 ha of heather moorland, 18 full-time jobs, 2,845 days of casual labour, 45 integrated farm management plans and £1.7m of private investment (Moorland Association 2002).

A contrasting approach to rural development is bottom-up rural regeneration. This endogenous approach is community-led, tapping into local resources. The classic example is LEADER and its associated Local Action Groups (LAGs), which now form the umbrella system for the Rural Development Regulation (see Chapter 9). LAGs also demonstrate another endogenous process, rural restructuring, where cross-cutting of power and association can force new combinations of actors from different contexts and political levels to seek common ground (Marsden and Smith 2005). In other words, private individuals (like farmers) have as much say as an employee of a state quango on how money is used locally.

The switch from exogenous to endogenous management of rural development can create tension between state-led RDAs and LAGs, as seen in England in 2008. This is probably a result of power transferring from the state to the community, so quangos feel they are losing control.

Summary

In this chapter we have investigated the social dimension in upland agriculture. We can now recognise that upland farming is unique because it heavily relies on the effective operation of local knowledge systems. These are made up of natural, human and social capital, which are inter-related. Central to the local knowledge system is the human capital of the farmer, who will draw on their knowledge and skills to operate the farm business to the best of their ability. Behavioural theories, however, have confirmed what we all know, that farmers (like all of us) have imperfect knowledge and their decision making is affected by personal characteristics and social circumstances. For upland farmers, demography seems to play a pivotal role in how a farm business develops; dependants and succession are particular issues now.

Upland farmers do not work in isolation, however: they draw on human and social capital from family, neighbours, the wider farming community and beyond. Unfortunately, economic decline in the sector has led to deterioration of much of the traditional social capital of the farming system, with an associated decline in human and natural capital. The public no longer understands the physical relationship between these three kinds of capital. It is implicit in the upland landscape, but the public sees that as entirely natural, and its understanding is not helped by agricultural support being mainly targeted at production.

However, society does more readily recognise the importance of cultural landscapes, of which upland farming systems are an example. So as a priority we need to identify which structures and processes are keystones in sustaining upland farming practices with the farming community being a pivotal role. It has been suggested here that certain elements of social capital (such as hefting) are essential, but they need to be supported by new income streams linked to business diversification (e.g. farmers' markets and public goods provision). If upland farming is to move from its position of economic marginality to become mainstream and sustainable, the fundamental importance of local knowledge systems needs to be acknowledged and explicitly supported by government. All too often social development is seen as implicit: solve the economic and environmental issues, which are tangible and quantifiable, and the intangible social challenges will resolve themselves.

What has become evident from this chapter is that upland agriculture is not just about individual farmers; it's about supporting the social and cultural context in which they operate. The LEADER philosophy has gone some way to addressing this; its assimilation into mainstream policy heralds the possibility of a more coherent approach to sustaining upland agriculture. Getting policy to this point has however been slow, painful and problematic. The next chapter investigates this by exploring how 'the hill farming problem' has been addressed by successive governments in the UK and how it fits into the wider contexts of the European Union and global agriculture. We can then appreciate that well-meaning agricultural policies may not 'deliver the goods' if they fail to consider the entire character of a farming system.

8

Politics and upland agriculture

The difficulties faced by upland agriculture between 1880 and 1940 culminated in the sector being labelled as problematic. The financial returns of hill farming began to lag behind those of other sectors of British agriculture, marginalising it economically (Attwood and Evans 1961). Various theories why this happened were outlined towards the end of Chapter 3 and the effects on farming were considered in chapters 6 and 7. This chapter investigates how the British Government has responded to the challenges posed by upland agriculture since 1947, but first we consider briefly how governments support farmers.

Political intervention

According to Hill (2006), policies develop in four stages. First, there is a phase of problem formulation: understanding what the problems are, what evidence there is for them and why they matter. This stage can be affected by changing priorities within and without government and, in a democracy, individuals and pressure groups can influence this process. We have seen that the problem of upland agriculture is, broadly, that it lags behind the economy of other farming sectors. Its marginal nature, alongside rural depopulation and environmental damage (both caused by a cost–price squeeze), has been recognised by successive British governments since the 1900s and latterly by European administrators.

The second stage is policy formulation, identifying policy objectives, which may be any combination of those listed in Table 8.1. Each aim is assessed to see whether it conflicts with others and whether the programme's effectiveness can be measured. Hill (2006) notes that monitoring is not always welcome because testable objectives leave little room for policies to flex with changes in the policy environment. Pivotal here is who is responsible for informing and developing policy. Traditionally this is done from the top down by government ministers and civil servants – specifically departments of agriculture.

Since 2000 there has been a shift in philosophy, as agriculture has become part of an overall rural development strategy in Europe rather than being the driver (see below and Chapter 9 for more on this). Other 21st-century developments reflect a more bottom-up rural planning approach. Taken at face value, the Upland Entry-Level Scheme (part of the ESS) suggests a high level of consultation between the UK Government and the upland sector, since proposals were submitted to a Hill Farmers Panel for comments on practical application (M. Edwards, pers. comm., Dec. 2009). However, much of the debate seems to have been within agencies and not with farmers, unless the farmer happened to represent an agency.

Table 8.1: Objectives of Agricultural Policy

Main Goals	*Objectives*	*Options*
UTILITY	Protect the environment	• *Conserve rural landscapes* • *Conserve soil resources* • *Maintain ecological diversity*
	Contribute to economic growth	• *Improve the balance of payments in international trade* • *Ensure efficient allocation of resources to and within agriculture*
	Contribute to economic stability	• *Stabilise domestic food and fibre prices* • *Stabilise cost of agricultural support programmes*
	Provide strategic security of food supply through self-sufficiency	• *Counteract political or military unrest* • *Anticipate developing world food shortage*
EQUITY	Provide equitable renumeration for those employed in agriculture	• *Reduce income disparities by farm by:* *Location* *Type* *Size*
	Contribute to the social stability of rural areas	• *Reduce instability of farm incomes* • *Maintain rural population densities through either:* *Equal welfare and educational facilities in rural areas* *Economic opportunity in rural areas*

(Adapted from: Bowler, 1979:9)

The third stage is policy implementation, typically by one of five instruments (Ilbery 1985):

- price support,
- production control,
- marketing boards,
- grant aid,
- structural reform.

Price support protects farmers by guaranteeing payments (or prices) for their products. In the UK uplands the classic example has been the Sheep Annual Premium (see below). The EU uses intervention buying, whereby governments buy produce to keep prices artificially high for farmers. Such policies encourage overproduction, creating surpluses as well as environmental damage. Production quotas are the usual way to control surpluses created by price support. Classic examples are the 1984 milk quotas (Halliday 1987) and arable set-aside (Mansfield 1994). These rarely work because farmers switch to non-controlled enterprises to maintain their gross margin, except there are limited opportunities for this in livestock-dominated uplands.

Marketing boards are set up to act as 'middlemen' between farmers and consumers/ manufacturers, preventing farm business receiving low prices as a result of low bargaining power. Eventually UK marketing boards controlled almost all sales of their product and were seen as too restrictive. Many were disbanded in the 1980s and 1990s, to be replaced by farmer-owned businesses or commercial enterprises. For example the Milk Marketing Board, set up in 1933, was disbanded in 1995 and replaced by private businesses such as Milk Marque and Milk First; with the exception of the Wool Board.

Grant aid is extensively used. Farmers can apply for money for certain inputs, to reduce production costs or to increase productivity. There have been many such grants in the UK uplands, for building restoration, pest control, ploughing up and much else. Finally, structural reform aims to improve internal economies of scale. Farmers have received payments to amalgamate farms, retire early or retrain outside agriculture, as with the EU Cessation of Farming Regulation 1096/88 (Dessylas 1987). Unfortunately the effect of these mechanisms is often counter-intuitive and can lead to rural depopulation, subsequent loss of services and other undesirable effects, as we shall see below.

It is the government or civil service's remit to demonstrate why one mechanism is preferred to another or use a combination of tools to counterbalance each other. The choice is determined by factors like expense, administration costs, political acceptability of alternatives and the value placed on wildlife. They also need to determine who will run them (which level of government, quango or even independent project).

The final phase of policy development is to measure the impact, to ascertain whether the instrument has been effective in achieving its policy objectives. It is important to ensure monitoring takes place (through controls or baseline audits) as schemes unfold and that monitors have enough data to make a judgement. Concepts like additionality, deadweight, side and multiplier effects are picked up here. A good example is monitoring of the Environmentally Sensitive Areas (ESA) scheme from its inception in 1986. Every three to five years, each ESA has been monitored to judge whether it has met its environmental, wildlife and heritage objectives (e.g. MAFF 1997, 1998). The entire programme was then reviewed in 2003 as part of the English Rural Development Programme (ERDP) Mid Term Review (Slater 2003) in anticipation of Round 2 of the Rural Development Regulation (see below).

Evolution of upland agricultural policy, 1947 to 2011

British governments have often intervened in upland agriculture since the mid-19th century. In Chapter 4, we discussed such interventions from 1815 to 1946:

- The Corn Laws – to protect Britain from cheap grain imports;
- Corn Production Act 1917 – to encourage home-produced food, led to 1.2m hectares of permanent grassland being ploughed;
- Creation of the Forestry Commission 1919;
- Agriculture Act 1920 – guaranteed prices for oats and wheat, but repealed in 1921 (the 'Great Betrayal');
- Agricultural Marketing Acts 1931/1933 – to help farms develop economies of scale and get a fair price at the farm gate;
- Ploughing-up campaign 1939 to 1945 – arable land increased by 2.1m hectares, notably in northern England, Devon, central Wales and Dumfries and Galloway;
- Hill Sheep and Hill Beef subsidies of 1941 and 1942 – to mitigate loss of animals in the Second World War.

From 1947, we can identify five phases of political intervention in UK upland agriculture:

- 1947–72 – before accession to the European Community, generally a period of policies encouraging agricultural production;
- 1972–84 – control of agriculture through the Common Agricultural Policy (CAP), which continues to support expansion of production;
- 1984–99 – global markets begin to influence EU policy, production controls develop, higher profile for environmental conservation of farmland;

- 2000–06 – EU agricultural policy shifts away from production towards rural development;
- 2006 onwards – second round of the EU Rural Development Programme (see chapter 9).

Before accession: 1947 to 1972

Preparing for post-war Britain, the government knew that agricultural production would need to go on expanding (food rationing lasted until 1952), also housing, so there would be pressure on rural land. The Scott Committee was asked to:

> Consider the conditions which should govern building and other constructional development in country areas consistent with the maintenance of agriculture, and in particular: the factors affecting the location of industry, having regard to economic operation, part time and seasonal employment, the well being of rural communities and the preservation of rural amenities. (Scott 1942: 2)

The report predicted pressure of development from outside the agricultural industry and recommended that farming and forestry be exempt from planning controls. One of the report's more controversial statements, in retrospect, was:

> the beauty and pattern of the countryside are a direct result of cultivation of the soil and there is no antagonism between use and beauty. (ibid: para. 14, p. 4)

It failed to recognise the potential for farming to expand and develop through technological innovation, though Professor Dennison expressed this concern in a minority report. The Scott Report paved the way for the Agriculture Act 1947 and the Town and Country Planning Act 1947, and indirectly influenced the National Parks and Access to the Countryside Act 1949. The Agriculture Act had two aims:

- to secure a modest standard of living for farmers,
- to produce an adequate food supply at moderate cost.

This was a watershed in British agriculture.

Farmers were offered a guaranteed price for their produce; if this was not met at sale, the exchequer made up the difference with a deficiency payment (Bowers and Cheshire 1983). In parallel, the Hill Farming Act 1946 formalised the wartime subsidies for hill sheep and hill cows. It came just before the worst winter for fifty years – 3 million sheep were lost in hill areas. Recovery of the national flock inhibited any measurement of the act's effect for the next three years (Attwood and Evans 1961).

Some farmers saw the loss of the national flock quite differently. As part of the war effort, upland farmers were required to plough up intake land and plant potatoes. The manure they would normally use on the inbye hay meadows was diverted to increase potato yields. Thus hay production suffered, which meant that ewes had much less nutrition before lambing, compounded by the poor weather (pers. comm. H. Wilson, 28/03/09). Stock could not go to the lowlands for overwintering, because most of this land had been ploughed for crops.

The idea of the Hill Farming Act was to assist rehabilitation of hill land and thus increase livestock production. Ensuring a supply of store lambs for lowland farms was recognised as fundamental, but this legislation was not designed to raise production, merely to support farm incomes (Bowler 1979). Deficiency payments for hill sheep were paid in lean years when farmers found sale prices were below the cost of production. They thus fluctuated, being 2s. 6d. in 1940 and 8s. per head in 1942. Rates were higher for pure-bred hill breeds than for cross-breeds. By 1949 2,000 grants had also been paid under the act to improve 846,000 acres (342,000 ha) of hill land. Unfortunately the definition of hill land left out much of it (including the whole of Exmoor) covering only the higher areas grazed by sheep (over 427m asl). Luckily the Exmoor anomaly was ended after lobbying by the NFU (Lowe *et al.* 1986). A £3 subsidy per head of all beef cattle, paid from 1943, had little effect on the size of the upland herd, but expansion did occur in the Welsh Marches and Devon (Bowler 1979).

The 1951 Livestock Rearing Act rectified various flaws in the Hill Farming Act. Marginal uplands now came under the scheme, bringing in lower suckler-cow production areas, lowering the 'official' definition of uplands to about 244m. This marginal land had to be above 183m but was identified on a farm-by-farm basis by local MAFF officials. Complaints about the eligibility of land were only resolved after 1963, when independent arbitration was introduced (Bowler 1979). Between 1951 and 1963 another £20m (the Livestock Rearing Land Improvement Scheme, sometimes called the Livestock Rearing Grant) was allocated to initiatives for farm improvements. Unfortunately, as Figure 8.1 shows, the scheme did not work the way the government expected, as many farmers focused on improving their immediate surroundings (Farm buildings), benefiting the farm family, not business efficiency (Evans and Attwood 1961). Adoption rates varied, with farmers in Cumberland, Westmorland and Denbighshire most committed, in contrast to farmers in Flint, Monmouth and Staffordshire (Bowler 1979).

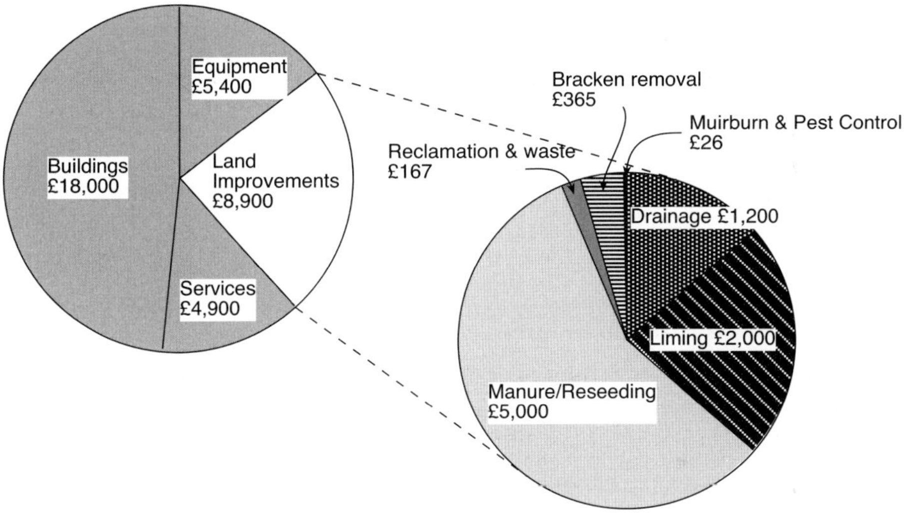

Figure 8.1 Livestock Rearing Grant Expenditure 1951 and 1956
Adapted from data derived from Attwood & Evans, 1961.

Farm improvement grants under the 1952 Agriculture (Ploughing Grants) Act offered farmers £12 per acre to plough up grass that had been a sward for over 12 years. Fields that were steep, uneven or had scrub were particularly eligible. Land was reclaimed in hill areas, but surprisingly the trend in south-west Britain was the reverse (Coppock 1971). Other upland farmers used the grant for sward renewal to improve productivity, which led to the loss of many upland hay meadows and permanent pastures (see Chapter 5; Rodwell *et al.* 1992).

The period of subsidy payment per head of cattle was extended by this act and by the 1956 Hill Farming Act. The payment was much higher than in 1943 (£10 per head per annum) and this stimulated growth in the national and average herd of beef breeding cattle. There were regional variations, as Bowler (1979) showed: in the northern uplands of England, marginal dairy farms converted to beef herds. In contrast, though the upland national flock increased in numbers, the amount of subsidy allocated to sheep fell between 1947 and 1956 as sheep became more profitable and farmers developed cross-bred flocks. In fact the hill sheep subsidy was paid (when sale prices fell below target levels) in only four years from 1951 to 1961 (Evans and Attwood 1961).

The government justified expanding the terms of the 1946 Hill Farming Act by stating that the 1956 act would:

1. help to arrest the decline in production of upland stock-rearing areas,
2. increase meat supply,
3. compensate upland farmers for poor financial support from existing government price guarantees,
4. improve financial returns for upland farmers,
5. compensate for the loss of land elsewhere for developments such as schools and afforestation.

The 1956 Hill Farming Act extended the Livestock Rearing Land Improvement Scheme (LRLIS). By 1957, some 11,500 schemes had been approved covering 2.5m acres on 9,700 farms (55% of the land, 39% of eligible farms) in England and Wales. Exmoor particularly suffered, losing 3,700ha of moorland between 1957 and 1966 (Sinclair 1966); the trend continued into the 1970s, leading to the Porchester Inquiry and the Wildlife and Countryside Act 1981 (Lowe *et al.* 1986). In Scotland 2,664 farms (23%) and 4.7m acres (91%) were enrolled (Attwood and Evans 1966). From 1956, upland and hill farmers received increasing subsidies for hill sheep and hill cattle. From 1961 to 1967, sheep were less profitable. This encouraged farmers to expand their flocks by drawing on more subsidy, so much so that in 1967 the government imposed a 1 ewe/1 hectare cap to prevent overstocking.

Despite fluctuations in subsidies, their impact on farm incomes cannot be underestimated. Indeed, Coppock commented:

> [Hill] Subsidies were paid on some 450,000 cows in 1965, or 52 percent of the breeding herd, but the proportion varied greatly, from 83 percent in Scotland and 71 percent in Wales to 24 percent in England. Hill cow and calf subsidies on such farms may nearly equal the revenue from sales of cattle and may even exceed the gross margin. (Coppock 1971: 189)

That trend was echoed later by Bowler:

> Published data, however, suggest that approximately 30–50 percent of the gross margin per hill cow has been provided by the subsidy and few upland farms would have made profit without it. (Bowler 1979: 79)

> [S]tudies have shown that the subsidy accounted for 75–100 percent of the net profit of hill flocks ... support for hill sheep appears to have done no more than prevent a further deterioration in income level. (ibid: 85)

In 1967 the Hill Land Improvement Scheme resurrected the LRLIS, this time funding mainly fencing, reseeding of pastures and manuring, whereas the earlier LRLIS put most money into farm buildings. By 1971 these hill grants accounted for 17% of net farm income in England and Wales, 32% in Scotland and 50% in Northern Ireland (EDCO 1973).

Governments from 1945 to 1972 tried to raise production levels and incomes of the upland farming sector with a variety of subsidies and grants (figures 8.2 and 8.3). Some of these initiatives had more success than others; some were responsible for less desirable results, such as loss of semi-natural habitat by overstocking. Nevertheless, incomes on upland and hill farms still remained below those of other farms (Bowler 1979) and the sector lagged behind the rest of the industry. What then could the EU's Common Agricultural Policy achieve that national political intervention had not?

After accession: 1973 to 1984

The EU had operated the Common Agricultural Policy (CAP) since its inception in 1957, under Article 39 of the Treaty of Rome (Table 8.2). UK accession in 1973 meant that British agriculture

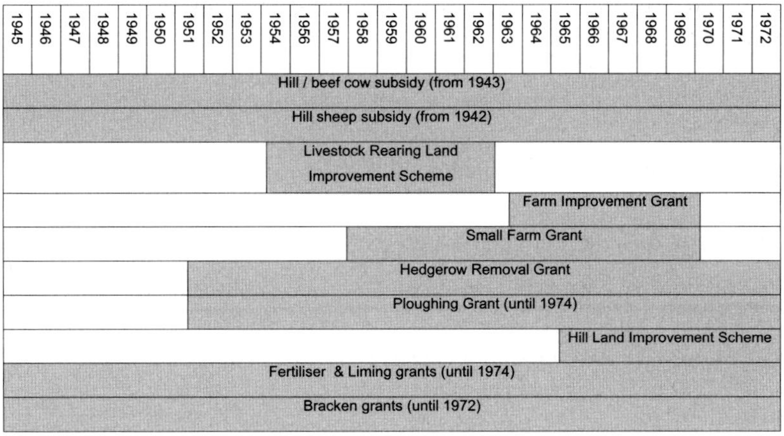

Figure 8.2 UK Agricultural Grants and Subsidies 1945 to 1972

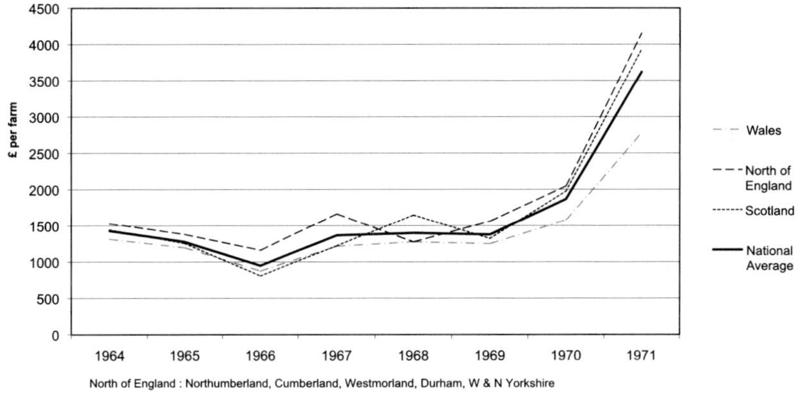

Figure 8.3 Net Farm Incomes on Upland Farms 1964 to 1971
Adapted from data derived from EDCA, 1973.

would now operate within this framework. The CAP was founded on three main principles (Ackrill 2000):

- a single market – goods could circulate free of tariff among member states;
- Community preference – protection against non-EU imports, first choice to partner markets;
- financial solidarity – a common system sharing the financing of the policy.

CAP objectives were much in line with those of the national 1947 Agriculture Act; neither mentioned environmental aims. This was eventually rectified by the 1987 Single European Act, which required an environmental protection component in all EU policies. Policy was implemented – unlike pre-1973 – by the European Agricultural Guidance and Guarantee Fund (EAGGF or FEOGA: *Fonds Européen d'Orientation et de Garantie Agricole*). The entire system was protected from cheap non-EU produce by a series of import levies, reinforced by intervention buying and guaranteed prices. The EU bought up surplus production to raise internal EU product prices above external EU import prices; surpluses were sold off on the world market, stored or destroyed to avoid deflating EU farm incomes. Farmers were also guaranteed payment at a set

Table 8.2: Objectives of Article 39 of the Treaty of Rome

Article 39 of the Treaty of Rome:
1) To increase agricultural productivity by promoting technical progress and by ensuring the rational development of agricultural production and the optimum utilisation of the factors of production, in particular labour
2) To ensure a fair standard of living for the agricultural community in particularly increasing the individual earnings of person engaged in agriculture
3) To stabilise markets
4) To assure the availability of supply
5) To ensure that supplies reach consumers at reasonable prices
Reformulated in 1999 by the European Council as:
1) Increase competitiveness internally and externally in order to ensure that Union producers take full advantage of positive world market developments
2) Food safety and food quality, which are both fundamental obligations towards consumers
3) Ensure a fair standard of living for the agricultural community and contributing to the stability of farm incomes
4) The integration of environmental goals in the CAP
5) Promotion of sustainable agriculture
6) The creation of alternative job and income opportunities for farmer and their families
7) Simplification of EU legislation

level for products, regardless of production volume. The Guidance section helped develop agricultural structures, including farm efficiency. The EAGGF paid farmers directly or channelled funds via the member state as grants or subsidies for particular objectives.

For upland farmers, the greatest benefit came from the EAGGF Guidance section under European Council Directive 75/268/EEC on mountain and hill farming and farming in less favoured areas, ratified on 28 April 1975. One of the main aims of the Less Favoured Areas (LFA) Directive was to reduce disparity between agricultural regions of the EU:

> Article 39 of the Treaty provides that account shall be taken of the social structure of agriculture and of the structural and natural disparities between agricultural regions in working out the common agricultural policy. (p. 1)

Intended mainly as a social and environmental initiative, the LFA Directive relied on member states identifying areas in need of financial support under one of three criteria:

1. Mountain areas where there are considerable limitations on the use of land in terms of:
    a) altitude and difficult climatic conditions shortening the growing season,
    b) slopes too steep for machinery at lower altitudes,
    c) a combination of both;

2. Areas in danger of depopulation, where farming conserves the countryside, with the following characteristics:

   a) infertile land needing high capital inputs,

   b) productivity lower than the agricultural mean,

   c) low/falling population dependent on agriculture;

3. Areas with specific problems, such as poor infrastructure or the need to maintain agriculture to either protect the countryside or preserve tourist potential.

Member states had a year to designate LFAs, using criteria of their own choosing within the three in the directive. Depending how criteria were applied, the definition of LFAs could have substantially different outcomes for agriculture and the environment. In Chapter 2, Table 2.7 listed four member states' criteria. Note how France used simply Criterion 1; the UK used Criterion 2. Mowle and Evans (1990) see this as the origin of much of the overgrazing and overstocking in the UK uplands between the 1970s and 1990s. Criterion 2 shifted the Ministry of Agriculture's attention away from the physical limitations of hill and upland farming towards more general economic and social challenges (Wathern 1992). Conversely, we can argue that the early LFA Directive was wholly successful in achieving its objectives in the UK by supporting the maintenance of farming in depopulating areas. Unfortunately for the environment there were less desirable side effects, despite the focus in upland politics being the reduction of rough grazing through ploughing and re-seeding (see Chapter 5).

Drawing on pre-1972 UK hill farming policy, the EU Commission used the LFA Directive to provide farmers with financial assistance in four main ways:

- Article 4:

  Title II – compensation for permanent natural handicap

  Title III – farm development grant aid

- Article 11: aid for joint investment
- Article 12: national aid for farms

In the UK, land designated as LFA in 1975 now forms the core of the Severely Disadvantaged Area. Articles 4 and 12 were implemented as follows:

- Title II – the Hill Livestock Compensatory Allowance (HLCA),
- Title III – Farm and Horticulture Development Scheme (FHDS),
- Article 12 – sheep premiums (guaranteed prices and deficiency payments)

The FHDS (later the Agriculture and Horticulture Development Scheme: AHDS) led to many upland farmers erecting wintering sheds for beef stock, abandoning traditional stone barns and re-seeding extensive summer pasture. It effectively modernised agricultural production on many farms, increasing grass and stock productivity through matched funding (30%+). Unfortunately, as Wathern *et al.* (1986) noted, to obtain a grant, farmers had to show that 50% of their income was from farming, with a minimum £6200 net farm income. This eliminated many smallholdings and part-time operations typical of upland areas at the time. These smaller farmers were encouraged to apply for lesser grants from the Agriculture and Horticulture Grant Scheme. In 1981/2 the largest LFA farms in the scheme were typically awarded over £13,000; the smallest gained an average of £590.

The FHDS–AHDS did little for the conservation of upland ecosystems. Mowle and Evans (1990) note that, with grant aid of up to 70% for re-seeding, many farmers ploughed up heather moorland and hay meadows, switching them to improved pasture and silage fields respectively. Moorland gripping became commonplace (see Chapter 4). In the new overwintering barns, cow cubicles and self-feeding yards, straw bedding was replaced by slurry storage, which on some farms proved inadequate and led to pollution of groundwater. Wathern *et al.* went further in condemning the system:

> The manipulation of the capital grant system to fulfil purely agricultural objectives confirms MAFF's narrow interpretation of the directive ... No special measures have been introduced by MAFF to encourage the 'conservation of the countryside' despite it being one of the objectives of the directive. (Wathern *et al.* 1986: 209)

The Hill Livestock Compensatory Allowance paid farmers for each head of stock, as long as they had at least 3 ha of LFA land. Payments were restricted by stocking densities to a maximum of 0.9LU/ha in the SDA and 1.35LU/ha in the DA. As we saw in Chapter 5, even these stocking rates far exceed the ecological carrying capacity of most semi-natural rough grazing in UK uplands. Mistakenly, the HLCA has been continually blamed for environmental damage in uplands, but it was the Sheep Annual Premium (SAP) for finished ewe lambs, the Beef Annual Premium (BAP) and Suckler Cow Premium (SCP) that were ultimately responsible (Winter *et al.* 1998). These three schemes where paid as a variable rate year on year, not unlike the deficiency payments of 1947 to 1972 (Figure 8.3). The flaw of SAPs and SCPs was that stocking rates were unregulated, so farmers went on receiving subsidies however many animals they put on their land. Stock numbers in the uplands rose substantially (Figure 8.4) and these subsidies formed the bulk of upland farm support between 1973 and 1984 (Table 8.3). The environmental effect was overgrazing of fells and moorland (see Chapter 5).

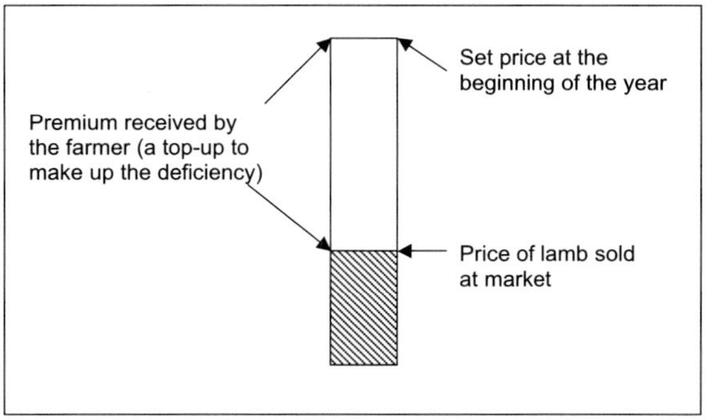

Figure 8.4 Sheep and Beef premiums until 1984

Table 8.3: Guidance Payment Proportions in LFAs

	sheep	*cattle*
HLCA	13.5 %	17.9%
Supplement to premiums	17.5%	16.2%
SAP/BAP	69%	66%

(Source: Winter *et al.* 1998)

But significant change was just around the corner. In the early 1980s, questions were raised about CAP operation. Seven challenges were identified (Moody and Neville 2004):

1. Cost was perhaps the most controversial issue for the EU public. EAGGF expenditure had no limit. In 1979–84, the CAP budget rose from 14m to 27m ECU, of which only 5% was allocated to guidance (Ackrill 2000). There were also concerns about how surpluses were dealt with.

2. Agriculture in the EU was declining in political importance and in its share of the workforce and GDP of member states.

3. Other sectors of the economy, such as infrastructure, were commanding greater attention especially after the accession of Spain and Portugal.

4. World trade politics was pushing for the EU to treat non-EU agricultural goods more fairly and remove import levies and subsidies.

5. Concern was rising over environmental damage caused by agriculture: pollution, landscape change, habitat deterioration and loss, overgrazing. For uplands the threat switched from intensification to decline in the remaining vegetation quality, caused by increased stocking densities and overgrazing (Winter *et al.* 1998).
6. Changing attitudes in West Germany, which with France had always played a pivotal role in formulating agricultural policy. Germans no longer felt the CAP was delivering what they wanted.
7. With the accession of central and eastern European countries with large agricultural sectors and low GDP, the EAGGF threatened to spin out of control completely.

Also people had begun to forget what it was like to go without food or be rationed.

Changes to the CAP: 1984 to 1999

In the light of these concerns, the EU started by slowing the rise in CAP costs and overproduction. Milk quotas were introduced in 1984. Farmers would be fined if they exceeded their 1981 production level plus 1% (Halliday 1988). The effect was immediate, stark and catatonic. Farmers were reported pouring milk down the drain as soon as it came from the cow, going out of business overnight or even committing suicide. Later, controls on land were introduced by extensifying production (EU Regulation 1760/87; Jones 1990) or diverting land out of production altogether (Set Aside: EU Regulation 1094/88; Ilbery 1992).

UK farmers opting for voluntary Set Aside had to take out of production 20% of their land that grew supported arable crops. Potatoes and fodder crops were excluded (Ilbery 1990), so initially upland farmers were little affected. Many commentators saw Set Aside as a mechanism to allay environmental concerns (e.g. Parish 1991, Wilson 1992, Sotherton *et al.* 1992) though these were not at first included. An environmental top-up to Set Aside, Countryside Premium, was piloted in 1989 and led to the extremely successful 1991 Countryside Stewardship Scheme (Mansfield 1994; Box XXXIII). Cross-compliance was introduced for the HLCA, which meant that farmers who wished to receive HLCA had to follow environmentally sustainable practices like supplementary feeding and reductions in overgrazing.

While arable overproduction was being tackled, the Less Favoured Areas were expanding. Boundaries were redefined under Directive 84/189 and an outer Disadvantaged zone was added, with less stringent criteria (Table 8.4). Inevitably this led to a significant jump in the stock population. Related grant schemes were re-aligned, with less funding for reclamation, improvement and drainage, but new money for replanting field boundaries, bracken control and heather reclamation (Ward *et al.* 1990). Once again, the need to match funding turned away the most needy: those with little capital, little labour or falling farm income.

Table 8.4: Defining the Disadvantaged Area in the UK Directive 84/189

criteria
Grassland >70% of Utilised Agricultural Area
Stocking Rates <1LU/ha
Farm rents <65% national average
Labour income <80% of national average
Population density <55/ sq km
Working population >30% in agriculture

(Source: Directive 84/189).

The other main developments for upland farmers were the Environmentally Sensitive Areas (ESA) and Countryside Stewardship Scheme (CSS). The ESA initiative originated in a local dispute in East Anglia between farmers trying to intensify production and what became the Broads National Park trying to save 4,000 ha of semi-natural grazing marshes and reedbeds

Box XXXIII: Development of the Countryside Stewardship Scheme (CSS)

The Countryside Stewardship Scheme developed out of the Countryside Premium Scheme in 1991. It was open to anyone owning or managing land, not solely farmers. In England it complemented ESAs in being available where ESA payments were not; this was not true of the parallel schemes that ran in Scotland, Wales and Northern Ireland (see Box XXXVII).

The scheme was designed and administered by the Countryside Commission from 1991 to 1996 and its aim was to "encourage positive conservation management to sustain or return to these [traditional management] practices, or adopt practical modern equivalents" (CoCo 1991: 1). It covered seven landscapes: chalk and limestone grassland; lowland heath; watersides; uplands; old meadows and pastures in Devon, Herefordshire and Worcestershire; historic landscapes. Each landscape had its own set of objectives; for uplands these were:

- Regenerate heather and other moorland vegetation on enclosed and agriculturally improved moorland
- Support and re-introduce traditional management on hay meadows and pastures
- Restore and protect characteristic landscape features
- Create and improve opportunities for people to enjoy the landscape and its wildlife

Because of limited funding, applicants competed to secure agreements, so only the best schemes were selected. Payments were a mixture of tiered annual rates and capital sums:

CSS payments: selected upland options, 1991 to 2006

option	*1991 rate (£/ha/yr)*	*1995 rate (£/ha/yr)*	*1998 rate (£/ha/yr)*	*2004 rate (£/ha/yr)*
Regeneration of suppressed heather on enclosed moorland	15 + 50 for first 5 years (Tier 1)	15 + 50 for first 5 years	15+ 50 for first 5 years	50 for first 300ha, 20 thereafter
Regeneration of heather moor on agriculturally improved land	50 + 50 for first 5 years (Tier 2)	50 + 50 for first 5 years	70 + 50 for first 5 years	70 + suppt of 50 for 10 years
Restoration and management of hay meadows Tier 3	80 (Tier 3)	80	80	150
Upland inbye	50	50	50	60
Heather burning	n/a	n/a	n/a	5
Bracken control: Mechanical Chemical	 50 85	 50 100	 80 50	 30 70

(Source: adapted from CoCo 1991, 1995; MAFF 1998; DEFRA 2004)

In 1996 the scheme passed into the hands of MAFF, employing the FRCA (Farming and Rural Conservation Agency) to implement it, with some differences introduced by MAFF. First, tiered payments were ended and a general pick-and-mix menu of options was provided instead. Second, each county was given a list of target landscapes and objectives to help make best use of the funds available (see Howgill example below). An applicant who offered land in line with these targets would score more and was more likely to be awarded an agreement than other applicants.

CSS targets and objectives for the Howgill Fells, 1998

KEY STEWARDSHIP OBJECTIVES Applicants should satisfy the following objective:
Grassland, moorland and rough grazing land – proposals should include the management of herb-rich meadows or rough grazing land to benefit the flora or breeding waders and combined with the restoration of field boundaries on a whole farm basis.
In addition to the above key objective, applications are likely to be enhanced if they include appropriate proposals for: • Appropriate management of waterside habitats • Small scale tree planting in gillsides • Provision of permissive public access with a proven demand

Source: MAFF 1998b.

Application became a two-stage process: a desk-based phase where points were tallied from the forms and a visit by a field officer for the best-scoring proposals. The project was then re-scored and a draft management agreement drawn up. New landscapes were added: countryside around towns; field boundaries (through adsorption of the Hedgerow Incentive Scheme); old orchards; new access. Management information became detailed, explaining why managers should carry out certain operations or employ stocking densities and timings (DEFRA 2004). In 2000 Countryside Stewardship, along with ESAs, became one of the six main schemes operating under England's Rural Development Programme. In 2002 it subsumed the Arable Stewardship Scheme.

The table below shows the amount of agreement land for the upland landscape options. In fact, regeneration of suppressed heath moor and upland rough grazing pastures were two of the top three options selected by all managers entering the scheme between 1991 and 2000. Despite this, Ecoscope (2003) calculated that this was under-representative of the broader habitat type found in England, as identified by CS2000 (see Chapter 2). Thus in 1999 CSS absorbed the Moorland Scheme and the upland options were expanded in an attempt to increase representation.

Countryside Stewardship Upland Landscape option: uptake by hectares, 1991 to 2000

	1991–1995	*1996–2000*	*totals*
Upland Hay Meadows	1,323	1,426	2,750
Regenerating suppressed heather moor <300ha	11,538	54,625	66,163
Regenerating heather on improved land	425	3,460	3,885
Upland in-bye pasture	3,692	6,860	10,552
Upland rough grazing pastures	3,480	28,483	31,963
Upland hay meadows & in bye pasture	2,721	1,691	4,412
TOTALS	*23,179*	*96,545*	*119,724*
Wall restoration (m)	251,857	1,160,560	1,412,417

(Source: Ecoscope 2003)

In 2003 the rules changed slightly for applications for upland landscapes. Payments would only be made if the land was registered as LFA and all applications were to be supported by an upland survey if 50% or more of the land was LFA or if any was SDA. The survey comprised of a series of maps recording baseline information about the land holding, existing designations and agreements, landscape features, access, wildlife assessment, heather and other dwarf shrub areas, archaeological and historical information (DEFRA 2003). By the time CSS was replaced by the Environmental Stewardship Scheme in 2006, over 500,000 hectares of land were under agreement on 15,600 land holdings, of which 379,000 ha were in the English SDA (DEFRA 2006). CSS had a budget of about £126 million in 2006/7.

(Box XXXIV). For upland farmers ESAs and CSS agreements offered huge potential to diversify income away from sheep and beef production, while avoiding agricultural intensification. Both ESA and CSS have been held up by various UK governments as beacons of effective management of the interface between agriculture and environment. A lot of land has been enrolled in both schemes, peaking in 2004 at 653,000 ha in English ESAs and 570,000 ha for CSS (DEFRA 2008) but these admirable figures hide issues of delivery and effectiveness, for farmers and environmentalists alike (Table 8.5).

Table 8.5: Critique of the Environmentally Sensitive Areas Scheme

aspect	*example*	*source*
Voluntarism Adoption rate varied from one ESA to another, perhaps due to • level of payment • size of eventual total grant Historical conflict between farmers and conservationists Management prescriptions were inappropriate	Dartmoor 47% v. Exmoor 74% Welsh v. English Lleyn £750 v. North Peak £5000 Cutting time for Pennine Dales hay meadows too late	Ecoscope 2003 Wilson 1997 Brotherton 1991
Length of Agreement Originally 5 years – too short for certain objectives to be met	Heather takes at least 7 years to regenerate	Hughes 1994
Slippage/Halo Effect Part-farm agreements led to farmers intensifying production on non-agreement land (did not occur in all ESAs)	Cambrian Mts only 17% of farmers in ESA reduced stocking densities cf. Pennine Dales where no halo effect existed	Hughes 1994 Saunders 1994
Environmental Benefits Overall effect is to maintain status quo rather than create or restore habitats Few opted for higher tiers (active restoration/creation): too time-consuming or didn't fit farm management Other ESAs do have added environmental benefit	Pennine Dales, Clun Lake District only 6% of eligible heather moorland entered Breadalbane	Saunders 1994, MAFF 1991 ADAS 1997 Ecoscope 2004 Slater 2003 Skerrat 1994
Farm Incomes Gross margins declined as a result of entering scheme	Cambrian Mts ESA dropped from £90/ha to £79/ha for Tier 2 land	Hetherington *et al.* 2002
Public Perception ESA activity has gone unnoticed by the general public Some features were valued more than others	Breadalbane Loch Lomond v. Stewartry wherelitter, woodlands and wildflowers valued more than walls	Skerrat, 1994 Gourlay and Slee 1998

Effectiveness has been patchy in the uplands: Ecoscope (2003) noted that in the Pennine Dales, Clun, North and South-West Peak over 75% of eligible land was entered into the scheme, so large blocks of landscape had contiguous management regimes. However, there is some evidence that most land has been entered in Tier 1 rather than the more complex and environmentally beneficial Tier 2 options (Tables 8.6 and 8.7). For example, the Tier 1 prescriptions for the Lake District required few farmers to change their daily management regime.

Table 8.6: Percentage of Upland ESA Agreement land in Tiers 1 and 2

ESA	*area eligible*	*area under agreement*	*area under agreement as % of eligible area*	*Tier 1 %*	*Tier 2 %*
Pennine Dales	51,000	38,476	75	95	5
North Peak	50,300	43,631	87	61	39
Exmoor	71,000	52,576	74	98	2
Lake District	205,100	146,269	71	94	6
South West Peak	30,500	24,922	82	73	27
Dartmoor	86,600	40,386	47	95	5
*Shropshire Hills**	32,900	22,719	69	99	1
*Clun**	18,400	15,848	86	90	10

*classified by Ecoscope (2003b) as lowland, but included as upland here. (Source: adapted from Ecoscope 2003b: 50.)

Table 8.7: Selected Uptake Statistics by Tier for Upland ESAs 1986 to 2001

ESA	*tier*	*description*	*rate*	*total ha in agreement*
		hay meadow options		
Pennine Dales	2A	Herb-rich meadows	250	1637
North Peak	MMS	Hay meadow supplement	80	302
SW Peak	2B	Traditional hay meadows	170	930
Lake District	2A	Meadows	200	671
Dartmoor	2A	Species-rich hay meadow	160	157
Shropshire Hills	MMS	Hay meadow management supplement	90	269
		moorland		
North Peak	1C 2A 2B	Moorland Moor enhancement (extensification) Moor enhancement (exclosure)*	26 48 90	20061 15747 1112
Exmoor	1D 1E 2A	Moorland Heather moor and coastal heath Reversion to heather moor/coastal heath	34 50 225	2418 9098 50
SW Peak	1D 2E MRS	Moorland Enhanced moorland Moorland regeneration supplement	32 62 30	1186 3016 60
Lake District	1D 1E 2B 2C	Fell without heather Heather fell Intermediate heather fell Enhanced heather fell	25 38 42 50	35769 26395 397 3427
Dartmoor	1E 2B 2BR PGS	Moorland Heather moor maintenance Enhanced moorland Summer purple moor-grass grazing	60 78 105 25	16174 389 1473 1246
Shropshire Hills	1D 02 LHS LLS	Moorland Heather moorland Off-wintering 100% stock from moor Off-wintering 25% stock from moor	60 95 16 4.5	3191 172 182 2272

* Exclosure: a stockproof boundary designed to protect a small area, e.g. woodland amid open fell.

(Source: adapted from Ecoscope 2003c.)

Box XXXIV: Evolution of Environmentally Sensitive Areas (ESA)

The story of how Environmentally Sensitive Areas developed is central to understanding why the politics of agriculture became more pro-environmental. In East Anglia, the Broads National Park Authority manages an area of semi-natural grasslands, reed beds and waters supporting a rich biodiversity and tourist industry (Moss 2001). Between 1960 and 1980 the farmers of this fenland area had taken advantage of agricultural grants and guaranteed prices to drain wetlands and convert them to intensive grain and oilseed production. In 1980, the 4,000 ha Halvergate Marshes were earmarked for drainage. This was the largest remaining block of semi-natural habitat, and the newly created Broads Authority was quick to recognise the danger. Thus began four years of political conflict between the Internal Drainage Board, MAFF and the farmers on one side and the Broads Authority, Nature Conservancy Council and Countryside Commission (CoCo) on the other.

A blow-by-blow account of strike and counterstrike can be found in Lowe *et al.* (1986). Suffice it to say here that the Halvergate Marshes were given a temporary reprieve in 1984, when the Authority agreed to pay £140/ha to the farmers for income forgone that year in not draining the marshes. In 1985, the warring parties agreed on a more permanent solution, which became known as the Broads Grazing Marsh Scheme (BGMS). This was the first initiative ever set up for farmers that used nature conservation legislation (Section 40 of the Wildlife and Countryside Act 1981) to support them in forgoing agricultural profit. As Potter (1988) noted, payments were designed to

> 'bolster declining livestock enterprises to reduce the pressure of conversion to arable'.

In return for a flat-rate payment, the farmers agreed to employ restraint in stocking densities, fertiliser and herbicide applications.

The BGMS was amazingly successful, with 89% of the land brought into the scheme and saved from drainage. The only hitch was lack of funding. To overcome this, MAFF lobbied the EU to fund it under the LFA Directive. It was agreed that up to 2% of agricultural land outside standard LFA areas could be provided with grant aid to encourage traditional farming practices – in other words, paying farmers not to intensify farming operations if that would be detrimental to the environment (Potter 1988). This eventually led to the EU-wide formulation of Regulation 797/85 (on improving efficiencies of agricultural structures), enabled by the UK Agriculture Act 1986; thus the Environmentally Sensitive Areas initiative came into being. The ESA scheme was the first EU environmental initiative to be funded from agricultural grant aid and is now seen as a political turning point in the history of agricultural support in the EU.

ESAs were, by definition, areas with nationally recognised environmental interest which was under threat from agricultural change. They were also areas that could be conserved by maintaining certain agricultural practices where traditional farming techniques would prevent or reverse environmental damage (Potter 1988). After consultation in 1985 between MAFF, the NCC, CoCo and English Heritage, a list of 66 candidate areas (10 in Scotland) was submitted (Whitby 1994; Lowe 1995). Of these, 36 were upland areas, of which 24 were designated (see map below), the first six in 1986 and the rest rolled out in groups between then and 1994. By the time the process finished, 2.2m ha of land was ESA, with a budget of £96m by 2006.

Although designation was compulsory under the EU regulation, farmers could choose whether to enter the scheme. By 2000, 18.5% of ESAs in England held agreements on SDA land and 5.5% on DA land (Slater 2003). Farmers could apply to the ESA office and, as long as they signed up to a management agreement, no one was turned away. Agreements were initially for five years and paid farmers to carry out certain land-management practices, for annual payments by the hectare or metre, or put in capital items in exchange, for a one-off payment. Farmers could opt for different levels of management, known as tiers. Tier I typically focused on protecting and maintaining landscapes, wildlife and related vernacular architecture. Tier II payments allowed some scope for restoration. Each ESA had its own prescriptions, tier structures and payment regimes. As the scheme matured, from 1992, agreements were extended to 10 years, more in line with ecological objectives.

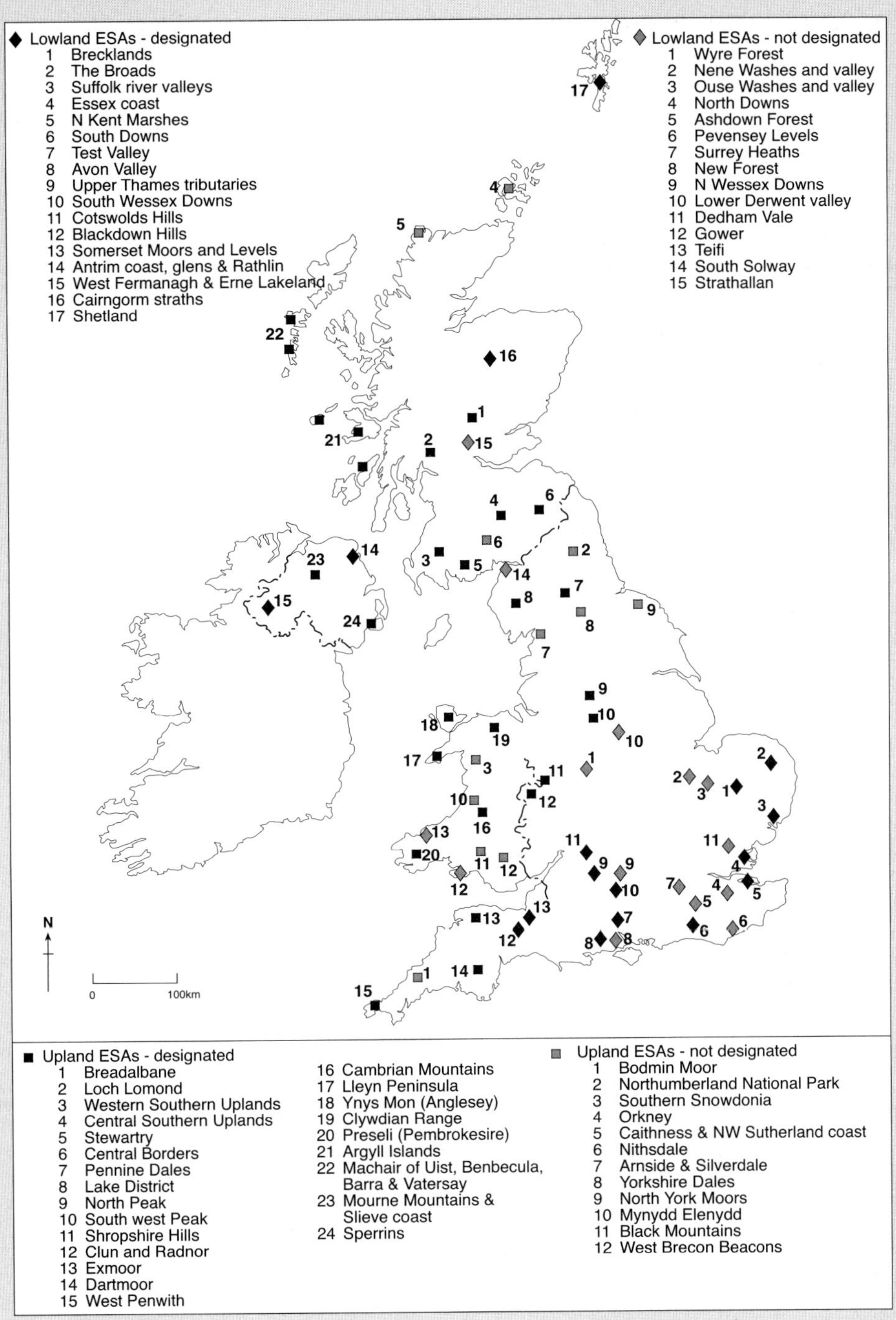
◆ Lowland ESAs - designated
1 Brecklands
2 The Broads
3 Suffolk river valleys
4 Essex coast
5 N Kent Marshes
6 South Downs
7 Test Valley
8 Avon Valley
9 Upper Thames tributaries
10 South Wessex Downs
11 Cotswolds Hills
12 Blackdown Hills
13 Somerset Moors and Levels
14 Antrim coast, glens & Rathlin
15 West Fermanagh & Erne Lakeland
16 Cairngorm straths
17 Shetland
◆ Lowland ESAs - not designated
1 Wyre Forest
2 Nene Washes and valley
3 Ouse Washes and valley
4 North Downs
5 Ashdown Forest
6 Pevensey Levels
7 Surrey Heaths
8 New Forest
9 N Wessex Downs
10 Lower Derwent valley
11 Dedham Vale
12 Gower
13 Teifi
14 South Solway
15 Strathallan
N
0
100km
■ Upland ESAs - designated
1 Breadalbane
2 Loch Lomond
3 Western Southern Uplands
4 Central Southern Uplands
5 Stewartry
6 Central Borders
7 Pennine Dales
8 Lake District
9 North Peak
10 South west Peak
11 Shropshire Hills
12 Clun and Radnor
13 Exmoor
14 Dartmoor
15 West Penwith
16 Cambrian Mountains
17 Lleyn Peninsula
18 Ynys Mon (Anglesey)
19 Clywdian Range
20 Preseli (Pembrokesire)
21 Argyll Islands
22 Machair of Uist, Benbecula, Barra & Vatersay
23 Mourne Mountains & Slieve coast
24 Sperrins
■ Upland ESAs - not designated
1 Bodmin Moor
2 Northumberland National Park
3 Southern Snowdonia
4 Orkney
5 Caithness & NW Sutherland coast
6 Nithsdale
7 Arnside & Silverdale
8 Yorkshire Dales
9 North York Moors
10 Mynydd Elenydd
11 Black Mountains
12 West Brecon Beacons

Tier 2 had more stringent stocking densities and temporal limitations on farming activities (MAFF 1997); cutting dates and stock utilisation timings for hay meadows were often inflexible (Clun, SW Peak, Lake District, Pennine Dales), failing to allow for year-on-year variation in weather patterns.

From the environmental perspective, Tier 2 often required restoration or creation of habitats. Ecoscope reported few instances of this being achieved in any ESA, not even the bird population levels (Ecoscope 2003: 59). On moorlands, many of the generic stocking rates in ESA prescriptions took no account of local variation in mosaic composition, grazing preferences, seasonal grazing impacts or variable physiological adjustments to de-stocking. Ecoscope concluded:

> These research findings indicate that it is not appropriate to use generic grazing rate prescriptions for the maintenance and enhancement of moorland habitats. (Ecoscope 2003: Chapter 10, p. 8)

To overcome stocking problems, a Moorland Management Plan was introduced, which allowed flexibility in grazing regimes. In other instances local variation was considered; for example, in the Pennines Dales ESA hay-cutting times were varied from the southern to the northernmost dales by a period of two weeks (Smith 1988, MAFF 1991).

Further issues arose from the lack of parity of data collection in line with other habitat monitoring and management regimes. Although English Nature (2001, quoted in Ecoscope 2003) was able to demonstrate that 62% of ESA agreements surveyed had not reached favourable condition, there was not enough detailed botanical surveying to identify the quality of priority habitats under the Biodiversity Action Plans (see Chapter 5). To rectify this Critchley *et al.* (2004) carried out a detailed survey of 164 Pennine Dales ESA Upland Hay Meadows. Their analysis revealed that the meadows had deteriorated if cutting times were before 15 July, if spring grazing continued beyond 15 May, if cattle were the sole grazing stock or if inorganic fertilisers were added. These results were helpful, leading English Nature to recommend that future agri-environment prescriptions about hay meadows should have a package of the reverse of these four management practices rather than providing a pick-and-mix menu typical of CSS.

This does not mean that ESAs have been a complete failure. They paved the way for development of effective agricultural grants that pay farmers to produce environmental goods. In the farming community there is evidence that many farmers joined the ESA scheme to increase farm income, despite concerns over de-stocking and heft management (IEEP 2004, Wilson 1997, Brotherton 1991). Walling, hedging and traditional farm-building renovations all provided work for local contractors, creating an economic multiplier effect in the wider community. Slater (2003) showed that ESA agreements were particularly important to upland farm businesses – 34% of England ESA agreements were on upland stock farms (only 12% of CSS) – because they maintained farm business incomes when stock prices were low in the late 1990s.

In much of the environmental research and monitoring literature on ESAs, the consensus is that they have maintained the *status quo* between agriculture and environment; if they had not existed, deterioration would have been much greater (Whitby 1994, Ecoscope 2003, Slater 2003). For instance, there were substantial reductions in stocking levels of sheep on some moorlands and fells under Tier I agreements (Edwards, pers. comm. 15/01/10). Also the geographical nature of the ESA scheme can be seen as a benefit, because it allowed for some tailoring of prescription design to customary practice and the evolution of large contiguous blocks of habitat. These practicalities fit well with developing whole-landscape ecosystem-management concepts, avoiding habitat fragmentation and all its faunal territorial issues (Sexton 2001). The ESA initiative also raised the profile of environmental considerations in agriculture and 'paved the way' for more demanding land-management regimes, like the Higher Level Scheme under Environmental Stewardship and the Sheep Wildlife Enhancement Scheme.

The Countryside Stewardship Scheme (CSS) has been similarly evaluated but to a lesser extent. Pre-1996 analysis was summarised by Ecoscope (1997, quoted in Ecoscope 2003). Adoption of CSS roughly met the targets set for the pilot by the Countryside Commission. Data analysed by Morris and Young (1997) showed that uplands were most popular landscape in the scheme (24,566 ha, 23% of the total). There was a regional pattern, with upland being the most popular option in the North Countryside Commission area (Cumbria, Northumberland, Tyne and Wear, Durham and Cleveland). This was due to lack of ESA designation on large tracts of upland in the Cheviots, North Pennines, Durham Dales and North York Moors.

Despite this, upland landscape agreements were actually under-represented in this period compared to broad habitat types. Comparative sampling analysis by Carey *et al.* (2002) found that, for Environmental Zone 3, dwarf shrub heath was grossly under-represented (8% CSS to 20% CS2000), acid grassland was exactly right (21% both CSS and CS2000) and improved grassland was over-represented (33% for CSS and 24% for CS2000). It also found overall landscape, wildlife and historic benefits from scheme land, except for re-creation of heather which had not matured enough to warrant a positive judgement (Ecoscope 2003). Of a sample of agreement holders, 82% stated that the environmental value of their land would/could have deteriorated if they had not entered CSS. A few sites had negative management issues like under- or over-grazing, or inappropriate cutting or burning regimes. There is little other independent critical analysis of CSS in this period, apart from a cirl bunting (*Emberiza cirlus)* survey in South Devon, which found the population had expanded by 83% on CSS land but only 2% on non-CSS land (Peach *et al.* 2001).

Less typical are socio-economic studies of farm business changes. An exception is a small (460-farm) postal survey by Harrison-Mayfield *et al.* (1998) which investigated the non-farm socio-economic effects of CSS, on- and off-farm. The main findings were:

- CSS had no real impact on farm labour: only 10% of farms had to make a change, usually for agreements with financial benefit (£40,000+).
- More respondents (43%) increased use of contractors/advisers, benefiting local businesses. These two changes in labour created 260 jobs nationally.
- There were changes in inputs, more in machinery purchases and fencing, less in fertiliser/crop-protection products, with negligible financial effect on input–output farm businesses, but a possible dissipated job-loss effect.
- Impacts on farm income varied +/−15% with increases more prevalent among farmers who committed less land to the scheme.
- Over a third had seen positive multiplier non-farming effects, including more visitors and renovation of attractive features.

A similar socio-economic assessment was made of CSS's Welsh equivalent, *Tir Cymen*, by ADAS (1995). This found that statutory whole-farm payments raised farm incomes by an average £1,500 a year and statutory habitat payments did so by £2,500 a year. Once the cost of capital items was factored in, farm incomes had risen by £1,500 a year, not to be ignored when gross margins at this time in Wales for LFA stock farms were around £10,000. Banks and Marsden (2000) recorded an improvement in stock quality, in terms of lambing rates and slaughter value, as stocking densities came down. A further improvement came from the restoration of field boundaries in the Gower area, which helped with overwintering of stock from upland Wales. There were also multiplier effects from off-farm and non-farm on-farm businesses.

While these initiatives were maturing, the agricultural agenda had moved on again. In 1992, the CAP underwent major reform. Named after the European Agriculture Commissioner of the time, the MacSharry reforms responded to overproduction and the Uruguay round of the GATT (General Agreement on Tariffs and Trade) talks (Le Heron 1993). The reforms aimed to:

- make farmers more competitive at the world level,
- bring production down to levels in line with market demand,
- focus support for farmers' incomes where it was most needed,
- encourage farmers to remain on the land,
- protect the environment and develop the natural potential of the countryside.

These were achieved by a combination of supply control, intervention-price reduction, compensation and agri-environment funding. The compensation was designed to soften lower subsidy incomes and would gradually be removed to bring farm businesses into line with non-EU farmers' prices. More effectively, the SAP (Sheep Annual Premium) and BAP (Beef Annual Premium) used quotas based on LUs per hectare of forage per farm unit. As an example, for the BAP this was 3.5LU in 1993, 3.0 in 1994, 2.5 in 1995 and just 2.0 in 1996. Farmers with a very low stocking density (less than 1.4LU/ha) received an enhanced premium. For sheep the quota was set at 1,000 ewes in the LFA per farm holding (Winter *et al.* 1998).

If these reductions were expected to decrease the national herd, then the EU were disappointed. A farm survey by Winter *et al.* (1998) showed that 27.1% of farmers increased their herd and 62.1% left it unchanged. More specifically the survey found that the new restrictions on stocking densities affected only 15% of upland farms. Of these, 86% compensated by acquiring access to more land, thus reducing their stocking density but retaining the same number of animals. Cattle re-stocking was seen as an environmental issue whereby habitats could still suffer from under- or inappropriate grazing (see chapters 4 and 5). The sheep sector responded by levelling off the national flock size, but Wales saw an expansion and Scotland a decline. Winter *et al.* (1998) concluded that there was little evidence that any LFA system created long-term environmental improvements in upland landscapes. In fact, it seemed the SAP quota system had encouraged a drift from lowlands to upland margins, which may well have reduced biodiversity there.

Regulation 2078/92 (on the Introduction and Maintenance of Agricultural Production Methods Compatible with the Requirements of the Protection of the Environment and the Management of the Countryside) provided matched funds for states to encourage environmentally-friendly forms of farming. In the UK this allowed current schemes, such as ESAs and CSS, to expand and a range of new initiatives to be developed (Table 8.8). The main Welsh, Scottish and Northern Irish agri-environment schemes in the UK (*Tir Cymen*, the Countryside Premium Scheme and

Table 8.8: Agri-Environmental Initiatives operating under Regulation 2078/92

	England	*Wales*	*Scotland*	*Northern Ireland*
Pre-existing expansions				
ESA	*	*	*	*
CSS	*			
Organic Aid	*	*		*
New schemes				
Nitrate-sensitive areas	*			
Tir Cymen		*		
Countryside Management Scheme				*
Countryside Premium Scheme			*	
Countryside Access Scheme	*	*		*
Habitat scheme	*	*	*	*
Moorland scheme	*			

Countryside Management Scheme respectfully) are outlined in boxes XXXV and XXXVII. All three had upland elements in their design, with some unique features.

Another scheme was the English Moorland Grant Scheme. It aimed to protect and improve upland heather moorland through maximum stocking densities. In summer (1 Mar. to 30 Sept; 1 Apr. to 31 Oct. in Somerset, Devon and Cornwall) these were set at a maximum of 0.23 LU/ha; in winter (1 Oct. to 29 Feb; 1 Nov. to 31 Mar. in Somerset, Devon and Cornwall) they were 0.15 LU/ha (MAFF 1995). Annual payments of £25 per ewe removed (£30 from 1996) were made for five years to compensate. Bracken management was funded at £120/ha and £110/ha for chemical and mechanical control respectively. A botanical monitoring programme was set up to measure improved ecological condition of the DSH. With only 17,785 ha of land enrolled by 1998, the scheme unfortunately failed to meet its objectives (Ecoscope 2003). A number of factors seemed to be responsible: lack of publicity, too low a level of compensation (£45 was seen as more realistic), too restrictive management (stocking densities were cripplingly low) and being limited to sheep (Soffe and Hetherington 1996/7). The monitoring found no change in vegetation condition, but it was executed too early in the plant community's ecological recovery to be of value. In 1999 the scheme closed and the management options were integrated into the CSS moorland options. The only really positive outcome was the introduction of the Moorland Line, but even this has issues (see Chapter 2).

Finally, in 1997 the Less Favoured Areas Directive was repealed and replaced by EC Council Regulation 950/97 on improving the efficiency of agricultural structures. The new regulation set out four main aims (Official Journal of European Council L142/4):

- To help restore the balance between production and market capacity;
- To help improve the efficiency of farms by reorganising their structures and promoting supplementary activities;
- To maintain a viable agricultural community and thus help develop the social fabric of rural areas by ensuring a fair standard of living for farmers and by offsetting the effects of natural handicaps in less-favoured areas;
- To contribute to safeguarding of the environment and preservation of the countryside, including long term conservation of natural farming resources.

Subsidies were limited to no more than 120 LU per farm and halved where LUs exceeded 60. Other clauses were more explicit about the role of LFAs in countryside conservation.

In effect, the 1992 reforms were a mechanism to apply the brakes slowly to the out-of-control CAP. It gave the EU a breathing space in which to demonstrate to observers that the EU was dealing with some of the worst of the inertia. It allowed time for the Commission to evaluate all that had gone before, with an eye to better management of agriculture in the future. At state level it allowed MAFF to evaluate the effectiveness of its agricultural support. Of particular interest are the HLCA evaluations for the four parts of the United Kingdom. Table 8.9 shows a similar trend in the effectiveness of HLCA in all parts of the UK. Though it now formed only about 10% of all support for LFA farms, it was seen as an important element in maintaining farming in upland areas alongside viable populations. There was varying evidence of its role in conserving the countryside but then, as Wilson *et al.* (1998) noted, the HLCA was never designed for this. In any case SAPs, with no cap on head counts, were more to blame for overstocking of fells. The other interesting conclusion from these evaluations is that they all agree that area payments were not the best option to replace headage; unfortunately this is what happened in 2001 as a result of further EU reform known as Agenda 2000.

A new approach: 2000 to 2006

Agenda 2000 was the main response to the Uruguay round of GATT. It continued the removal of production support (Pillar I), replacing it with non-production support (Pillar II) for desired

Table 8.9: HLCA Evaluations in the UK in Preparation for Agenda 2000

factor considered	*England*	*Northern Ireland*	*Scotland*	*Wales*
Current nature of hill farming sector	Decline in number of LFA holdings, notably beef and dairy; 9% of agric. labour force is in LFA	No off-farm income in 40% of LFA farms; 45% of SDA, 44% of DA farms are part-time; 66% of agric. labour force in LFA	Red meat consumption down; 90% of sheep/84% of sucklers in LFA; Highlands poorer than rest of LFA; 61% of labour force is in LFA	Sheep and cattle numbers have risen in last 30 years; 92% of sheep are in LFA; 68% of agric. labour force are in LAFA
HLCA changes	No longer dominant support mechanism		HLCA no longer dominant, now SAPs	Support cost up; HLCA no longer dominant, now SAP and BAP (10% of GM); switch to sheep; fewer larger units
Economic effect of HLCA	Larger farms get the lion's share	The smaller the farm, the greater the contribution to income; multiplier effects outside agriculture of 1000 jobs and £9m income	Multiplier effect of 1395 jobs and £8m income; farm incomes are comparable to outside the LFA; workforce declined but not as much as outside LFA	Larger farms get the lions share of HLCA payments. Multipler effect outside agriculture
Social effect of HLCA	Rise in LFA population but decline in agricultural population	Lack of off-farm income means HLCA is seen as a positive contribution to household income	Population up in rural areas; issues of succession on farms; incomers forcing house prices up and thus no affordable housing for young to stay	Slower depopulation in LFA than non-LFA; agricultural labour has declined; succession problems; decline in Welsh speaking and increase in transient in-migrants
Environmental effect of HLCA	Stock numbers have risen and stock are being left for longer on the hill Overgrazing is seen as an issue but not the sole effect of HLCA			Reclaimed moorland, agricultural improvement, stocking densities up/exceed ecological capacity; fall in structural diversity, bird nos., habitat quality/quantity; rise in soil erosion, acidification, bracken

Table continued over

Table 8.9: ***continued***

factor considered	*England*	*Northern Ireland*	*Scotland*	*Wales*
No HLCA	Would lead to decline in small/medium farms and faster structural change	Would impair net cashflow and servicing of overdrafts	Would lead to no hired labour Farm viability would suffer along with multiplier businesses	
Replacement options for HLCA: *A - area payments* *B - labour subsidy* *C - tighter stocking limits* *D - modulated headage*	Farmers did not favour options A and B nor an upland agri-environment scheme – but organisations favoured this	Option A would reduce financial support for farms	Keep HLCA, option A only favoured by 20% of farmers surveyed	Farmers favoured B, C or D; option A would have a detrimental effect for small and medium-sized holdings.
Overall conclusions	**Has kept up hill farming/viable population; 3rd aim not met; in the long term an upland AES would be good**	**Agri-environment schemes would not plug the deficit of an area-based scheme.**	**Has maintained hill farming and viable population; achievement of 3rd aim (countryside conservation) debatable**	**Local environmental variation means the effects of HLCS have been locally different**

(Sources: Drew Associates 1997; Midmore *et al.* 1998; Wilson *et al.* 1998; Caskie *et al.* 2001.)

aims – in this case, rural development and environmental management. This process, known as modulation, decoupled financial support from production (Moody and Neville 2004).

A main feature of decoupling was Regulation 1257/99 (On the support for Rural Development from the EAGFF), known colloquially as the Rural Development Regulation (RDR), articulating Pillar II. The use of EU agricultural funds to support rural development was completely new. Previously rural development was strictly the domain of regional and structural development funds from other sectors of the EU Commission (see Box XXXII). The RDR drew together a range of structural policies under a single umbrella (Ackrill 2000):

- Investment in farm businesses – to reduce production costs, improve quality, protect the environment, meet animal welfare needs or diversify farm businesses;
- Human resources – helping people under 40 enter farming, supporting those retiring early, providing training focused on farm outputs or on protecting the environment;
- Less Favoured Areas – payments per hectare for farmers in these areas;
- Forestry – supporting its economic, ecological and social functions, plus money for forest product processing and marketing;
- Processing and marketing of agricultural products - research into new commercial outlets and developing greater added value;

- Agri-environmental measures – annual hectarage payments compensating farmers for loss of income and management cost beyond normal practice;
- General development of rural areas ('Article 33 measures') – aiding integrated rural development, e.g. by preserving rural heritage, diversifying economic bases and promoting tourism.

The guarantee section of EAGGF (FEOGA), therefore, was re-routed to fund agri-environment initiatives, early retirement, afforestation and LFA payments. In addition to these four measures, member states were required to ensure continued environmental cross-compliance and modulation, and to draft Rural Development Programmes (RDPs) to deliver the RDR. The UK Government responded by producing four plans, one for each part of the UK, using a 10-stage process (Figure 8.5). In England for example, the initial SWOT[1] analysis identified three main trends in agriculture (www.defra.gov.uk/erdp, accessed 03/03/04):

- Economic change – agriculture's role in the economy was declining, its general labour force was contracting, with reduced inward investment. Populations were sparse and dispersed, with a large gap between skills and knowledge emerging.
- Environmental Change – environmental quality had declined, despite being valued, mainly because of agricultural change and a rural-to-urban shift of population. With no market for 'environmental goods' there was no incentive for farmers to protect the environment (despite ESAs and CSS).
- Social change – typified by poverty and social exclusion masked by general trends. The decline in services meant the population found it increasingly hard to access what was left.

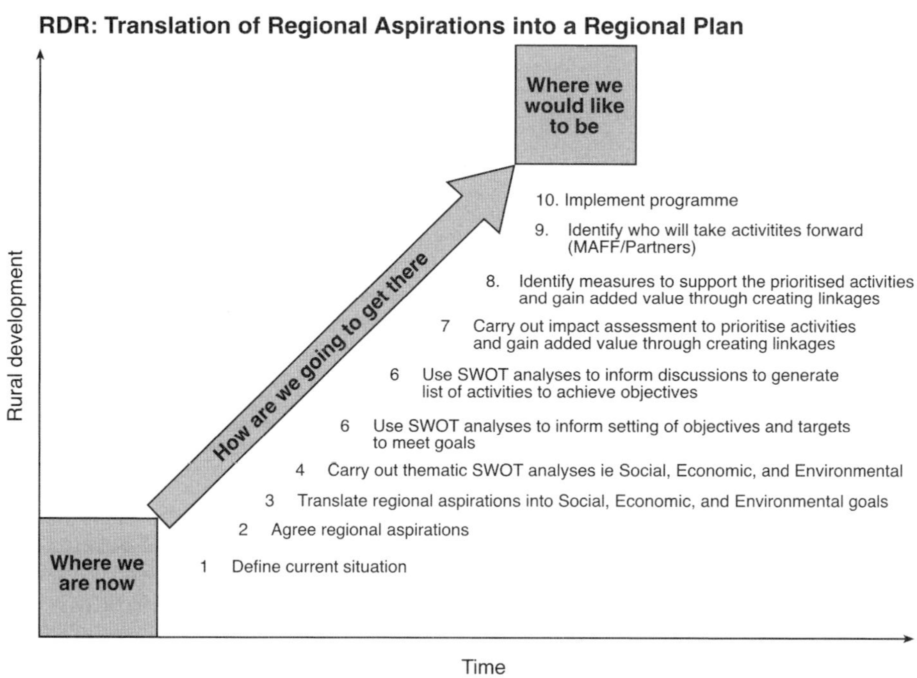

Figure 8.5 SWOT Analysis for England's RDP
DEFRA, 2000, '*Englands Rural Development Plan*'. This information is licensed under the terms of the Open Government Licence.

[1] SWOT means analysis by strengths, weaknesses, opportunities and threats.

Following this exercise, the UK Government identified two main priorities and operational objectives:

- Priority 1 – The creation of a productive and sustainable rural economy

  Operational Objective: to assist projects that will contribute to creating more diverse and competitive agricultural and forestry sectors and new jobs in the countryside, encourage the development of new products and market outlets with greater collaboration and provide targeted training to support these new activities

- Priority 2 – Conservation and enhancement of the rural environment

  Operational Objective: to increase significantly the area covered by schemes under these measures over the Programme period and to maintain sustainable management of an appropriate area of the Less Favoured Areas

RDP operational objectives were manifested as a series of initiatives (Table 8.10) which the government drew from extant agricultural, agri-environment and rural development schemes to form a coherent programme. Some were managed at national level (ESA, OFS, HFA), some operated in the national framework but had targeting at regional level (CSS, WGS/FWPS, ECS) and some were managed by regional groups with regionally agreed priorities and targets (PMG, RES, VTS). Funding varied, with agri-environment initiatives taking the largest slice (£1bn). Other schemes had regional budgets, like the VTS with £846,000 in the North West and £2.18m in the South West; these apportionments were governed by the earlier SWOT analyses. OFS, PMG, RES and VTS were discussed in Chapter 6; below we consider HFA, ESA and CSS.

Table 8.10: RDP Schemes 2000-2006

Initiative	*Funding (£m)*	*No. of Agreements*
Environmentally Sensitive Areas	347	9902
Countryside Stewardship	431	15433
Energy Crops	5.2	339
Voluntary Training Scheme	14.6	26925
Processing & Marketing Grant	33	237
Rural Enterprise Scheme	111	2891

Adapted from: archive/defra.gov.uk/rural/documents/rdpe/progdoc/annex34.pdf. Accessed: 15/10/11.

For upland agriculture the most important aspect of the RDP was the continuation of LFAs which, under Article 13, were designed to:

- ensure continued land use and thereby contribute to the maintenance of a viable rural community,
- maintain countryside,
- maintain and promote sustainable farming, in particular taking account of environmental protection requirements. (www.defra.gov.uk/erdp/docs/national, accessed 02/02/05)

The three main types of LFA continued (Articles 18, 19 and 20):

- Mountain areas – short growing seasons due to high altitude, steep slopes at lower altitude or a combination of the two; land north of 62nd parallel.
- Other LFAs (intermediate) – areas in danger of agricultural land abandonment and where conservation of the countryside is needed.

- Areas affected by Specific Handicaps – farming should be continued in order to conserve or improve the environment, maintain the countryside , preserve tourist potential or protect a coastline.

The crux was that member states could no longer provide headage payments. Instead, in line with SPS (see below), they could offer hectarage payments in LFAs. In England this measure was implemented via the Hill Farm Allowance scheme using area payments for land where sheep breeding and suckler-cow beef production occurred. The ERDP document itself stated that:

> the continuation of extensive suckler cow and sheep breeding is crucial to maintaining the environment and social fabric of the LFA .... It is traditional and extensive sheep and beef grazing that have largely moulded the environment, biodiversity and landscape of the English LFAs and contributes greatly to the social fabric of these areas. (ERDP 2000: Sect. 8, p. 136)

In one fell swoop DEFRA could turn overstocking into appropriate stocking. Looking at it another way, upland agricultural support had been decoupled from production and re-aligned as environmental management with rural development potentials. The other parts of the UK approached the issue in slightly different ways, drawing on previous experience (Box XXXV).

A final pot of money available from the EU was the 'National Envelope'. This equated to a payment of £22.66 in 2004 for beef cows; for sheep, it was used to fund the Sheep Wildlife Enhancement Scheme (SWES) in England. Administered by English Nature, SWES was designed to de-stock fells designated as SSSIs, allowing vegetation to reach favourable condition (see Box XXXVI). This supported the drive to meet their Public Service Agreement (PSA) set by the then UK Government (see Box XIX). All farmers were also required to follow a set of good farming practice conditions, known as Cross-compliance: for a farmer to get their HFA, they had to comply with a set of rules, updated in 2007 (Table 8.11).

One of the biggest concerns about the switch from HLCA to HFA was whether farmers would be put under undue hardship. Matthews *et al.* (2006) ran a modulation model for various farms in upland Wales and calculated an average decline of 5% in financial performance due to these CAP reforms. They also found a 15–18% decrease in direct payments resulted, partially offset by agri-environment grants. Although the model was run for only three farms, it showed that to maintain farm income upland farmers probably had no option but to adopt agri-environment schemes. Anecdotal evidence suggested that many upland farmers operating on commons lost substantial income once the HLCA switched to HFA (M. Edwards, pers. comm. 15/01/10). Up until the switch many commoners were using their full registered rights for stocking levels on their heft; under HFA, the payments were restricted to notional areas representing the proportion of grazing rights held on the common, irrespective of the actual area of common grazed.

In 2002, the EU Commission published the Mid Term Review (MTR), which was expected to be simply that, an evaluation of how the Agenda 2000 initiative was going, with perhaps a few modifications. Instead the MTR became infamous as a root-and-branch re-vamp in view of the forthcoming round of the World Trade talks and the imminent enlargement of the EU (Moody and Neville 2004). The official objectives of the Mid Term Review were to:

- enhance the competitiveness of EU agriculture,
- produce a market-orientated sustainable agriculture,
- strengthen rural development.

Member states were required to review RDPs in 2003. In England, the main aims were to:

- review the impact of the ERDP on rural areas since its inception in 2000,
- measure progress against the plan and national objectives,
- review the appropriateness of the strategy and its implementation.

Table 8.11: SPS Payments: Land Entitlements

cross-compliance requirement	*range*
Good Agricultural and Environmental Condition	Development of a Soil Protection Review (see Chapter 4), to include: • post-harvest management • waterlogging management • burning of crop residues • soil-erosion management Maintenance of habitats and landscape features, to include: • EIA • SSSI • scheduled monuments • public rights of way • overgrazing and unsuitable supplementary feeding • heather and grass burning • control of weeds • land not in agricultural production currently • stone walls • protection of hedgerows and watercourses • felling of trees and TPOs
Permanent Pasture	Defined as land: "used to grow grasses or other herbaceous forage either naturally or through cultivation, which has not been included in the crop rotation for five years or more" Land can be ploughed and re-seeded under this definition.
Statutory Management Requirements	Environmental legislation: • wild birds • groundwater • sewage sludge • nitrate-vulnerable zones • habitats Public & Animal Health: • pigs, goats and sheep ID & registration • cattle identification • sheep and goats registration [again!] • public health • animal & plant health • notification of diseases (RIDDOR) • animal welfare

(Sources: RPA 2005 c–f; RPA 2006; RPA 2007.)

Box XXXV: Direct hill farm support in the United Kingdom

Hill Farm Allowances (HFA) in England were made up of three components:

- *A basic payment per hectare* – no payments were made on land over 700 ha, and half rate applied for 350 to 700 ha. This allayed concerns raised in HLCA evaluations that larger farms would receive the lion's share. The payment varied with the type of land, supposedly to reflect how severe was the permanent natural handicap for sheep, but common land could technically be of any productive potential;
- *An enhancement* – 10% for meeting one criterion, 20% for meeting two or more;
- *A safety net* – ensuring that HFA payments did not fall below a set percentage of the HLCA in a particular year, reducing annually until withdrawn in 2004.

HFA, England

	£/ha			
	2001	*2004*	*2006*	*2008*
Land above the moorland line	13.02	11.27	9.39	14.12
Common land not above the moorland line	13.02	11.27	9.39	14.12
Other DA land	18.60	16.11	13.42	–
Other SDA Land	34.40	29.80	24.82	37.31

(Source: DEFRA website, accessed 10/10/08)

On top of the HFA payment, farmers received a headage payment per ewe (about £18 in 2004) or suckler cow (about £124 in 2004), which in line with WTO agreements gradually decreased year on year. These had quotas to dissuade farmers from overstocking, thus limiting national expenditure. They were based on how many stock the farmer had in the base year 1991/2.

The Less Favoured Areas Support Scheme (LFASS) in Scotland had three elements: area-based payments, environmental measures and a safety net. There were three rates for area-based payments, based on a farm classification system developed for SERAD (Scottish Executive Rural Affairs Department):

- *Moorland farms* – highest hill land with limited production potential, typically low stocking rates and large area;
- *Southern Upland farms* – best-quality land with highest stocking rates;
- *Northern Upland farms* – more disadvantaged than Southern Upland farms, for example because of a shorter growing season or further from markets.

The payments were divided again by land type, with improved pasture commanding a higher price than rough grazing. According to the Scottish Parliament (2001) these were supposed to reflect differences in carrying capacity between moorland and the other two farm types. To avoid overcompensating farmers with large land areas, there were minimum stocking rates for each farm type. Farmers could receive 10% enhanced payments for having low-intensity cattle grazing. As with the HFA there were cross-compliance requirements with environmental legislation, including restrictions on hedge trimming, boundary removal, inappropriate supplementary feeding and over- or under-grazing. The safety net was meant to avoid financial hardship as farmers switched between schemes; it operated by comparing HLCA and LFASS payments and ensuring LFASS did not fall below a set percentage of HFA in a particular year. On top of this, Scottish livestock farmers continued to receive SCPS, BSPS and SAPS until 2004.

LFASS, Scotland

farm type	*rate per hectare (£)*	
	Improved Pasture	*Rough Grazing*
Moorland Farms	30.40	6.10
Southern Upland farms	39.40	7.90
Northern upland farms	45.00	9.00
10% enhancement	2.50 < 0.65LU all land classes 3.00 <0.65 LU moorland farms only	

(Source: SERAD website, accessed 10/10/08)

In Wales *Tir Mynydd* was developed with the objective of supporting and maintaining livestock production to avoid land abandonment and rural depopulation. Like the other LFA schemes it uses area-based payments, but with a 60/40 weighting between SDA and DA land. For 140–640 ha payments were reduced by 35%; over 640 ha they reduced by 70%. Land used for milk production was ineligible for *Tir Mynydd* payments. In 2007 about £23 million was paid out. From 2010 *Tir Mynydd* was subsumed in the new *Glastir* (see Chapter 9).

***Tir Mynydd*, Wales**

	rate per hectare (£)
SDA	28
DA	19
10% Enhancement (1 of) **20% enhancement (2+ of)**	mixed stock enterprise registered organic under UKROFS or in conversion >2% of farm is under cropping stocking rate <1.2 LU/ha sheep are removed from commons for 3+ months/yr >2% of farm under forage, with 1 ha fenced woodland farm is member of Farm Assurance scheme

(Source: Welsh Assembly website, accessed 10/10/08)

Northern Ireland has the Less Favoured Area Compensatory Allowances Scheme (LFACAS). It has two parts: an area-based payment for SDA or DA land, and a 'cattle bonus' for those farmers who have at least 25% of the LUs as suckler cows or heifers. Stocking densities must be at most 0.2LU/ha – this is very low, equating to one breeding female farmed deer over 6 but less than 27 months old; a suckler cow over 24 months equals 1 LU (DARD 2010). The usual cross-compliance regulations exist (GAECs and SMRs) in order for farmers to receive LFACAS. Particular financial rules apply to conacre land, a traditional method across Ireland of renting land for a single year.

LFACAS, Northern Ireland

		rate per hectare (£)
SDA		40
DA		20
SDA bonus of £10/ha DA bonus of £5/ha	}	Farmers maintaining 25% or more of the LUs as suckler cows for environmental and socio-economic benefit

(Source: DARDNI website, accessed 10/10/08)

Box XXXVI: The Sheep Wildlife Enhancement Scheme

Initially, land-management agreements for designated sites (SSSIs and NNRs) were based on profit foregone and paid farmers and land managers not to do certain activities that would have a negative impact on ecological value. This created a great deal of tension between farmers and conservationists, often culminating in acts of deliberate destruction by farmers (Pacione 1995). By the 1990s it was realised that a more positive approach might yield better results. Paying farmers for proactive management, though not the cheapest system to operate (Falconer and Saunders 2002), has reduced tension and encouraged co-operation between land managers and nature conservationists.

This was tested by the Wildlife Enhancement Scheme (WES), a pilot programme in the Devon and Cornwall culm grasslands and the Pevensey Levels, later extending into the uplands of Craven limestone grasslands (Yorkshire Dales) and White Peak (Derbyshire). Each WES area had its own objectives; for example, the Craven Dales focused strongly on the improvement of CG9 and CG10 upland calcareous grasslands through modified grazing regimes (£60/ha) and stock exclusion (£90/ha). Some 3,600 ha of land was enrolled, about two thirds of the total SSSI land (Hooper 1996). Both parties benefited from the scheme: various butterflies and plants re-emerged as grazing pressure reduced, and commoners reported improved livestock condition.

In contrast, the WES on the North York Moors SSSI was a joint project between Natural England, the North York Moors National Park and the owners/occupiers of the central moorland area. The NYM WES was specifically designed to encourage farmers to continue grazing moorland on a site that has very few graziers still operating (pers. comm. R. Pickering, July 2007). The objectives – extending beyond those of the Craven Dales – include a range of habitats (blanket bog, grassland, woodlands, DSH) and management of moorland for the benefit of wildlife and the community, with increased public understanding. Payment rates vary, for instance stocking density restrictions of £7/ha for the first 200 ha, foddering payments at £5/ha for first 200 ha and gathering payments at £1.50/ha.

The WES was effective because each scheme was tailored to local conditions and challenges. In 2003, it was refined as the Sheep Wildlife Enhancement Scheme (SWES), which drew on funding from Article 33 (National Sheep Envelope) of the English Rural Development Programme. The SWES has been used as part of the larger Sustainable Grazing Initiative (SGI) in the Cumbrian uplands (Johnston *et al.* 2005). The aim was to maintain or restore the ecological quality of designated sites throughout the county. It is noteworthy that agreements were not restricted to SSSIs and NNRS, but applied across a whole fell or common so as to balance grazing levels over a massif. By pooling money from different grant schemes (SWES, WES, ESA, CSS and various capital funds), it was possible to allocate £8.7m between 2002 and 2012. In contrast to WES (mainly annual payments for agreed stock management), SWES uses front-loaded capital payments for stock reductions in the first year, followed by four consecutive annual payments for additional management such as off-wintering or shepherding.

The scheme began by focusing on fells that had been badly hit by FMD, working with the various commoners' associations: 'farmers needed to re-establish their business on a sound economic footing and we wanted to ensure that they were able to do this in a way that also encouraged the recovery of important wildlife habitats' (Johnston *et al.* 2005: 8). Between 2000 and 2005 the area in favourable condition rose from 10% to 56%, solely because of changes in grazing management, an increase of 29,000 ha. As with many complex schemes, various practical heft-management problems did emerge, but on the whole the SGI team must be congratulated on working around the plethora of different scheme criteria to achieve a more holistic fell-management approach, which is desperately needed.

Thus the entire programme was evaluated, scheme by scheme and regionally. Finch and Slater (2003) recommended seven changes to the CSS, most importantly its integration with other ERDP schemes to avoid contradictions, better integration with BAP and HAP targets and a more landscape-based approach in relation to contiguous holdings. Similar issues were raised for ESA (see Chapter 8). With respect to the HFA Mid Term Review, Evans (2003) believed that it helped to provide continued support for the hill farming sector, though there were few economic arguments to justify its continuation. However, it did seem to raise more questions than it answered. For example, the HFA supported hill farming families in LFAs, but these were now a minority of the population of these areas; was the HFA thus a disproportionate response? The aims of HFA were unclear; were they socio-economic, environmental or both? He was able to demonstrate that it was more effective at controlling overgrazing than the old HLCA and that its ability to provide environmental public goods was high. In contrast, it was difficult to link directly the HFA to specific environmental outputs due to problems with measuring outcomes, hence additionality[2] was low. Other problems were the high level of deadweight,[3] lack of understanding among farmers of cross-compliance (see below) and the need for better integration with other schemes, though it was complementary.

The MTR concluded that four changes were needed: a greater move towards area-based payments; a ceiling of €300,000 on any farmer's support payments; greater modulation; and cuts in cereal support prices. Along with its own review, the UK Government's response focused on two far-reaching changes derived from EU CAP legislation:

Pillar I – Single Farm Payment Scheme (SPS)

Pillar II – national agri-environment schemes

Pillar 1: Single Farm Payment Scheme

In 2005, SPS replaced all livestock and crop support payments operating under EU Pillar I. This decoupled subsidy from production completely. Instead UK farmers would be paid by the area of land they farmed, not by how it was used. As long as they could be classified as a farmer under Article 299 of the Treaty of Rome, there was no requirement to undertake any agricultural activity.

Areas of land declared as being farmed in 2005 are known as entitlements, of which there are two types:

- Normal – standard payments will be made per hectare of land;
- Set Aside – land laid down as set aside at the time of the entitlement claim, but still classified as farmland under IACS.

If a person holds an entitlement, they can claim SPS provided the land falls into one of the categories listed in Table 8.12 and it is "land which has not been set-aside during this period whether under AAPS or SPS" (Rural Payments Agency 2005b).

Thus it does not cover land taken out of production under other agri-environmental initiatives (Table 8.12) that produce grass-like swards. In the UK the scheme is administered nation by nation. In England, the land is sub-divided again, for payment purposes, thus: outside the upland SDA, in the upland SDA, or moorland within the upland SDA (i.e. above the moorland line).

Farmers are further required to:

> Demonstrate that they are keeping their land in good agricultural and environmental condition and complying with a number of specified legal requirements relating to the environment, public and plant health and animal health and welfare. (RPA 2005a, para. 1)

The extremely detailed and comprehensive requirements are divided into three sets, all of which must be met to receive SPS or any other direct support payments the farmer may be entitled to (RPA 2005 c–f):

- Maintenance of the land in Good Agricultural and Environmental Condition (GAEC), which focuses on soil management and protection, and maintenance of habitats and landscape features. GAECs do not however, take precedence over land entered into agri-environment schemes where the management prescriptions require greater activity.
- Maintenance of permanent pasture at the 2003 national level.

[2] Additionality: unforeseen benefits from a scheme which are beyond the original objectives and targets.

[3] Deadweight: how far a policy and associated funding pay for outcomes that would happen anyway.

Table 8.12: A Summary of Cross Compliance Requirements 2005

eligible	*not eligible*
Arable crops (+ energy crops, protein crops, flax and hemp)	Countryside Stewardship land under options R3, R7, R8, WM2
Potatoes and other roots	Arable Stewardship land under options 4A, 4B, 5
Forage maize or forage rape	ESA Broads Scheme Tier 4B ESA Upper Thames Tier 3c
Temporary grass (leys)	Habitat scheme 1B
Fruit (+ strawberries) – not top fruit	Christmas trees
Vegetables	Tree nursery stock
Greenhouses	Peat production
Multi-annual crops	Forest
Cut flowers & bulbs	Non-agricultural use
Turf	Non-grazeable orchards
Existing set-aside	
Fallow in good agricultural and environmental condition	
Hop yards	
Permanent pasture	

(Source: RPA 2005a.)

- Compliance with Statutory Management Requirements (SMRs) derived from domestic and EU legislation; for example, under the Wild Birds section, farmers are required to comply with any SPA designation and rules, comply with the Wildlife and Countryside Act 1981 and not take game birds in the closed season.

Cross-compliance is enforced through inspection by the Rural Payments Agency (RPA) working with relevant specialist agencies. Farmers demonstrating non-compliance are fined up to 5% reduction in the SPS per infringement. Repeated infringements accrue a 20% penalty, possible criminal action or removal from the SPS altogether. Given the voluminous requirements stipulated in cross-compliance it seems almost impossible for farmers to comply with every part of the regulations. The other issue is that the RPA is not provided with a large enough budget to visit every farm, so a sample is selected at random within each region. GAECs are an interesting component of cross-compliance as they are, in effect, an attempt to introduce baseline environmental management standards into an industry that has previously not been subject to minima. They are typically portrayed as the base of a pyramid on which all other agri-environment schemes sit (Figure 8.6).

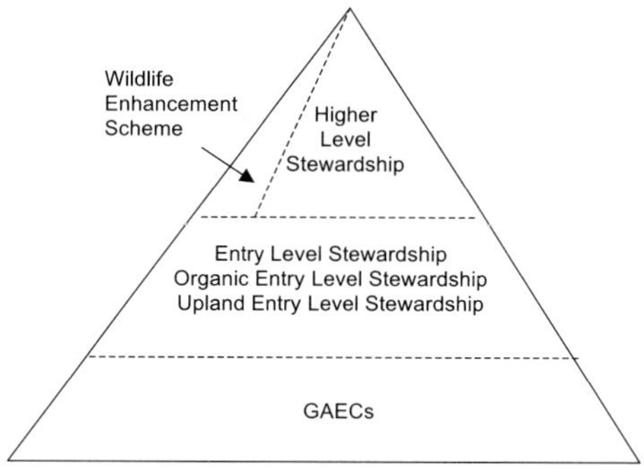

Figure 8.6 GAECs and other Agri-Environmental schemes in England 2005

Rural Development Service, 2005c, *Entry Level Stewardship Handbook – terms and conditions and how to apply.* RDS: London PB10355. This information is licensed under the terms of the Open Government Licence.

Under SPS in England, a farmer receives multiples of a flat rate based on the number of entitlements they have claimed for that year. This flat rate is derived from two sources:

- The Ceiling – a limited pot of money, to be apportioned around the regions of England. This accounted for 10% of SPS flat rate in the first year, 2005.
- The Historic Reference Amount (HRA) – based on the farmer's claims for direct support in 2000–05. This made up 90% of the flat rate in 2005.

The HRA is reducing over eight years to reach zero in 2012, when the SPS is to be 100% ceiling money. The idea is to wean farmers gradually off subsidies based on production rather than land. In Wales and Scotland, historical payments have been retained until 2012, though this is likely to change with the CAP reforms in 2013 (see Chapter 9).

Since its inception, SPS in Europe has come under mounting criticism on eight points (Bureau and Witzke 2010):

- Food security – decoupling provides no incentive to farmers to grow more food despite growing global food security challenges (see Chapter 9);
- Obstacles to competitiveness and bureaucratic burden – cross-compliance has put unfair restrictions on EU farmers that non-EU farmers do not have;
- Limited environmental record – farm systems that supply the highest biodiversity face high compliance costs and lower payments compared to lower-diversity farm systems; also, agricultural output is intrinsically linked to public goods, so decoupling means the latter will decline;
- Transfer-inefficiency – little reaches the actual producer, particularly the labourer; more goes to the landowner;
- Unfair distribution within the EU – most money goes to the largest farm businesses, though some states counter this view by saying that SPS is propping up marginal businesses unnecessarily;
- Unfair distortions for third countries – supported EU agriculture unfairly competes with developing countries, limiting their ability to export, particularly if SPS still does this despite GATT;
- High budgetary burden – SPS is expensive, however where CAP had accounted for 80% of the entire EU budget previously, it is now down to only 32%;
- Reformed CAP neglects rural vitality – any decoupling measure will reduce agricultural activity, creating rural depopulation, a phenomenon which seems to vary regionally (examples cited in Bureau and Witzke 2010).

There are also specific regional issues. In England for example, common land is particularly contentious. To be paid entitlement for SPS on a common, the farmer must first have the legal right to use the land. In 2005, this right was established by the Commons Registration Act 2005 (see Chapter 3). The effect was that some commoners could not claim SPS because their interests were not on the 1965 register, but had been legally established between 1965 and 2005 (Federation of Cumbria Commoners website, accessed 06/08/08). Negotiations to resolve this continue. Once legal right of use is established, payments are claimed for any right of common related to grazing (including horses). Area is calculated from the area of the common, the proportion of rights the applicant has and the number of commoners eligible to claim. An owner can claim any entitlement that becomes surplus and is not claimed by registered commoners. Other problems have occurred with misclassification of the type of land the common is (e.g. SDA non-moorland, yet is above the moorland line) and the use by commoners of local names for pieces of land rather than those that appear on the 1965 register (Cumbria Commoners website, accessed 06/08/08). There are also issues where a commoner is not a farmer and thus has no entitlements.

Pillar II: Agri-environment initiatives

Pillar II funding concentrated on combining environmental management and rural development, and different parts of the UK opted to address this in different ways (Box XXXVII). In England, a new Environmental Stewardship Scheme (ESS) replaced ESA and CSS. This was brought in early to offset the reduction in direct support that SPS was considered to bring.

ESS was piloted in four areas (for four farming systems) at entry level in 2003; the upland area was around Barnard Castle, Co. Durham. Boatman *et al.* (2004) found that upland farmers were keen to enter the pilot, but many thought their land did not have enough suitable features to meet the points target. Investigation revealed widespread lack of knowledge and misunderstandings. Many upland farmers had not fully read and understood the scheme booklet (only 30–50% had read the handbook, compared to 90% on average). Boatman *et al.* therefore recommended that there be upland events to demonstrate the range of options. The RDS later published guidance notes listing specific possibilities for farmers in each Joint Character Area. For example, in the Howgill Fells (JCA 018) the guidance suggested access, species-rich meadows, broadleaved woods, wall restoration and protection of Scheduled Ancient Monuments (RDS 2005b).

With various changes the scheme was rolled out to the rest of England in 2005. The initiative is, in fact, four separate grants :

- Entry Level Stewardship (ELS) – a 'broad and shallow' mechanism to get the majority of farmers and landowners involved, with a relatively low level of management commitment across the whole farm;
- Organic Entry Level Stewardship (OELS) – subsuming the Organic Farming Scheme, with similar aims to ELS, but with additional payments to support organic management;
- Higher Level Stewardship (HLS) – a 'narrow and deep' scheme to target farmed landscapes with the highest conservation potential, resulting in fewer applications but of greater environmental value;
- Upland Entry Level Stewardship (UELS) – this replaced HFA in 2010 and rewards farmers for delivering landscape and environmental benefits in LFAs (see Chapter 9).

The design of ELS and OELS was particularly influenced by the need to make them straightforward enough for farmers to apply without professional help. In practice, many farmers have employed agents and consultants anyway.

The overall aim of the Entry Level Scheme is to:

> Encourage large numbers of farmers and land managers across England to deliver simple yet effective environmental management that goes beyond the Single Payment Scheme (SPS) requirement to maintain Good Agricultural and Environmental Condition. (RDS 2005a: 5)

Its objectives cover wildlife conservation, maintaining and enhancing landscape quality and character, historic environment protection and resource protection. Voluntary agreements run for five years and land managers receive a flat rate payment of £30/ha per year. Until 2007, there was a special rate of £8/ha per year for parcels of land in the LFA of 15 ha or more, since limited to land above the moorland line. Every farmer who enters the scheme is given a points target (30 points per hectare, 8 points for LFA land). Each option has a certain number of points attached to it; for example, management of rush pastures (EL4) is worth 60 points per hectare and unenclosed moorland rough grazing (EL6) is worth 5 points. Points scores relate to the increase in practical management required and the impact this may have on farm management, on the basis of income foregone.

Box XXXVII: Agri-environmental initiatives in Scotland, Wales and Northern Ireland, 1986 to 2007

ESAs were launched in Scotland in 1987. Running parallel from 1996 was the Countryside Premium Scheme. Similar to CSS in England, it was discretionary, competitive and not area-based but targeted on environmental enhancement, rather than paying for the *status quo* (CRER 2002). With the Rural Development Regulation in 2000, the two initiatives were re-vamped and blended together, with changes to avoid undesirable aspects of earlier schemes:

- Tier 1 no longer selected under ESA, which had allowed farmers to make few changes to farm management,
- Removal of a pick-and-mix menu, which again farmers were using to minimise changes in farming practice,
- Declaration of farm ownership now required in environmental audits and plans,
- No more loosely worded management prescriptions that could be interpreted in various ways,
- Geographical targeting was no longer seen as fundamental to protecting the Scottish landscape because vulnerable areas were not location-specific. (CRER 2002)

The Rural Stewardship Scheme, unveiled in 2001, aimed to encourage land managers (farmers, crofters and common grazing groups) to use environmentally friendly practices to maintain certain habitats and landscapes. Successful applicants had to comply with basic environmental standards on farming. Of the 23 management options, five were based on moorland landscapes (RRS website, accessed 1/08/08):

- Moorland management to enhance biodiversity
- Stock disposal for heather and mosaic moor enhancement
- Muirburn and heather swiping to produce heather age mosaics
- Bracken eradication on moorland
- Capital works for black grouse breeding areas and drystone walling

Anyone applying under these options was required to produce a moorland management plan reporting on the moor's condition, its existing management and proposals for change. Selection was discretionary, based on the range of activities proposed and how they helped meet local and national BAPs (see Box XXI). Managers were contracted for five years and received both annual and capital payments. By the end of 2006, 461,000 hectares were under agreement on 6,200 farm holdings (DEFRA 2007).

Wales chose a different approach. In 1986 the Countryside Commission launched the pilot *Tir Cymen* ('tidy land') in three areas: Meirionnydd, Dinefwr and the Gower. There were mandatory whole-farm and annual habitat-specific payments, and optional extras. The ten-year agreements accounted for 89,000 ha by 1999 (DEFRA 2004). In 1999 *Tir Cymen* and the ESA were merged as *Tir Gofal* ('land in care'), operated throughout Wales by the Countryside Council for Wales. By 2007 it covered 364,000 ha (DEFRA 2008). *Tir Gofal* was designed to:

- benefit wildlife on agricultural land through restoration and creation of habitats, and positive management of existing ones,
- protect characteristic rural landscapes and promote and restore related landscape features,
- protect and enhance the historic environment by compatible farming practices,
- integrate delivery of public access to the countryside.

Applications were subject to a discretionary scoring system, needing to reach 100 points made up of whole farm, habitat enhancement/creation and capital works per hectare/metre. Farmers entered into 5-year contracts with 17 mandatory and 26 voluntary management prescriptions. The table below shows the upland prescription rates for 2006:

Upland prescriptions and payment rates for *Tir Gofal*, 2006

prescription	*mandatory or option*	*payment rate/ha/yr (£)*
Whole farm	M	up to 3475
Upland heath	M	50

Unimproved unenclosed grassland:
    <200ha
    200–500ha
    500+ ha
	M	37.50
		30
		15

Unimproved grassland commons: <200 ha >200 ha	M	37.50 30
Unimproved neutral grassland	M	80
Unimproved limestone grassland	M	110
Upland marshy grassland	M	50
Raised and blanket bog	M	35
Restoration of semi-improved to unimproved grassland	O	105
Restoration of heath vegetation on acid grassland	O	95
Restoration of heath vegetation on improved farmland	O	215

Capital works:
    Drystone walling
    Heather cutting
    Heather burning
    Heather restoration: seed & mulch
    Bracken control: mechanical
    Bracken control: chemical
	O	22.50/sq. m.
		60
		130
		235
		60
		120

(Source: Welsh Assembly Government website, accessed 01/08/08)

In 2005, *Tir Cynnal* ('constructed land') was introduced, filling a similar niche to ELS in England. It was designed to encourage farmers to protect wildlife habitats, traditional landscapes, ancient monuments and historic features while reducing pollution and maintaining rights of way. It provided environmental grants to small and medium-sized farm businesses that found the entry standards of *Tir Gofal* too demanding or whose production was too intensive. These whole-farm agreements ran for five years with annual payments per hectare, which decreased as the area increased (the first 20 ha received £45/ha/yr, whereas every hectare over 200 ha received only £5/ha/yr). By 2010, *Tir Cynnal* covered 200,000 ha of Wales.

Five ESAs covering 20% of the land area of Northern Ireland were developed from the late 1980s and were considered successful in raising environmental awareness among the farming population, having 148,000 ha under agreement by 2002 (DARD 2006, DEFRA 2008). The programme was then re-launched, aligning it more closely with the Countryside Management Scheme (CMS). This was introduced in 1999 to allow farmers outside the ESAs to join in enhancing biodiversity, landscape and heritage in Northern Ireland. Administered by DARD (the Department of Agriculture and Rural Development), both initiatives used a whole-farm ten-year agreements combining mandatory and optional activities related to habitat and landscape.

There were several features unique to ESAs and CMS in Northern Ireland which warrant brief attention. First, the two schemes were identical in management, except that ESAs were geographically specific. Second, all farmers had to manage *any* identified habitats on their farm in line with set prescriptions to maintain and increase its extent *and* in line with BAP targets. For example, if moorland (defined as dry or wet heath, blanket bog and degrade heath) existed, prescriptions of grazing-season length and stocking density had to be applied, the latter varying by animal and habitat type (DARD 2006); peat cutting for domestic use was limited to 0.1 ha per year and the removal of western gorse (*Ulex gallii*) was not permitted. Third, once an application was received, DARD staff completed an audit, mapping farm habitats, offering waste-management advice and suggesting habitat options. Particularly innovative was that payments were the same for both schemes, and farmers had to attend agri-environment training workshops and implement a farm waste development plan. By 2006, ESA agreements covered 131,000 ha and CMS 317,000 ha (DEFRA 2008). The CMS programme has continued in the new RDP (2007–13), retaining many of its former features and becoming the main Axis 2 agri-environment scheme for Northern Ireland, equivalent to ESS in England (DARD 2007).

This increased complexity was an issue raised in evaluation of the pilot study because managers were opting for simpler activities (Boatman 2004). Applicants are provided with (RDS 2005a):

- Personalised application form – which records option details in tabular format and allows the manager to select an agreement start date from one of four points in the year;
- Environmental Information Map (EIM) – which shows existing features on their land of historic, landscape or wildlife interest;
- Farm Environmental Record Map (FER) – this is used to record existing farm features like landscape, buildings, habitats and soil erosion risk areas using a set of coloured pencils;
- Options map – on which to draw what they would like to do, using the same pencils.

Once completed, the form and maps are returned to the RDS and, if the points target is appropriately met, an agreement letter is issued.

The aim of the Higher Level Scheme is to:

> deliver significant environmental benefits in high priority situations and areas. (RDS 2005e: 5)

HLS has five main objectives: wildlife conservation, maintaining and enhancing landscape quality and character, historic environment protection, natural resource protection and promoting the public's access to and understanding of the countryside. Each of these has regional and specific JCA targets. HLS has two subsidiary objectives of flood management and conservation of genetic resources. Payments are higher than for ELS because the active management is more complex and demanding (e.g. restoration or creation projects) and the impacts on farm management systems will be greater and require more changes. Other differences are that HLS agreements are for ten years, are not necessarily whole-farm agreements and are competitive, selecting the best projects for the most efficient use of limited funds.

To decide which management options are best for a farm there is a formal assessment, a Farm Environment Plan (FEP). Completing it is time consuming and complex, and requires a level of environmental knowledge beyond most farmers (Table 8.13), so most are completed by land agents or consultants. FEPs are, in effect, a detailed audit of environmental features on the farm, followed by habitat management planning. If the application is successful, there is financial assistance for producing an FEP, based on area; for instance, less than 5.5 ha gets £395, over 3000 ha gets £3,350 (Natural England 2008). A scoring system identifies the best applications; payments vary depending on the management complexity. For moorland and upland rough grazing there are five main options with three supplementary prescriptions (Table 8.14). Payments are substantially lower than those of the lowland management options, but are designed this way to take into account the lower incomes foregone in upland farming systems.

Table 8.13: Summary of Components in a Farm Environment Plan

main part	*details*
1 Farm and farmer details	Farm holding information such as field combinations, areas and numbers
2 Environmental Features Data Sheet	A baseline audit of the quantity and quality of the features.The whole lot is coded in the FEP manual. It also asks for the HLS management options
3 Farm Scale environmental features and potential for creation of new features	This section is designed to cover those issues that are best considered on a farm scale rather than an individual field basis. It covers: • Landscape – JCA areas • HLC (Historic Landscape Characterisation) if it exists • Field Boundaries • Biodiversity • Access • Natural resource Protection • Flood and Coastal Management • Genetic conservation • Woodlands and orchards
4 Other relevant information	Eg SSSI, significant features of interest
5 Summary of FEP findings and management proposals	
6 Record of Consultations and FEP check list	

(Source: Natural England 2008d)

By 2008, 4.3m hectares of land had been enrolled in ELS, representing 51% of the farmland in England. In comparison, 178,500 ha of land were under HLS agreements. Evaluation by CSL (2007) showed that only 12.7% of ELS farms are in LFA areas (16,749 farms in the SDA, 8,196 farms in the DA). Of course, much upland was (and still is) covered by earlier ESA, CSS or WES agreements. As these 'come to term', farmers will hopefully switch to the new schemes.

ELS agreement holders tend to be arable farms in eastern England with the largest units (47% were over 150 ha). Cereal farms have the most ELS options in the scheme (8.6) and LFA stock farms the fewest (4.8). Farms over 150 ha have twice as many ELS options as small farms (under 50 ha). Boundary options are the most popular, the top two being hedgerow management (EB1 at 55.9%, EB2 at 62.8%). However, of the 61 ELS options, only eight reached a 30% adoption rate (Figure 8.7) and only one option, EB1, was above 30% of LFA stock farmers. As Figure 8.7 demonstrates, there is little similarity of options between LFA and non-LFA farms. Although this is understandable for the more specific upland types EL1 to EL6, other options are more generic, such as producing a range of land management plans to address set environmental farm issues or boundary management activities. This can be interpreted in various ways. Perhaps LFA farmers already do these activities or do not deem them relevant to their land, or they consider the practices too restrictive (such as those relating to fertiliser use) or simply have yet to recognise the opportunity, as Boatman *et al.* (2004) found in the pilot. Of course, EM4 would be fairly rare in the uplands and option EK2 is very similar to EL2. In fact the EM options were withdrawn in 2008 because they were actually already part of cross-compliance under EU rules (Frances Dalton, pers. comm., 9/9/2009).

Table 8.14 : Some Upland options available under Higher Level Stewardship

option	*code*	*overview*	*payment £/ha*
main prescription			
Maintenance of moorland	HL9	Grazing to an agreed stock calendar depending on vegetation mosaic and condition	20/40
Restoration of moorland	HL10	Grazing to an agreed stock calendar, grip blocking, temporary fencing to exclude or reduce grazing	40
Creation of upland heathland	HL11	Site preparation through various processes, establishment of dwarf shrub heath species, control of grass or weeds	60
Maintenance of rough grazing for birds	HL7	Targeted at breeding waders on allotments, intakes, newtakes and semi-improved fields. Stocking density limited to 0.4 to 1.0 LU/ha from 31/3 and 20/6, with densities higher at other times to control sward height	80
Restoration of rough grazing for birds	HL8	As above plus grip blocking to create flushes or wet marshy grassland	80
Restoration of species-rich semi-natural grassland	HK7	Scrub clearance, invasive weed control, relevant seed introduction along with grazing/cutting but no ploughing, re-seeding or new drainage	200
Creation of species-rich semi-natural grassland	HK8	Ley grassland conversion, type controlled by environmental conditions of site using local seed source. Cut year 1 for tillering and then grazed/cut where appropriate	280
supplements			
Supplement for management of heather, gorse and grass by burning, cutting or swiping	HL12	To aid rotational vegetation management where significant change is needed	7
Seasonal livestock exclusion	HL15	To encourage moorland restoration by keeping stock off	10
Moorland re-wetting	HL13	To support re-wetting and help reduce diffuse pollution run-off	10
Native breeds at risk grazing	HR2	To encourage use of appropriate native breeds to help achieve other options	<70
Haymaking	HK18	To support the re-establishment or continuation of the process where it is uneconomic	75
Bracken control	HR5	To cover the costs of follow-up management of bracken control	35
Small fields	HR6	To compensate for the higher levels of management required, fields must be less than 2ha in size	35

(Source: Natural England, 2008c)

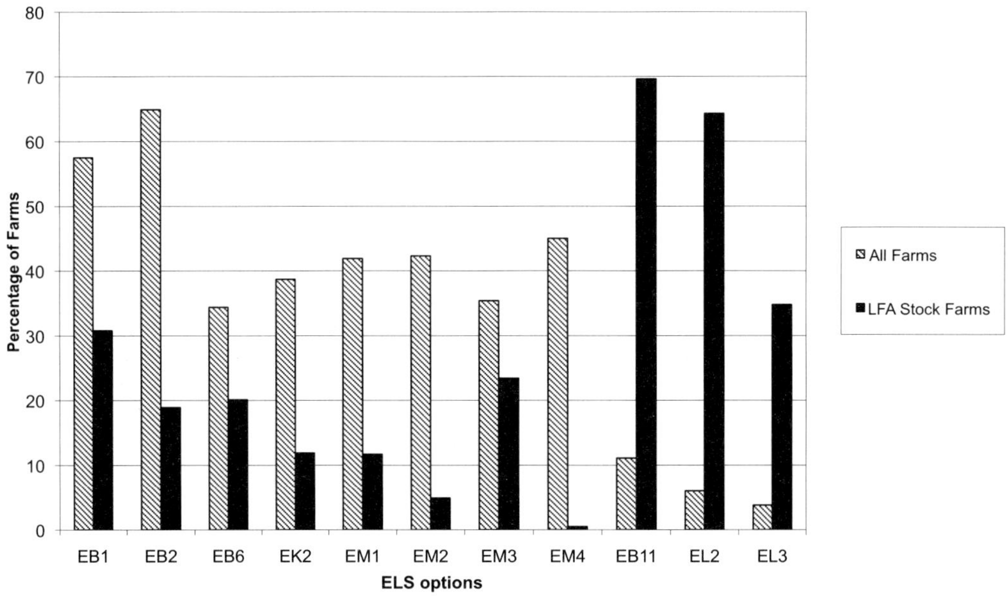

Key:
EB1 – Hedgerow management both sides
EB2 – Hedgerow management one side
EB6 – Ditch management
EK2 – Permanent grassland with low inputs
EM1 – Soil management plan
EM2 – Nutrient management plan
EM3 – Manure management plan
EM4 – Crop protection management plan
EB11 – Stone wall protection and maintenance
EL2 – Manage permanent inbye grassland with low inputs
EL3 – Managed inbye pasture and meadows with very low inputs

Figure 8.7 ELS Options Adopted by at least 30% of Farmers 2006
Adapted from data derived from Boatman ND *et al.*, 2004.

These issues were picked up in an ESS review by Natural England, which administers and manages the scheme. Among its recommendations were:

- the introduction of capital item payments (non-existent in the ELS);
- a more comprehensive advice system for ELS applicants to improve quality and appropriateness of management;
- option changes, making prescriptions more or less restrictive and adding new ones (eg an ELS option to encourage hay making in uplands);
- a possible split menu approach for ELS, so land managers would have to choose X amount from list A or B, to create more coherent environmental management in a single application;
- some form of geographical targeting of the ELS to help with regional biodiversity targets;
- simplifying the FEP in the HLS. (Natural England 2008)

Evaluating the environmental effects of the scheme is essential to meet the objectives of the ELS, but so is the financial impact on farmers. Wallis and Jones (2008) completed a detailed analysis of seven Teesdale upland farms. They noted how the ELS, under EU rules, is not allowed to provide farmers with a profit; the payments are there to offset profit forgone, the costs of application and managing prescriptions, and labour time lost from farming. Though these rules are strictly applied by DEFRA, in practice it is feasible to make a marginal profit because of the variable nature of farming and individuals. Wallis and Jones showed that all farms made a net financial gain from joining the ELS, when the costs of direct revenue change, farmers' own time and environmental work were calculated. On average the resulting profit margin was over half the ELS payment. Further detailed analysis by prescription showed no overall problems of profit forgone, though it was an issue in hedge and wall management, inbye management and de-stocking.

The period 2000 to 2006 was dominated by the move away from production support to one of rural development and environmental stewardship. For upland farmers, direct support came from SPS and HFA (and its national equivalents), which do not encourage overproduction and overstocking. Upland farmers could also draw on a range of environmental and rural economic initiatives (see Chapter 6). A particularly positive step was the drive to increase use of Joint Character Areas to inform scheme targets, which helped to reinforce the need for varied management practices, massif by massif. The Mid Term Review led to re-packaging of environmental schemes into a single coherent project, but it failed to address the continued reliance of upland farmers on HFA and SPS payments. The lack of a threshold for SPS payments or a scaling approach has exacerbated the restructuring of upland holdings, with larger units becoming even larger.

Summary

The politics of upland agriculture have rumbled on as long as there has been political intervention in UK agriculture. Responses to the agricultural depression from 1870 to 1940 shaped British agricultural policy, designed to solve economic and social hardship in uplands, for the rest of the 20th century, alongside farm restructuring to increase production. Deficiency payments and improvement grants started with the Hill Farming Act 1946 and continued after EU accession in 1973 through the Less Favoured Areas Directive. Pressures inside and outside the EU forced a radical rethink of agricultural strategy in the late 1980s. By the end of the 1990s production was no longer king, despite much of the CAP still focusing on Pillar I and production. It took another decade before this was really addressed at farm level, with the protectionist and productivist policies of the 1970s and 1980s being dismantled.

The shift to rural development and environmental management, as the central philosophy of EU policy, has meant that livestock production is no longer valued as the core farming activity of uplands. Many farmers may see this as a betrayal of trust in their primary role as producers of food; but upland farmers, in the midst of HNV landscapes, have greater opportunities than their lowland cousins to exploit this new policy regime. They may not be able to continue to farm at the same level, so simple economic and social survival means they need to re-invent their businesses. This will allow them to maintain their lifestyle and continue stock production, albeit at lower intensity. These new opportunities are the subject of the final chapter.

9

A future for upland agriculture

This book has focused on the relationship between upland agriculture and the environment. The association that has developed has not been straightforward, and it has left upland farming on the margins in all senses.

The upland farming system has been fundamentally determined by the physical environment. Working within harsh climatic, topographical and soil conditions, enterprise choice is always restricted to livestock of some kind. These environments are also easily damaged because carrying capacities are typically lower than in lowlands, so good stock management is central to sustainable farming. Extensive grazing regimes have gradually emerged as farmers have developed special management processes like transhumance, hefting and intercommoning. These have allowed low-intensity exploitation of the uplands, underpinned by strong social capital, which has persisted in a fragmentary nature until today. The product has been a series of highly prized HNV landscapes in the sub-alpine zone, merging into limited natural alpine vegetation. This is the absolute advantage uplands have, but these landscapes are the product of livestock management, in which uplands have only comparative advantage.

Responses to pressure

From the mid-18th century, pressure to increase production has encouraged inappropriate management practices which have had negative effects on the upland ecology. In hindsight we can recognise a tipping point in the 19th century, when the biophysical capacity of uplands began to be exceeded, typified by the Highland Clearances.

Increased agricultural competition from the lowlands, both here and abroad, better returns from other land uses and poor land management together pushed upland farming to the economic margins. This was amplified by inadequate tenancy arrangements and demographic push–pull factors, which created economic and social deprivation. Compounding these issues were agricultural management challenges posed by the land itself: poor drainage, poaching, infertility and erosion. The combined effects pushed upland farming to the edges of the the agricultural sector as its comparative advantage dwindled, labelling it a ‘problem’.

Successive governments, national and EU, have attempted to alleviate the economic and social consequences of this decline. Throughout the 20th century, intervention in upland agriculture focused on raising farm efficiency and profit margins, in the hope that this in turn would maintain a viable upland population. An amalgam of agricultural policies was tried as a strategy to support production and re-structuring of uneconomic units. For a time, subsidies did support farming incomes. Unfortunately, the result was inappropriate management and overstocking. Many biophysical carrying capacities of the system were exceeded; a symptom of which was, and still is, declining biodiversity quality and quantity.

This strategy protected the sector from the worst excesses of market forces, though upland farms continued to suffer from low gross margins, inherent to the physical and economic marginality of their businesses. Farmers responded with a range of re-structuring options, such as intensifying production or seeking other sources of income on- or off-farm, but these options were less viable in the uplands because of inherent physical limitations, the reduced labour force (to lessen overheads) and remoteness. The mounting economic and social hardship which has developed as a result has started to drive farmers from the land. This has weakened some aspects of farm management, notably hefting, and with it social capital, which is fundamental to these systems. For those who remain, upland farming has become a less attractive lifestyle for their children, creating a looming problem of succession.

Parallel to this developing economic and social crisis in upland farming has been an escalating political and public crisis for agriculture in general. Overproduction and related disposal costs have become a huge issue for the EU. Global political pressure driven by GATT (the General Agreement on Tariffs and Trade) and its successor, the WTO (World Trade Organization), has demanded elimination of the inflated protectionist market. A series of public health scares and livestock welfare issues have forced a re-evaluation of how food is produced, processed and marketed. These issues coalesced in the launch of Agenda 2000 and the first RDR (Rural Development Regulation).

As a result, general agricultural support has switched focus from production to rural development and environmental management. In the academic literature this re-alignment has been viewed in two ways: either as restructuring agriculture from productionism to post-productivism (Box XXXVIII) or as a move towards multifunctionality (see below). There is no doubt that the face of how we farm and what we produce is changing; determining what is driving that change is more difficult. The driver(s) responsible will vary from farm to farm, as we have seen throughout this book. For some farmers the driver may be family circumstances, for others the cost of production or reliance on subsidies. Any change may be a response, a cause or both; such is the complexity of agricultural change. Whatever the reasons for the individual business, the effect has been to push European agriculture towards non-food production. Farmers will react to this situation in various ways, as Lobley *et al.* (2002, 2004) showed in their restructuring spectrum (Table 6.7).

Upland farmers, already on the physical and economic margins, may see the shift away from production-led farming as catastrophic. It will be if upland farming continues to view livestock production as its *raison d'être*, slowly boxing itself into a corner. Yet there are opportunities that upland farmers can exploit, that perhaps their lowland counterparts cannot. In this final chapter we investigate these new opportunities and consider threats that might stop upland agriculture from flexing enough to exploit them. We start by looking at the current policy framework for upland farming and then move on to consider the possibilities.

Box XXXVIII: Theories of post-productivism

Over the last few years a debate has emerged in agricultural change research that focuses on the topic of post-productivism (e.g. Shucksmith 1993; Winter 1997; Bowler and Ilbery 1998). At its simplest the concept of post-productivism is viewed as the antithesis of production, that is, a move away from production to less intensive activity.

The academic literature focuses on two particular issues, which seem to have exercised researchers. First, what is the character of post-productivism in agriculture, and second, more fundamentally, whether it exists at all with respect to agricultural landscapes. There have been two notable attempts to describe the character of post-productivism. First, Ilbery and Kneafsey (1997, quoted in Evans *et al.* 2002) identified the following features:

- a move from quantity of production to quality of food,
- a rise in alternative farm enterprises,
- more effort by the state to develop traditional farming systems, using agri-environment initiatives to encourage sustainable farming,
- an increase in environmental regulations,
- the restructuring of government support away from production.

Later work by Ilbery and Bowler (1998) suggested three bipolar aspects of the transition from production to post-productivism: intensification to extensification; concentration to dispersion and specialisation to diversification.

Using these two typologies, there is evidence to indicate that upland agriculture could be moving towards post-productivism. The use of Environmentally Sensitive Area and Countryside Stewardship agreements (Morris and Young 1997) demonstrates Ilbery and Kneafsey's third element; environmental cross-compliance in the new Single Payment Scheme in the UK (Rural Development Service 2005) illustrates the fourth element and the switch from headage to hectarage payments for upland livestock (Drew Associates 1997, MAFF 2000) exemplifies the fifth element. With respect to the bipolar transitions, Walford (2002) suggests that hill farms may be more responsive to adopting post-productivist behaviour.

Empirical evidence of the three bipolar processes is less convincing, because few studies have specifically investigated upland farming systems and post-productivism, a notable exception being a beef cattle study by Evans *et al.* (2003). In the wider agricultural debate more headway has been made in the search for evidence of post-productivism. For example, Walford (2002, 2003) tried to determine the existence of post-productivism in south-east England and found that, by and large, farmers were using agri-environment initiatives and schemes such as set-aside as a management tool to support agricultural production, rather than as a move away from production. Slippage has also occurred where farmers simply extensify production on scheme land, but intensify on land not under agreement (Potter 1988, Saunders 1994).

Evans *et al.* (2002), in their comprehensive critical review of post-productivism, reject many of the features identified by Ilbery and Kneafsey (1997, in Evans *et al.* 2002). For example, all food, whether small-scale or mass-produced, has undergone quality improvements, that the growth of pluriactivity on farms has started to decline, and that in some places specialisation of agricultural activity has increased (Evans 2000, cited in Evans *et al.* 2003). Evans *et al.* (2002) go on to say that many EU policies – like the Extensification, Agri-Environment or Rural Development regulations – are not there because of genuine concern for environmental amelioration, but to reduce CAP budgetary demands and address GATT free trade agreements. The mounting empirical evidence disputing post-productivism has led some researchers to the view that the concept is, in fact, a mythical beast (e.g. Walford 2003).

Underpinning this is the theoretical argument put forward by Wilson (2001) that the relationship between production and post-productivism is non-linear: that is, the latter does not necessarily follow the former, as the terminology and characterisation of Ilbery and Kneafsey (1997, in Evans *et al.* 2002) and Ilbery and Bowler (1998) suggested. Wilson recommends that instead we should be explaining the recent changes in agriculture as moves to a 'multifunctional agricultural regime', creating patterns of uneven development (in line with the earlier work of Marsden *et al.* 1986). Others believe that the whole concept of post-productivism is a cul-de-sac and that researchers should be looking toward other human geography theories to explain agrarian change (Evans *et al.* 2002).

The current policy environment

Drawing on experiences from the first RDR, the MTR and the rural development programme LEADER+, the EU ratified Council Regulation (EC) No. 1290/2005 on financing the common agricultural policy. It allowed money to be channelled from agriculture directly to rural development. The new Rural Development Regulation for Pillar II, Council Regulation (EC) No. 1698/2005 on the European Agriculture Fund for Rural Development (EAFRD), links up with the European Agricultural Guarantee Fund (EAGF) for Pillar I. The EAGF component provides joint funding by the EU and the member states to support:

- refunds for exporting farm produce to non EU countries,
- intervention measures,
- direct payments to farmers,
- international promotions for farm produce,
- restructuring measures for the sugar industry,
- various veterinary activities,
- promotion of farm produce by the EU collective,
- conservation of the genetic resource of farming,
- farm survey systems (e.g. IACS),
- fisheries market issues. (Europa website, accessed 07/08/08)

The most important feature of the EAGF for UK farmers is that SPS continues.

Under Pillar II, the EAFRD has three main objectives:

- increasing the competitiveness of agriculture and forestry,
- supporting land management and improving the environment,
- improving the quality of life and encouraging diversification of economic activities. (European Communities 2006)

New strands of this package include: access to public services, coverage of all EU territory, cohesion, stakeholder participation, partnership and simplification. To achieve all this, the LEADER principle has been integrated into the EAFRD (see Chapter 7) using an axis approach (Figure 9.1). Underlying this is a range of measures (Table 9.1), many presenting opportunities for upland farmers, with 212 designed specifically for them. Britain's budgetary allocation is 1.9 billion euros (about £1.6bn) with each part of the UK delivering its rural development programme with slightly different objectives in mind (Table 9.2). Which measures are adopted and offered also varies. For example, in England DEFRA chose not to apply several measures at all:

- 112 Setting up young farmers
- 113 Early retirement
- 131 Adapting to Community standards
- 132 and 133 Participating in food quality schemes
- 126 Restoring agricultural production

Their justification for not including 112 and 113 was based on a report by ADAS (2004) and entry/exit trends in 2005, which found no evidence of major barriers to entry and no market failure in the processes of entry and exit for farmers. DEFRA (2007b) therefore concluded that financial support for entry and exit would probably be deadweight (it would happen anyway, regardless of financial support) and that the real cause, low income, could not be addressed by these measures. In contrast, the Scottish RDP (SRDP 2007) recognised the value of a young entrant's scheme, because new farmers often lacked cash for capital outlays at start-up.

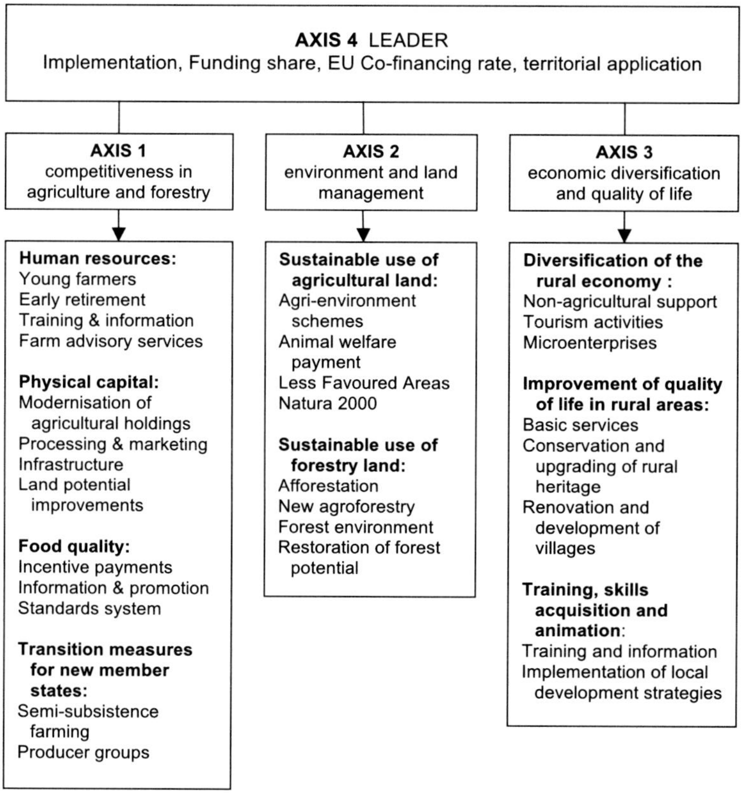

Figure 9.1 The Axes Approach of the EAFRD 2005
Adapted from DEFRA, 2007a.

Actual delivery also varies. In England the RDP operates as follows:

- Axis I and III delivered by Rural Development Agencies and individual projects are considered against single measures;
- Axis II delivered through the HFA and ESS (see Chapter 8);
- Axis IV varies from one Rural Development Agency to another via a 'Leader' method with LAG coverage geographically uneven and even delivery is measured in different ways.

In contrast, the Scottish RDP has identified five main themes it wishes to address: revitalising communities, rural community capacity, new markets and products, progress in the rural economy and conservation of the rural environment. To achieve these within the axes and measures, three main cross-cutting elements have been designed:

- Rural Development Contracts – made up of a combination of:
  1. Land Management Contract Menu Scheme (LMCMS) which has set minimum payments based on the amount of land a person manages:

     £109.50/ha – first 10 ha

     £43.80/ha – next 90 ha

     £1.46/ha – next 900 ha

     £0.15/ha – above 1,000 ha

Table 9.1 The Measures under EAFRD

measure			North West	England	Scotland	Wales	Northern Ireland
	code	*title*					
AXIS I	111	Vocational training and information actions	X	X	X	X	X
	112	Setting up young farmers			X		
	113	Early retirement					
	114	Use of advisory services	X	X	X	X	
	115	Setting up of management, relief and advisory services	X	X	X		
	121	Modernisation of agricultural holdings	X	X	X	X	X
	122	Improving the economic value of forests	X	X	X		
	123	Adding value to agricultural and forestry products	X	X	X	X	X
	124	Co-operation for the development of new products	X	X	X	X	X
	125	Infrastructure related to development and adaptation of agriculture and forestry	X	X	X		
	126	Restoring agricultural production					
	131	Adapting to community standards					
	132	Participation in food quality schemes			X		
	133	Participation in food quality schemes					
AXIS II	212	Payments to farmers in areas with handicaps, other than mountain areas		X	X	X	X
	214	Agri-environment payments		X	X	X	X
	216	Non productive investment		X	X	X	
	221	First afforestation of agricultural land		X	X	X	X
	223	First afforestation of non-agricultural land		X	X	X	
	225	Forest-environment payments		X	X		X
	227	Forestry support for non-productive investments		X	X	X	
AXIS III	311	Diversification into non-agricultural activities	X	X	X	X	
	312	Support creation and development of micro-enterprises	X	X	X	X	
	313	Encouragement of tourism	X	X	X	X	
	321	Basic services	X	X	X	X	X
	323	Conservation and upgrading of rural heritage	X	X	X	X	
	331	Training and Information	X	X	X	X	
	341	Skills acquisition, animation and implementation of local development strategies			X	X	

(Sources: NWDA 2007, DEFRA 2007, SRDP 2007.)

Table 9.2: Budgetary allocations and RDP Objective in the UK 2007-2013

nation	*strategies employed*	*budget (million euros)*			
		Axis 1	*Axis 2*	*Axis 3*	*Axis 4*
England	To build profitable, innovative and competitive farming, food and forestry sectors To improve the environment and countryside To enhance opportunity in rural areas, in a way that harnesses and builds on environmental quality	447.5	4,182	334.3	219.5
Northern Ireland	Focus on vocational training, farm modernisation and supply-chain improvements Support for LFAs, agri-environment and forestry Diversification of the rural economy, cultural heritage and promotion of tourism	35.5	187.5	No information	99.9
Scotland	Support business viability, add value to the rural economy and increase market facilitation Integrated approach to meet environmental objectives Encourage private enterprise, improve services and infrastructure at local level, support Scotland's culture Support capacity building and innovation	305.9	1,468	247.7	107.5
Wales	Stimulating a dynamic and innovative agriculture sector Encourage sustainable production methods with a view to improving the environment Improving the quality of life in rural areas and encouraging diversification of the rural economy Building capacity and innovation in rural areas	121.9	721.6	93.9	46.6

(Source: www.Europa.eu.com, accessed July 2008.)

After this, land managers can take on additional management prescriptions (Land Manager Options) of which there are 21, covering environmental, economic and social themes. For upland farmers, specific options include:

Improving rush pasture for wildlife– £100/ha

Management of moorland grazing – £1.30/ha

Biodiversity cropping on inbye – £2.70/ha

2. Rural Priorities initiative, which has 74 measures with environmental, economic and social themes. Particularly useful for upland farmers are:

Muirburn and heather swiping – £67/ha

Off wintering of sheep – £9/ha

Away wintering of sheep – £21/ha

- LFASS – continued in order to service Measure 212 (see Chapter 8)
- LEADER – to deliver Axis III mainly, but to contribute to Axis I and II.

Understanding the way these rural development programmes differ teaches important lessons in how to approach general farming support. Whereas the Scottish RDP recognises the financial handicaps for new entrants, the English model sees support for new entrants as irrelevant; yet we know that new entrants and succession are particular issues in some uplands (Chapter 7). To meet this need for greater flexibility, a bottom-up LEADER approach would be beneficial; individual LAGs would then have been able to support entry initiatives where appropriate.

A second contrast is the approach to land management. Whereas the English RDP has gone down the tried and tested route of (what is to all intents and purposes) a biodiversity management scheme (ESS), the Scottish programme has stepped outside this box and developed the cross-cutting Rural Development Contract. The English version is still too sectoral, so development is fragmented rather than integrated. It again begs the question whether the rigid approach to axes and measuring delivery is designed to aid departmental accounting, rather than agriculture, or whether it simply shows lack of vision, which could be symptomatic of power relations between state and community, the philosophical difference between development and regeneration.

Finally, though the English and Scottish RDPs both provide explicitly for upland farmers through the HFA and LFASS respectively; the Scottish RDP goes one step further by acknowledging actively the economic disadvantage of small farms by implementing a graded payment regime. Even though the LFASS seems far more integrated, it has already been criticised for its overly complex approach, making it hard for farmers to work out exactly what they can and cannot do.

There are more specific changes unfolding for upland farmers. Re-alignment of the purpose of LFAs means that financial aid is now referred to as Natural Handicap Payments. These are designed to prevent farmland being abandoned by giving compensation for specific disadvantage that requires additional costs (CEC 2009b). For upland farmers in Britain this will have two effects: the re-definition of LFA land (see below) and the subsuming of HFA, *Tir Mynydd* and LFASS into broader schemes providing other public goods beyond food. This change started in England in 2010 with roll-out of the Upland Entry Level Scheme (UELS), which provides broad but shallow environmental grants for upland farms (Table 9.3). In line with the ELS, it works by allocating points for areas or features to be managed for five years (www/defra.gov.uk/rural/uplands/support.htm, accessed March 2009). Prescriptions are divided into requirements and options (examples in Table 9.4). The UELS also offers a supplementary payment of £5/ha/yr for commons and shared grazing, an important step forward as it recognises the complexity of heft and common management.

At the development stage, it was acknowledged that each English upland is unique, with its own issues and challenges, so consideration was given to varying the structure of UELS from upland to upland (DEFRA 2008b). This was rejected on three counts. First, ELS is a national scheme, and a regional UELS would conflict with national prescriptions; second, the administrative burden would increase for Natural England and the RPA; third, there would be "significant" IT difficulties. Frankly, these excuses are rather weak, given that ESA and CSS both had 'regional' prescriptions fitting Joint Character Area targets, as HLS does. Admittedly, the administrative burden would probably be greater, as previous schemes have shown (Chapter 8), but cynically we could see this as simply a way of reducing programme costs and missing an opportunity to tackle the problems of individual massifs, which is needed if uplands are to become sustainable. There is said to be some flexibility, but it remains to be seen whether it is enough to accommodate the intricacies of stocking densities of upland moorland mosaics (see Chapter 5).

For Wales, *Tir Mynydd* is being withdrawn in 2012, along with other Axis 2 schemes, and replaced by one scheme known as *Glastir* ('green land') (Table 9.5). It contains no specific LFA payment, though there is a 20% premium for all LFA farmers entering the first *Glastir* objective. A noteworthy difference is that UELS payments are limited to just the SDA in England, whereas in Wales there is no differential between DA and SDA. This is a result of the re-evaluation of LFA criteria in England (see below). In Wales (and Scotland) such land is almost all SDA, with

Table 9.3 The English Upland Entry Level Scheme

Overview: *a straightforward approach to supporting the good stewardship of the countryside for farmers with land in Severely Disadvantaged Areas (SDA). The successor to the hill Farm Allowance. A voluntary scheme with agreements for 5 years*	
Award System:	**Points target set for total farm area as follows:** SDA land below moorland line — 62 points/ha SDA land above moorland line, < 15 ha. — 62/ha SDA land above moorland line, >15 ha — 23 /ha non SDAbelow moorland line — 30 points/ ha non SDAabove moorland line, < 15ha — 30/ ha non SDAabove moorland line, > 15 ha — 8/ha for each point, £1 is awarded per year
Compulsory elements : *where applicable*	**UX1 Moorland commons shared grazing (2 or more graziers sharing) gain a £5 per ha supplementary payment** • Maintain a hefted flock • Establish and maintain a commoners group or association • Maintain an active graziers record **UX2 moorland (SDA) land below the moorland line (x 15 points/ha)** • No supplementary feeding • No fertiliser or herbicides with 6m of watercourse • Retain traditional features • Hedgebanks maintained using traditional materials • Employ local style hedge management • No supplementary feeding in native woodland • Retain existing areas of native scrub • No boulder or rock outcrop removal • Prevent the spread of bracken • Collect all plastic waste **UX3 Upland Grassland and arable SDA (x 11points/ha)** • Avoid over and under grazing (detailed guidance provided by breed, timing and occurrence of blanket bog/ peat) • Maintain wetlands including peat bogs, other mires and hillside flushes • Block all drains and grips bar along tracks & boundaries • Supplementary feeding to avoid damaging notable habitats • Legal burning must for the Heather & Grass Burning Code • Do not apply fertilisers or manures. • Do not plough, cultivate, re-seed or harrow. • Controlling bracken by harrowing is permitted • Retain and protect native woodland **Completion of Farm Environment Record (x 3 points/ha)**
Optional element:	A combination of ELS and UELS specific options, the latter of which can be seen on Table 9.4

(Source: Natural England 2010b, Part 5, accessed: 24/6/11)

Table 9.4: Presciption Requirements and Options under UELS

prescription	*score/ payment*
requirements (grassland and arable below moorland line)	
No supplementary feeding within 6m of watercourse No fertiliser applications within 6m of watercourse Any maintenance of drystone walls undertaken must be completed in the local style Any maintenance of hedge/hedgebanks undertaken must be completed in the local style No supplementary feeding in native woodland except during periods of extreme weather Retaining existing areas of scrub by not supplementary feeding nor applying fertilisers, herbicides or pesticides No removal of large boulders or stone outcrops Prevent the spread of bracken beyond the current areas where a tractor can get Remove discarded plastic bags and wrappings deposited on the land	11 points/ha
requirements (open moorland)	
Avoid over- and under-grazing, and maintain a minimum stocking rate of 0.05LU/ha between 1/6 and 30/9 Maintain wetlands, including peat bogs, mires and hillside flushes Manage supplementary feeding sensitively Follow heather- and grass-burning code No fertiliser applications, cultivations or harrowing Retain and protect native woodland – no supplementary feeding except in extreme weather	15 pts/ha
requirements (additional for common and shared grazing)	
All sheep must consist of hefted self-maintained flocks Establish and maintain a commoners' association or group	£5/ha/yr
options (examples)	
Stone wall restoration	30pts/m
Sheep fencing around small woodlands	50pts/100m
Maintaining visibility of archaeological features on moorland	53pts/feature
Cattle grazing on upland grasslands and moorland	30pts/ha
Haymaking	60pts/ha
Management of upland grasslands for birds	37pts/ha

(Source: http://www/defra.gov.uk/rural/uplands/support.htm accessed March 2009)

minimal amounts of simple DA; thus re-evaluation would have little effect. *Glastir* is the first scheme to explicitly build in ecosystem service elements, such as carbon management and energy efficiency, whilst addressing regional differences and also acknowledging variations in stocking density on commons. *Glastir* is interesting as it includes carbon reduction activities for land and buildings.

Whilst new developments in Pillar II offer some financial reward, they are perceived as a particular threat by upland farming communities. This will indeed be the case if upland farmers continue to rely on livestock production as their core business (as shown in Figure 9.2) and exploit little else their farm has to offer. Yet the new agenda means new opportunities.

Table 9.5 Glastir

Overall Objectives: *Managing soils to help conserve carbon stocks and reduce soil erosion.* *Improving water quality and reducing surface run-off.* *Managing water to help reduce flood risks.*	*Conserving and enhancing wildlife and biodiversity.* *Managing and protecting landscapes and the historic environment.* *New opportunities to improve access and understanding of the countryside.*
Three elements:	
All Wales Element (AWE)	A whole farm land management scheme which is open to all farmers and land managers for a 5 year agreement. A compulsory 'whole farm code' element gaining £33.60/ ha for LFA land A set of management options with regional variations, the latter of which 'score' 10% higher when selected Operates through a point system scored by no. of eligible Ha x 28 Payments per year = up to 20ha £15/ha, 20 to 50 ha £8/ha and 50 to 100ha £2.75/ha A competitive part farm scheme intended to deliver significant improvements to a range of habitats, species, soils and water that might also require changes to current agricultural practices.
Targeted Element	Options include: • *Carbon management*eg No lime on improved land or semi-improved grassland over peat soils (£20/ha/yr) • *Water management* eg Rivers – Management of land adjacent to sensitive rivers (£96.60/ha/yr) • *Biodiversity* eg Upland limestone grassland (£61.82/ha/yr), Montane heath (£64.10/ha/yr), a wide range of individual species needs such as medicated grit for black grouse (£94.60) • *Historic landscape and environment* eg drystone wall restoration (£26.50/m2) • *Access* eg permissive access area (£447/ha)
Common land Element	Designed to provide support for the delivery of environmental benefits on common land. Paid at £33.60/ha in LFA Series of management options available to the grazing association not on a points system: *Option 1* - Management of stocking density min. & max calculated based on the common's proportion of habitats as well as recommended stocking rate *Option 2* – Closed period stock free for 3 months between 1/11 and 31/3
Additional:	
Agricultural Carbon Reduction and Efficiency Scheme (ACRES)	A capital grant scheme aimed at improving resource and business efficiency, and reducing the carbon equivalent emissions of agricultural and horticultural holdings, such as new technology and equipment to promote the efficient use of energy, water and manure/slurry. **Heat generation** – such as heat-recovery systems; biomass boilers; ground/air source heat pumps; solar hot water panels. **Energy efficiency** – a list of approved capital items which have a demonstrated energy efficiency saving – including variable speed drives, plate heat exchangers, ventilation/temperature controls, energy efficient lighting systems. **Water efficiency** – including rainwater harvesting equipment and water recycling systems e.g. fixed pumps, UV filtration systems, pipe work and storage tanks. **Manure/slurry efficiency** – Includes capital items aimed at expanding storage capacity on-farm to enable better timing of applications to meet crop growth requirements. Includes clean water separation, floating covers and slurry/manure storage. Paid at a rate of up to 40% against eligible expenditure (50% for young farmers) (from £5K to £50K capital expenditure)

(Source: http://wales.gov.uk/docs/drah/publications/100915glastirgeneralguidanceen.pdf., accessed: 24/6/11)

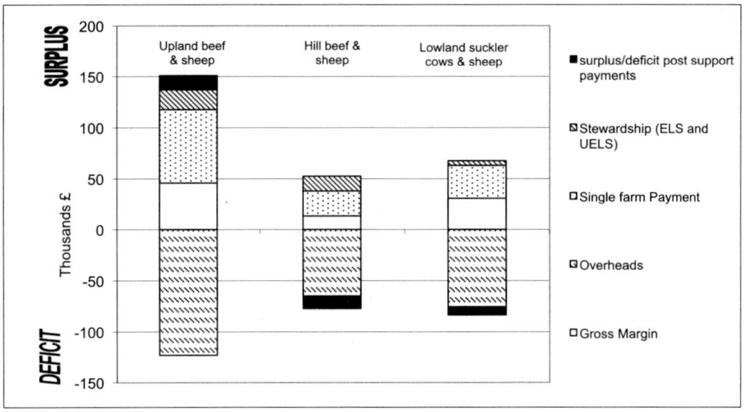

Figure 9.2 Income sources and profitability of livestock farms
Adapted from HSBC Agriculture (2011) *Forward Planning 2011* HSBC Bank Plc: Hemel Hempstead.

Emerging opportunities

Historically the uplands provided people with a range of resources, which were exploited on a subsistence level. In Chapter 3 we saw that upland communities took a range of products from the hills, not just livestock and cereals. For them the landscape was multifunctional.

Today multifunctionality is making a re-appearance, this time as an approach to agricultural restructuring. It can be seen as a way of maintaining Western farming even when exposed to the full rigour of global market forces. The term was first applied in the forestry sector, with forests and woods being shown to have social as well as economic value (e.g. Buttoud 2002). When applied to agriculture, the term evolved simultaneously in policy and academic thinking to explain how farming provides a range of products beyond foodstuffs. In the EU it was seen, rather cynically, as a mechanism to justify continued agricultural support when the WTO were pushing for its complete removal in the early 1990s (Potter and Tilzey 2003). The term was officially ratified by the EU in the 1996 Cork Declaration, from which the basis of Pillar 2 emerged (Lowe *et al.*, cited in Wilson 2007).

Multifunctionality has no agreed definition, but Tilzey (2003) proposed that agricultural multifunctionality is a concept that tries to capture the multiple functions of agriculture in addition to food and fibre production, such as its contributions to cultural heritage, environmental protection and rural community. Marsden and Sonnino (2008) took this further and identified three dimensions to the term 'multifunctionality':

- as a palliative to the cost–price squeeze through various survival strategies which encompass a combination of agricultural and non-agricultural incomes within the farm household;
- as a landscape approach where agriculture has lost its central role in society and is replaced by its landscape value only;
- through the provision of rural development whereby the farm is interconnected with its rural location in many integrated ways.

Wilson (2007) provides a detailed examination of agricultural multifunctionality; interestingly he demonstrates that it sits within a funnel of decision making (Figure 9.3). The funnel has two axes:

- strength of multifunctionality, from weak to strong;
- the ability of a farm business to adopt multifunctionality, controlled by a range of opportunities and constraints.

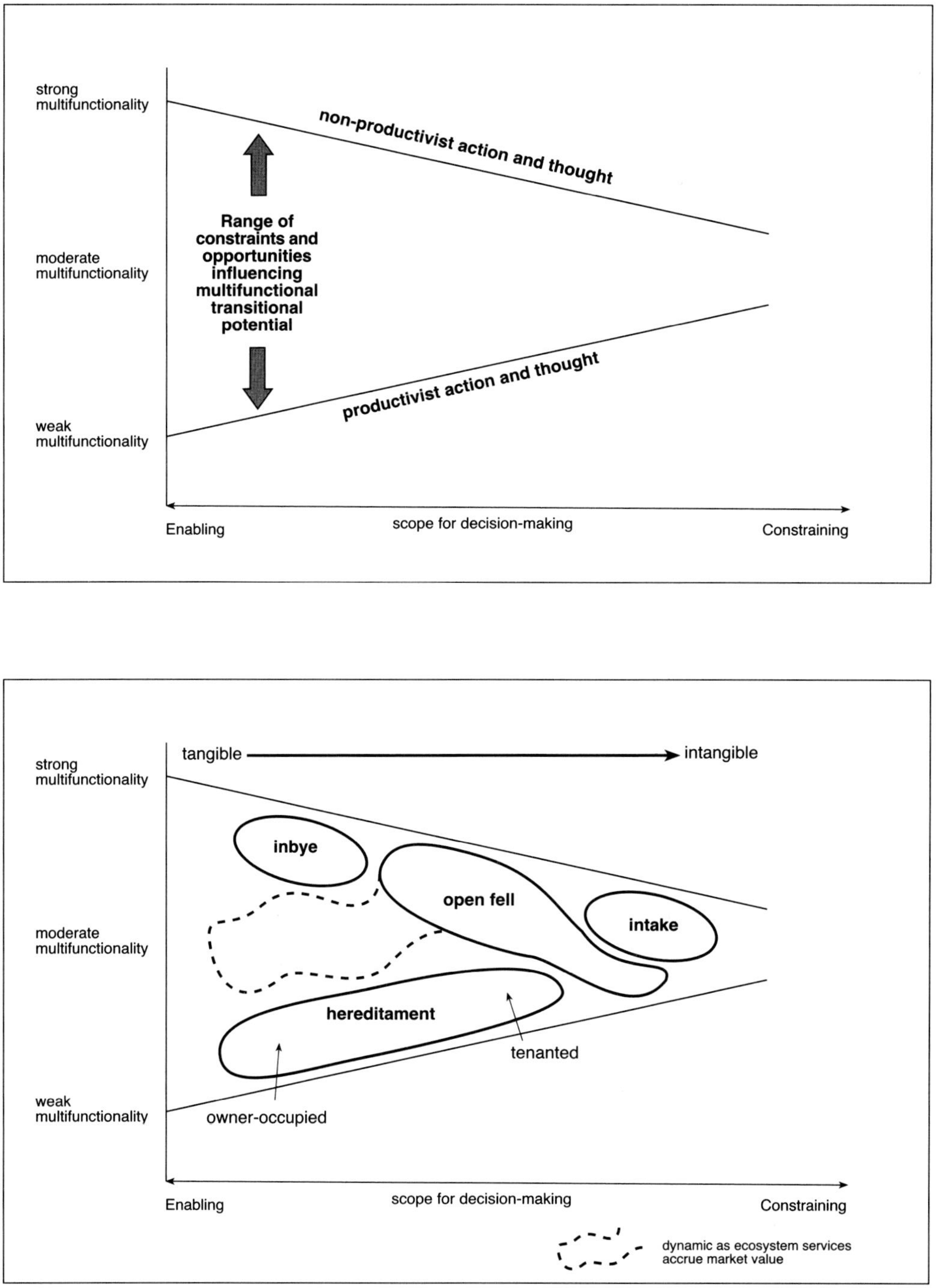

Figure 9.3 Wilson's Concept of Multifunctionality
Reproduced from, Wilson GA, 1987, *'Multifunctional Agriculture A Transitional Theory Perspective'* with kind permission from CABI, Wallingford, UK.

As the constraints increase, so a farmer's ability to develop a multifunctional approach to their business decreases, hence the funnel shape. Within the funnel Wilson has placed different farm types to form a loose spatial classification. Upland farming sits towards the constraint end of the funnel, where multifunctionality is moderately to highly possible, based on the HNV of these environments, but fairly constrained. He cites a number of empirical studies from Austria and Greece, which support this location within the funnel. He also describes a study by Knickel *et al.* (2004) suggesting that 80 to 90% of European upland farms are pluriactive, which is not typical of UK uplands, as we saw in Chapter 6.

Wilson recognises that many exceptions to his funnel pattern can be found, but he sees that the lowland/upland divide does exist. However, he does not really explain why upland farms are constrained, which suggests he is playing the obvious 'economically marginal' card; and he is suggesting that pluriactivity and multifunctionality are the same, which they are not. Pluriactivity is about increasing income on and off farm; multifunctionality is about goods and services that are by-products of farming. There are also particular marketing constraints for upland farm businesses, related to close proximity to competition and spheres of influence in upland valleys, which were discussed in Chapter 6.

So, while HNVs are generally more available to upland farmers (Figure 5.2), there is a plethora of other constraints that force this farm type closer to the funnel's neck in relation to other multifunctional opportunities. There are other, better ways of looking at Wilson's funnel in uplands:

1. Massif by massif – because each massif has unique multifunctional characteristics to exploit. For instance, the Orton Fells of south east Cumbria were shown in Chapter 6 to have less tourism potential than the Lake District; but Lake District farmers are more risk-averse than farmers in the Pennines.
2. As a series of temporal waves – typical of Lake District valleys, where the initial adopter corners the market before any competitors can. Only once the first adopter moves up to a bigger geographical scale can another adopter move into the vacated niche.
3. Option by option – some forms of diversification are less available than others, depending on individual circumstances. However, exploitation of general HNV should be less constrained than other options in uplands.
4. By land type – Figure 9.3 shows conceptually how this might work for upland farm land types. The hereditament has most opportunities, whereas the intake has the least (see Table 6.8) because of lack of access and utilities. The fell has the most potential, particularly as goods and services develop market values (Table 7.5).

Wilson (2007) adds a long-term temporal dimension to his conceptual discourse, but only on a European scale. He considers the development of farming in western Europe from 1500 forwards into the 22nd century, identifying three phases (with a small overlap between the first two):

- 1500 to 1980 – typified by a shift from strong to weak multifunctionality, (very characteristic of upland farming as chapters 3 and 6 showed).
- 1950 to 1990 – the productionist trough, when the primary function of agriculture was to produce food; (during this window, upland agriculture became tarnished with being economically marginal, a burden as the 'hill farming problem' and as a result, highly subsidised).
- 1990 onwards – a broadening of multifunctional potential, based on how society's view of agriculture will change.

Wilson's thoughts about multifunctionality tell us three things of practical value for upland agriculture. First, HNV is a crucial option to exploit (as Chapter 5 showed). Second, Western capitalist agriculture is moving out of an era focused on production into one of multifunctionality, which should allow hill farms to be valued more for their HNVs. Third, as a result of these two, the hill farming problem should disappear, putting this sector back into mainstream agriculture. One could argue, of course, that if HNVs are the panacea then 25 years of agri-environment and related schemes should mean we no longer have a hill farming problem – but we do. This suggests there are other factors at work that prevent upland agriculture from thriving; we explore the nature of these below.

Another way of looking at multifunctionality in upland agriculture is to see it as a method of rural development – one of Marsden and Sonnino's dimensions. A crass interpretation would be to see anything that increases the income base of farmers as rural development. Instead Marsden and Sonnino (2008) suggest an activity needs to meet three criteria to qualify as multifunctional:

1. The activity needs to add income and employment opportunities to the agricultural sector;
2. It should respond and contribute to the needs and expectations of society at large;
3. There should be a radical redefinition and reconfiguration of rural resources in and beyond the farm business.

Pluriactivity and diversification on upland farms are therefore obvious examples of the first criterion and Chapter 6 has been shown they are possible for some upland farm businesses. They may be limited by unique geographical and temporal qualities of upland massifs and/or by the individual circumstances of a farm business, as discussed in Chapters 6 and 7. The second criterion broadly depends on the changing perception of the role of upland agriculture. Chapter 7 showed evidence that the public have yet to make the link between various upland landscape attributes and the act of farming. We also know that upland agriculture is still perceived as over subsidised, despite the realignment of funds from Pillar I to Pillar II.

However, both phases of Pillar II between 2000 and 2013 should drive forward multifunctionality, giving upland farming goods and services a distinct comparative advantage over other types of farming. Upland farmers could really benefit from this, for example in the production of ecosystem services, which emerged as a concept in the 1990s (Box XXXIX), though it began to gain political validity only in about 2005. Ecosystem services can be provided by uplands in copious amounts beyond the traditional environmental goods. Bonn *et al.* (2009) identify eighteen main ones, of which many are derived from agricultural practice directly or indirectly (Table 9.6). Some are a product of continuous management of upland landscapes for millennia (e.g. biodiversity and cultural heritage); others have emerged from the needs of modern urban society (e.g. health benefits and tourism). The word 'service' is useful here because its connotation of benefit to society gives it a more positive spin. Society can view this as paying for something we need in line with Marsden and Sonnino's second criterion. It is here that Wilson's recognition of upland HNVs becomes important.

Ecosystem services also provide evidence of the redefinition of rural resources in the farm context – part of Marsden' and Sonnino's third criterion, reconfiguration of rural resources in and beyond the farm. This re-alignment seems to be occurring with the changing financial support for upland farmers from HFA to UELS and other agri-environmental schemes. Non-farm on-farm diversification is another facet of this, where upland farmers are re-structuring the use of some of their farm resources and assets. However, the exploitation of agri-environment schemes and pluriactivity on upland farms is not universal, as we saw in Chapters 6 and 8, and as Figure 9.3 suggests. As with all diversification on farms, the aim is to choose the right combination in line with the resource base. Thus some upland farms may be better off focusing on carbon sinks as an ecosystem service; these can have a higher grazing density than biodiversity, as long as soil erosion does not occur. So management strategies to maintain multiple services on the same piece of land may coincide, but they are not paramount.

Table 9.6: Upland Ecosystem Services

ecosystem service	*role of farming*
Provisioning	
Food	Continued supply of livestock
Fibre	Sustainable exploitation of quarries and mines
Minerals	Afforestation and woodland maintenance
Energy Provision	Micro generation & turbine location
Fresh water	Halting soil erosion and pollution
Regulating	
Carbon storage & sequestration	Maintaining active mire complexes
Air quality	Halting soil erosion
Water quality	Appropriate grazing regimes
Flood risk prevention	Retaining vegetation
Wildfire risk prevention	
Cultural	
Recreation, tourism & education	Maintaining access and egress across land
Field sports & game management	Providing appropriate vegetation through sensitive grazing
Landscape aesthetics	Maintaining field structures
Cultural heritage	Continuing practice and traditions
Biodiversity	
Health Benefits	
Supporting	
Nutrient cycling	Appropriate grazing and general farm management
Water cycling	Halting soil erosion
Soil formation	Limiting pollution of water courses
Habitat provision	

(Source: adapted from Bonn *et al.* 2009.)

The corollary is that upland rural development derived from agricultural multifunctionality will be uneven at best, with different upland farms, farm clusters (particularly if intercommoning) or massifs representing the various dimensions of the term. The pragmatic value of knowing which type of multifunctional agriculture a particular upland farm or area should be adopting is that rural development support can be modified and targeted to the local needs of those farms, while also addressing the needs of the local or regional environment. It is this that should underpin the RDP process, LEADER and LAGs.

For upland agriculture there needs to be an acceptance that the opportunities for upland farms can be one of three multifunctional types: palliative (Type 1), landscape (Type 2) or rural development (Type 3). These are all *equally valid* and should be *equally valued* in sustaining upland agriculture and resolving the hill farming problem. Of course, some farm businesses may adopt a composite model to suit their resource base. Type 1 has been typical of most upland farms in their increasingly marginal economic role since the 1950s. It is often perceived as more acceptable to farming families because the farm unit remains central; however, it has the most constraints because each farm competes with its neighbour for on- and off-farm non-farm income sources, exacerbated by the way upland farms are grouped in the valley bottoms. It also has negative connotations from past subsidy regimes.

We could argue that all upland farms should be multifunctional in a Type 2 sense by exploiting their HNVs (as Wilson suggested), driven by society's growing demands for a less production-based agriculture. But, as noted above, it does not seem to have resolved the hill farming problem – yet. Perhaps agri-environment policy needs to evolve further by integrating, more fully, the entire range of ecosystem services into policy and practice. The Rural Development Contracts

Box XXXIX: Ecosystem services

Ecosystem services are those ecological processes that provide a service for the benefit of people. It is based on one of the main outputs of the Convention on Biological Biodiversity, which recognised the validity of applying an ecosystem approach to resource management. The Convention on Biodiversity defines the ecosystem approach as:

> a strategy for the integrated management of land, air, water and living resources that promotes conservation and sustainable use in an equitable way, and which recognises that people with their cultural and varied social needs, are an integral part of ecosystems. (quoted in DEFRA 2007b, http://biodiv.org/decisions/default.asp?lg=0&m=cop-05&d=06)

In 2000, Kofi Annan, then UN Secretary General, called for an assessment of the state of global ecosystems, identifying actions needed to conserve and sustain their use. There were four main conclusions of the Millennium Ecosystem Assessment (MEA):

- Ecosystems have changed a great deal in the last 50 years due to increased human needs for food and fibre, resulting in biodiversity loss;
- Whilst human well-being has gained, it has been at a cost of ecosystem degradation, which in turn has had a positive feedback effect on human well-being;
- This trajectory of degradation will continue for the next 50 years at least;
- To solve this and to retain ecosystem services, our policies, institutions and practices need to change to accommodate practices that balance need against loss.

In 2007, the UN Convention on Biological Diversity proposed twelve principles to implement its ecosystem service philosophy:

1. The objectives of management of land, water and living resources are a matter of societal choices.
2. Management should be decentralised to the lowest appropriate level.
3. Ecosystem managers should consider the effects (actual or potential) of their activities on adjacent and other ecosystems.
4. Recognising potential gains from management, there is usually a need to understand and manage the ecosystem in an economic context.
5. Conservation of ecosystem structure and functioning, in order to maintain ecosystem services, should be a priority target of the ecosystem approach.
6. Ecosystems must be managed within their limits of their functioning.
7. The ecosystem approach should be undertaken at the appropriate spatial and temporal scales.
8. Recognising the varying temporal scales and lag effects that characterise ecosystem processes, objectives for ecosystem management should be set for the long term.
9. Management must recognise that change is inevitable.
10. The ecosystem approach should seek the appropriate balance between, and integration of, conservation and use of biological diversity.
11. The ecosystem approach should consider all forms of relevant information, including scientific and indigenous and local knowledge, innovations and practices.
12. The ecosystem approach should involve all relevant sectors of society and scientific disciplines. For upland agriculture, principles 2, 4, 6, 8 and 11 are particularly pertinent.

The management concepts of the ecosystem approach and ecosystem services are now employed as the underlying framework of UK biodiversity conservation (DEFRA 2007c). The government requires all areas of government, statutory organisations and stakeholders to embed these into their strategies. They also need to encourage behavioural change, gather more evidence of status and trends; develop adaptations to climate change; evaluate policies and outcomes against biodiversity targets; fulfil national and international obligations; and have a better balance between biodiversity, ecosystem functions, goods and services (DEFRA 2007c).

Although there is no agreed classification of ecosystem services, DEFRA (2007c) have adopted that designed by the MEA:

- *Provisioning services* – those products obtained from ecosystems;
- *Regulating services* – benefits created through the regulation of ecosystem services;
- *Cultural services* – non-material benefits that we can acquire through spiritual enrichment, cognitive development, recreation and so on;
- *Supporting services* – those necessary for the production of all other ecosystem services.

The emphasis is on valuing ecosystem services that contribute to human welfare, which gives them a policymaking focus. However, valuing some of these services is tricky as they fall under the category of externalities (see Box XXIX), so CVM and WTP have been used as initial tools for this, along with Choice Experiment (see Hanley and Colombo 2009). Despite these valuation issues, Natural England has adopted the ecosystem services approach as a strategic framework, to achieve a 'healthier upland environment' through their *Vital Uplands Vision* (Natural England 2009a: 1; Natural England 2009b). This, in turn, has led to various pilot implementation projects in three upland catchments in the Lake District, Pennines and South West (see Chapter 2).

introduced by Scotland and the Welsh *Glastir* go some way to achieving this, but application complexity issues mean we are not quite there yet; nor do they provide, for instance, market values for carbon or water quality.

What is probably needed is a parametric shock (see Chapter 7) that will force farmers to re-evaluate and re-structure their knowledge and skill base, in line with the broader 'ecosystem service' concept and so slide left across Wilson's funnel. How this shock occurs could be institutionally led, as proposed by Bureau and Witzke (2010), who considered that some form of direct payment scheme is needed for public goods/ecosystem services post-2013. They proposed two payments: a flat rate for all farms meeting set criteria in a particular region (such as a water catchment or massif) and a variable payment to farmers contracted to produce specific public goods and services over one to fifteen years. These issues are currently under review for the CAP Reforms and UK developments due to roll out in 2013 (Box XL). On the other hand, the decision may be taken out of the bureaucrats' hands as the global environment takes us down other paths (see below).

Chapters 7 and 8 showed that upland farmers do adopt agri-environmental grants, but only on their terms – that few adjustments to farm management are required, that it doesn't really detract from main farming operations, that it is not essential for them to know environmental 'stuff', that it doesn't undermine their *raison d'être.* In this situation, society may benefit from the landscape goods or ecosystem services provided, but upland farmers do not have to re-define their role completely. The re-definition comes, as Marsden and Sonnino suggest, in the move to Type 3 multifunctionality, where rural development becomes central and upland agriculture is merely one piece of the jigsaw. I would argue that, for most upland farmers as individuals, this is a step too far; it is, as *Star Trek* puts it, 'The Undiscovered Country'. It is simply too far removed from why they farm in the first place. As one upland farmer put it, "we are not glorified park keepers, we are stockmen". This view is underlined by the fact that any diversification away from farming is usually managed by the partner, reinforcing the farmer's production focus.

Box XL: CAP Reform, 2013 onwards

The European Union and its member states have been preparing for CAP Reform in 2013 for several years. The date was chosen because it forms a natural break in policy operation at the end of the (second) Rural Development Regulation 1698/05. Some specific issues form the basis of these reforms:

- food security, climate change and peak oil,
- environmental management,
- expansion of the EU,
- WTO negotiations (the Doho round),
- the importance of agriculture to the vitality of rural areas.

Comprehensive studies have been undertaken in preparation for the Commission's proposals at various institutional levels. Two notable agricultural studies, which will affect future developments in upland agriculture in the UK, are:

- Bureau and Witze (2010), who investigated changes to the operation of SPS for the European Parliament;
- Vrolijk *et al.* (2010) who analysed the effects of reduced farm payments on farm viability for DEFRA in the UK.

The main proposal put forward by the European-wide team led by Bureau and Witze focused on the provision of public goods through direct CAP payments. They recognised that, due to the importance of SPS for farm incomes, transitional payments would be needed as SPS was phased out, because rapid dismantling of Pillar I would result in many farm bankruptcies. Member states could vary the value of the transitional arrangements through flat-rate hectarage payments linked to labour levels. The public goods and service element could be tailored to regional conditions and needs in member states (often referred to as 'subsidiarity').

The work of Vrolijk *et al.* investigated the effects of complete abolition of decoupled payments on farm viability – in other words, abolishing SPS and any other area payments. They considered farm viability in various ways: by entire states, by state regions (in the UK's case, Wales, Scotland, Northern Ireland, Northern England, South West England and South East England) and farm type. The survey collected data only for those farms considered by member states to be 'commercial' (34% in the UK); as a result many UK upland farms were not represented. For the UK, around 25% of dairy farms would become unviable and 30 to 40% of grazing livestock farms.

In November 2010, the European Commission finally issued its draft proposals for its post-2013 agricultural agenda – 'The CAP towards 2020: Meeting the food, natural resources and territorial challenges of the future' (European Commission, COM(2010)672 final). This identifies three main objectives:

- viable food production,
- sustainable management of natural resources and climate action,
- balanced territorial development.

To achieve these, the European Commission sees three ways forward:

- adjust the policy gradually, focusing on equity of direct payments across the EU,
- overhaul it, mainly targeting measures to be more sustainable but also efficient,
- begin a far-reaching overhaul, focusing on environmental and climate change objectives at regional level.

At the time of writing, no final decision has been made, although the first option is more likely simply because those member states with most political clout have a greater vested interest in agriculture than others.

A further two documents have recently been released by the UK Government:

- The UK National Ecosystem Assessment,
- The government White Paper 'The Natural Choice: securing the value of nature'.

Both of these will have wide-ranging repercussions on agriculture and its relationship with the environment. The UK National Ecosystem Assessment (UK NEA) is the first survey to be published by any state in response to the UN Millennium Ecosystem Assessment (Box XXXIX). The report comprises a summary document and a large technical report containing chapters on various broad habitats, including one on mountain, moorland and heath (UK National Ecosystem Assessment 2011a). The main finding is that, while living standards and technology have improved since 1950, this has been at the expense of ecosystem services. Although some improvements have been made, 30% of ecosystems are still in decline and climate change will exacerbate these trends unless we respond now. A particular challenge is to find ways of intensifying agriculture sustainably to increase food production while ensuring our environmental impacts are lessened.

The UK NEA also attempted to calculate the economic value of various market and non-market services (see Box XXIX). These types of economic value assessment derived a figure of £430m p.a. for biodiversity pollination, an average of £845m p.a. for terrestrial biodiversity, £90m p.a. for green timber production and £30bn p.a. for recreation and tourism. Agriculture produced mixed fortunes from minus £50/ha p.a. to plus £75/ha pa depending on geographical location, which is exemplified by a scenario they ran for Wales by switching rural land use from agriculture to multi-purpose forestry (UK NEA 2011a: 43–4). A Foresight report (2010: 27, 42; cited in UK NEA 2011b) calculated the value of UK agriculture's non-production ecosystem services to be £1.74bn p.a; on the downside, they demonstrated agriculture's negative impact on ecosystem services at £2.57bn p.a.

The White Paper (HM Government 2011) is a broad policy response to the findings of the UK NEA with regard to nature. The key aim is to 'mainstream' nature across all areas of UK life by:

- facilitating more local action to protect nature,
- creating a green economy,
- strengthening connections between people and nature,
- demonstrating leadership within the EU and internationally to protect and enhance natural resources.

To achieve this, the government want to develop an integrated approach to management to create a 'resilient ecological network' through Local Nature Partnerships (LNPs), Nature Improvement Areas (NIAs) and reforms in the planning system. Further actions revolve around valuing natural capital in an economic sense (see chapters 5 and 7) by making all businesses trade and develop their supply chains in a more sustainable way, reconnecting people with nature through public health initiatives, removing barriers to outdoor learning for school children and providing more volunteering opportunities.

Within the White Paper's objectives, there are certain key points that upland farmers need to bear in mind and exploit. First, the development of LNPs should focus attention on the unique local character of areas and the processes required to maintain each, for example more structured and targeted support under agri-environmental schemes in line with farming systems. Second, NIAs will be used as a vehicle to promote integrated catchment management (see Chapter 2) and whole valley planning, which in turn will require greater co-operative working by land managers and stakeholders to address landscape issues, not just on a farm-by-farm basis. Third, planning reforms will have a presumption in favour of sustainable development, for example the use of green technologies in buildings and processes as well as offsetting the impacts of development on biodiversity.[1] There will be more support for low-carbon energy production and more investigation of the impacts of bioenergy

[1] Biodiversity offsetting – conservation activities designed to deliver biodiversity benefits in compensation for losses in a measurable way (UK NEA 2011:22), for example, building a new home may require offsetting the impact of the production of the building materials.

(see Chapter 6) production on the environment, food production and other sectors. Other opportunities may arise from demands for more woodland creation and management or greater protection for soil and peat deposits, along with more money for the Catchment Sensitive Farming initiative and a new Catchment Restoration Fund.

From this summary it is clear that, while the EU and the UK Government recognise the need to continue supporting agriculture, the emphasis is shifting towards reconciling food production and the environment. At the same time food security is paramount, but not at the expense of biodiversity. The next few years are going to be crucial for upland farming, as the sector will be shaped for the coming generation by these new visions and policies.

Realising farm assets

Fundamental to the three criteria recognised by Marsden and Sonnino (2000) is a farmer's ability to realise all the assets available to their business that can sustain their livelihood. We have seen that top-down financial aid to the upland farming sector has really failed to make a substantial difference for many.

Consequently other approaches are needed, ones that adopt radical methods of delivery. LEADER has gone some way to doing this, but it still requires the individual to actively recognise the potential of their farm and its resource base to evolve a sustainable business. To this end, we need a more person-centred approach to upland farming development, to help farmers identify and value all of their assets. This individual approach would also be a means to embrace local upland variations. One way to do this is the Sustainable Livelihoods Approach advocated by various international development organisations. This has six key objectives (DFID 2000: 1.2):

- Improved access to education, information, technologies and training, better nutrition and health
- More supportive and cohesive social environment
- More secure access to, and management of, natural resources
- Better access to basic and facilitating infrastructures
- More secure access to financial resources
- A policy and institutional environment that supports multiple livelihood strategies and promotes equitable access to competitive markets for all

This approach was designed to eliminate poverty in poorer countries, yet many of these objectives resonate with the plight of upland farming systems and communities in the UK and elsewhere, as we have seen. The process emphasises that people should be central: they should be the ones to identify what makes them vulnerable, what constrains their choices and how these factors interact, to help them realise all their assets. This information should then be used to develop a strategy – a Sustainable Livelihoods Framework (SLF) – for the individual to achieve their own livelihood outcomes (Figure 9.4). We can apply this framework to upland farming to formulate suitable strategies.

A central concern of the SLF is to allow farmers and their families to identify their limiting factors. We can divide these into two types: vulnerabilities that affect farming livelihoods, cannot be controlled by farmers themselves and derive from outside the sector; and constraints specific to the exploitation of individual farm and farming community assets. Although every farming situation will be different, there are general vulnerabilities and constraints, affecting many upland farmers and their businesses, that have been voiced by farmers themselves in recent years (Figure 7.11). In a way, we could argue that vulnerabilities become constraints as soon as they begin to affect farmers at an operational level. However, there are some vulnerabilities that farmers have

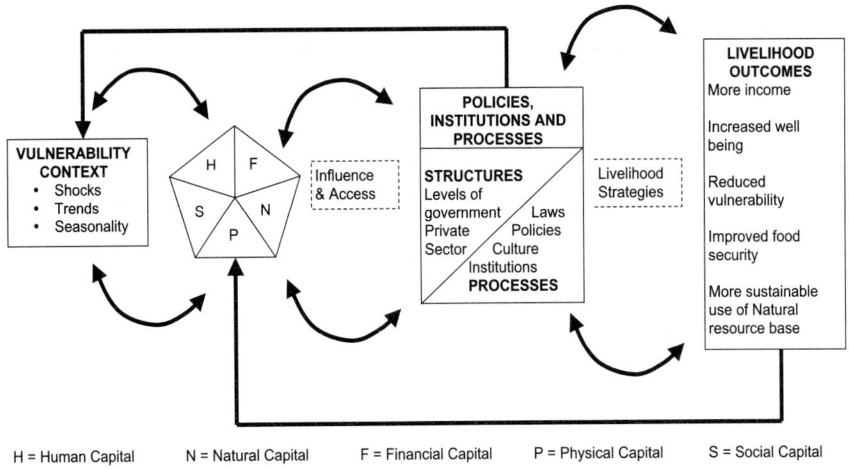

Figure 9.4 Sustainable Livelihoods Framework
Department of International Development (1999) *Sustainable Livelihoods: guidance sheets.* DFID, London. This information is licensed under the terms of the Open Government Licence (www.dfid.gov.uk/sustainable-livelihoods, accessed: 14/11/10.

yet to recognise as constraints, or perhaps they recognise them but (just like the rest of society) do not wish to acknowledge them because of their enormity. We now explore both realised and potential vulnerabilities and constraints.

The vulnerability context

Vulnerability is essentially about risk, uncertainty and insecurity (DFID 2000). As we saw in Chapter 6, vulnerability is closely linked to the concept of resilience. Making a farm business less vulnerable should therefore increase its resilience.

We can divide vulnerability into shocks, which cannot be predicted, seasonal risks and long-term trends, which are inevitable and can usually be seen coming. Unpredictable shocks are typically outbreaks of disease and there can be no more stark example than Foot-and-Mouth in the UK in 2001 (see Chapter 1). Seasonal risks disrupt management systems and put stress on other farm resources. For example, flooding in West Cumbria in 2009 rendered much inbye land useless for winter grazing after river gravel was dumped on grassland. Another example is a long period of snow cover, which increases supplementary feeding costs. Trends making upland farming vulnerable at present are those

- external to the EU, like peak oil, climate change and food security, or
- internal to the EU –like over-reliance on direct support payments, LFA redesignation and enlargement of the EU.

Peak oil

'Peak oil' means the point when global oil production reaches its maximum. There are various estimates of exactly when this will be, but the consensus is that it will occur by 2025 or 2030, possibly much sooner. The main impact for farmers is the combined effect of rising electricity and fuel costs (Table 9.7).

In the UK the price of red diesel (only for use in tractors) rose from 18p/litre in 2003 to 56.5p/litre in 2008 and heating oil (kerosene) went up from 16p to 49.7p. The cost of fertilisers stayed steady between 1995 and 2003 at about £120/tonne, but by 2008 it had increased fivefold

Table 9.7: Rising fuel costs

Variable Costs	*England*			*Scotland*			*Wales*	
	LFA stock	*Lowland stock*	*Cereal*	*LFA stock*	*Lowland stock*	*Cereal*	*LFA stock*	*Lowland stock*
Concentrates	62	88	16	98	128	33	88	14
Roughage & grazing	0	0	0	46	47	3	0	0
Sundry livestock expenses (incl. haulage)	49	65	13	80	87	18	54	87
Seeds	1	5	31	7	8	35	2	9
Fertiliser and lime	16	23	71	61	58	91	32	51
Sundry crop expenses (incl. spraying)	4	11	94	13	12	79	5	19
Contract and casual labour	15	33	42	40	62	51	nk	nk
Total variable costs	146	226	267	345	402	310	180	281
Gross margin	**336**	**544**	**673**	**543**	**503**	**649**	**409**	**635**
Fixed costs	292	470	528	424	342	497	353	541
Net farm income	**44**	**74**	**145**	**118**	**161**	**151**	**56**	**94**

Note: nk = not known.

(Source: Beaton 2009.)

to £620/tonne (Agro Business Consultants 2008). These prices are linked because artificial fertilisers and feed concentrates rely on petro-chemicals for production. A report by DEFRA (2005) calculated that LFA livestock farm costs would go up by £1.17/ha/annum if oil prices were to go up by the predicted 73% increase in real terms by 2025. Though other types of farm would suffer more in real terms per hectare (dairy costs up by £4.18/ha, cereal by £2.62), LFA farms with their lower profit margins might find this disproportionately burdensome.

In response to rising fuel costs, upland farmers could adopt transition farming, whereby they reduce their fossil fuel dependency. First, they could stop using feed concentrates, which would save money in the short term, though it would mean more reliance on forage production. The easiest way to do this would be to increase silage or improve intake/fell land; unfortunately, silaging uses fuel and improvement involves fuel initially and also loss of biodiversity. The other option would be to increase the amount of extensively grazed land. This is feasible if land is being abandoned, but has logistical challenges as we discussed in chapters 6 and 7. Second, they could stop using artificial fertilisers, saving money but again reducing forage production capacity, leading to the same challenges as for concentrates. It could be partially offset by greater utilisation of FYM and by climate change. Upland farmers would need to return to a mixed livestock system, which would require changes in infrastructure and farm management, but would be beneficial ecologically (see Chapter 5).

Another aspect of transition farming is to exploit alternative fuels such as:

- biomass crops – short rotation coppice, *miscanthus* and forestry woodfuel;
- biofuels – liquids produced from oilseeds, sugar beet or cereals;
- electricity supply – from wind, water and solar energy.

These sources sit well with the UK Government's Non-Fossil Fuels Obligation (NFFO) to produce 20% of electricity from renewable energy sources by 2020, but upland farmers are at various disadvantages to their lowland neighbours. First, upland is less adaptable to biomass and biofuel crops; for example, it is too cold and wet for *miscanthus*. Heaton *et al.* (1999) compared short rotation coppice (SRC) with HLCA and SAP sheep production in the Welsh uplands over 250m asl. They found that SRC would be an economic option for farmers only if subsidies for sheep ended. However, knowing the reduced rates offered by *Tir Mynydd* in 2009 for sheep in uplands and the proposed *Glastir* rates for 2012 (£33.60/ha), SRC has become more feasible as part of the enterprise mix on upland farms. It also provides other ecosystem services for sustainable catchment management (Chapter 2). A final option would be to encourage local production and consumption – people eating what is grown locally, reducing our dependence on oil – though for upland areas in capitalist economies this would lead to more restricted diets, which the public might not be willing to accept, being used to non-seasonal products on supermarket shelves.

A second challenge is the aesthetic impact of non-fossil fuel production. SRC could distort upland landscape character (NE 2009); wind turbines are an even more contentious issue, since the windiest parts of Britain correspond to many uplands with HNV (Countryside Commission 1991). Most wind turbine projects have been set back, cancelled or curtailed by local and national opposition (McLaren 2007; Cowell 2010). They appear more positive from an upland farmer's perspective: they do not preclude grazing and provide 'free' farm electricity, reducing electricity costs on farm by as much as 60% (Soil Association website, accessed 21 July 2009). They also bring in welcome rental from turbine companies and enhance agricultural land values (RICS 2004). Plenty of information and advice on energy use reduction in farm buildings can already be obtained from bodies like the National Farmers Union and the Carbon Trust.

Climate change

The second main vulnerability that upland farmers will have to deal with is the impact of climate change. Over the last 4,000 years the British climate has remained relatively stable, though intermittently less conducive to agriculture (see Chapter 3), but in the late 20th century researchers became concerned that industrial activity was changing our climate more rapidly than those natural fluctuations of the past. Particular blame was placed on rising use of fossil fuels and changes in land use, especially more agriculture and less forest cover.

Pollutants from economic activity have led to changes in composition of the atmosphere in two ways:

- Global warming – there has been an increase in chemicals that enhance the atmosphere's insulating properties (notably carbon dioxide and methane). Temperature rises encourage ice to melt, changing ocean salinity. In turn this changes the direction of currents. Because the global atmosphere and the oceanic system are interrelated, general climatic patterns begin to change and regions become wetter or drier, warmer or cooler. Other effects are rising sea levels and increased storminess.

- Ozone depletion - chemical pollutants are breaking down the natural ozone in our atmosphere (e.g. chlorofluorocarbons). Ozone reflects harmful ultraviolet rays from the sun back into space. Depletion of the ozone layer means more UV reaches the Earth's surface, damaging living cells of plants and animals.

Various computer models have been developed to try to predict the rate and character of the effects of such changes on human activity. However, as with all predictive modelling tools, output quality is only as good as that of the input. The first global climate models were fairly primitive, allowing sceptical governments to dismiss the scientists' claims as scaremongering. As our understanding of the global atmosphere-oceanic system has improved, so these models have become more refined and sophisticated, producing more valuable predictive modelling (as well as more trend evidence as years have gone by). Coupled with global treaties, notably the Kyoto Accord, bodies like the Inter-governmental Panel on Climate Change (IPCC) and reports such as 'The Economics of Climate Change' (Stern 2008), governments have been forced to take the threat more seriously.

Parry and Carter (1991) showed how a crop–climate simulation model could plot future enterprise changes in northern England. A mean annual temperature increase of 3.7°C would shorten growing seasons by two months and reduce yields by one third. In fact, Holden and Adamson (2002) demonstrated a real temperature increase of 0.7°C between the 1950s and 1990s on the Moorhouse NNR in the North Pennines. As a result, land at about 500m asl would be close to the optimum for growing winter wheat (currently not grown at all). The increase in the soil moisture deficit suggests cereals would flourish in uplands, rather than grass production. However, a climate model developed by Brouwer and Chadwick (1991) suggests the UK would experience a drop of 2°C in the mean annual temperature and an increase in rainfall by 0.8mm per day. The overall effect would be to lengthen the growing season in the north-west and south-east by three months.

These scenarios suggest agricultural production based on grass will no longer be central to upland farming and there will be a switch to arable enterprises. Pepin (1995) corroborated this by calculating increases in annual accumulated temperatures above 6°C, enough for cereal production in the northern Pennines, eliminating all marginal and sub-marginal land to 800m asl. Unfortunately, many of these models do not take account of the effects of climate change on other environmental parameters, such as soil, which will probably limit cereal production to areas with more than 30cm depth.

Anecdotal evidence suggests that changes in weather patterns are already causing management challenges for upland farmers in parts of Britain. Earlier supplementary feeding in northern Cumbria and housing cattle earlier in barns in the Forest of Bowland are two such examples (H. Martin, pers. comm., 2007). Agricultural research has identified several trends (West and Gawith 2005):

• the potential to grow new crops • longer growing seasons positive • enhanced yields	positive
• more/different pests and diseases • increased need for irrigation • damage to crops and soil from intense winter rainfall • difficulty accessing fields due to increased winter rain • reduced yields with no irrigation	negative

These predictions of climate change are based on further models developed by organisations like the Hadley and Tyndall Centres (DEFRA 2002). They describe climate scenarios for four levels of greenhouse-gas emissions and three time slices: 2020, 2050 and 2080. All the scenarios predict higher temperatures with changing patterns of precipitation leading to hotter, drier summers and wetter, stormier, milder winters. These trends vary for the five main upland areas:

South West:	Loss of competitive advantage from more conducive climate
Wales:	Decreased yields, more animal heat stress
North West:	Increased soil erosion, more animal heat stress
North East:	Increased soil erosion, increase in growing season and more arable farming
Scotland:	Increased demand for water in the east

Climate change issues specific to extensive grazing can be divided into forage and stock management. Forage concerns include grassland productivity, seasonal shifts, palatability and weed invasion. Scenarios suggest that in the short term grassland productivity will increase, though limited by nutrient availability. However, long-term climatic warming may warrant increased use of fertilizers, pushing more intensive livestock production up fell, probably leading to more poaching, damage to soil structure and eventually soil erosion. Seasonal shifts are also apparent in lengthening growing seasons, which increase opportunities for early winter grazing. Again this could create problems with soil erosion, but a more likely issue is overgrazing of plant species that are usually left to recover at this time of year. An indirect spin-off of lengthened growing seasons is a drop in the need to off-winter stock. This could save money in off-wintering fees and allow farmers to fatten livestock, adding to the individual animal's marketable value, but it would also require farmers to re-adjust their farm management systems to accommodate more stock on inbye and intake, reducing business for recipient farms. This could disrupt flock and suckler herd rotations, as well as affect early spring flushes of fodder, which would normally occur before the stock arrived. In turn, it could also force the need for more supplementary feed from concentrates and more conversion to silage fields.

If farmers did adapt their management to keeping stock on the farm all year, the effect of this would cascade across the farm stratification system by drying up the movement of stock up and down chain. This would undermine the viability of some lowland stock farms. It might not, of course, be such a challenge because lowland farms would probably be suffering from increased drought conditions anyway and thus be unable to support the usual influx of upland stock. These farms might then convert to arable, the corollary being an even more acute dichotomy emerging between lowland arable and upland stock systems in the UK. With such a scenario, it could be that true hill farmers would benefit more than upland farmers, because hill farmers could find themselves shifting away from the open fell–intake system to the triparite system found on upland farms. Upland farmers would probably be in a less enviable position, as described above.

However, other predicted changes in the fodder system could play against such a scenario unfolding, because there is evidence that palatable species will decrease and thus grassland nutrition will be lower in the long term (Glynn 2005). As well as the obvious stress on animals' health and greater probability of overgrazing, agriculturally viable land would contract. The effect could be further lowering of appropriate stocking densities, making many hill and upland farms uneconomic even at their current size. Farmers would be forced to sell up or extend their grazing lands over larger and larger areas (ranching). The only way this could happen is by combining hefts on fells into larger units, which would ultimately bring with it practical gathering and personnel problems. A shift uphill does not, however, mean more agricultural land, as has happened previously. The new land will probably not compensate for losses at lower altitudes, compounded by inappropriate soil types.

In stock management, the main problems are expected to be health and transport (DEFRA 2002). Stock health will be affected as temperatures rise because disease will increase and severe cold snaps diminish, which normally kill off bacteria, viruses and zoonoses. Upland stock will also suffer from more heat stress because many are adapted to cooler environments. This will require more animal shelters, creating extra expense for farmers and problems of landscape aesthetics. Animal transporters will require more ventilation, adding yet again to cost of production. There may also be a decline in reproductive efficiency, making upland farms less efficient and increasing the cost–price squeeze.

Farm management regimes will have to be adapted to these new conditions. Upland farmers may be able to increase silage production, but they will spend more on disease control. Wetter winters will require more erosion control on steeper slopes and bring more catastrophic events. With changing conditions more capital will be needed, as well as new practical management skills to tackle new challenges. However, this debate may be academic because land may not be needed if the national flock and herd decrease due to wider policy and public dietary changes.

Food security

The World Food Summit in 1996 defined food security as the situation when all people have access to sufficient, safe, nutritious food to maintain a healthy and active life (WHO website, accessed 09/03/2010). Food security rests on:

- Food availability – sufficient quantities of food are available constantly;
- Food access – people have sufficient resources to obtain a nutritious diet of appropriate foodstuffs;
- Food use – people know about nutrition and act appropriately, and have adequate water and sanitation.
- Other considerations are food quality and safety (Food Security website, accessed 09/03/2010). Food security has become a major issue because several challenges are coalescing.

By 2050 it is estimated that the global population will have reached 9 billion (in 2009 it was 6 billion); as a result the FAO (cited in DEFRA 2010a) suggest we may need 70% more food than we now produce. A second concern is the way our diet has become more meat-based and less vegetable/cereal-based. Meat and dairy products take greater inputs to produce than cereals and vegetables. For every 1kg of meat we produce, we need to supply 3kg of cereal for feed (DEFRA 2010a). Alongside higher transport and energy costs, this will force agricultural prices to rise by 20% in the next ten years. Also, biofuel production has reduced arable land area, exacerbated by changing precipitation and temperature regimes. These food-security problems suggest that nations would be better off producing their own food. The UK, for instance, could be self-sufficient if meat consumption fell slightly and we stopped using cereals as animal feed. Complete self-sufficiency, however, would be unwise: it would not protect against poor harvests or new diseases, nor meet the full range of dietary needs.

In response to this looming crisis the UK Government commissioned and published a series of documents on food security. *Food 2030* lays out government strategy for twenty years (DEFRA 2010b). It is based on a comprehensive assessment of our food security, under six heads (DEFRA 2010a):

- Global food availability
- Global resource sustainability
- UK food availability and access
- UK food chain resilience
- Household affordability and access
- Food safety and confidence

The UK now supplies about 50% of its food needs, with another quarter coming from five other countries (Netherlands, Spain, France, Germany and Ireland). The UK produces 80% of its meat, 84% of its dairy and poultry, 50% of cereals and 19% of fresh fruit and vegetables (DEFRA 2010a). The general view is that the British livestock industry is in a strategic position to negate some concerns about food security because 80% of animals do not rely on concentrates and grain for feed. Thus upland meat production could regain some of its comparative advantage lost historically (see chapters 3 and 7). Climate change could also increase arable cultivation and forage production on inbye land, though this would be at the expense of valley-bottom semi-natural habitats and floodplain management (Natural England 2008e).

The challenges of peak oil, climate change and food security are therefore inter-related (Shiva 2009). At farm level, upland farmers need to adopt some form of transition farming to cut costs, but that may have an impact on ecology and landscape. The practicalities of extensifying even further may also preclude the most economic businesses from opting for it. Livestock production affects climate change and is in turn affected by it. The need to cut meat intake in the West may reduce production, but the developing comparative advantage of upland forage-grazed stock over grain-fed lowland stock should put upland farmers in a strong position, as long as the public are prepared to pay more for meat. But there are other vulnerabilities that upland farmers need to be aware of.

Over-reliance on direct support payments

Pillar I payments are gradually being replaced by Pillar II. In 2004 the SPS averaged 77% of TIFF on UK farms. From 2012 onwards the amount of money available under the Single Farm Payment in the UK will probably shrink, as decoupling increases and the EU enlarges (Box XL; and see below).

A report by ADAS UK (2008) on the potential impact of complete decoupling (no more SPS) and the removal of trade barriers on English agriculture estimated that as much as 15% of land could go out of production, most probably in upland areas, and there could be a 25% decline in cattle and 35% decrease in sheep. These trends have been corroborated by Vrolijk *et al.* (2010), who predict that 50% of livestock farms in the UK will be worse off after complete decoupling, moving from positive to negative net income while farmers will also have to compensate for structural change. The ADAS report concluded that such a situation would result in extensification, benefiting biodiversity and soil erosion, but might lead to undergrazing, which can be a disadvantage to biodiversity depending on your site objectives. ADAS also foresaw a collapse in cross-compliance as production increased to offset loss of SPS. It would only be maintained by farmers operating voluntary ESS, so non-adoption of ESS would reduce the environmental quality of farmed land. These specific changes were corroborated by a 2006 farmer survey in the Exmoor National Park. Lobley *et al.* (2006, cited in Spedding 2006) found that introducting an upland rewards scheme to replace subsidies would lead to about 17% of the land being abandoned and 60% of farms reducing or stopping livestock farming.

The move from HFA (an entitlement) to UELS (a choice) in 2010 caused much consternation among farmers. In 2008 in England, 350,000 ha – 27% of farmed SDA land – was not in any agri-environment scheme: ESA, CSS or ESS (DEFRA 2008b). Logically, this land should be entered into UELS fairly quickly to replace income lost from the removal of HFA, to be followed over the next five years by another 50% of the English SDA, as CSS and ESA agreements end, and farmers switch to UELS. Because a number of upland farmers will be ineligible for UELS, as they currently hold agri-environment agreements, the government has agreed to introduce transitional payments (cost about £9m, tapering to £2m in 2014) to help replace HFA. These will be about £9.30/ha, with farmers averaging £3,300. DEFRA (2008b) also estimated additional costs for farmers joining UELS at 7 to 14% of costs, at most £3.30/ha more. It may not sound much but, as shown in Table 9.7, rising costs of fuel will increase the burden. Anecdotal evidence is already emerging that upland farmers are worse off under UELS than they were with either ESA or HFA (Farm Business North West Event 10/2/11).

Redefinition of LFAs

Long overdue, LFA re-delineation is desperately needed. As we saw in chapters 2 and 8, various attempts have been made to clarify their definition. Under the three current criteria, 31% of agricultural land in the EU27 is classified as intermediate, whereas only 16.4% are Mountain LFAs (CEC 2009b).

Following non-agreement on reclassification in 2005, a 2009 EU project is trialling a set of biophysical criteria and thresholds based on climate, soil and terrain parameters (Table 9.8). To be classified as LFA under these proposals, an area must meet at least one of the criteria at the threshold. A report by Wright *et al.* (2006) investigated the potential of the Scottish LCA in identifying land under these new criteria. They found that 94% of current LFA land falls into Class 4.1 or poorer (down to 7), suggesting that the LCA could be used successfully for the new LFA definition. The situation is probably similar in Wales, given the little DA area that exists. It is England that would probably suffer the most.

This has already happened because in 2008, in preparation, the English LFA was cut back to the core SDAs, a reduction of 25%, yet many peripheral upland areas beyond the SDA would fit at least one of the criteria listed in Table 9.8 (see Chapter 4). DEFRA put forward three arguments for these changes. First, DA land is typically of better agricultural quality and therefore more flexible. Second, SDA farmers rely more on subsidies than DA farmers and the

> gap will continue to widen as a result of the CAP reform because the DA is subsumed within the lowland for the purpose of defining SPS regions and therefore receives a higher flat create payment than the SDA. (DEFRA 2007b: 5-39)

Third, SDA land has higher landscape and biodiversity value. DEFRA identify 16% of SDAs are SSSI compared to only 1% of DA. Supporting their claim is initial analysis by the Commission for European Communities (2009c), which does not identify the UK as a state where farm incomes would be affected negatively if the new LFA criteria were applied and if land that was now intermediate were to fall outwith the proposed thresholds. More important is the proportion farm by farm of income derived from LFA payments in relation to overall farm income. The average for the UK is 18% which is in the 'not overly reliant' category, but this figure hides regional disparities. The challenge is whether the 2012 definition increases financial hardship for upland rather than hill farmers. Other European countries like Poland or Slovakia are more at risk in this instance.

Enlargement of the EU

The expansion of the EU to include central and eastern European countries poses new challenges for the CAP. A number of the countries that joined in 2004 have extensive uplands (Table 9.9), large farming workforces and a disproportionately large part of their GDP coming from food production. Agricultural support for these states takes the form of direct payments to encourage modernisation and restructuring or for rural development. Several transitional measures exist under the EARFD, supporting semi-subsistence farming and the formation of producer groups in the new member states. Some of these countries could put themselves in direct competition for various sectors that have comparative advantage in the uplands, for instance the Czech Republic and Croatia with larger organic production and Poland with cheaper milk (DEFRA 2005).

A second group of countries are now candidates (Macedonia, Turkey and Croatia), while the western Balkan states of Bosnia-Herzegovina, Albania, Montenegro, Serbia and Kosovo are potential candidates. All of these have substantial mountainous and upland landscapes, many of which still operate traditional transhumance systems (Table 9.9). The agricultural structure in

Table 9.8: Biophysical Criteria under review by EU for new LFA Designation

criterion	*definition*	*threshold*
climate		
Low temperature	Length of growing period – number of days below 5°C	Less than or equal to 180 days OR
	Accumulated daily average temperature above 5°C number of degree days	Less than or equal to 1500 degree days
Heat stress	Number and length of continuous periods (days) within the growing period that exceeds the threshold	One or more period of 10 days or more with daily maximum above 35°C
soil		
Drainage	Areas waterlogged for a significant duration of the year	Poorly drained soil
Texture and Stoniness	Relative abundance of clay, silt, organic matter and coarse material fractions in the top soil	More than 15% topsoil is coarse material OR Unsorted, coarse or medium sand, loamy coarse sand OR Heavy clay (over 60% clay) OR Organic OR Vertisols or with vertic (clay) properties OR Rock outcrop, boulder within 15cm of the surface
Rooting depth	Depth (cm) from soil surface to coherent hard rock or hard pan	Less than 30cm
Chemical properties	Presence of salts, exchangeable sodium and gypsum in the topsoil	Salinity more than 4 dS/m OR Sodicity more than 6 exchangeable Na % OR Gypsum more than 15%
soil & climate		
Soil moisture balance	Days within the growing period (above 5°C) where precipitation and water available in profile exceeds half potential evapotranspiration	Less than or equal to 90 days
terrain		
Slope	Change of elevation with respect to planimetric distance	Greater than or equal to 15% (1:6.6)

(Source: CEC 2009c.)

these countries is based on small, family-run, subsistence-level operations with up to 80% of farms under 5 ha (www.europa.com.eu, accessed 28 June 2009). To join the EU, these states are required to meet a range of political, economic and social criteria on which the EU was originally founded. The CAP is one of the more difficult negotiations for all these candidate states, but entry is based on them embracing modernisation, rural development and environmental management.

For any farmers in the current EU-27, continued enlargement of the Union beyond 2013 will increase pressure as limited agricultural funds are shared round (DEFRA 2005; Box XL). Financial support under Pillar II for existing members will have to decline to compensate. In the

Table 9.9: EU Enlargement

country	*agriculture as % of GDP*	*average farm size (ha)*	*mountain area as % of whole*
candidate or proposed candidate			
Albania	22.6	1.1	81
Bosnia–Herzegovina	10.4	2	79
Croatia	6.9	2.4	40
Kosovo	30.0	0.5 to 2	*no data*
Macedonia	13.2	2.5 to 2.8	90
Montenegro	11.3	5	94
Serbia	16.6	<10	53
Turkey	11.6	6	78
accession in 2007			
Bulgaria	7.5	6.2	49
Romania	12.4	3.5	38
accession in 2005			
Poland	4.6	12.3	5
Czech Republic	2.6	89.3	33
Hungary	3.3	6.8	5
Slovakia	2.7	28.1	60
Slovenia	2.4	6.5	76
(UK) (Comparator)	(0.9)	(76.4)	(54)
EU27	**1.6**	**12.6**	**29**

(Sources: www.europa.eu, accessed 28/07/09; EEA 1992; EEA 2010.)

UK this means a loss of £73m in 2013 (about 3% of the current budget). Funds could be channelled to states or even territories within states that do not fit the EU average. The CAP may be forced to adapt to new members, rather than the other way round, which could have a profound affect on LFA status and opportunities (www.europa.com.eu, accessed 28 June 2009). On the other hand, in many of the candidate states a high proportion of the population work in agriculture for low wages; in view of this, there could be a migration of workers from east to west across the Union which could benefit upland farm businesses elsewhere.

The vulnerabilities external to the upland farming sector are therefore daunting. They demonstrate how farming is no longer a local or even a national matter, but one controlled and influenced by global issues. As if these were not enough, the upland farmer also faces a number of constraints in the sector itself.

Reducing constraints, realising assets

Constraints affect an upland farmer's ability to exploit any emerging agenda. We have identified and discussed at length in this book a number of constraints:

- Declining human capital
- Lack of financial capital
- Poor market valuation of non-market goods
- Lack of appropriate new knowledge and skills
- Inadequate public understanding
- Decline of local knowledge systems and related social capital

As chapters 6 and 7 showed, there are several issues in upland agriculture related to declining human capital in the workforce. Depopulation can be dealt with in two ways. One solution is to accept it and adjust the farm management system to 'ranchification', not unlike systems in the Australian outback or North American prairies. There are three problems with this approach for uplands. First, practical farm management systems may not be adaptable enough across fell tops and commons. Second, there could be detrimental effects on the ecology through less controlled stock behaviour. Finally, allowing depopulation is simply not politically acceptable in rural development policy.

The other solution is to endeavour to revive interest in upland farming as a vocation. The FFTS described in Chapter 8 was a pilot for this and is now being followed in a number of upland areas in Britain. However, numbers are small and it may all be too little, too late. Related issues – the lack of affordable housing, low service levels and poor incomes – will also need addressing if we are to retain those who do want to work in the sector. Planning regulations for conversions on hereditaments will need relaxing as well, something which the LDNPA is now considering (pers. comm., R. Leafe, Hill Farming Matters conference Jan 2010).

The majority of upland farmers are effective stockmen, able to produce increasingly high-quality meat for the table. More farm businesses, often operated by other family members, are exploiting premium niche markets. To increase the diversity of products coming from farms, there needs to be more support for business and entrepreneurial innovation to combat the lack of appropriate knowledge and skills. Particular areas in need of support are environmental goods, micro-generation, energy self-sufficiency and new agendas like carbon storage. Some organisations can provide these skills already, for example, the National Farmers Network or Farm Business Link. Attention also should be paid to the knowledge and skills that young upland farmers receive in their formal education at Level 2, the main qualificatory level of LFA farmers. The agricultural content is high and of good quality, but more rounded business and environmental skills are needed – farming is no longer just that. Changes in curricula to reflect the new agenda (LBCNC 2009) and broader discipline-based teaching teams in agricultural colleges would help to address this.

More informal mechanisms should be rolled out for the current farming workforce, as has happened in Wales (*Tir Gofal*). A recent survey looking at the UELS found that 38% of upland farmers did not understand how to calculate their points target until they had received one-to-one advice (CSL 2008, cited in DEFRA 2008b). Perhaps initial training should become a requirement of receiving SPS, ESS or UELS grants, as in Northern Ireland. Under the current NICMS, participants must take three approved two-hour outreach training courses on cross-compliance, field-boundary management, farm nutrients and waste management. These are run by the College of Agriculture, Food and Rural Enterprise, which also provides voluntary workshops in related environmental skills (DARD 2007). Although DEFRA have allocated £1m from 2010 to help UELS applicants, details of how this will be delivered are yet to be published, and obviously £1m doesn't go very far. Hopefully it will be a combination of farm- and college-based activities, to try and avoid too much top-down delivery, so despised by farmers.

Policy initiatives of the last 30 years have gone some way to addressing the lack of financial capital on upland farms, through a combination of production subsidies and agri-environment initiatives. We know that these subsidies are unpopular with the public and that the EU now employs more positive support mechanisms, providing multifunctional products, landscape or otherwise. All upland farmers should therefore be encouraged to exploit the HNV element of their farms, and hopefully new schemes like UELS and *Glastir* will accommodate this by being more directed to upland circumstances than current equivalents. However, this should be achieved through an SLF approach rather than top-down diktats.

As we have seen, ever since the first agri-environmental initiatives in 1984, financial support for environmental management has always been and continues to be on an income-foregone basis; that is, the grant paid should be equivalent to the money a farmer would have got from farming the land, instead of extensifying or directing the land out of agriculture. There are problems with this approach. First, there is a negative perception among farmers that they are being paid NOT to farm, but to do other things that are not mainstream to their core business. Second, there is a supposition at both EU and national levels that the mechanism of income-foregone is enough recompense, supported by research showing that farmers are not out of pocket and some may make a small profit (Chapter 8). There is, nevertheless, copious evidence from past agri-environmental schemes that non-adoption is because payment rates are too low to compensate for the work required. This was clearly demonstrated by the pathetic adoption rate of the Moorland Scheme, low participation in certain ESAs and in certain options within ESAs, particularly in the higher tiers where more management was demanded.

These systems do not, however, allow farmers to accrue financial capital in any large amount, making diversification away from farming rather difficult. Inherently risk-adverse, the farming community avoids debt and loans in general. Sale of land to subsidise a new enterprise is almost unheard of because, if the new activity failed, the farm would be 'worse off'. This is why grant schemes such as RES and the PMG were beneficial, but again they did require matched funding. Even the new RDPE continues to use this system, precluding many financially marginal farms from deriving benefit from this scheme.

We also lack a method to value non-market goods adequately. If we are to embrace multifunctionality and ecosystem services as part of the new agenda for upland agriculture, we need to ensure that we have parity of value. Production of provisioning services is obviously the easiest to value because they operate in an economic market. Regulating services sit at the other end of the spectrum where we have no basis yet to value them financially for society. If we could do this, say, for the maintenance of one hectare of peat bog for carbon storage and sequestration, we could develop service grants for upland farmers. Thus, how much peat does a farmer need to maintain to offset the use of a petroleum-based car for a year? In other words, what is the cost to society of not maintaining the peat? The cost aspect could suggest a solution, because it is easier to calculate the cost downstream of flooding (the clean-up) or not flooding (erection of flood defences and catchment management) based on experience. However, these types of costs can be hard to calculate if we do not have perfect knowledge of the ecosystem itself.

Somewhere in between lie the cultural and supporting services. The financial value of these non-market goods can be derived in two ways:

1. income foregone,
2. non-market valuations.

Historically and currently biodiversity, landscape and heritage grants are valued by the first method (e.g. ESA, ESS payment rates) and, as we have seen, are flawed. The non-market techniques calculate their financial value as the public view them (WTP and choice experiment Box XXIX). However, relying on the public to value ecosystem services could be misleading. First, it will depend on people's knowledge of the importance of the service to their everyday life; thus ignorance could over- or underestimate its worth. Second, there is inadequate understanding among the public in that they do not recognise certain landscapes as even a product of upland agricultural activity (e.g. Moran *et al.* 2004). Positive contact between the public and farming community is limited to informal voluntary experiences such as agricultural shows, chance meetings whilst out walking and short food-supply chains. At the same time, the public do appreciate the negative aspects of the industry, such as food scares, diseases like BSE and FMD, and animal-welfare issues. This is underpinned by the portrayal of agriculture on television, in the newspapers and on the Internet. To overcome this, we need to go out and meet the public on their own terms, use venues and contexts with which they are familiar. In one such project under way in Cumbria, supported by RDPE, information about the post-2013 agenda is going out to the public through libraries, extant social gatherings and third-party goods packaging (University of Cumbria 2011).

The last issue to address is the decline in local knowledge systems and related social capital. We can argue that all the other constraints stem from this one, the ability to keep people farming in a context that relies on a unique skill set. We identified in Chapter 7 a number of keystone processes and structures that need maintaining if we are to perpetuate this farming system. Heft management, the tripartite land types, social networks and connections, together with a sense of place, all need to be maintained in some way. Many of these come from within communities, so the solution to the hill farming problem needs to come from there. However, they cannot do it alone, due to all the constraints above. Alleviating these should empower the upland communities themselves and should ultimately be the responsibility of government, if the philosophy of rural development is socially acceptable. Nevertheless, history has shown that top-down policies, strategies and initiatives do not really work in an environment that is so geographically variable. The more the institutions tighten their grip, the more farming systems and their communities will slip through their fingers.

The answer is that institutions need to invest in bottom-up regeneration that places trust back in communities on a stakeholder–partnership basis. A starting point would be to employ an SLF approach by asking farmers and farming communities to identify their assets in five main ways: physical, financial, human, environmental and social (Figure 9.5). The rub is how to help farmers, their families and farming communities to recognise the range and depth of what they have individually to draw upon, and then provide specific support based on their unique situation. These types of project are only just beginning to trickle into the upland arena. The National Farmers Network (Ponder and Hindley 2009) worked with 16 farming families in the Peak District to identify their ability to build greater resilience and sustainability into their livelihoods. Another example is an evolving

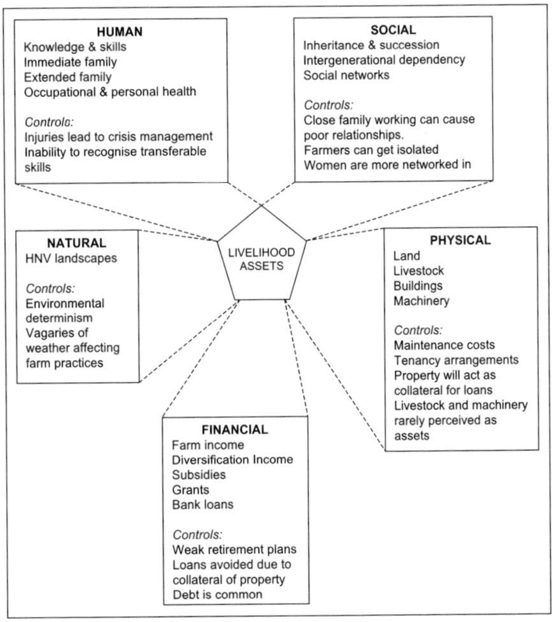

Figure 9.5 Identifying Upland Farm Assets
Based on the framework proposed by DFID, 1999.

post-2013 whole-valley planning exercise supported by the RPDE LAGs in Cumbria, whereby communities identify their assets and develop sustainable options (University of Cumbria 2011 website). In adopting this approach, short-term costs may rise steeply, but in the long run the bill will be less, as upland farming communities build their resilience to fit their own circumstances.

Support, therefore, needs to be driven from the bottom up, at the point of consumption (the farm, farmer, farming community), and it needs to pay for all services on an equal footing, returning upland farming to its multifunctional roots from which it has strayed. Unfortunately the first parameter may be seen as undesirable, because it means institutions and governments are going to have to let go control and power, the second is difficult because it is expensive and time-consuming, and the third is still out of reach given our limited understanding of non-market valuations. The alternative, of course, is to continue relying on the same tools we have used since the 1940s. If we do, our upland inheritance will seriously risk being poorer socially, culturally and environmentally.

Concluding Remarks – what can be done?

Some form of intervention is therefore a necessary evil. For the uplands, full engagement within the farming community needs to occur. Farmers will then be in a position to take better decisions about what is best economically, socially and environmentally for their individual businesses and to help them work more collaboratively in order to achieve effective catchment-scale management.

Our first objective is to develop a secure economic basis for upland farmers. This can be achieved initially by providing fair market value for their core products of meat, wool and milk through a mix of commodity and premium marketing. The next step is to expand financial value to other ecosystem services produced by upland agriculture, such as carbon and habitat. Whilst provisioning services need to be market driven with no centralised support, these other ecosystem services such as carbon storage need a quantitative allocation system by farm per unit price. These rewards must be more than just income foregone to demonstrate that society financially values all forms of ecosystem service production equally. For example, feed in tariffs for carbon neutral farming similar to those beginning to operate for renewable energy micro-generation. For habitat management, it should not just be the cost of the act of habitat management, but the financial benefit gained from the management act in other ways (such as the reduction in cost of reservoir dredging downstream of soil erosion that has been halted by the act of management).

Second, we need to devise an approach to support social rural regeneration in uplands. The fundamental action here is to provide basic local services in remote rural areas that allow farmers and their families to remain, such as schools, shops, health care, post offices and community facilities. Practical examples of this include full access to broadband, appropriately focused training and provision of community social events. Probably the hardest of these is the training element as we need to break out of providing that which is within a farmers' comfort zone towards that which is required, even to the point of not using the word 'training', but perhaps whole farm planning. The first step is to focus on enhancing individual farm income by improving the flock/herd, farming enterprise mix, taking advantage of contemporary farm diversifications (for example, electricity micro-generation) and general awareness raising of what delivering ecosystem services can do for farm incomes. These topics can be addressed in a variety of forums and so could include schools and colleges, women's groups (such as the Rural Farming Women's Network), Young Farmers Groups and auction marts. After that, farm-by-farm asset identification and development needs to be rolled out based on the Sustainable Livelihoods Framework. This would be a joint exercise between the farmer, farm family and an 'off-farm facilitator' team.

A second social action is to deal with farm succession. Currently the typical process is for those with a farming background to go into farming, often, but not always via inheritance. A minority are from outside the farming community. For the industry to be sustainable there needs to be avenues into farming from the general populous like any other type of employment. This could be achieved through starter farms with short tenancies, financial support for first herds/ flocks, formal training funding and a paid farmer mentor service.

Finally, we need to support appropriate environmental management on uplands to achieve national biodiversity targets. The approach here is to develop prescriptions which revolve around what farmers are good at (the production of sheep and cattle) along with the ancillary related management tasks, such as wall and hedge maintenance, ditching, production of fodder and the extensive management of marginal agricultural land (the non application of fertilisers and pesticides, lack or reversal of drainage, timing of grazing, etc.). An essential element of this approach is to ensure that biodiversity experts actually understand how the upland farming system operates and the constraints and vulnerabilities farm businesses operate. This will allow for an adaptive approach to management prescription formulation farm by farm and not through fixed 'one size fits all' approach, which is one of the main 'bug bears' raised by farmers currently.

The varied individual solutions to each of the three main objectives will also create synergies between them. For example, by increasing the financial income of farms to at least if not more than the average annual wage through a combination of ecosystem service prices and tariffs, more people from outside the sector would consider farming as a career. In turn, appropriate training and subsequent management behaviours would work towards nationally agreed upland biodiversity targets. The overall effect is that the whole is greater than the sum of the parts.

We need our uplands more than ever, for food, for biodiversity and for our survival. Working with farmers, towards pluralistic benefits, secures farm livelihoods. In turn the environment profits through positive low cost management and the public, as the taxpayer, can continue to enjoy High Nature Value landscapes.

References

Ackrill R (2000) *The Common Agricultural Policy.* Sheffield Academic Press: Sheffield.

ADAS (1996) *Socio Economic Assessment of Tir Cymen.* Research Report to Countryside Council for Wales. ADAS: Aberystwyth.

ADAS (1997) *Lake District ESA – report of environmental monitoring, 1993 to 1996.* ADAS: London.

ADAS Consulting (2005) *The Economics of Extensive Livestock Grazing post CAP Reform – analysis of ADAS Farmers' Voice 2004 survey.* ADAS Consulting: Wolverhampton.

ADAS Consulting and University of Reading (2003) 'Evidence assessment to inform the review of the Organic Farming Scheme.' ADAS: London. Accessed at www.statistics.gov.uk/esg/evaluation/ofs/whole (10/11/2003).

ADAS UK and SAC Commercial (2008) *Estimating the Environmental Impacts of Pillar I Reform and the Potential Implications for Axis II Funding.* ADAS UK: Wolverhampton.

Adger WN (2000) 'Social and ecological resilience: are they related?' *Progress in Human Geography,* 24: 347–64.

Adkins L and Adkins RA (1998) *The Handbook of British Archaeology.* Constable: London.

Agate E (2003) *Hedging – a practical handbook.* BTCV: Wallingford, 2nd edn.

Agate E (2005) *Footpaths – A practical handbook* BTCV: Wallingford, 3rd edn.

Agro Business Consultants (2008) *The Agricultural Budgeting and Costing Book.* ABC: Melton Mowbray, 68th edn [known as 'The Green Book'].

Aitchison J (1990) 'The commons and wastes of England and Wales, 1958 to 1989' *Area.* 22(3): 272–7.

Aitchison J, Crowther K, Ashby M and Redgrave L (2000) *The Common Lands of England: a biological survey.* University of Aberystwyth: Aberystwyth.

Aitchison J and Gadsden G (1992) 'Common land', Ch. 8 in Howarth W and Rodgers CP, *Agriculture, Conservation and Land Use – law and policy issues for rural areas.* University Press of Wales: Cardiff.

Ajzen I and Fishbein M (1980) *Understanding Attitude and Predicting Social Behaviour.* Prentice Hall: Englewood Cliffs NJ.

Alderson L and Small R (1997) 'Rare breeds, natural choices'. *Enact,* 5(4): 4–7.

Anderson I (2001) *Foot and Mouth Disease 2001: lessons to be learned,* Inquiry Report. The Stationery Office: London.

Anderson P (2002) 'Diversity from *Molinia* moorlands'. *Enact,* 10(4): 15–18.

Anderson P and Radford E (1994) 'Changes in vegetation following reduction in grazing pressure on the National Trust Kinder Scout Estate, Peak District, Derbyshire, England.' *Biological Conservation,* 69: 55–63.

Anderson P, Tallis P. and Yalden D (1997) *Restoring moorland. Peak District moorland erosion study: Phase 1 report,* ed Phillips J, Tallis J and Yalden D. Peak Park Joint Planning Board, Bakewell.

Anderson P and Yalden D (1981) 'Increased sheep numbers and the loss of heather moorland in the Peak District, England.' *Biological Conservation,* 20: 195–213.

Ardron P, Gilbert O and Rotherham I (1997) 'Factors determining contemporary upland landscapes: a re-evaluation of the importance of peat cutting and associated drainage, and the implications for mire restoration and remediation' in Tallis JH, Meade R and Hulme PD, *Blanket Mire Degradation: causes, consequences and challenges*, pp. 38–41. English Nature: Wakefield. [Proceedings of the Blanket Mire Degradation Conference, 9–11 April 1997, University of Manchester.]

Armstrong H (1993) 'Taking stock of sheep.' *Enact*, 1(4): 7–9.

Armstrong HM, Gordon IJ, Grant SA, Hutchings NJ, Milne JA and Sibbald AR (1997a) 'A model of the grazing of hill vegetation by sheep in the UK: I. The prediction of vegetation biomass.' *Journal of Applied Ecology*, 34: 166–85.

Armstrong HM, Gordon IJ, Hutchings NJ, Illius AW, Milne JA and Sibbald AR (1997b) 'A model of the grazing of hill vegetation by sheep in the UK: II. The prediction of offtake by sheep.' *Journal of Applied Ecology*, 34: 186–207.

Armstrong HM and Milne JA (1995) 'The effects of grazing on vegetation species composition' in Thompson DBA, Hester AJ and Usher MB, *Heaths and Moorland: cultural landscapes*, pp. 162–71. HMSO: Edinburgh.

Arnstein SP (1969) 'A ladder of citizen participation.' *American Institute of Planners*, 35: 216–24.

Astill G and Grant A (1992) *The Countryside in Medieval England.* Blackwell: Oxford.

Aston M (2000) *Monasteries in the Landscape.* Tempus: Stroud.

Atherden M (1976) 'The impact of late prehistoric cultures on the vegetation of the North York Moors.' *Transactions of the Institute of British Geographers*, 1 (NS): 284–300.

Atherden M (1992) *Upland Britain: a natural history*. Plantlife and Manchester University Press: Manchester.

Atkin MA (1994) 'Land use and management in the upland desmesne of the De Lacy estate of Blackburnshire c.1300.' *Agricultural History Review*, 42(1): 1–19.

Attwood EA and Evans HG (1961) *The Economics of Hill Farming.* University of Wales Press: Cardiff.

Audsley E, Annetts JE, Pearn K and De Baets A (2000) 'Integrated impacts on agriculture', Chapter 6 in *REGIS – Regional Climate Change Impact and Response Studies in East Anglia and North West England.* DEFRA: London; UKCIP: Oxford.

Avery BW (1980) *System for Soil Classification for England and Wales (Higher Categories).* Technical Monograph No. 14, Soil Survey: Harpenden.

Avery BW (1990) *Soils of the British Isles.* CAB International: Wallingford.

Backshall J (1999) 'Managing bracken in the English uplands.' *Enact*, 7(2): 7–9.

Backshall J, Manley J and Rebane M (2001) *The Upland Management Handbook.* English Nature: Peterborough.

Bacon J (1999) 'Back to purple with mean machines.' *Enact*, 7(2): 4–5.

Baker ARH (1973) 'Changes in the later middle ages' in Darby HC (ed.) *A New Historical Geography of England*, pp. 186–247. Cambridge University Press: Cambridge.

Balfour AJ (1942) *Report of the Committee on Hill Sheep Farming in Scotland.* Dept of Agriculture for Scotland: London.

Ball DE, Radford GL and Williams WM (1983) *A land character data bank for Great Britain.* Institute of Terrestrial Ecology: Bangor.

Ballantyne CK and Harris C (1995) *The Periglaciation of Great Britain.* Cambridge University Press: Cambridge.

Banks J and Marsden T (2000) 'Integrating agri-environment policy, farming systems and rural development: Tir Cymen in Wales' *Sociologia Ruralis*, 40(4): 466–80.

Barclay C (2001) *Farming after Foot and Mouth.* Research Paper 01/67, Science and Environment Section: House of Commons Library.

Bardgett RD and Marsden JH (1992) *Heather Condition And Management in the Uplands of England and Wales.* English Nature: Peterborough.

Bardgett RD, Marsden JH and Howard DC (1995) 'The extent and condition of heather in moorland in the uplands of England and Wales.' *Biological Conservation*, 71: 155–61.

Barrett J (1997) 'Moor gripping in the uplands.' *Enact*, 5(1): 16–17.

Barry RG, Chorley RJ and Serreze M (2009) *Atmosphere, Weather and Climate.* Routledge: London, 9th edn.

Bateman D and Ray C (1993) 'Pluriactivity in the less favoured areas of nature, extent and variability', Ch. 3 in Bateman D, Hughes G, Midmore P, Lampkin N and Ray C, *Pluriactivity and the Rural Economy in the LFAs of Wales.* Department of Economy and Agricultural Economics, University of Aberystwyth: Aberystwyth.

Battershill MRJ and Gilg AW (1997) 'Socio-economic constraints and environmentally sensitive farming in the southwest of England.' *Journal of Rural Studies*, 13(2): 213–28.

Bayfield NG (1979) 'Recovery of four montane heath communities on Cairngorm, Scotland, from disturbance by trampling.' *Biological Conservation*, 15: 165–79.

Bazeley K and Hayton A (2007) *Practical Cattle Farming*. Crowood Press: Marlborough.

Beaton C (ed.) (2007) *The Farm Management Handbook 2007/8: the UK reference for farm business management*, Scottish Agricultural College: Ayr, 28th edn.

Beaufoy G (2007) 'EU labelling of geographical origin: good, bad or irrelevant for HNV farming?' *La Canada*, No. 21: 4–5.

Beckett JV (1982) 'The decline of the small landowner in eighteenth and nineteenth century England: some regional considerations.' *Agricultural History Review*, 30: 97–111.

Beckett JV (1990) *The Agricultural Revolution*. Historical Association Studies, Basil Blackwell: Oxford.

Bell M and Walker MJC (1992) *Late Quaternary Environmental Change: Physical and Human Perspectives*. Longman Scientific and Technical: Harlow

Beniston M (2000) *Environmental Change in Mountains and Uplands*. Arnold: London.

Bennett K, Carroll T, Lowe P and Phillipson J (2002) *Coping with Crisis in Cumbria: the consequences of foot and mouth disease*. Centre for Rural Economy: University of Newcastle on Tyne.

Bennett K, Phillipson J, Lowe P and Ward N (2001) *The Impact of the Foot and Mouth Crisis on Rural Firms: a survey of microbusiness in the North East of England*. Centre for Rural Economy: University of Newcastle on Tyne.

Bessière J (1998) 'Local development and heritage: traditional food and cuisine as tourist attractions in rural areas.' *Sociologia Ruralis*, 38(1): 21–34.

Bibby CJ (1988) 'Impacts of agriculture on upland birds' in Usher MB and Thompson DBA (eds), *Ecological Change in the Uplands*, pp. 223–36. Blackwell: Oxford.

Bibby JS, Douglas HA, Thomasson AJ and Robertson JS (1991) *Land Capability Classification for Agriculture*. Macaulay Land Use Research Institute: Aberdeen.

Bilton L (1957) 'The Chillingham herd of wild white cattle.' *Transactions of the Natural History Society of Northumberland*, 12: 137–60.

Birkeland PW (1999) *Soils and Geomorphology*. Oxford University Press: Oxford, 3rd edn.

Birks HH (1988) 'Long-term ecological change in the British uplands' in Usher MB and Thompson DBA, *Ecological Change in the Uplands*, pp. 37–56. Special Publication No. 7 of the British Ecological Society, Blackwell Scientific: Oxford.

Birse EL (1971) '*Assessment of Climatic Conditions in Scotland: 3. The bioclimatic sub-regions*.' Soil Survey of Scotland: Macaulay Institute for Soil Research, Aberdeen.

Blunt D (1993) 'BioEngineering – new products and services.' *Landscape Design*, Extra No. 37: 4–6.

Boatman ND, Deppe C, Garthwaite D, Gregory S and Jones NE (2004) *Evaluation of the Pilot Entry Level Agri-Environment Scheme: final report*, Centre Science Laboratory: York.

Bond J (2004) *Monastic Landscapes*, Tempus: Stroud.

Bonn A, Allott T, Hubacek K and Stewart J (2009) *Drivers of Environmental Change in Uplands*. Routledge: London.

Bonn A, Rebane M and Reid C (2009) 'Ecosystem services: a new rationale for conservation of upland environments', Ch. 25 in Bonn A, Allott T, Hubacek K and Stewart J (eds), *Drivers of Environmental Change in Uplands*, pp. 448–74. Routledge: London.

Bowers JK and Cheshire P (1983) *Agriculture, the Countryside and Land Use: an economic critique*. Methuen: London.

Bowler IR (1976) 'Regional agricultural policies: experience in the United Kingdom.' *Economic Geography*, 52(3): 267–80.

Bowler IR (1979) *Government and Policy: a spatial perspective*. Longman: London.

Bragg OM (2002) 'Hydrology of peat-forming wetlands in Scotland.' *Science of the Total Environment*, 294: 111–29.

Bragg OM and Tallis JH (2001) 'The sensitivity of peat-covered upland landscapes.' *Catena*, 42: 345–60.

Braithwaite D, Grooby L and Newborn D (1997) 'Hebridean sheep.' *Enact*, 5(4): 8–9.

Brassley P (1996) 'Silage in Britain, 1880 – 1990: the delayed adoption of an innovation.' *Agricultural History Review*, 44(1): 63–87.

Bridges EM (1984) *World Soils*. Cambridge University Press: Cambridge.

Britton AJ and Pearce I (2004) *Studies into the Condition and Conservation of Montane Heath and Summit Heath Vegetation in Wales.* Final Report to CCW.

Britton AJ, Pearce ISK and Jones B (2005) 'Impacts of grazing on montane heath vegetation in Wales and implications for the restoration of montane areas.' *Biological Conservation*, 125: 515–24.

Brotherton I (1991) 'What limits participation in Environmentally Sensitive Areas?' *Journal of Environmental Management*, 32: 241–9.

Brouwer FM and Chadwick MJ (1991) 'Future land use patterns in Europe', Ch. 3 in Brouwer FM, Thomas AJ and Chadwick MJ, *Land use changes in Europe – processes of change, environmental transformations and future patterns*, pp. 60–74. Kluwer Academic Publishers: London.

Brown I (2004) *Discovering a Welsh landscape: archaeology in the Clwydian Range.* Windgather Press: Macclesfield.

Brown R and Robinson R (n.d.) *Bracken Management Handbook – Integrated Bracken Management: a guide to best practice.* Rhone-Poulenc Agriculture.

Browne D and Hughes S (1993) *The Archaeology of the Welsh Uplands.* Royal Commission on the Ancient and Historical Monuments of Wales: Aberystwyth.

Buckingham H, Chapman J and Newman R (1999) 'The future of hay meadows in the Peak District National Park.' *British Wildlife*, 10(5): 311–18.

Bullock CH and Kay J (1997) 'Preservation and change in the upland landscape: the public benefits of grazing management.' *Journal of Environmental Planning and Management*, 40: 315–34.

Bullock DJ (2005) 'Large herbivores in upland Britain: what can the past tell us about the future?' Paper at *The role of Large Herbivores in Shaping the Upland Landscapes of Britain – what does the science of herbivore ecology tell us?* Battleby: Perth, 16 February.

Bullock P (1971) 'The soils of the Malham Tarn area.' *Field Studies*, 3(3): 381–408.

Bunce RGH and Barr CJ (1988) *Rural Information for Forward Planning.* Symposium 21. Institute of Terrestrial Ecology: Grange-over-Sands, Cumbria.

Bunce RGH, Barr CJ, Clarke RT, Howard DC and Lane AMJ (1996a) 'ITE Merlewood land classification of Great Britain.' *Journal of Biogeography*, 23: 625–34.

Bunce RGH, Barr CJ, Clarke RT, Howard DC and Lane AMJ (1996b) 'Land classification for strategic ecological survey.' *Journal of Environmental Management*, 47: 37–60.

Bunting E and Guelke L (1979) 'Behavioural and perception geography: a critical appraisal.' *Annals of the Association of American Geographers*, 69(3): 448–61.

Bureau J-C and Witzke H-P (2010) *The Single Payment Scheme after 2013: new approach – new targets.* Directorate General for Internal Policies, Policy Dept B: Structural and Cohesion Policies, Agriculture and Rural Development. IP/B/AGRI/IC/2009_038.

Burel F (1995) 'Ecological patterns and processes in European landscapes.' *Landscape and Urban Planning*, 31(1–3): 1–2.

Burgess C (1985) 'Population, climate and upland settlement', Ch. 10 in Spratt D and Burgess C (eds), *Upland Settlement in Britain: the second millenium* BC *and after*, BAR British Series 143. British Archaeological Reports: Oxford.

Burgess J, Clark J and Harrison CM (2000) 'Knowledges in action: an actor network analysis of a wetland agri-environment scheme.' *Ecological Economics*, 35: 119–32.

Burton J (1994) *Monastic and Religious Orders in Britain, 1000–1300*, Cambridge University Press: Cambridge.

Burton R (2004a) 'Reconceptualising the "behavioural approach" in agricultural studies: a socio-psychological perspective.' *Journal of Rural Studies*, 20(3): 359–71.

Burton R (2004b) 'Seeing through the "good farmer's" eyes: towards developing an understanding of the social symbolic value of "productivist behaviour".' *Sociologia Ruralis*, 44(2): 195–215.

Burton R, Mansfield L, Schwarz G, Brown K and Convery I (2005) *Social Capital in Hill Farming.* Report to the International Centre for the Uplands: Hackthorpe: Cumbria.

Burton R, Rigby D and Young T (1999) 'Analysis of the determinants of adoption of organic horticultural techniques in the UK.' *Journal of Agricultural Economics*, 50(1): 48–63.

Burton R, Schwarz G, Brown KM, Convery IT and Mansfield L (2008) 'The future of public goods provision upland regions: learning from hefted commons in the Lake District, UK', Ch. 17 in Bonn A, Allott T, Hubacek K and Stewart J, *Drivers for Environmental Change in Uplands*, Routledge: London.

Buttoud G (2002) 'Multipurpose management of mountain forests: which approaches?' *Forest Policy and Economics*, 4: 83–7.

Caldwell JR (1994) 'Farming systems' in Arntzen CJ and Ritter EM, *Encyclopedia of Agricultural Science*, Vol. 2, pp. 128–38. Academic Press: London.

Campbell C, Tipping R and Cowley D (2002) 'Continuity and stability in past upland land uses in the western Cheviot Hills, southern Scotland.' *Landscape History*, 24: 111–19.

Capstick M (1978) 'Economic, social and political structures in the uplands of Cumbria', Ch. 46 in Tranter R (ed.), *The Future of Upland Britain*, Vol. 2, Centre for Agricultural Strategy: Reading.

Carey PD, Barnett CL, Greenslade PD, Hulmes S, Garbutt RA, Warman EA, Myhill D, Scott RJ, Smart SM, Manchester SJ, Robinson J, Walker KJ, Howard DC and Firbank LG (2003) 'A comparison of the ecological quality of land between an English agri-environment scheme and the countryside as a whole.' *Biological Conservation*, 108: 183–97

Carr S and Tait J (1990) 'Farmer's attitudes to conservation.' *Built Environment*, 16(3): 218–31.

Carroll T and Lowe P (2001) 'Introduction' in Bennett K, Carroll T, Lowe P and Phillipson J, *Coping with Crisis in Cumbria: the consequences of foot and mouth disease*. Centre for Rural Economy, University of Newcastle: Newcastle upon Tyne.

Caseldine C and Hatton J (1993) 'The development of high moorland on Dartmoor: fire and the influence of Mesolithic activity on vegetation change' in Chambers FM (ed) *Climate Change and Human Impact on the Landscape*, pp. 119–32. Chapman & Hall: London.

Caskie P, Davis J and Wallace M (2001) 'Targeting disadvantage in agriculture.' *Journal of Rural Studies*, 17: 471–9.

Castree, N (2004) 'Differential geographies: place, indigenous rights and "local" resources.' *Political Geography*, 23: 133–67.

Centre for Rural Economics Research *see* CRER.

Centre for Rural Research (2002) *Farm Diversification in England 2002: Final Report*. University of Exeter, on behalf of DEFRA.

Chadwick L (2003) *The Farm Management Handbook 2002/3*. SAC: Ayr.

Chambers FM (1988) 'Archaeology and the flora for the British Isles: the moorland experience' in Jones M (ed.), *Archaeology and the Flora of the British Isles*, pp. 107–15. Oxford University Committee for Archaeology: Oxford.

Chambers F (ed.) (1993) *Climate Change and Human Impact on the Landscape*. Chapman & Hall: London.

Chandler TS and Gregory S (1976) *The Climate of the British Isles*. Longman: London.

Chapman JL and Reiss MJ (1998) *Ecology: Principles and Applications*. Cambridge University Press: Cambridge.

Clayden B and Hollis JM (1984) *Criteria for Differentiating Soil Series*. Technical Monograph No. 17, Soil Survey: Harpenden.

Clements FE (1916) *Plant Succession*. Carnegie Institute: Washington.

Clifton SJ, Ranner DS and Ward L (1995) *The conservation of juniper in Northumberland*. Research Report No. 152, English Nature: Peterborough.

Coed Cymru (2005) *Woodland Owners Handbook*. Coed Cymru: Newtown, Powys.

Cole L, Kernon T and Knightsbridge R (1996) *Identifying and describing farm character and structure in the Natural Areas – a guidance manual*. Research Report No. 206, English Nature: Peterborough.

Cole L, Kernon T and Knightsbridge R (1997) *Characterisation of Farming in Natural Areas*. Research Report No. 205. English Nature: Peterborough.

Coleman R (1981) 'Footpath Erosion in the English Lake District.' *Applied Geography*, 1: 121–31.

Collins EJT (1978) *The economy of upland Britain, 1750–1950: an illustrated review*. CAS Paper 4 University of Reading: Reading.

Commission of the European Communities (2009a) *Towards a better targeting of the aid to farmers in areas with natural handicaps*. Communication for the Commission of the European Parliament, the Council, the European Economic and Social Committee and the Committee of the Regions. SEC449, COM2009 161.

Commission of the European Communities (2009b) *Towards a better targeting of the aid to farmers in areas with natural handicaps. Technical Annex*. Communication for the Commission of the European Parliament, the Council, the European Economic and Social Committee and the Committee of the Regions. SEC449, COM2009 161.

Commission of the European Communities (2009c) *Towards a better targeting of the aid to farmers in areas with natural handicaps. Impact Assessment*. Communication for the Commission of the European Parliament, the Council, the European Economic and Social Committee and the Committee of the Regions. SEC449, COM2009 450.

Conway VM (1954) 'Stratigraphy and pollen analysis of southern Pennine blanket peats.' *Journal of Ecology*, 42(1): 117–32.

Coppock JT (1971) *An Agricultural Geography of Britain.* G. Bell: London.

Coulson JC and Butterfield JEL (1978) 'The animal communities of upland Britain,' Ch. 37 in Tranter R (ed.) *The Future of Upland Britain: 2*, Centre for Agricultural Strategy: Reading.

Countryside Commission (1978) *Upland Land Use in England and Wales*. Countryside Commission and Institute of Terrestrial Ecology: Cheltenham.

Countryside Commission (1984a) *A Better Future for the Uplands*. Countryside Commission: Cheltenham.

Countryside Commission (1984b) *Upland Landscape Change: a review of statistics*. Countryside Commission: Cheltenham.

Countryside Commission (1991a) *Application Pack for Countryside Stewardship* Countryside Commission: Cheltenham.

Countryside Commission (1991b) *Wind Energy Development and the Landscape*. Countryside Commission: Cheltenham.

Countryside Commission (1995) *Countryside Stewardship: Handbook and Application Form*. Countryside Commission: Cheltenham.

Countryside Commission (2000) *Countryside Character Areas: North West Region* Countryside Commission: Cheltenham.

Courtney FM and Trudgill ST (1984) *The Soil: an introduction to soil study*, Hodder & Stoughton: Sevenoaks, 2nd edn.

Cowell R (2010) 'Wind power, landscape and strategic spatial planning – the construction of "acceptable locations" in Wales.' *Land Use Planning*, 27(2): 222–32.

Cox CB and Moore PD (1993) *Biogeography: An Ecological and Evolutionary Approach*. Blackwell Scientific: Oxford.

CRER (2002) *Economic Evaluation of Agri-Environment Schemes*. Centre for Rural Economics Research and CJC Consulting, Report for DEFRA.

Critchley CNR, Fowbert JA, Wright B and Parkin AB (2004) *Upland Hay Meadows in the Pennine Dales Environmentally Sensitive Area: vegetation change between 1987and 2002 and its relation with management practices and soil properties.* Report to DEFRA. ADAS: Newcastle.

Crofts A and Jefferson RG (1999) *The Lowland Grassland Management Handbook.* English Nature/The Wildlife Trusts: Peterborough.

Crofts R (2004) 'Connecting the pieces: Scotland's integrated approach to the natural heritage', Ch. 11 in Bishop K and Phillips A, *Countryside Planning: new approaches to management and conservation*, Earthscan: London.

Croston D and Pollott G (1994) *Planned Sheep Production*. Blackwell Scientific: Oxford, 2nd edn.

CSL (2007) *Evaluation of the Operation of Environmental Stewardship Scheme on behalf of DEFRA*. CSL:York MA01028.

Cullingworth JB and Nadin V (2006) *Town and Country Planning in the UK*. Routledge: London, 14th edn.

Cumbria County Council (1995) *Cumbria Landscape Classification*, Cumbria County Council Economy and Environment Department: Carlisle.

Cumbria County Council (2001) *Cumbria Landscape Strategy: a vision for Cumbria's future landscapes with guidelines for managing change*. Cumbria County Council Economy and Environment Department: Carlisle.

Cumulus Consultants, Institute of Environmental European Policy and Countryside Research Unit (2005) *Assessment of the Impact of CAP Reform and other key policies on upland farms and land use implications in both Severely Disadvantaged and Disadvantaged Areas of England.* Report to DEFRA: Cambridge.

Cundill PR (1976) 'Late Flandrian vegetation and soils in Carlingill Valley, Howgill Fells.' *Transactions of the Institute of British Geographers*, 1 (NS): 301–9.

Curry N (2009) 'Leisure in the landscape: rural incomes and public benefits,' Ch. 15 in Bonn A, Allott T, Hubacek K and Stewart J, *Drivers of Environmental Change in Uplands.* Routledge: London.

Dalton RT (1971) 'Peas for freezing: a recent development in Lincolnshire agriculture.' *East Midlands Geographer*, 5: 133–41.

DARD (2006) *Agri-environment Scheme Explanatory Booklet: Environmentally Sensitive Areas Scheme and Countryside Management Scheme*. DARD: Cookstown, NI.

DARD (2007) *Countryside Management Scheme 2007-2013.* DARD: Cookstown, NI.

DARD (2010) *Less Favoured Area Compensatory Allowances (LFACA) Scheme 2011 – explanatory notes.* DARD, LFACA Section: Derry/Londonderry.

Dark P (2004) *The Environment of Britain in the First Millenium AD.* Gerald Duckworth & Co: London.

Darvill T (1986) *The Archaeology of the Uplands: a rapid assessment of archaeological knowledge and practice*, with Royal Commission on the Historical Monuments of England. London: Council for British Archaeology.

Daugstad K, Ronningen K and Skar B (2006) 'Agriculture as an upholder of cultural heritage? Conceptualisations and value judgements – A Norwegian perspective in an international context.' *Journal of Rural Studies*, 22: 67–81.

Davies AL (2007) 'Upland agriculture and environmental risk: a new model of upland land-use based on high spatial-resolution palynological data from West Affric, NW Scotland.' *Journal of Archaeological Science*, 34: 2053–63.

Davies B, Finney B and Eagle D (2001) *Resource Management: Soil.* Farming Press: Ipswich.

Davies R (2005) 'Predation and the profitability of Grouse moors.' *British Wildlife*, 17: 339–47.

Davis J, Caskie P, Gilbreath G and Wallace M (1998) *A socio economic evaluation of the HLCA Scheme in Northern Ireland.* CRS, Queens University: Belfast.

De La Warr, Earl (1944) *Report of the Committee on Hill Sheep Farming in England and Wales*, Cmnd 6498, Ministry of Agriculture and Fisheries: London.

DEFRA (2000) *England's Rural Development Plan*, 1. HMSO: London. Accessed at: www.defra.gov/erdp (3/3/2004).

DEFRA (2001a) *Task Force for the Hills.* DEFRA: London.

DEFRA (2001b) *England LEADER+ Programme 2000 to 2006.* DEFRA: London, PB5706.

DEFRA (2003a) *Classification of Land to reflect the Environmental Value and Difficult Farming Conditions of Upland Grazing Land.* Project LS3405. DEFRA: London prepared by the University of Reading

DEFRA (2003b) *Sites of Special Scientific Interest: encouraging positive partnerships – code of guidance.* DEFRA: London PB6576.

DEFRA (2003c) '*Upland Farm Survey for the Countryside Stewardship Scheme: guidance notes*' DEFRA: London

DEFRA (2004a) *Agriculture in the UK 2003.* The Stationery Office: London.

DEFRA (2004b) *Organic Food and Farming.* DEFRA: London.

DEFRA (2004c) *The Countryside Stewardship Scheme 2004: payment summary sheet.* DEFRA: London.

DEFRA (2005a) *Controlling Soil Erosion: incorporating former advisory leaflets on grazing livestock, wind, outdoor pigs and the uplands.* DEFRA: London.

DEFRA (2005b) *A Study of Long-Term Trends affecting the Farming Industry*, Appendices. DEFRA: London.

DEFRA (2006a) *Farm Practices Survey.* DEFRA: London. Accessed at: www.defra.gov.uk/evidence/statistics (20/6/2007).

DEFRA (2006b) *Agricultural Census 2005.* DEFRA: London. Accessed at: www.defra.gov.uk/evidence/statistics/foodfarm/landuselivestock/junesurvey/results.htm (20/6/2007).

DEFRA (2007a) *Rural Development Programme for England 2007 – 2013.* Accessed at: www.defra.gov.uk/rural/rdpe/progdoc.htm (July 2009).

DEFRA (2007b) *An Introductory Guide to Valuing Ecosystem Services.* PB12852. DEFRA: London.

DEFRA (2007c) *Agriculture in the United Kingdom 2006.* HMSO: London. Accessed at: www.defra.gov.uk/evidence/statistics/foodfarm/general/auk/documents/AUK-2007.pdf (26/10/2007).

DEFRA (2007d) *Convention on Biological Diversity. COP 5 Decision V/6* Accessed at: http://biodiv.org/decisions/default.asp?lg=0&m=cop-05&d=06) (26/10/2007)

DEFRA (2008a) *Agriculture in the United Kingdom 2007.* HMSO: London.

DEFRA (2008b) *Impact Assessment of Upland Entry Level Stewardship (Uplands ELS): Summary Intervention and Options.* DEFRA: London. Accessed at: www.defra.gov.uk/rural/uplands/support.htm (July 2009).

DEFRA (2009a) *Farm Practices Survey 2009 – Uplands and Other Less Favoured Areas (England).* National Statistics, DEFRA: London.

DEFRA (2009b) *Agricultural Census 2009.* DEFRA: London. Accessed at: www.defra.gov.uk/evidence/statistics/foodfarm/landuselivestock/junesurvey/results.htm (12/11/2009).

DEFRA (2009c) *The Upland Entry Level Environmental Stewardship Scheme.* Accessed at: ww2.defra.gov.uk/food-farm/land-manage/stewardship/ (24/03/2009).

DEFRA (2010a) 'UK *food security assessment detailed analysis'*, August 2009, updated January 2010. Accessed at: www.defra.gov.uk/foodfarm/food/pdf/food-assess100105.pdf (12/6/2010).

DEFRA (2010b) *Food 2030: how we get there.* Accessed at: www.defra.gov.uk/foodfarm/food/pdf/food2030strategy-summary.pdf (5/1/2011).

Department of Agriculture and Rural Development *see* DARD.

Department of Environment, Food and Rural Affairs *see* DEFRA.

Department of International Development (1999) *Sustainable Livelihoods: guidance sheets.* DFID, London.

Dessylas D (1987) 'The Commission's proposals adapting agriculture to new market conditions – preservation of the countryside' in Baldock D and Conder D (eds), *Removing Land from Agriculture – the implications for farming and the environment*, pp. 31–4. CPRE, IEEP and WWF(UK): London.

Dimbleby GW (1961) 'The ancient forest of Blackamore.' *Antiquity*, 35: 123–8.

Dodds GW, Appleby MJ and Campbell L (1996) *A Management Guide to Birds of Upland Farmland.* RSPB: Sandy, Bedfordshire.

Dodge RE (1911) 'Geography and agriculture.' *Annals of the Association of American Geographers*, 1: 144.

Dodgshon RA (1988) 'The ecological basis of Highland peasant farming 1500 to 1800 AD' in Birks HH, Birks HJB, Kaland PE and Moe D, *The Cultural Landscape: past, present and future*, pp. 139–52. Cambridge University Press: Cambridge.

Dodgshon RA (2004) 'The Scottish Highlands before and after the Clearances: an ecological perspective' in Whyte ID and Winchester A, *Society, Landscape and Environment in Upland Britain*, pp. 67–78. Society for Landscape Studies Supplementary Series No. 2. Society for Landscape Studies: Whitby.

Dodgshon RA and Olsson GA (2006) 'Heather moorland in the Scottish Highlands: the history of a cultural landscape, 1600-1800.' *Journal of Historical Geography*, 32: 21–37.

Dower J (1945) *National Parks in England and Wales.* HMSO: London, Cmnd 6378.

Drew Associates (1997) *Economic Evaluation of the Hill Livestock Compensatory Allowances Scheme in England.* Drew Associates and University of Exeter: Exeter.

Drewitt AL and Manley V (1997) *The Vegetation of the Mountains and Moorlands of England – national assessment of significance.* Research Report No 218. English Nature: Peterborough.

Duchafour P (1982) *Pedology*, trans. T Paton. George Allen & Unwin: London.

Duffey E, Morris MG, Sheail J, Ward LK, Wells DA and Wells TCE (1974) *Grassland Ecology and Wildlife Management.* Chapman & Hall: London.

Dumayne-Peaty L (1999) 'Late Holocene human impact on the vegetation of south eastern Scotland: a pollen diagram from Dodgen Moss, Berwickshire.' *Review of Palaeobotany and Palynology*, 105: 121–41.

Dumayne-Peaty L and Barber KE (1994) 'The impact of the Romans on the environment of northern England: pollen data from three sites near Hadrian's Wall' *The Holocene*, 4: 165–73.

Dunford NB and Feehan J (2001) 'Agricultural practices and natural heritage: a case study of the Burren Uplands, Co Clare.' *Tearmann*, 1(1): 19–34.

EBLEX (2008) *Understanding cattle and carcases for better returns.* EBLEX: Huntingdon.

Ecoscope Applied Ecologists, CPM Environmental Planning & Design and CJC Consulting (2003a) *Review of Agri-Environment Schemes – monitoring information and R and D results.* Final Report, Part A, chs 1–9. Ecoscope Applied Ecologists: Cambridge.

Ecoscope Applied Ecologists, CPM Environmental Planning & Design and CJC Consulting (2003b) *Review of Agri-Environment Schemes – monitoring information and R and D results.* Final Report, Part B, chs 10–14. Ecoscope Applied Ecologists: Cambridge.

Ecoscope Applied Ecologists, CPM Environmental Planning & Design and CJC Consulting (2003c) *Review of Agri-Environment Schemes – monitoring information and R and D results.* Final Report: Technical Appendix. Ecoscope Applied Ecologists: Cambridge.

EDCA (1973) *UK Farming and the Common Market: hills and uplands.* National Economic Development Office: London.

Edmond H and Crabtree JR (1994) 'Regional variation in Scottish pluriactivity: the socio-economic context for different types of non-farming activity.' *Scottish Geographical Magazine*, 110(2): 76–84.

Edwards K (1988) 'The hunter-gatherer/ agricultural transition and the pollen record in the British Isles' in Birks HH, Birks HJB, Kaland PE and Moe D (eds), *The Cultural Landscape: past, present and future*, pp. 255–66. Cambridge University Press: Cambridge.

Edwards KJ (1993) 'Models of mid-Holocene forest farming in north-west Europe' in Chambers FM (ed.), *Climate Change and Human Impact on the Landscape*, pp. 137–56. Chapman & Hall: London.

Elliot J, Temple ML, Clinton S, Tiffin AL, Rees E and Standen J (2003a) *Evidence Assessment to Inform the Review of the Processing and Marketing Grant Scheme – report to Department of Farming and Rural Affairs.* ADAS: Wolverhampton; and University of Reading: Reading.

Elliot J, Temple ML, Clinton S, Tiffin AL, Rees E and Standen J (2003b) *Evidence Assessment to Inform the Review of the Organic Farming Scheme – report to Department of Farming and Rural Affairs.* ADAS: Wolverhampton; and University of Reading: Reading.

Ellis S and Mellor A (1995) *Soils and Environment.* Routledge: London.

Ellison W, Boyd DA and Church BM (1955) 'The progress of improvement of upland and hill farms in England and Wales.' *Journal of the Royal Agricultural Society*, 116: 34–49.

Emery FV (1973) 'England circa 1600' in Darby HC (ed.), *A New Historical Geography of England*, pp. 248–302. Cambridge University Press: Cambridge.

Emery M (1986) *Promoting Nature in Cities and Towns.* Croom Helm: Beckenham.

English Nature (1998) *State of Nature: the upland challenge.* External Relations Team, English Nature: Peterborough.

English Nature (2003) *The Wildlife Enhancement Scheme for the North York Moors SSSI*, Estate Managers' Version. English Nature: Peterborough.

English Nature (2004) *Sheep and Wildlife Enhancement Scheme 2004 – sustaining wildlife and sheep farming.* English Nature: Peterborough.

ENTEC *see* Cole L, Hernon T and Knightsbridge R (1997).

EUNIS (2008) http://eunis.eea.europa.eu/habitats-annex1-browser.jsp (11/09/2008).

Europa (2008) www/europa/eu/agriculture/eagf/pillar1 (accessed 7/8/2008).

European Commission *see* Commission of the European Communities.

European Court of Auditors (2003) 'Special Report No. 4/2003 concerning rural development: support for less-favoured areas, together with the Commission's replies.' *Official Journal of the European Union*, C151, 46.

European Environment Agency (2010) *Europe's Ecological Backbone: recognising the true value of our mountains.* EEA: Copenhagen EEA Report No 6/2010.

Evans D (1992) *A History of Nature Conservation in Britain.* Routledge: London.

Evans J (2003) *The Mid-Term Evaluations of the England Rural Development Programme (ERDP): Hill Farm Allowance* ADAS Consulting: Wolverhampton.

Evans N (1997) 'Something old, new borrowed, new and blue: the marriage of agriculture and conservation in England', Ch. 15 in Ilbery B, Chiotti Q and Rickard T, *Agricultural Restructuring and Sustainability: a geographical perspective.* CAB International: Wallingford.

Evans N and Ilbery BW (1992) 'Farm-based accommodation and the restructuring of agriculture: evidence from three English counties.' *Journal of Rural Studies*, 8: 85–96.

Evans N, Morris C and Winter M (2002a) 'Conceptualising agriculture: a critique of post-productivism as the new orthodoxy.' *Progress in Human Geography*, 26(3): 313–32.

Evans N and Yarwood R (1995) 'Livestock and landscape.' *Landscape Research*, 20(3): 141–6.

Evans N and Yarwood R (2000) 'The politicization of livestock: rare breeds and countryside conservation.' *Sociologia Ruralis*, 40(2): 228–48.

Evans N, Gaskell P and Winter M (2003) 'Re-assessing agrarian policy and practice in local environmental management: the case of beef cattle.' *Land Use Policy*, 20: 231–42.

Evans R (1977) 'Overgrazing and soil erosion on hill pastures with particular reference to the Peak District.' *Journal of British Grassland Society*, 32: 65–76.

Evans R (2005) 'Curtailing grazing – induced erosion in a small catchment and its environs, the Peak District, Central England.' *Applied Geography*, 25: 81–95.

Falco S di and Rensburg T van (2006) *Cooperation and Conservation in the Management of Common Property Resources: the case of Irish commonage.* Paper presented at 3rd World Congress, Environmental and Resource Economists, 3–7 July.

Falconer K and Saunders C (2002) 'Transaction costs for SSSIs and policy design.' *Land Use Policy*, 19: 157–66.

Farr E and Henderson WC (1986) *Land Drainage.* Longman Handbooks in Agriculture. Longman: London.

Federation of Cumbria Commoners (2009) website accessed at: www.cumbriacommoners.org.uk/ (11/10/2009).

Felton M and Marsden J (1990) *Heather Regeneration in England and Wales – a feasibility study for the Department of the Environment.* NCC: Peterborough.

Fernie J and Pitkethly AS (1985) *Resources: Environment and Policy.* Harper & Row: London.

Fielding AH and Howarth PF (1998) *Upland Habitats.* Routledge: London.

Finch C and Slater J (2003) *The Mid Term Evaluation of the England Rural Development Programme (ERDP): Countryside Stewardship Scheme.* ADAS Consulting: Wolverhampton.

Firbank LG, Barr CJ, Bunce RGH, Furse MT, Haines-Young R, Hornung M,

Howard DC, Sheail J, Sier A and Smart SM (2003) 'Assessing stock and change in land cover and biodiversity in GB: an introduction to Countryside Survey 2000.' *Journal of Environmental Management*, 67: 207–18.

Fleming A (1983) 'The prehistoric landscape of Dartmoor: wider implications.' *Landscape History*, 6: 5–19.

Folke C (2006) 'Resilience: the emergence of a perspective for social-ecological system analyses.' *Global Environmental Change*, 16(3): 253–67.

Food from Britain (2008) Accessed at: www/foodfrombritain/ (18/8/2008) [now subsumed in the Food and Drink Federation].

Forestry Authority (1994a) *Upland Birchwoods: the management of semi natural woodlands.* Forestry Authority: Edinburgh.

Forestry Authority (1994b) *Upland Ashwoods: the management of semi natural woodlands.* Forestry Authority: Edinburgh.

Forestry Authority (1994c) *Upland Oakwoods: the management of semi natural woodlands.* Forestry Authority: Edinburgh.

Forestry Authority (1994d) *Native Pinewoods: the management of semi natural woodlands.* Forestry Authority: Edinburgh.

Forestry Commission (2006) *Polmaddy Settlement.* Forestry Commission: Newton Stewart.

Forestry Commission (2008) *English Woodland Grant Scheme.* Accessed at: www.forestry.gov.uk/ewgs (11/11/2010).

Forestry Commission (2011) *National Forest Inventory*. Accessed at: www.forestresearch.gov.uk/inventory (6/4/2011).

Forestry Commission Wales (2008) *Better Woods for Wales.* Accessed at: www.forestry.gov.uk/forestry/INFD-7XWF7V (11/11/2010).

Found WC (1974) *A Theoretical Approach to Rural Land Use Patterns.* Edward Arnold: London.

Fowler PJ (1978) 'Later prehistory' in Limbrey S and Evans JG (eds), *The Effect of Man on the Landscape: the lowland zone*, pp. 1–12. CBA Research Report: Cambridge.

Fowler PJ (1983) *The Farming of Prehistoric Britain*. Cambridge University Press: Cambridge.

Frame J (1992) *Improved Grassland Management*. Farming Press: Ipswich.

Franks J, Lowe P, Phillipson J and Scott C (2003) 'The impact of foot and mouth disease on farm businesses in Cumbria.' *Land Use Policy*, 20: 159–68.

Fuller RJ and Warren MS (1993) *Coppiced Woodlands: their management for wildlife.* JNCC: London, 2nd edn.

Fuller RJ, Ward W, Hird D and Brown AF (2002) 'Declines of ground-nesting birds in two areas of upland farmland in the south Pennines of England.' *Bird Study*, 49: 146–52.

Funnell D and Parrish R (2001) *Mountain Environments and Communities.* Routledge: London.

Fyfe RM, Brown AG and Rippon SJ (2004) 'Characterising the late prehistoric, "Romano-British" and medieval landscape, and dating the emergence of a regionally distinct agricultural system in south west Britain.' *Journal of Archaeological Science*, 31: 169–174.

Galbraith H and Kinnes L (1988) 'Pressures on ptarmigan populations.' *Game Conservancy Annual Review*, 19: 60–64.

Gallopin GC (2006) 'Linkages between vulnerability, resilience and adaptive capacity.' *Global Environmental Change*, 16(3): 293–303.

Gasson R (1973) 'Goals and values of farmers.' *Journal of Agricultural Economics*, 24: 521–37.

Gasson R (1992) 'Farmers' wives – their contribution to the farm business.' *Journal of Agricultural Economics*, 43(1): 74–87.

Gasson R (1998) 'Educational qualifications of UK farmers: a review.' *Journal of Rural Studies*, 14(4): 487–98.

Gasson R and Errington F (1993) *The Farm Family Business.* CAB International: Wallingford.

Gearey BR, Charman DJ and Kent M (2000) 'Palaeoecological evidence for the prehistoric settlement of Bodmin Moor, Cornwall, southwest England: Part II: land use changes from the Neolithic to the present.' *Journal of Archaeological Science*, 27: 493–508.

Gerrard J (2000) *Fundamentals of Soils.* Routledge: London.

Gilbert D (2006) *Montane Scrub Restoration Project – guidance for the restoration of montane scrub.* Highland Birchwoods and Scottish Natural Heritage: Munlochy.

Gilbert D and Anderson P (1998) 'Montane and submontane habitats', Ch. 7 in *Habitat Creation and Repair.* Oxford University Press: Oxford.

Gilbert D, Horsfield D and Thompson DBA (1997) *The Ecology and Restoration of Montane and Subalpine Scrub Habitats in Scotland.* SNH: Edinburgh.

Gillmor DA (1986) 'Behavioural studies in agriculture: goals, values and enterprise choice.' *Irish Journal of Agricultural Economics and Rural Sociology*, 11: 19–33.

Gimingham CH (1975) *An Introduction to Heathland Ecology*. Oliver & Boyd: Edinburgh.

Gimingham C (ed.) (2002) *The Ecology, Land Use and Conservation of the Cairngorms.* Packard: Chichester.

Glassock RE (1973) 'England circa 1334' in Darby HC (ed.), *A New Historical Geography of England*, pp. 136–85. Cambridge University Press: Cambridge.

Glynn S (2005) *Climate Change in the Northwest and its Impacts: a summary document.* Sustainability North West: Manchester.

Godwin HH (1985) *History of the British Flora: a factual basis for phytogeography.* Cambridge University Press: Cambridge.

Gordon JE, Brazier V, Thompson DBA and Horsfield D (2001) 'Geo-ecology and the conservation management of sensitive landscapes in Scotland.' *Catena*, 42: 323–32.

Gordon JE, Thompson DBA, Haynes VM, Brazier V and MacDonald R (1998) 'Environmental sensitivity and conservation management in the Cairngorm Mountains, Scotland.' *Ambio*, 27(4): 335–44.

Gorham E (1998) 'Acid deposition and its ecological effects: a brief history of research.' *Environmental Science and Policy*, 1: 153–66.

Grace J and Unsworth MH (1988) 'Climate and microclimate of the uplands' in Usher MB and Thompson DBA, *Ecological Change in the Uplands*, pp. 137–50. Special Publication No. 7 of the British Ecological Society. Blackwell Scientific: Oxford.

Graham H, Mason R and Newman A (2009) *Literature Review: historic environment, sense of place, and social capital*, commissioned for English Heritage by International Centre for Cultural and Heritage Studies (ICCHS), University of Newcastle.

Grant SA and Armstrong H (1993) 'Grazing ecology and the conservation of heather moorland: the development of models as aids to management.' *Biological Conservation*, 2: 79–94.

Grayson B (1997) 'Does conservation grazing work?' *Enact*, 5(4): 19–22.

Grayson B and Small R (2007) 'Can it pay? The economics of conservation grazing.' *Conservation Land Management*, 5(4): 12–15.

Green B (1985) *Countryside Conservation*, 2nd edn. George Allen & Unwin: London.

Greene JP (1992) *Medieval Monasteries.* Leicester University Press: Leicester.

Gregory J and Essery R (2006) 'Mountain climate' in Kempe N and Wrightham M (eds), *Hostile Habitats: Scotland's mountain environment*, pp. 10–21. Scottish Mountaineering Trust and Scottish Natural Heritage. [Distributed by] Cordee: Leicester.

Grieve IC (2001) 'Human impacts on soil properties and their implications for the sensitivity of soil systems in Scotland.' *Catena*, 42: 361–74.

Griffith M (1951) 'Reclamation of hill and moorland' in Stapledon RG, *Reclaiming Land For Agriculture*, pp. 23–57. Crosby Lockwood & Son: London.

Grigg D (1984) *An Introduction to Agricultural Geography*. Hutchinson: London.

Grigg DB (1987) 'Farm size in England and Wales from early Victorian times to the present.' *Agricultural History Review*, 35: 179–90.

Grigg D (1989) *English Agriculture: an historical perspective.* Basil Blackwell: Oxford.

Grubb PJ, Green HE and Merrifield RCJ (1969) 'The ecology of chalk heath: its relevance to the calcicole-calcifuge and soil acidification problems.' *Journal of Ecology*, 57: 175–212.

Hagerstrand T (1967) *Innovation Diffusion as a Spatial Process.* University of Chicago Press: Chicago.

Haines-Young RH, Barr CJ, Black HIJ, Briggs DJ, Bunce RGH, Clarke RT, Cooper A, Dawson FH, Firbank LG, Fuller RM, Furse MT, Gillespie MK, Hill R, Hornung M, Howard DC, McCann T, Morecroft MD, Petit S, Sier ARJ, Smart SM, Smith GM, Stott AP, Stuart RC and Watkins JW (2000) *Accounting for Nature: assessing habitats in the UK countryside.* DETR, London.

Haines-Young R, Barr CJ, Firbank LG, Furse M, Howard DC, McGowan G, Petit S, Smart SM and Watkins JW (2003) 'Changing landscapes, habitats and vegetation diversity across Great Britain.' *Journal of Environmental Management*, 67: 267–81.

Hall CM and McArthur S (1998) *Integrated Heritage Management: principles and practice.* The Stationery Office: London.

Hall C, McVittie A and Moran D (2004) 'What does the public want from agriculture and the countryside? A review of evidence and methods.' *Journal of Rural Studies*, 20: 211–25.

Hall SJG (1979) 'Studying the Chillingham wild cattle.' *Ark*, 6: 72–9.

Halliday J (1988) 'Dairy farmers take stock: a study of milk producers' reactions to quota in Devon.' *Journal of Rural Studies*, 4(3): 193–202.

Hanley N, Colombo S, Mason P and Johns H (2007) 'The reform of support mechanisms for upland farming: paying for public goods in severely disadvantaged areas of England.' *Journal of Agricultural Economics*, 58(3): 433–53.

Hanley N and Colombo S (2009) 'The economic value of landscapes in the uplands of England', Ch. 18 in Bonn A, Allott T, Hubacek K and Stewart J, *Drivers of Environmental Change in Uplands*, pp. 323–38. Routledge: London.

Hardcastle and Nisbet (2008) *Drystone Walling Techniques and Traditions.* Drystone Walling Association: Westmorland Showground, Cumbria.

Hardin G (1968) 'The tragedy of the commons.' *Science*, 162: 1243–8.

Harrison-Mayfield L, Dwyer J and Brookes G (1998) 'The socio-economic effects of the Countryside Stewardship Scheme.' *Journal of Agricultural Economics*, 49(2): 157–70.

Harrison SJ and Harrison DJ (1988) 'The effect of elevation on the climatically determined growing season in the Ochil Hills.' *Scottish Geographical Magazine*, 104(2): 108–15.

Hart K and Wilson GA (2000) 'United Kingdom: from agri-environmental policy shaper to policy receiver?' Ch. 6 in Buller H, Wilson G and Holl A, *Agri-environmental Policy in the European Union.* Ashgate: Aldershot.

Harvey RJ (1993) *Three Peaks Project: project report 1986-1992.* YDNPA: Grassington.

Havins PJN (1976) *The Forests of England.* Robert Hale & Co: London.

Hawkins-Byass N (1999) 'Aerial bracken control.' *Enact*, 7(2): 14–18.

Hay R (1998) 'Sense of place in developmental context.' *Journal of Environmental Psychology*, 18: 5–29.

HC [House of Commons] (2003) Debate, 27 January 2003, col. WA137.

Healey M and Ilbery BW (1990) *Location and Change: perspectives on economic geography.* Oxford University Press: Oxford.

Heathwaite AL and Gottlich K (eds) (1993) *Mires: process, exploitation and conservation.* Chichester: Wiley.

Heaton RJ, Randerson PF and Slater FM (1999) 'The economics of growing short rotation coppice in the uplands of mid-Wales and an economic comparison with sheep production.' *Biomass and Bioenergy*, 17: 59–71.

Heddle RG and Ogg WG (1936) 'Irrigation experiments on a Scottish hill pasture.' *Journal of Ecology*, 24(1): 220–37.

Hester A and Brooker R (2007) 'Threatened habitats marginal vegetation in upland areas' in Hester R and Harrison RM (eds), *Biodiversity Under Threat*, pp. 107–34. Royal Society of Chemistry: Cambridge.

Hey D (2002) 'Moorlands', Ch. 8 in Thirsk J (ed.), *Rural England: an illustrated history of the landscape.* Oxford University Press: Oxford.

Higham N (1986) *The Northern Counties until AD 1000.* Longman: London.

Higham N (1987) 'Landscape and land use in northern England: a survey of agricultural potential c.500BC–AD1000.' *Landscape History*, 9: 35–44.

Hill B (2006) *An Introduction to Economics: concepts for students of agriculture and the rural sector.* CAB International: Wallingford.

HM Government (2011) *The Natural Choice: securing the value of nature.* The Stationery Office: Norwich.

Hobhouse A (1947) *Report to the National Park Committee (England and Wales).* Cmnd 6628. HMSO: London.

Hodder KH, Bullock JM, Buckland PC and Kirby KJ (2005) *Large Herbivores in the Wildwood and in Modern Naturalistic Grazing Systems.* Research Report No. 648. Natural England: Peterborough.

Hodgson JM (1974) *Soil Survey Field Handbook: describing and sampling soil profiles.* Technical Monograph No. 5. Soil Survey of England and Wales: Harpenden.

Holden J and Adamson JK (2002) 'The Moorhouse long term upland temperature record: new evidence of recent warming.' *Weather*, 57: 119–27.

Holl K and Smith M (2007) 'Scottish upland forests: history lessons for the future.' *Forest Ecology and Management*, 249: 45–53.

Hollin CS (1973) 'Resilience and stability of ecological systems.' *Annual Review of Ecology and Systematics*, 4: 1–23.

Holloway L and Kneafsey M (2000) 'Reading the space of the farmers' market: a preliminary investigation from the UK.' *Sociologia Ruralis*, 40(3): 285–99.

Holman IP, Loveland PJ, Nicholls RJ, Shackley S, Berry PM, Rounsevell MDA, Audsley E, Harrison PA and Wood R (2002) *REGIS – Regional Climate Change Impact Response Studies in East Anglia and North West England*. DEFRA: London.

Hoogeveen Y, Petersen J-E, Balazs K and Higuero I (2004) *High Nature Value Farmland: characteristics, trends and policy challenges*. European Environment Agency: Luxembourg.

Hope D, Picozzi N, Catt DC and Moss R (1996) 'Effects of reducing sheep grazing in the Scottish Highlands.' *Journal of Range Management*, 49(4): 301–10.

Hopkins J (1996) 'Achieving the aims of the EC Habitats Directive: links with reform of agricultural policies in mountain areas' in Poole A, Pienkowski M, McCracken DI, Petrelli F, Bredy C and Deffeyes C, *Mountain Livestock Farming and EU Policy Development*, pp. 117–24. European Forum on Nature Conservation and Pastoralism: Isle of Islay. [Proceedings of 5th European Forum on Nature Conservation and Pastoralism, Valle d'Aosta, Italy, 18–21 Sep.]

Hopkins JJ and Buck AL (1995) *The Habitats Directive, Atlantic Biogeographical Region*. JNCC Report, No. 247. [Report of Atlantic Biogeographical Workshop, Edinburgh, 13–14 Oct. 1994.]

Hooper S (1996) *The Wildlife Enhancement Scheme in the Uplands.* English Nature Research Report No. 199. English Nature: Peterborough.

Horsfield D and Thompson D (1996) *The Uplands: guidance on terminology regarding altitudinal zonation and related terms*. SNH Information and Advisory Note No. 26. Scottish Natural Heritage: Edinburgh.

Hoskins WG and Stamp LD (1964) *The Common Lands of England and Wales*. New Naturalists Series. Collins: London.

Hossell JE, Jones PJ, Marsh JS, Parry ML, Rehman T and Tranter RB (1996) 'The likely effects of climate change on agricultural land use in England and Wales.' *Geoforum*, 27(2): 149–57.

Howkins A (2003) *The Death of Rural England: a social history of the countryside since 1900.* Routledge: London.

Hubbard CE (1984) *Grasses – a guide to their structure, identification, uses and distribution in the British Isles*, 3rd edn. Penguin: London.

Hudson G (1988) 'Effects of latitude, oceanicity and geology on alpine and sub-alpine soils in Scotland' in Usher MB and Thompson DBA, *Ecological Change in the Uplands*, pp. 125–35. Special Publication No. 7 of the British Ecological Society. Blackwell Scientific: Oxford.

Hughes G (1994) 'ESAs in the context of a "culturally sensitive area": the case of the Cambrian Mountains.' Ch. 7 in Whitby M (ed.), *Incentives for Countryside Management: the case of environmentally sensitive areas*, pp. 135–52. CAB International: Wallingford.

Hughes J and Huntley B (1988) 'Upland hay meadows in Britain – their vegetation, management and future' in Birks HH, Birks HJB, Kaland PE and Moe D, *The Cultural Landscape: past, present and future*, pp. 91–110. Cambridge University Press: Cambridge.

Hulme M, Jenkins GJ, Lu X, Turnpenny JR, Mitchell TD, Jones RG, Lowe J, Murphy JM, Hassell D, Boorman P, McDonald R and Hill S (2002) *Climate Change Scenarios for the United Kingdom: The UKCIP02 Scientific Report.* Tyndall Centre for Climate Change Research, University of East Anglia: Norwich.

Hulme PD and Birnie RV (1997) 'Grazing-induced degradation of blanket mire: its measurement and management' in Tallis JH, Meade R and Hulme PD (eds), *Blanket Mire Degradation: causes, consequences, and challenges*, pp. 163–73. Proceedings of the Mire Research Group. British Ecological Society: London.

IEEP (2004a) *An Assessment of Hill Farming in England: on the economic, environmental and social sustainability of the uplands and more widely*. Vol. 1: Main Report. Institute for European Environmental Policy, Land Use Consultants and GHK Consulting: London.

IEEP (2004b) *An Assessment of Hill Farming in England: on the economic, environmental and social sustainability of the uplands and more widely*. Vol. 2: Literature Review and Consultations. Institute for European Environmental Policy, Land Use Consultants and GHK Consulting: London.

IEEP (2004c) *An Assessment of Hill Farming in England: on the economic, environmental and social sustainability of the uplands and more widely.* Vol. 3: Reports of Case Studies. Institute for European Environmental Policy, Land Use Consultants and GHK Consulting: London.

Ilbery BW (1977) 'Point score analysis: a methodological framework for analysing the decision making process in agriculture.' *Tijdschrift voor Economische en Sociale Geografie*, 68: 66–71.

Ilbery BW (1978) 'Agricultural decision making: a behavioural perspective.' *Progress in Human Geography*, 2: 448–66.

Ilbery BW (1983) 'Goals and values of hop farmers.' *Transactions of the Institute of British Geographers*, 8: 319–41.

Ilbery BW (1984) *Agricultural Geography: a social and economic analysis.* Oxford University Press: Oxford.

Ilbery BW (1988) 'Farm diversification and restructuring of agriculture.' *Outlook on Agriculture*, 17: 35–9.

Ilbery BW (1990) 'Adoption of the arable Set Aside scheme in England' *Geography*, 76: 259–63.

Ilbery BW (1991) 'Farm diversification as an adjustment strategy on the urban fringe of the West Midlands.' *Journal of Rural Studies*, 7: 207–18.

Ilbery BW (1992) 'Agricultural policy and land diversion in the European Community' in Gilg A, *Progress in Rural Policy and Planning*, Vol. 2, pp. 153–66. Belhaven: London.

Ilbery BW and Bowler IR (1998) 'From agricultural productivism to post-productivism.' Ch. 4 in Ilbery BW (ed.), *The Geography of Rural Change.* Longman: London.

Ilbery BW and Kidd J (1992) 'Adoption of the Farm Woodland Scheme in England.' *Geography*, 77(4): 363–7.

Ilbery BW and Kneafsey M (1998) 'Product and place: promoting quality products and service in the lagging rural regions of the European Union.' *European Urban and Regional Studies*, 5: 329–41.

Ilbery BW and Kneafsey M. (1999) 'Niche markets and regional speciality food products in Europe: towards a research agenda'. *Environment and Planning A*, 31(12): 2207–22.

Ilbery BW and Kneafsey M (2000) 'Registering regional speciality food and drink products in the United Kingdom: the case of PDOs and PGIs.' *Area*, 32(3): 317–25.

Ilbery BW, Maye D, Kneafsey M, Jenkins T and Walkley C (2004) 'Forecasting food supply chain developments in lagging rural regions: evidence from the UK.' *Journal of Rural Studies*, 20: 331–44.

Ilbery BW and Maye D (2005) 'Food supply chains and sustainability: evidence from specialist food producers in the Scottish/English borders.' *Land Use Policy*, 22: 331–44.

Ilbery BW and Maye D (2006) 'Retailing local food in the Scottish-English Borders: a supply chain perspective.' *Geoforum* 37 p352–367.

Ilbery BW and Stiell B (1991) 'Uptake of the Farm Diversification Scheme in England.' *Geography*, 75: 69–73.

Ingrouille M (1995) *Historical Ecology of the British Flora.* Chapman & Hall: London.

IPCC (2000) *Climate Change and Agriculture in the United Kingdom.* IPCC: Oxford.

Jarvis RA, Bendelow VC, Bradley RI, Carroll DM, Furness RR, Kilgour INL and King SJ (1984) *Soils and their Use in Northern England.* Bulletin No. 10. Soil Survey of England and Wales: Harpenden.

Jerram R (1992) 'Montane lichen and moss heath in the Lake District.' Contract report for English Nature: Bowness-on-Windermere. Unpublished.

JNCC (1995) 'Purple Moor Grass and Rush Pasture Habitat Action Plan' in *UK Biodiversity Group Tranche 2 Action Plans – Vol. II: Terrestrial and freshwater habitats* (December 1995, Tranche 1, II, p. 253).

JNCC (1998a) 'Montane Habitat Action Plan' in *UK Biodiversity Group Tranche 2 Action Plans – Vol. II: Terrestrial and freshwater habitats* (December 1998, Tranche 2, II, p. 16).

JNCC (1998b) 'Woolly willow Species Action Plan' in: *UK Biodiversity Group Tranche 2 Action Plans – Vol. I: Vertebrates and vascular plants* (June 1998, Tranche 2, I, p. 221).

JNCC (1999a) 'Upland heathland Habitat Action Plan' in *UK Biodiversity Group Tranche 2 Action Plans – Vol. VI: Terrestrial and freshwater species and habitats* (October 1999, Tranche 2, VI, p. 217).

JNCC (1999b) *UK Biodiversity Group Tranche 2 Action Plans – Vol. VI: Terrestrial and freshwater species and habitats* (October 1999). JNCC: London.

JNCC (2007) *Second Report by the UK under Article 17 on the Implementation of the Habitat Directive from January 2001 to December 2006.* JNCC: Peterborough. Accessed at www.jncc.gov.uk/article17.

JNCC (2008) *UK Biodiversity Action Plan Priority Habitat Description: mountain heaths and willow scrub.* Accessed at: www.jncc.gov.uk/page-5220 (23/11/2008).

Jobey G (1985) 'Unenclosed settlements of the Tyne–Forth – a summary' in Spratt D and Burgess C (eds) *Upland Settlement in Britain: the second millenium BC and after*, pp. 177–94. British Series 143. British Archaeological Reports: Oxford.

Johnston JR, Gregory K and Smith DM (1994) *The Dictionary of Human Geography*, 3rd edn. Blackwell: Oxford.

Johnston J, Webb S and Hunt D (2005) *English Nature's Sustainable Grazing Initiative in Cumbria – a summary of English Nature's work in the Cumbrian uplands from 2003 to 2005.* English Nature: Kendal.

Joint Nature Conservation Committee *see* JNCC.

Jones GE (1963) 'The diffusion of agricultural innovations.' *Journal of Agricultural Economics*, 15: 49–59.

Jones GE (1975) *Innovation and Farmer Decision Making.* D203 Agriculture. Open University: Milton Keynes.

Jones A (1990) 'New directions for West German agricultural policy – the example of Schleswig-Holstein.' *Journal of Rural Studies*, 6(1): 9–16.

Jones P, Mortimer S, Park J, Strange A and Griffiths G (2003) *Classification of Land to Reflect the Environmental Value and Difficult Farming Conditions of Upland Grazing Land.* Report for DEFRA project LS34055. Centre for Agricultural Strategy, University of Reading: Reading.

Kapos V, Rhind M, Edwards M, Price MF and Ravilious C (2003) 'Developing a map of the world's mountain forests' in Price MF and Butt N (eds), *Forests in Sustainable Mountain Development: a state of knowledge report for 2000*, pp. 10–17. CAB INTERNATIONAL Publishing: Wallingford.

Kennedy F (2002) *The Identification of Soils for Forest Management.* Forestry Commission: Edinburgh.

Kirby K (2005) 'Was the wildwood closed forest or savannah – does it matter?' Paper given at conference on 'The role of large herbivores in shaping the upland landscapes of Britain – what does the science of herbivore ecology tell us?' [Battleby, Perth, Scotland, 16 Feb. 2005]

Kirby KJ, Saunders GR and Whitbread AM (1991) 'The National Vegetation Classification in nature conservation surveys – a guide to the use of the woodland section' *British Wildlife*, 3(2): 70–80.

Knightsbridge R and Kernon T (1997) *Characterisation of Farming in Natural Areas.* Research Report No. 205. English Nature: Wakefield.

Korner C. (2002) 'Mountain biodiversity, its causes and function: an overview' in Korner C and Spehn EM (eds), *Mountain Biodiversity: a global assessment*, pp. 3–20. Parthenon Publishing: London.

Kristensen SP, Thenail C and Kristensen L (2004) 'Farmers' involvement in landscape activities: an analysis of the relationship between farm location, farm characteristics and landscape changes in two study areas in Jutland, Denmark.' *Journal of Environmental Management*, 61: 301–18.

Kuznesof S, Tregear A and Moxey A (1997) 'Regional foods: a consumer perspective.' *British Food Journal*, 99(6): 199–206.

Lake District World Heritage Project (2009) website accessed at: www.lakeswhs.co.uk/ (2/2/2011).

Land Based Colleges National Consortium (2009) website accessed at: www.lbcnc.org.uk/moodle/ (23/12/2009).

Latacz-Lohmann U and Laughton R (2000) 'Farmers' markets in the UK – a study of farmers' perceptions.' *Farm Management*, 10(10): 579–89.

Le Heron R (1993) *Globalized Agriculture: political choice.* Pergamon Press: Oxford.

LEADER + (2000) *LEADER+ Cumbria Fells and Dales Development Plan 2000 to 2006.* Accessed at: www.defra.gov.uk (23/3/2004).

LEADER+ (2001) *England LEADER+ Programme Development Plan 2000 to 2006.* Accessed at: www.defra.gov.uk (23/3/2004).

Lindberg K, McCool S and Stankey G (1997) 'Rethinking carrying capacity.' *Annals of Tourism Research*, 24(2): 461–5.

Linton D (1968) 'The assessment of scenery as a natural resource.' *Scottish Geographical Magazine*, 84: 219–38.

Lobley M, Errington A, McGeorge A, Millard N and Potter C (2002) *Implications of Changes in Structures of Agricultural Businesses.* Final Report. DEFRA: London.

Lobley M, Potter C, Butler A, Whitehead I and Millard N (2005) *The Wider Social Impacts of Changes in the Structure of Agricultural Businesses.* University of Exeter and Imperial College: Exeter.

Long D and Williams J (2007) *Juniper in the British Uplands: the Plantlife Juniper Survey results.* Plantlife International: Stirling.

Lowe JL and Walker MJC (1984) *Reconstructing Quaternary Environments.* Longman: London.

Lowe P, Buller H and Ward N (2002) 'Selling the next agenda? British and French approaches to the Second Pillar for the CAP.' *Journal of Rural Studies*, 18(1): 1–17.

Lowe P, Cox G, MacEwen M, O'Riordan T and Winter M (1986) 'Ploughing into the Halvergate Marshes', Ch. 11 in *Countryside Conflicts: the politics of farming, forestry and conservation*. Temple Smith: Aldershot.

McClatchey J (1996) 'Spatial and altitudinal gradients of climate in the Cairngorms – observations from climatological and automatic weather stations.' *Botanical Journal of Scotland*, 48: 31–49.

MacDonald A (1996) 'Cutting of heather as an alternative to muirburn'. Information and Advisory Note No. 58. Scottish Natural Heritage: Battleby, Perth.

MacDonald A (2004) 'New developments at the treeline.' *Look to the Hills*, 12 (March). Accessed at: www.jncc.gov.uk/pdf/ulcn_newsletter12.pdf.

MacDonald A, Stevens P, Armstrong H, Immirzi P and Reynolds P (2001) *A Guide to Upland Habitats: surveying land management impacts*. Vol. 1: *Background information and guidance for surveyors*. Scottosh Natural Heritage: Edinburgh.

McEachern C (1992) 'Farmers and conservation: conflict and accommodation in farming politics.' *Journal of Rural Studies*, 8(2): 159–71.

McInerney JP and Turner MM (1991) 'Farm diversification: enterprises and incomes.' *Farm Management*, 7(11): 525–34.

McEvoy D, Handley JF, Cavan G, Aylen J, Lindley S, McMorrow J and Glynn S (2006) *Climate Change and the Visitor Economy: challenges and opportunities for England's Northwest*. Sustainability Northwest: Manchester; UKCIP: Oxford.

MacEwan A and MacEwan M (1982) *National Parks: conservation or cosmetics*? George Allen & Unwin: London.

MacEwan A and MacEwan M (1987) *Greenprints for the Countryside: the Story of Britain's National Parks*. Allen & Unwin: London.

MacEwen M and Sinclair G (1983) *New Life for the Hills: policies for farming and conservation in the uplands*. Council for National Parks: London.

MacGowan F and Doyle G (1997) 'Vegetation and soil characteristics of damaged Atlantic blanket bogs in the west of Ireland' in Tallis JH, Meade R and Hulme PD (eds), *Blanket Mire Degradation: causes, consequences and challenges*, pp. 54–63. English Nature: Wakefield. [Proceedings of the Blanket Mire Degradation conference, University of Manchester, 9–11 April].

MacKenzie NA (2000) *Low Alpine, Subalpine and Coastal Scrub Communities in Scotland*. Highland Birchwoods: Munlochy.

Mackey EC, Shewry MC and Tudor GJ (1998) *Land Cover Change: Scotland from the 1940s to the 1980s*. The Stationery Office: Edinburgh

Mackney D, Hodgson JM, Hollis JM and Staines SJ (1983) *Legend for the 1:250000 Soil Map of England and Wales*. Soil Survey of England and Wales: Harpenden.

McLaren Loring J (2007) 'Wind energy planning in England, Wales and Denmark: factors influencing project success.' *Energy Policy*, 35(4): 2648–60.

Maclean M (1992) *New Hedges for the Countryside*. Farming Press: Ipswich.

McLeish A (1992) *Geological Science*. Nelson: London.

McNally S (2001) 'Farm diversification in England and Wales – what can we learn from the farm business survey?' *Journal of Rural Studies*, 17: 247–57.

McVittie A, Moran D, Smyth K and Hall C (2005) *Measuring Public Preferences for the Uplands*. Final report. International Centre for the Uplands: Hackthorpe, Cumbria.

MAFF (1970) *Modern Farming and the Soil. Report of the Agricultural Advisory Council on Soil Structure and Soil Fertility*. HMSO: London [the Strutt Report].

MAFF (1983) *Agricultural Land Classification of England and Wales: the distribution of the grades*. Note No. TN/RP/01 TFS 846.

MAFF (1988) *Agricultural Land Classification of England and Wales: revised guidelines and criteria for grading the quality of agricultural land*. MAFF: London.

MAFF (1991) *The Pennine Dales ESA Report of Monitoring*. MAFF: London.

MAFF (1995) *The Moorland Scheme Information Pack*. PB2058. MAFF: London.

MAFF (1997) *The Organic Aid Scheme Information Pack*. PB1825. MAFF: London.

MAFF (1998a) *Countryside Stewardship Application Pack*. MAFF: London.

MAFF (1998b) 'Howgill Fells' in: *Countryside Stewardship in Cumbria*. Countryside Stewardship Scheme: MAFF: London.

MAFF (1999a) *Agricultural Land Classification*. FRCA: London.

MAFF (1999b) *Countryside Stewardship Upland Survey Guidance for Applicants* MAFF: London.

MAFF (1999c) *Success with Marketing Diversified Enterprises: concepts and applications*. MAFF: London.

MAFF (2000) *Processing and Marketing Grant – introductory booklet*. PB5268. MAFF: London.

Mallik AU, Gimingham CH and Rahman AA (1984) 'Ecological effects of heather burning. I. Water infiltration, moisture retention and porosity of surface soil.' *Journal of Ecology*, 72: 767–76.

Mallon D, Wheeler P, Whiteley D and Yalden DW (2003) 'Mountain hares in the Peak District.' *British Wildlife*, 15(2): 110–16.

Manning RE (2007) *Parks and Carrying Capacity: commons without tragedy*. Island Press: London.

Mansfield LT (1994) '*Land diversion and environmental management in eastern England*'. University of Coventry, unpublished PhD thesis.

Mansfield LT (2005) 'Change in upland agriculture and its effects on the environment: a case study from Cumbria, United Kingdom.' Working Group 16, European Rural Sociological Society Conference, Kesthely, Hungary, 22–27 August.

Mansfield, LT (2008) 'The Cumbria Hill Sheep Initiative: a solution to the decline in the upland hill farming community in England?' in Robinson G (ed.), *Sustainable Rural Systems – sustainable agriculture and rural communities*, pp. 161–84. Ashgate: Aldershot.

Mansfield LT (in press) 'Hill farming identities and connections to place' in Convery I, Corsane G and Davis P (eds) (2012), *Making Sense of Place: multidisciplinary perspectives*. The Boydell Press: Woodbridge [bibliographic details to be confirmed].

Mansfield LT and Martin H (2004) *The Fell Farming Traineeship Scheme: Final Evaluation*. On behalf of Cumbria Fells and Dales LEADER+: Penrith, Cumbria.

Mansfield LT, Martin H and Brown G (2005) 'Sustaining an upland labour force in Cumbria: the Fell Farming Traineeship Scheme.' *Journal of Farm and Rural Development*, 20: 43–61.

Mansfield LT (2011) *'Farmers' views on vital uplands: Bassenthwaite catchment, Cumbria'* A pilot survey commissioned by Natural England, NW Region.

Mardon D (1997) 'Eight years of montane scrub restoration at Ben Lawers NNR' reprinted in Gilbert D (2006) *Montane Scrub Restoration Project – guidance for the restoration of montane scrub*, pp. 65–74. Highland Birchwoods and Scottish Natural Heritage: Munlochy.

Mardon D (2003) 'Conserving montane willow scrub on Ben Lawers NNR.' *Botanical Journal of Scotland*, 55(1): 189–203.

Mardon D (2007) 'Conserving montane scrub – an uphill struggle.' *Conservation Land Management*, 5(2): 4–7.

Marks HF and Britton DK (1989) *A Hundred Years of British Food and Farming: a statistical survey*. Taylor & Francis: London.

Marland A (1989) 'An overview of organic farming in the UK.' *Outlook on Agriculture*, 18(1): 24–7.

Marrs RH, Le Duc MG, Mitchell RJ, Goddard D, Paterson S and Pakeman RJ (2000) 'The ecology of bracken: its role in succession and implications for control.' *Annals of Botany*, 85 (supplement B): 3–15.

Marsden T (1998) 'Theoretical approaches to rural restructuring: economic perspectives', Ch. 2 in Ilbery BW (ed.), *The Geography of Rural Change*. Longman: London.

Marsden T, Munton R and Ward N (1992) 'Incorporating social trajectories into uneven agrarian development: farm businesses in upland and lowland Britain.' *Sociologia Ruralis*, 32(4): 408–30.

Marsden T, Munton R, Ward N and Whatmore S (1996) 'Agricultural geography and the political economy: a review.' *Economic Geography*, 72(4): 361–75.

Marsden T, Munton RJC, Whatmore S and Little J (1986) 'Towards a political economy perspective of capitalist agriculture: a British perspective.' *International Journal of Urban Regional Research*, 10: 498–521.

Marsden T and Smith E (2005) 'Ecological entrepreneurship: sustainable development in local communities through quality food production and local branding.' *Geoforum*, 36(4): 440–51.

Marsden T and Sonnino R (2008) 'Rural development and the regional state: denying multifunctional agriculture in the United Kingdom.' *Journal of Rural Studies*, 24: 422–31.

Mason WL, Hampson A and Edwards C (eds) (2004) *Managing the Pinewoods of Scotland*. Forestry Commission: Edinburgh.

Matthews KB, Wright IA, Buchan K, Davies DA and Schwarz G (2006) 'Assessing the options for upland livestock systems under CAP reform: Developing and applying a livestock systems model within whole-farm analysis.' *Agricultural Systems*, 90: 32–61.

Messerli B and Ives J (1997) *Mountains of the World*. Parthenon: New York.

Messerli B and Messerli P (2007) 'From local projects in the Alps to global change programmes in mountain areas: the development of interdisciplinary and transdisciplinary in the last 25 years' in Price MF (ed.), *Mountain Area Research and Management – integrated approaches*, pp. 24–48. Earthscan: London.

Midmore P, Sherwood A-M, Hounsome B, Hughes G, Jenkins T, Roughley G and Russell S (1998) *LFA Policy in Wales: a review of the socio-economic and environmental effects of the HLCA Scheme*. University of Wales: Aberystwyth.

Midmore P, Sherwood A-M and Roughley G (2001) 'Policy reform and the sustainability of farming in the uplands of the United Kingdom: conflicts between environment and social support.' *Journal of Environmental Policy and Planning*, 3: 43–63.

Miles J (1987) 'Effects of man on upland vegetation' in Bell M and Bunce RGH (eds), *Agriculture and Conservation in the Hills and Uplands*, pp. 7–17. Symposium No 23. Institute of Terrestrial Ecology: Grange-over-Sands, Cumbria.

Miller GR, Geddes C and Mardon DK (1994) 'Response of the alpine gentian (*Gentiana nivalis*) and other montane species to protection from grazing.' *Scottish Natural Heritage Review*, No. 32.

Miller GR, Miles J and Heal OW (1984) *Moorland Management: a study of Exmoor*. Institute of Terrestrial Ecology: Cambridge.

Milligan AL, Putwain PD, Cox ES, Ghorbani J, Le Duc MG and Marrs RH (2004) 'Developing an integrated land management strategy for the restoration of moorland vegetation on *Molinia caerulea* dominated vegetation for conservation purposes in upland Britain.' *Biological Conservation*, 119: 371–85.

Milne JA and Hartley SE (2001) 'Upland plant communities – sensitivity to change.' *Catena*, 42: 333–43.

Milward R and Robinson A (1980) *Upland Britain*. David & Charles: Newton Abbot.

Ministry of Agriculture and Fisheries (1934) 'Bracken.' Advisory leaflet No. 190.

Ministry of Agriculture, Fisheries and Food *see* MAFF.

Mitchell BW (1989) *Geography and Resource Analysis*. Longman Scientific and Technical: London.

Mitchell BW (2004) *Resource and Environmental Management*, 2nd edn. Prentice Hall: London.

Moody J and Neville W (2004) *Mid Term Review: a practical guide to the Common Agricultural Policy*. [Self-published].

Moore PD (1973) 'The influence of prehistoric cultures upon the initiation and spread of blanket bog in upland Wales.' *Nature*, 241: 350–3.

Moore PD (1988) 'The development of moorlands and upland mires' in Jones M (ed.), *Archaeology and the Flora of the British Isles*, pp. 116–22. Oxford University Committee for Archaeology: Oxford.

Moore PD (1993) 'The origin of blanket mire, revisited' in FM Chambers (ed.), *Climate Change and Human Impact on the Landscape*, pp. 217–25. Chapman & Hall: London.

Moore PD and Merryfield DL (1974) 'Prehistoric human activity and blanket peat initiation on Exmoor.' *Nature*, 250: 439–41.

Moore PD and Webb MJ (1978) *An Illustrated Guide to Pollen Analysis*. Hodder & Stoughton: London.

Moore-Colyer RJ (1998) 'Farming in depression: Wales between the wars, 1919 to 1939.' *Agricultural History Review*, 46(2): 177–96.

Moorland Association (2002) *Northern Uplands Moorland Regeneration Project*. The Moorland Association: Lancaster.

Morgan RJC (1995) *Soil Erosion and Conservation*. Longman: London.

Morgan R and Munton R (1971) *Agricultural Geography*. Methuen: London.

Morin PJ (1999) *Community Ecology*. Blackwell Science: Oxford.

Morris C and Potter C (1995) 'Recruiting the new conservationists: farmers' adoption of agri-environment schemes in the UK.' *Journal of Rural Studies*, 11: 51–63.

Morris C and Young C (1997) 'Towards environmentally beneficial farming? – an evaluation of the Countryside Stewardship Scheme' *Geography*, 82(4): 305–16.

Moseley MJ (1979) *Accessibility: the rural challenge*. Methuen: London.

Moss B (2001) *The Broads*. New Naturalist Library. Harper Collins: London.

Mowle A and Evans S (1990) 'Conserving the rural environment – reconciling farmer and nature on less favoured areas' in Bowers J (ed.), *Agricultural and Rural Land Use: into the 1990s*. ESRC: Swindon.

Muir R and Muir N (1989) *Fields*. Macmillan: London.

Munton RJC (1977) 'Financial institutions: their ownership of agricultural land in Great Britain.' *Area*, 9: 29–37.

Nagy L, Grabher G, Korner C and Thompson DBA (2003) *Alpine Diversity in Europe.* Ecological Studies 67. Springer-Verlag: Berlin.

National Assembly for Wales (2000) *Rural Development Plan for Wales 2000–2006.* Food and Farming Development Division, National Assembly for Wales: Cardiff.

National Trust for Scotland (2006) 'The Ben Lawers treeline woodland and montane scrub project'. Information notes.

Natural England (2008a) *Higher Level Stewardship Part C Options, Capital Items and Management Conditions Handbook*, 2nd edn. Natural England: Peterborough.

Natural England (2008b) *Organic Entry Level Stewardship Scheme Handbook*, 2nd edn. Natural England: Peterborough.

Natural England (2008c) *Higher Level Stewardship Part A Application Handbook*, 2nd edn. Natural England: Peterborough.

Natural England (2008d) *Environmental Stewardship: Higher Level Stewardship. Part B. Farm environment plan (FEP) features manual: technical guidance on the identification, condition assessment and recording of HLS FEP features*, 2nd edn. Natural England: Peterborough.

Natural England (2008e) *Responding to the Impacts of Climate Change on the Natural Environment: the Cumbria High Fells – a summary.* Natural England: Peterborough.

Natural England (2008f) *Environmental Stewardship Review of Progress.* DEFRA and Natural England: London. Accessed at: http://collections.europarchive.org/tna/20081027092120/http://defra.gov.uk/erdp/schemes/es/es-report.pdf (15/10/2008).

Natural England (2009a) *Mapping Values: the vital nature of our uplands: an atlas linking environment and people.* NE209. Natural England: Peterborough.

Natural England (2009b) *Vital Uplands: a 2060 vision for England's upland environment.* NE210. Natural England: Peterborough.

Natural England (2010a) *National Character Areas.* Accessed at: www.naturalengland.org.uk/ourwork/landscape/englands/character/areas/default.aspx (24/10/2010). [previously called Joint Character Areas]

Natural England (2010b) *Entry Level Stewardship: environmental stewardship handbook*, 3rd edn [Part 5, Section U: Uplands ELS]. NE226. Natural England: Peterborough. Accessed at: http://naturalengland.etraderstores.com/NaturalEnglandShop/NE226 (24/6/2011).

Nature Conservancy Council (1984) *Nature Conservation in Great Britain.* Nature Conservancy Council: Shrewsbury.

Neville-Rolfe E (1984) *The Politics of Agriculture in the European Community.* Policy Studies Institute/European Centre for Policy Studies: Brussels.

Newby HH (1979) *Green and Pleasant Land? Social change in rural England.* Hutchinson: London.

NORDREGIO (2004) *Mountain Areas in Europe: analysis of mountain areas in EU states, acceding and other European countries.* NORDREGIO, European Union: Brussels.

NWDA (2007) *Rural Development Programme for England: the implementation plan for England's Northwest 2007–2013.* North West Development Agency: Manchester.

NYMNPA (2006) 'Farming in the North York Moors National Park.' North York Moors National Park Education Service: Helmsley.

O'Donovan G (2001) 'Productivity, grazing pressure and phenology of a limestone grassland in the Burren National Park, Co. Clare.' *Tearmann*, 1(1): 35–54.

Office of Official Publications of the European Communities (1990) *Upland Areas Basic Data and Statistics.* Economic and Social Committee, Office of Official Publications of the European Communities: Luxembourg.

Oom S, Sibbald AM, Hester AJ, Miller DR and Legg CJ (2008) 'Impacts of sheep grazing a complex vegetation mosaic: relating behaviour to vegetation change.' *Agriculture, Ecosystems and Environment*, 124: 219–28.

Overton M (1996) *The Agricultural Revolution in England: the transformation of the agrarian economy, 1500 to 1850.* Cambridge University Press: Cambridge.

Pacione M (1984) *Rural Geography.* Paul Chapman: London.

Pain DJ, Hill D and McCracken DI (1997) 'Impact of agricultural intensification of pastoral systems on bird distributions in Britain, 1970 to 1990.' *Agriculture, Ecosystems and Environment*, 64: 19–32.

Pakeman RJ and Hay E (1996) 'Heathland seedbanks under bracken *Pteridium aquilinium* (L.) Kuhn and their importance for re-vegetation after bracken control.' *Journal of Environmental Management*, 47: 329–39.

Pakeman RJ, Hulme PD, Torvell L and Fisher JM (2003) 'Rehabilitation of degraded dry heather [*Calluna vulgaris* (L.) Hull] moorland by controlled sheep grazing.' *Biological Conservation*, 114: 389–400.

Pakeman RJ, Marrs RH, Howard DC, Barr CJ and Fuller RM (1996) 'The bracken problem in Great Britain: its present extent and future changes.' *Applied Geography*, 16(1): 65–86.

Palmer SCF (1997) 'Prediction of the shoot production of heather under grazing in the uplands of Great Britain.' *Grass and Forage Science*, 52: 408–24.

Palmer SCF, Gordon LF, Hester AL and Pakeman RJ (2004) 'Introducing spatial grazing impacts into the prediction of moorland vegetation dynamics.' *Landscape Ecology*, 19: 817–27.

Pankhurst M (2007) 'Keskadale: an oak wood on the edge.' *British Wildlife*, 19(3): 187–92.

Parrish T (1991) 'Is Set Aside benefiting wildlife?' *Farming and Wildlife Advisory Group Eastern Region Newsletter*, No. 3 (Summer): 5.

Parry ML (1973) 'Secular climate change and marginal agriculture.' *Transactions of the Institute of British Geographers*, New Series 1(1): 1–13.

Parry ML (1975) 'Secular climatic change and marginal agriculture.' *Transactions of the Institute of British Geographers*, 64: 1–14.

Parry ML (1985) 'Upland settlement and climatic change: the medieval evidence', Ch. 2 in Spratt D and Burgess C (eds), *Upland Settlement in Britain: the second millenium BC and after*, BAR British Series 143. British Archaeological Reports: Oxford.

Parry M, Bruce A and Harkness C (1981) 'The plight of British moorlands.' *New Scientist* (28 May 1981): 550–1.

Parry M, Carson I, Rehman T, Tranter R, Jones P, Mortimer D, Livermore M and Little J (1999) *Economic Implications of Global Climate Change on Agriculture in England and Wales*. Research Report No. 1. Jackson Institute, University College: London.

Parry ML and Carter TR (1991) 'Climatic changes and land use potential in Europe', Ch. 9 in Brouwer FM, Thomas AJ and Chadwick MJ (eds), *Land Use Changes in Europe – processes of change, environmental transformations and future patterns*. Kluwer Academic: London.

Passmore J (1980) *Man's Responsibility for Nature*, 2nd edn. Duckworth: London.

Peach WJ, Lovett LJ, Wotton SR and Jeffs C (2001) 'Countryside Stewardship delivers cirl buntings (*Emberiza cirlus*) in Devon, UK.' *Biological Conservation*, 101: 361–73.

Peak District National Park Authority (2001) *Peak District National Park Biodiversity Action Plan*. Accessed at: www.peakdistrict.gov.uk/bap.pdf.

Pearce-Higgins JW and Yalden DW (1997) 'The effect of resurfacing the Pennine Way on recreational use of blanket bog in the Peak District National Park, England.' *Biological Conservation*, 82: 337–43.

Pearsall WH (1955) *Mountains and Moorlands*. New Naturalist Series. Collins: London.

Peattie R (1936) *Mountain Geography: a critique and field study*. Harvard: Cambridge MA.

Pienkowski MW, Bignal EM, Galbraith CA, McCracken DI, Stillman RA and Boobyer MG (1996) 'A simplified classification of land-type zones to assist the integration of biodiversity objectives in land use policies.' *Biological Conservation*, 75: 11–25.

Pennington W (1969) *The History of British Vegetation*. Unibooks, English University Press: London.

Pepin NC (1995) 'The use of GCM scenario output to model effects of future climatic change on the thermal climate of marginal maritime uplands.' *Geografiska Annaler*, 77A: 167–85.

Perren R (1995) *Agriculture in Depression, 1870 to 1940*. Cambridge University Press: Cambridge.

Perry PJ (1974) *British Farming in the Great Depression 1870 to 1914: an historical geography*. David and Charles: Newton Abbot.

Petit S, Stuart RC, Gillespie MK and Barr CJ (2003) 'Field boundaries in Great Britain: stock and change between 1984, 1990 and 1998.' *Journal of Environmental Management*, 67: 229–38.

Phillips A and Clarke R (2004) 'Our landscape from a wider perspective', Ch. 4 in Bishop K and Phillips A (eds), *Countryside Planning: new approaches to management and conservation*, pp. 49–62. Earthscan: London.

Phillips CJC (2001) *Principles of Cattle Production*. CAB INTERNATIONAL Publishing: Wallingford.

Plantlife (2004) *Juniperus communis species dossier for Plantlife*. Accessed at: www.plantlife.org.uk (14/07/2004).

Plantlife (2005) *Managing upland for juniper*. Plantlife Scotland: Stirling.

Pollock ML, Milner JM, Waterhouse A, Holland JP and Legg CJ (2005) 'Impacts of livestock in regenerating upland birch woodlands in Scotland.' *Biological Conservation*, 123: 443–52.

Ponder V and Hindley A (2009) *Farming Lives: using the sustainable livelihoods approach in the Peak District farming community*. National Farmers Network and Oxfam: Nottingham.

Poore MED and McVean DN (1957) 'A new approach to Scottish mountain vegetation.' *Journal of Ecology*, 45: 401–39.

Porter K (2004) 'The natural area experience', Ch. 6 in Bishop K and Phillips A, *Countryside Planning: new approaches to management and conservation*. Earthscan: London.

Poschlod P and WallisDeVries MF (2002) 'The historical and socioeconomic perspective of calcareous grasslands – lessons from the distant and recent past.' *Biological Conservation*, 104: 361–76.

Potter CA (1988) 'Environmentally Sensitive Areas in England and Wales: an experiment in countryside management.' *Land Use Policy* (July): 301–13.

Potter C, Barr C and Lobley M (1996) 'Environmental change in Britain's countryside: an analysis of recent patterns and socio-economic processes based on the Countryside Survey 1990.' *Journal of Environmental Management*, 48: 169–86.

Potter C and Burney J (2002) 'Agricultural multifunctionality in the WTO – legitimate non trade concern or disguised protectionism?' *Journal of Rural Studies*, 18: 35–47.

Potter C and Lobley M (1992) 'The conservation status and potential of elderly farmers: results from a survey of England and Wales.' *Journal of Rural Studies*, 8: 133–43.

Potter C and Lobley M (1996) 'The farm family life cycle, succession paths and environmental change in Britain's countryside.' *Journal of Agricultural Economics*, 47(2): 172–90.

Potter C and Tilzey M (2003) 'Agricultural multifunctionality, environmental sustainability and the WTO: resistance or accommodation to the neoliberal project of agriculture?' *Geoforum*, 38(6): 1290–1303.

Pred A (1967) 'Behaviour and location: part 1.' *Land Studies in Geography*, B 28.

Pretty J and Ward H (2001) 'Social capital and the environment.' *World Development*, 29(2): 209–27.

Price MF and Butt N (eds) (2001) *Forests in Sustainable Mountain Development: a state of knowledge report for 2000.* CAB International Publishing: Wallingford.

Pringle D (1994) *The First 75 years: a brief account of the history of the Forestry Commission, 1919–1994.* Forestry Commission: Edinburgh.

Putnam R. (1993): *Making Democracy Work: civic traditions in modern Italy*. Princeton University Press: Princeton NJ.

Raistrick A (1983) *Malham and Malham Moor.* Dalesman: Clapham, Yorkshire.

Rare Breeds Survival Trust (2008) *Breeds at Risk*. Accessed at: www.rbst.org.uk/breedsatrisk (18/11/2008).

Ratcliffe DA (1964) 'Mires and bogs' in Burnett JH (ed.), *The Vegetation of Scotland*, pp. 426–78. Oliver & Boyd: Edinburgh.

Ratcliffe DA (ed.) (1977) *A Nature Conservation Review*. Cambridge University Press: Cambridge.

Ratcliffe DA and Thompson DBA (1988) 'The British uplands: their ecological character and international significance' in Usher MB and Thompson DBA (eds), *Ecological Change in the Uplands*, pp. 9–36. Special Publication No. 7 of the British Ecological Society. Blackwell Scientific: Oxford.

Rawes M and Welch D (1969) 'Upland productivity of vegetation and sheep at Moor House National Nature Reserve, Westmorland England.' *Oikos Supplement*, 11: 7–72.

Ray C (1998) 'Territory, structures and interpretation – two case studies of the European Union's LEADER I programme.' *Journal of Rural Studies*, 14(1): 79–87.

Ray C (2000) 'The EU LEADER programme: rural development laboratory.' *Sociologia Ruralis*, 40(2): 163–71.

Reading *see* University of Reading.

Reid C (2009) *Nature Conservation Law*, 2nd edn. W. Green/Sweet & Maxwell: Edinburgh.

Rees J (1990) *Natural Resources: allocation, economics and policy*, 2nd edn. Routledge: London.

RICS (2004) *Wind Farm Research: impact of wind farms on the value of residential property and agricultural land.* Royal Institute of Chartered Surveyors: London.

RICS (2006) *RICS rural land market survey: Great Britain, second half 2006.* Royal Institute of Chartered Surveyors: London.

Rieder P and Wyder J (1997) 'Economic and political framework for sustainability of mountain areas', Ch. 5 in Messerli B and Ives J (eds), *Mountains of the World*. Parthenon: New York.

Riley M (2004) 'Silent meadows: the uncertain decline and conservation of hay meadows in the British Landscape.' *Landscape Research*, 30(4): 437–58.

Rizov M (2004) 'Rural development and welfare implications of CAP reforms.' *Journal of Policy Modelling*, 26: 209–22.

Roberts RA (1949) 'The historical development of hill farming in Wales.' *Agricultural Progress*, 24: 25–9.

Roberts C and Roberts D (1980) *A History of England*, Vol. 1: *Prehistory to 1714*. Prentice Hall: New York.

Robertson I and Richard P (2003) *Studying Cultural Landscapes*. Arnold: London.

Robinson G (2005) *Geographies of Agriculture – globalisation, restructuring and sustainability*. Prentice Hall: London.

Robinson M (1985) 'The hydrological effects of moorland gripping – a reappraisal of the Moor House research.' *Journal of Environmental Management*, 21: 205–11.

Robinson R (1999) 'Bracken clearance.' *Enact*, 7(2): 10–14.

Rodwell J (1993) 'The National Vegetation Classification.' *British Wildlife*, 6: 266–8.

Rodwell JS (ed.) (1991a) *British Plant Communities*, Vol. 1: *Woodland and scrub*. Cambridge University Press: Cambridge.

Rodwell JS (ed.) (1991b) *British Plant Communities*, Vol. 2: *Mires and heaths*. Cambridge University Press: Cambridge.

Rodwell JS (ed.) (1992) *British Plant Communities*, Vol. 3: *Grasslands and montane communities*. Cambridge University Press: Cambridge.

Rodwell JS (ed.) (1995) *British Plant Communities*, Vol. 4: *Aquatic communities, swamps and tall-herb fens*. Cambridge University Press: Cambridge.

Rodwell JS (ed.) (2000) *British Plant Communities*, Vol. 5: *Maritime communities and vegetation of open habitats*. Cambridge University Press: Cambridge.

Rogers EM (1962) *Diffusion of Innovations*. Macmillan: New York.

Roseland M (2000) 'Sustainable community development: integrating environmental, economic, and social objectives.' *Progress in Planning*, 54: 73–132.

Rotherham ID, Egan D and Ardon PA (2004) 'Fuel economy and the uplands: the effects of peat and turf utilisation on upland landscapes' in Whyte ID and Winchester A, *Society, Landscape and Environment in Upland Britain*, pp. 99–110. Society for Landscape Studies Supplementary Series No. 2. Society for Landscape Studies: Whitby.

Roworth P and Morris T (1998) 'Flailing Crompton's moor grass.' *Enact*, 6(3): 4–5.

Royal Institution of Chartered Surveyors *see* RICS.

Rural Development Service (2005a) *Organic Entry Level Stewardship Handbook – terms and conditions and how to apply*. RDS: London PB10381.

Rural Development Service (2005b) *Environmental Stewardship Targeting Statements 2005* [Upland Joint Character Areas only]. RDS: London.

Rural Development Service (2005c) *Entry Level Stewardship Handbook – terms and conditions and how to apply.* RDS: London PB10355.

Rural Development Service (2005d) *Higher Level Stewardship Handbook – terms and conditions and how to apply.* RDS: London PB10382.

Rural Development Service (2005e) *Higher Level Stewardship – farm environment plan, guidance booklet.* RDS: London PB10383.

Rural Payments Agency (2005a) *Single Payment Scheme: Cross compliance Soil Protection Review*. RPA/DEFRA: London PB11160.

Rural Payments Agency (2005b) *Single Payment Scheme: Handbook and Guidance for England 2005*. RPA/ DEFRA: London.

Rural Payments Agency (2005c) *Single Payment Scheme: information for farmers and growers in England.* February 2005 update. RPA: London PB10525.

Rural Payments Agency (2005d) *Single Payment Scheme 2005 Application Form.* RPA: London PB10525A.

Rural Payments Agency (2005e) *Single Payment Scheme: Cross Compliance Handbook for England, 2005 edn.* RPA: London PB10222A.

Rural Payments Agency (2005f) *Single Payment Scheme: Cross Compliance Guidance for the Management of Habitats and Landscape Features.* RPA: London PB10222C.

Rural Payments Agency (2005g) *Single Payment Scheme: Cross Compliance Guidance for Soil Management.* RPA: London PB10222B.

Rural Payments Agency (2007) *Hill Farm Allowance: Explanatory Booklet 2008.* RPA/DEFRA: London.

Russell T and Harris A (2008) 'Swaling – an ancient tool for modern management.' *Conservation Land Management*, 6(3): 14–17.

Ryder ML (1964) 'The history of sheep breeds in Britain.' *Agricultural History Review*, 12(1): 1–12.

Ryder ML (1984) 'Medieval sheep and wool types.' *Agricultural History Review*, 32(1): 14–28.

Rymer L (1976) 'History and ethnobotany of bracken.' *Botanical Journal of the Linnean Society*, 73: 151–76.

Sansom AL (1999) 'Upland vegetation management: the impacts of overstocking.' *Water Science and Technology*, 39(12): 85–92.

Saul GR (1992) 'Potential systems for increasing income from wool in hill and upland sheep flocks.' *Agricultural Systems*, 39(3): 273–87.

Saunders C (1994) 'Single tier system with many farms partly outside the ESA the case study of the Pennine Dales', Ch. 3 in Whitby M (ed.), *Incentives for Countryside Management: the case study of environmentally sensitive areas.* CAB International: Wallingford.

Scotia Seeds (2011) *Northern Hay Meadow Mix*. Accessed at: www.scotiaseeds.co.uk (6/4/2011).

Scott A, Christie M and Midmore P (2004) 'Impact of the 2001 foot-and-mouth disease outbreak in Britain: implications for rural studies.' *Journal of Rural Studies*, 20: 1–14.

Scott K (2003) 'Grazing issues on Warcop Fell, Cumbria'. Unpublished dissertation. University of Central Lancashire: Penrith.

Scott M (2000) *Montane Scrub*. Scottish Natural Heritage: Edinburgh.

Scott M (2005) 'Montane scrub – the challenge of the "wee trees".' *British Wildlife*, 318–25.

Scott, Mr Justice (1942) *Report of the Committee of Land Utilisation in Rural Areas* Cmnd 6378. HMSO: London.

Scottish Natural Heritage (2002a) *Natural Heritage Futures: an overview.* SNH: Battleby, Perth.

Scottish Natural Heritage (2002b) *Natural Heritage Futures: Cairngorms Massif.* SNH: Battleby, Perth.

Scottish Natural Heritage (2002c) *Natural Heritage Futures: Hills and Moors.* SNH: Battleby, Perth.

Scottish Parliament (2001) 'Support for Agriculture in Less Favoured Areas (LFA).' Research Note RN01/37. The Information Centre: Edinburgh.

Scottish Parliament (2007) *Scottish Rural Development Plan 2007–2013.* Scottish Parliament: Edinburgh.

Scottish Parliament (2010a) *Less Favoured Area Support Scheme 2010 to 2013: explanatory notes.* Rural Payments and Inspections Directorate. Accessed at: www.scotland.gov.uk/Resource/Doc/310944/0098118.pdf (5/11/2010).

Scottish Parliament (2010b) *Economic Report on Scottish Agriculture, 2010 edn.* Accessed at: www.scotland.gov.uk/Publications/2010/06/16110738/87 (5/11/2010).

Selby JA (1980) 'Field afforestation in Finland and its regional variations.' *Communicationes Instituti Forestalis Fenniae*, 99(1): 27–38.

Selby JA (1987) 'On the operationalization of Pred's behavioural matrix.' *Geografiska Annaler*, 69 Series B: 81–90.

Selman P (2004) 'Community participation in the planning and management of cultural landscapes.' *Journal of Environmental Planning and Management*, 47(3): 365–92.

Sexton WT (2001) 'Ecosystem management: expanding the resource management tool kit'. *Landscape and Urban Planning*, 40: 103–12.

Shaw S, Wheeler B and Backshall J (1997) 'Review of effects of burning and grazing of blanket bogs – conservation issues and conflicts' in Tallis JH, Meade J and Hulme PD (ed.), *Blanket Mire Degradation: causes, consequences and challenges*, pp. 174–82. Macaulay Land Use Research Institute: Aberdeen.[Proceedings of the Blanket Mire Degradation Conference, 9-11 April 1997, University of Manchester]

Shepherd M, Pearce B, Cornack B, Phillips L, Cuttle S, Bhogal A, Costigan P and Unwin R (2003) *An Assessment of the Environmental Impacts of Organic Farming – a review for DEFRA funded project OFO405.* DEFRA: London.

Shiva V (2009) *Soil, not Oil: climate change, peak oil and food insecurity.* Zed Books: London.

Shoard M (1980) *The Theft of the Countryside.* Temple Smith: London.

Shucksmith M (1993) 'Farm household behaviour and the transition to post-productivism.' *Journal of Agricultural Economics*, 44: 466–78.

Shucksmith M (2000) 'Endogenous development: social capital and social inclusion: perspectives from LEADER in the UK.' *Sociologia Ruralis*, 40(2): 208–18.

Silvester RJ (2000) 'Medieval upland cultivation on the Berwyns in north Wales.' *Landscape History*, 22: 47–60.

Silvester RJ (2004) 'The commons and the waste: use and mis-use in Mid Wales' in Whyte ID and Winchester A, *Society, Landscape and Environment in Upland Britain*, pp. 53–66. Society for Landscape Studies Supplementary Series 2. Society for Landscape Studies: Whitby.

Simmons IG (1993) 'Vegetation change during the Mesolithic in the British Isles: some amplifications', Ch. 10 in Chambers FM (ed.), *Climate Change and Human Impact on the Landscape.* Chapman & Hall: London.

Simmons IG (2005) *The Moorlands of England and Wales: an environmental history 8000BC to AD2000.* Edinburgh University Press: Edinburgh.

Simmons IG and Ines JB (1996) 'The ecology of an episode of prehistoric cereal cultivation on the North York Moors, England.' *Journal of Archaeological Science*, 23: 613–18.

Simon HA (1957) *Models of Man: social and rational*. Wiley: New York.

Sinclair G (1966) *Can Exmoor Survive*? The Exmoor Society: Dulverton.

Skerrat SJ (1994) 'Itemised payment systems within a scheme: the case of Breadalbane', Ch. 6 in Whitby M (1994) (ed.), *Incentives for Countryside Management: the case of environmentally sensitive areas*, pp. 105–34. CAB International: Wallingford.

Slater J (2003) *The Mid term Evaluation of the England Rural Development Programme (ERDP) Environmentally Sensitive Areas Scheme.* ADAS Consulting: Wolverhampton.

Slee B (1989) *Alternative Farm Enterprises: a guide to alternative sources of income for the farmer*, 2nd edn. Farming Press: Ipswich.

Smart SM, Clarke RT, van der Poll HM, Robertson EJ, Shield ER, Bunce RGH and Maskell LC (2003) 'National-scale vegetation change across Britain: an analysis of sample-based surveillance data from the Countryside Survey of 1990 and 1998.' *Journal of Environmental Management*, 67: 239–54.

Smith AG (1981) 'The Neolithic' in Simmons IG and Tooley MG (eds), *The Environment in British Prehistory*, pp. 125–209. Duckworth: London.

Smith LP (1984) *The Agricultural Climate of England and Wales: areal averages 1941 to 1970.* HMSO: London.

Smith RS (1988) 'Farming and the conservation of traditional meadowland in the Pennine Dales Environmentally Sensitive Area' in Usher MB and Thompson DBA (eds), *Ecological Change in the Uplands*, pp. 183–99. Special Publication No. 7 of the British Ecological Society. Blackwell Scientific: Oxford.

Smithson P, Addison K and Atkinson K (2002) *Fundamentals of the Physical Environment.* Routledge: London.

Soffe RJ and Hetherington SP (1996/7) 'The Moorland Scheme: examining the response.' *Farm Management*, 9(8): 392–9.

Soil Association (2009) *Transition Farming.* Accessed at: www.soilassociation.org/Farmersgrowers/Adviceandsupport/Transitionfarming/tabid/149/Default.aspx (21/07/2009).

Sotherton NW, Boatman ND and Robertson PA (1992) *The Use of Set Aside Land for Game Conservation.* BCPC Monograph No. 50. Game Conservancy Trust: Fordingbridge.

Spedding A (2006) '*Hill farm reform – possible impacts*.' RuSource Briefing 348. Arthur Rank Centre: Norwich.

SQW Ltd (2003) '*ERDP Mid Term Review evaluation – sub regional case study reports*.' On behalf of DEFRA: London.

Stamper P (1992) 'Woods and parks', Ch. 7 in Astill G and Grant A (eds), *The Countryside in Medieval England.* Blackwell: Oxford.

Stapledon RG (1937) *The Hill Lands of Britain*. Faber & Faber: London.

Stapledon RG (1949) 'The hill lands of England and Wales.' *Agriculture*, 53(3): 99–101.

Stapledon RG (1951) *Reclaiming Land for Agriculture*. Crosby Lockwood & Son: London.

Starr C (2005) *Woodland Management: a practical guide*. Crowood: Ramsbury.

Staughton E (2004) 'Beyond enclosure: upland common land in England and Wales since 1800' in Whyte ID and Winchester A (eds), *Society, Landscape and Environment in Upland Britain*, pp. 89–98. Society for Landscape Studies Supplementary Series 2. The Society for Landscape Studies: Whitby.

Stenseke M (2009) 'Local participation in cultural landscape maintenance: lessons from Sweden.' *Land Use Policy*, 26: 214–23.

Stern, Lord (2006) '*The economics of climate change*'. HM Treasury: London. Accessed at: http://webarchive.nationalarchives.gov.uk/+/http:/www.hm-treasury.gov.uk/sternreview_index.htm (27/6/2011).

Stevenson AC and Thompson DBA (1993) 'Long term changes in the extent of heather moorland in upland Britain and Ireland: palaeoecological evidence for the importance of grazing.' *The Holocene*, 3(1): 70–76.

Stirling Council (2002) *Inland Rock and Montane Zone Habitat Action Plan.* Accessed at: www.stirling.gov.uk/montane_ap.pdf (10/06/2007).

Stoneman R and Brooks S (1997) *Conserving Bogs: the management handbook*. The Stationery Office: London.

Swanwick C (2002a) *Landscape Character Assessment Guidance for England and Scotland*. Countryside Agency: Wetherby/Scottish Natural Heritage: Edinburgh.

Swanwick C (2002b) *Making Sense of Place: landscape character assessment – summary guidance for England and Scotland*. Countryside Agency: Wetherby/Scottish Natural Heritage: Edinburgh.

Swanwick C (2004) 'The assessment of countryside and landscape character in England: an overview', Ch. 7 in Bishop K and Phillips A (eds), *Countryside Planning: new approaches to management and conservation*. Earthscan: London.

Sydes C and Miller GR (1988) 'Range management and nature conservation in the British uplands' in Usher MB and Thompson DBA (eds), *Ecological Change in the Uplands*, pp. 323–37. Special Publication No. 7 of the British Ecological Society. Blackwell Scientific: Oxford.

Tallis JH (1964) 'Studies on the southern Pennines peats. I. The general pollen record.' *Journal of Ecology*, 52: 323–47.

Tallis JH (1997) 'The southern Pennine experience: an overview of blanket mire degradation' in Tallis JH, Meade J and Hulme PD (eds), *Blanket Mire Degradation: causes, consequences and challenges*, pp. 7–15. Macaulay Land Use Research Institute: Aberdeen. [Proceedings of the Blanket Mire Degradation Conference, 9 -11 April 1997, University of Manchester]

Tallis JH (1998) 'Growth and degradation of British and Irish blanket mires.' *Environmental Reviews*, 6: 81–122.

Tansley AG (1939) *The British Islands and their Vegetation*. Cambridge University Press: Cambridge.

Tarrant J (1974) *Agricultural Geography*. David & Charles: Newton Abbot.

Taylor C (1987) *Fields in the English Landscape*. Alan Sutton: Gloucester.

Thirsk J (1987) *England's Agricultural Regions and Agrarian History 1500 to 1750*. Macmillan: Basingstoke.

Thomas C (1992) 'A cultural-ecological model of agrarian colonisation in upland Wales.' *Landscape History*, 14: 37–50.

Thomasson A (1975) *Soil and Field Drainage*. Technical Monograph No. 7. The Soil Survey: Harpenden.

Thompson DBA and Brown A (1992) 'Biodiversity in montane Britain: habitat variation, vegetation diversity and some objectives for conservation.' *Biodiversity and Conservation*, 1: 179–208.

Thompson DBA, Galbraith H and Horsfield D (1987) 'Ecology and resources of Britain's mountain plateaux: land use conflicts and impacts' in Bell M and Bunce RGH (eds), *Agriculture and Conservation in the Hills and Uplands*, pp. 22–31. Institute of Terrestrial Ecology: Grange-over-Sands, Cumbria.

Thompson DBA, Macdonald AJ, Marsden JH and Galbraith CA (1995) 'Upland habitat management in Great Britain: a review of international importance, vegetation change and some objectives for nature conservation.' *Biological Conservation*, 71: 163–78.

Tichelar M (2004) 'The Scott Report and the Labour Party: the protection of the countryside during the Second World War.' *Rural History*, 15(2): 167–87.

Tilzey M (2000) 'Natural areas, the whole countryside approach and sustainable agriculture.' *Land Use Policy*, 17: 279–94.

Tinsley H (1976) 'Cultural influences on Pennine vegetation with particular reference to North Yorkshire.' *Transactions of the Institute of British Geographers*, New Series 1: 310–22.

Tinsley HM (1981) 'The Bronze Age' in Simmons IG and Tooley MG (eds), *The Environment in British Prehistory*, pp. 210–49. Duckworth: London.

Tipping R (1997) 'Pollen analysis: late Iron Age and Roman agriculture around Hadrian's Wall' in Gwilt A and Haselgrove C (eds), *Reconstructing Iron Age Societies*, pp. 239–47. Oxford University: Oxford.

Tipping R (1999) 'Towards an environmental history of the Bowmont Valley and the northern Cheviot Hills.' *Landscape History*, 20: 41–50.

Toghill P (2000) *The Geology of Britain: an introduction*. Swanhill Press: Shrewsbury.

Tranter RB (1978) *The Future of Upland Britain*. CAS Paper No. 2. Centre for Agricultural Strategy: Reading.

Truscott AM, Mitchell RJ, Palmer SJF and Welch D (2004) 'The expansion of native oakwoods into conifer cleared areas through planting.' *Forest Ecology and Management*, 193: 335–43.

Tucker HG (2004) 'The burning of uplands and its effect on wildlife.' *British Wildlife* (April): 251–57.

Turner J (1981) 'The Iron Age' in Simmons IG and Tooley MG (eds), *The Environment in British Prehistory*, pp. 250–81. Duckworth: London.

UK *see also* HM Government.

UK Biodiversity Steering Group (1995) *Biodiversity: the UK Steering Group report*, vols 1 and 2. London: HMSO.

UK Biodiversity Steering Group (1998) *Tranche 2 Action Plans: terrestrial and freshwater habitats*, Vol. 2. London: HMSO.

UK LEADER+ (2004) *The Essentials of LEADER+: a guide to practical approaches* LRDP: London.

UK LEADER+ (2004) website. Accessed at: www.defra.gov.uk (23/4/2004).

UK National Ecosystem Assessment (2011a) *The UK National Ecosystem Assessment: synthesis of key findings*. UNEP-WCMC: Cambridge.

UK National Ecosystem Assessment (2011b) *The UK National Ecosystem Assessment: technical report.* UNEP-WCMC: Cambridge.

UNESCO (1972) *Convention concerning the Protection of the World Cultural and Natural Heritage.*

UNESCO (1998) *Operational Guidelines for the Implementation of the World Heritage Convention.*

University of Cumbria (2011) 'Post 2013 sustainable land management for Cumbria' http:/www.post2013 @cumbria.ac.uk

University of Reading (2003) *Classification of Land to reflect the Environmental Value and Difficult Farming Conditions of Upland Grazing Land.* Project LS3405. DEFRA: London.

Usher M and Thompson DBA (1993) 'Variation in the upland heathlands of Great Britain: conservation importance.' *Biological Conservation*, 66: 69–81.

Van der Veen M (1992) *Crop Husbandry Regimes: an archaeobotanical study of farming in northern England 1000BC–AD500.* Sheffield Archaeological Monographs No. 3. J.R. Collis Publications, University of Sheffield: Sheffield.

Varjo U (1979) 'Productivity and fluctuating limits of crop cultivation in Finland.' *Geographia Polonica*, 40: 225–33.

Ventatachalam L (2004) 'The contingent valuation method: a review.' *Environmental Impact Assessment Review*, 24: 89–124.

Vera FWM (2000a) *Grazing Ecology and Forest History.* CAB International: Wallingford.

Vera FWM (2000b) 'The dynamic European forest.' *Arboricultural Journal*, 26: 179–211.

Vera FWM (2005) 'The effects of large herbivores on vegetation dynamics in temperate Europe.' Paper given at: 'The role of large herbivores in shaping the upland landscapes of Britain – what does the science of herbivore ecology tell us?' [Conference at Battleby, Perth, Scotland: 16 February].

Verdant Engineering (2011) *'Generating Income from Renewable Energy'* CAfS training evening, Alston Town Hall 13 September 2011.

Vogler J (2000) *The Global Commons: environmental and technological governance.* Wiley: London.

Vrolijk HCJ, de Bont CJAM, Blokland PW and Soboh RAME (2010) *Farm Viability in the European Union: assessment of the impact of changes in farm payments*. LEI Report 31964. DLO Foundation: Wageningen, Holland.

Walford N (2001) 'Patterns of development in tourist accommodation enterprises on farms in England and Wales.' *Applied Geography*, 21: 331–45.

Walford (2002) 'Agricultural adjustment: adoption of and adaptation to policy reform measures by large-scale commercial farmers.' *Land Use Policy*, 19: 243–57.

Walford N (2003) 'Productivism is allegedly dead, long live productivism. Evidence of continued productivist attitudes and decision-making in South East England.' *Journal of Rural Studies*, 19: 491–502.

Wall T (1993) 'Burning issues.' *Enact*, 1(4): 12–14.

Wallis JR and Jones JVH (2008) 'The financial impact of Entry Level Scheme participation on farms in the Teesdale Less Favoured Area.' *Journal of Farm Management*, 13(6): 453–64.

Walmsley DJ and Lewis GJ (1984) *Human Geography – behavioural approaches.* Longman: London.

Warburton D (2004) 'Community involvement in countryside planning in practice', Ch. 16 in Bishop K and Phillips A (eds), *Countryside Planning: new approaches to management and conservation.* Earthscan: London.

Warburton J, Holden J and Mills AJ (2004) 'Hydrological controls of surficial mass movements in peat.' *Earth Science Review*, 67: 139–56.

Ward SD and Evans DF (1976) 'Conservation assessment of British limestone pavements based on floristic criteria.' *Biological Conservation*, 9: 217–33.

Ward N and McNicholas K (1998) 'Objective 5b of the structural funds and rural development in Britain.' *Journal of the Regional Studies*, 32(4): 369–74.

Ward N, Marsden T and Munton R (1990) 'Farm landscape change: trends in upland and lowland England.' *Land Use Policy* (October): 291–302.

Warr *see* De La Warr.

Warren C (2002) *Managing Scotland's Environment.* University Press: Edinburgh.

Wathern P (1992) 'Less favoured and environmentally sensitive areas: a European dimension to the rural environment' in Howarth W and Rodgers P (eds), *Agriculture, Conservation and Land Use: law and policy issues for rural areas*, pp. 186–208. University of Wales: Cardiff.

Wathern P, Young SN, Brown IW and Roberts DA (1986) 'The EEC Less Favoured Areas Directive – implementation and impact on upland land use in the UK.' *Land Use Policy*, 3: 205–12.

Watt AS (1955) 'Bracken versus heather – a study of plant sociology.' *Journal of Ecology*, 43(2): 490–506.

Webb S and Glading P (1998) 'The ecology and conservation of limestone pavement in Britain.' *British Wildlife*, 10(2): 103–13.

Webster D (1992) *Understanding Geology.* Oliver & Boyd: Sevenoaks.

Webster R and Oliver MA (1990) *Statistical Methods in Soil and Land Resources Survey.* Oxford University Press: Oxford.

Welch D, Scott D and Thompson DBA (2005) 'Changes in the composition of *Carex bigelowii–Racomitrium lanuginosum* moss heath on Glas Maol, Scotland, in response to sheep grazing and snow fencing.' *Biological Conservation*, 122: 621–31.

Welsh Assembly (1999) *Welsh Rural Development Plan 2000-2006.* Welsh Assembly: Cardiff.

Werkman BR and Callaghan TV (2002) 'Responses of bracken and heather to increased temperature and nitrogen addition, alone and in competition.' *Basic and Applied Ecology*, 3: 267–76.

West CC and Gawith MJ (eds) (2005) *Measuring Progress: preparing for climate change through the UK Climate Impacts Programme*. UKCIP Technical Report. UKCIP: Oxford.

Westmacott R and Worthington T (1974) *New Agricultural Landscapes.* Countryside Commission: Cheltenham.

Whatmore S (1991) *Farming Women*. Macmillan: London.

Whitby M (eds) (1994) *Incentives for Countryside Management: the case of environmentally sensitive areas*. CAB International: Wallingford.

Whitby M and Lowe P (1994) 'The political and economic roots of environmental policy in agriculture'. Ch, 1 in Whitby M (ed.), *Incentives for Countryside Management: the case of environmentally sensitive areas*, pp. 1–24. CAB International: Wallingford.

White P and Lovett J (1999) 'Public preferences and willingness to pay for nature conservation in the North York Moors National Park.' *Journal of Environmental Management*, 55(1): 1–33.

White R (2002) *The Yorkshire Dales – a landscape through time*. Great Northern: Ilkley.

WHO (2010) *Food Security Crisis.* Accessed at: www.who.int/foodcrisis/en/ (9/3/2010).

Whyte ID (2000) 'Patterns of parliamentary enclosure of waste in Cumbria: a case study from north Westmorland.' *Landscape History*, 22: 77–89.

Whyte ID (2002) *Transforming fell and valley – landscape and parliamentary enclosure in north west England*. Centre for North West Regional Studies, University of Lancaster: Lancaster.

Whyte ID (2006) 'Parliamentary enclosure and changes in landownership in an upland environment: Westmorland c.1770–1860.' *Agricultural History Review*, 54(2): 240–56.

Wiggering H, Dalchow C, Glemnitz M, Helming K, Muller K, Schultz A, Stchow U and Zander P (2006) 'Indicators for multifunctional land use – linking socio-economic requirements with landscape potentials.' *Ecological Indicators*, 6: 238–49.

Wigginton MJ (1999) *British Red Data Books 1: Vascular Plants.* Joint Nature Conservation Committee: Peterborough.

Williams JM (2006) *Common Standards Monitoring for Designated Sites: first six year report*. JNCC: Peterborough.

Williamson T (2002) *The Transformation of Rural England – farming and the landscape 1700 to 1870*, esp. Ch. 5, pp. 115–38. University of Exeter Press: Exeter.

Willis K and Garrod GD (1993) 'Valuing landscape: a contingent valuation approach.' *Journal of Environmental Management*, 37: 1–22.

Willis K and Garrod GD (1994) 'The ultimate test: measuring the benefits of ESAs', Ch. 9 in Whitby M (ed.), *Incentives of Countryside Management: the case of environmentally sensitive areas.* CAB International: Wallingford.

Wilson G (1997) 'Factors influencing farmer participation in the Environmentally Sensitive Areas Scheme.' *Journal of Environmental Management*, 50: 67–93.

Wilson G (2001) 'From productivism to post-productivism .... and back again? Exploring the (un)changed natural and mental landscapes of European agriculture.' *Transactions of the Institute of British Geographers*, 26(1): 77–102.

Wilson G (2007) *Multifunctional Agriculture: a transition theory perspective.* CAB International: Wallingford.

Wilson O (1992) 'Land ownership and rural development in the North Pennines: a case study.' *Journal of Rural Studies*, 8(2): 145–58.

Wilson RM, Busque J, Fawcett RH, Jones JW, Morse DJ, Tait EJ and Wright JA (1997) *Economic Evaluation of the Hill Livestock Compensatory Allowances Scheme in Scotland.* Institute of Ecology and Resource Management, University of Edinburgh: Edinburgh.

Winchester AJL (2000) *The Harvest of the Hills: rural life in northern England and the Scottish Borders 1400 to 1700.* Edinburgh University Press: Edinburgh.

Winchester A (2004) 'Moorland forests of medieval England' in Whyte ID and Winchester A, *Society, Landscape and Environment in Upland Britain*, pp. 21–36. Society for Landscape Studies Supplementary Series No. 2. The Society for Landscape Studies: Whitby.

Winter M (1997) 'New policies and new skills: agricultural change and technology transfer.' *Sociologia Ruralis*, 37: 363–81.

Winter M, Gaskell P, Gasson R and Short C (1998a) *The Effects of the 1992 Reform of the Common Agricultural Policy on the Countryside of Great Britain: final report (March 1998)*, Vol. 1: Project overview and summary findings. CCX47(i). Countryside Commission: Cheltenham.

Winter M, Gaskell P, Gasson R and Short C (1998b) *The Effects of the 1992 Reform of the Common Agricultural Policy on the Countryside of Great Britain: final report (March 1998)*, Vol. 2: Main findings. CCX47(ii). Countryside Commission: Cheltenham.

Winter M, Gaskell P, Gasson R and Short C (1998c) *The Effects of the 1992 Reform of the Common Agricultural Policy on the Countryside of Great Britain: final report (March 1998)*, Vol. 3: Appendices. CCX47(iii). Countryside Commission: Cheltenham.

Winter M, Gaskell P and Short C (1998) 'Upland landscapes in Britain and the 1992 CAP reforms.' *Landscape Research*, 23(3): 273–88.

Wolpert J (1964) 'The decision process in spatial context.' *Annals of the Association of American Geographers*, 54: 537–8.

Woodcock N (1996) *Geology and Environment in Britain and Ireland.* UCL Press: London.

Woodin SJ (1988) 'Acidic deposition and upland conservation: an overview and the way ahead' in Usher MB and Thompson DBA (eds), *Ecological Change in the Uplands*, pp. 355–63. Special Publication No. 7 of the British Ecological Society. Blackwell Scientific: Oxford.

Woods M (2005) *Rural Geography.* Sage: London.

Woodside R and Crow J (1999) *Hadrian's Wall: an historic landscape*. The National Trust: London.

World Health Organization *see* WHO.

Wright JA (1998) *Economic Evaluation of the Hill Livestock Compensatory Allowance Scheme in Scotland.* IERM, University of Edinburgh: Edinburgh.

Wright IA, Birnie RV, Malcolm A, Towers W and McKeen M (2006) *The Potential Use of the Land Capability for Agriculture Classification for Determining Support to Disadvantaged Areas of Scotland.* Macaulay Institute: Aberdeen.

Wynn G, Crabtree B and Potts J (2001) 'Modelling farmer entry into the environmentally sensitive area schemes in Scotland.' *Journal of Agricultural Economics*, 52(1): 65–82.

Yallop AR, Thomas G, Thacker J, Brewer T and Sannier C (2005) *A History of Burning as a Management Tool in the English Uplands. 1: Estimates of the areal extent of management burning in English uplands.* Research Report No. 667. English Nature: Peterborough.

Yarwood R and Evans N (1998) 'New places for 'Old Spots': the changing geographies of domestic livestock animals.' *Society and Animals*, 6(2): 137–65.

Young R and Simmonds T (1995) 'Marginality and the nature of later prehistoric upland settlement in the north of England.' *Landscape History*, 17: 5–16.

Zuckerman S (1957) *Forestry, Agriculture and Marginal Land – a report by the Natural Resources (Technical) Committee.* HMSO: London.

Legislation

National

Heather and Grass burning Regulations 1986.

Hill Farming Act 1946 ch73: 1056–1086.

Hill Livestock Compensatory Allowance Regulation 1996.

Livestock Rearing Act 1951.

Wildlife and Countryside Act 1981.

Wildlife and Countryside (Amendments) Act 1986.

European Union

Council Directive concerning the Community list of less-favoured farming areas within the meaning of Directive 75/268/EEC (United Kingdom).

79/409/EEC Council Directive of 2 April l979 on the conservation of wild birds

Council Directive 84/169 of 28 February 1984 concerning the Community list of less favoured farming areas within the meaning of Directive 75/268/EEC (UK)

Council Directive 92/43/EEC on the conservation of natural habitats and of wild fauna and flora

Council Directive (EC) 2078/92 on agricultural production methods compatible with the requirements of the protection of the environment and the maintenance of the countryside.

Council Directive 2003/85/EC on community measures for the Control of Foot and Mouth Disease.

Council Regulation EEC 797/85 of 12 March 1985 on improving the efficiency of agricultural structures

Council Regulation 1760/87 of 15 June 1987 amending Regulations (EEC) No. 797/85, (EEC) No. 270/79, (EEC) No. 1360/78 and (EEC) No. 355/77 as regards agricultural structures, the adjustment of agriculture to the new market situation and the preservation of the countryside

Council Regulation 1094/88 Set Aside amending Regulation (EEC) No. 797/85 and (EEC) No. 1760/87 as regards the set aside of arable land and the extensification and conversion of production

Council Regulation 1096/88 of 25 April 1988 establishing a community scheme to encourage the cessation of farming.

Council Regulation1098/88 of 25 April 1988 amending Regulation No.136/66/EEC

Council Regulation 2078/92 of 30 June 1992 on agricultural production methods compatible with the requirements of the protection of the environment and the maintenance of the countryside

EU Council Regulation 2081/92 on the protection of geographical indicators and designations of origin for agricultural products and foodstuffs.

Council Regulation (EC) No 950/97 on improving the efficiency of agricultural structures.

Council Regulation (EC) No 1257/1999 on support for rural development from the European Agricultural Guidance and Guarantee Fund (EAGGF) amending and repealing certain Regulations.

Council Regulation (EC) No 1783/2003 amending Regulation (EC) No 1257/1999 on support for rural development from the European Agricultural Guidance and Guarantee Fund (EAGGF).

Council Regulation (EC) No 817/2004 laying down detailed rules for the application of Council Regulation (EC) No 1257/1999 on support for rural development from the European Agricultural Guidance and Guarantee Fund (EAGGF).

Council Regulation (EC) No 1698/05 on support for rural development by the European Agricultural Fund for Rural Development.

Council Regulation EC 1290/05 on the financing of the Common Agricultural Policy.

Council Regulation 510/2006 on the protection of geographical indicators and designation of origin for agricultural products and foodstuffs.

European Landscape Convention ETS No. 176, Accessed at: http://conventions.coe.int/Treaty/en/Treaties/Html/176.htm on 19/1/11

Proposal on the consideration, on behalf of the EU of the Protocol on Mountain farming attached to the Alpine Convention 2006/0059(CNS) Com 2006 170 Final Brussels 25/4/06.

Index